TIME-DOMAIN MEASUREMENTS IN ELECTROMAGNETICS

TIME-DOMAIN MEASUREMENTS IN ELECTROMAGNETICS

Edited by

Edmund K. Miller

Van Nostrand Reinhold Electrical/Computer Science and Engineering Series

VNR VAN NOSTRAND REINHOLD COMPANY

New York

Library of Congress Catalog Card Number: 86-5662
ISBN 0-442-26384-8

Manufactured in the United States of America

Published by Van Nostrand Reinhold Company Inc.
115 Fifth Avenue
New York, New York 10003

Van Nostrand Reinhold Company Limited
Molly Millars Lane
Wokingham, Berkshire RG1 2PY, England

Van Nostrand Reinhold
480 Latrobe Street
Melbourne, Victoria 3000, Australia

Macmillan of Canada
Division of Gage Publishing Limited
164 Commander Boulevard
Agincourt, Ontario MIS 3C7, Canada

15 14 13 12 11 10 9 8 7 6 5 4 3 2 1

Library of Congress Cataloging-in-Publication Data

Miller, E. K. (Edmund Kenneth), 1935-
Time domain measurements in electromagnetics.

(Van Nostrand Reinhold electrical/computer science and engineering series)
Bibliography: p.
Includes index.
1. Electromagnetic measurements. I. Title.
II. Series.
TK277.M55 1986 621.38′043 86-5662
ISBN 0-442-26384-8

This book is dedicated to my parents, Edmund W. and Viola L. Miller, who saw me through my BS degree; to my wife, Patricia A. Denn Miller, who helped make the PhD degree a reality; and to my children, Kerry A. and Mark C. Miller, who made it all worthwhile.

Van Nostrand Reinhold
Electrical/Computer Science and Engineering Series
Sanjit Mitra—Series Editor

HANDBOOK OF ELECTRONIC DESIGN AND ANALYSIS PROCEDURES USING PROGRAMMABLE CALCULATORS, by Bruce K. Murdock

COMPILER DESIGN AND CONSTRUCTION, by Arthur B. Pyster

SINUSOIDAL ANALYSIS AND MODELING OF WEAKLY NONLINEAR CIRCUITS, by Donald D. Weiner and John F. Spina

APPLIED MULTIDIMENSIONAL SYSTEMS THEORY, by N. K. Bose

MICROWAVE SEMICONDUCTOR ENGINEERING, by Joseph F. White

INTRODUCTION TO QUARTZ CRYSTAL UNIT DESIGN, by Virgil E. Bottom

DIGITAL IMAGE PROCESSING, by William B. Green

SOFTWARE TESTING TECHNIQUES, by Boris Beizer

LIGHT TRANSMISSION OPTICS, Second edition, by Deitrich Marcuse

REAL TIME COMPUTING, edited by Duncan Mellichamp

HARDWARE AND SOFTWARE CONCEPTS IN VLSI, edited by Guy Rabbat

MODELING AND IDENTIFICATION OF DYNAMIC SYSTEMS, by N. K. Sinha and B. Kuszta

COMPUTER METHODS FOR CIRCUIT ANALYSIS AND DESIGN, by Jiri Vlach and Kishore Singhal

HANDBOOK OF SOFTWARE ENGINEERING, edited by C. R. Vick and C. V. Ramamoorthy

SWITCHED CAPACITOR CIRCUITS, by Phillip E. Allen and Edgar Sanchez-Sinencio

SOFTWARE SYSTEM TESTING AND QUALITY ASSURANCE, by Boris Beizer

MODERN DC-TO-DC SWITCHMODE POWER CONVERTER CIRCUITS, by Rudolf P. Severns and Gordon E. Bloom

ESTIMATION AND OPTIMUM CONTROL OF SYSTEMS, by Theodore F. Elbert

DIGITAL TELEPHONY, by Bernhard Keiser and Eugene Strange

DATA COMPRESSION TECHNIQUES AND APPLICATIONS, by Thomas J. Lynch

DIGITAL TRANSMISSION SYSTEMS, by David R. Smith

FLAT-PANEL DISPLAYS AND CRTS, edited by Lawrence E. Tannas, Jr.

MICROCOMPUTER SYSTEMS, by Ivan Flores and Christopher Terry

COMPUTER ARCHITECTURE, 3/e, by Caxton Foster and Thea Iberall

RADIOWAVE PROPAGATION IN SPACE COMMUNICATIONS SYSTEMS, by Louis J. Ippolito, Jr.

SEMICONDUCTOR POWER ELECTRONICS, by Richard G. Hoft

LONG-WAVELENGTH SEMICONDUCTOR LASERS, by G. P. Agrawal and N. K. Dutta

APPLIED RELIABILITY, by Paul A. Tobias and David C. Trindade

TIME DOMAIN MEASUREMENTS IN ELECTROMAGNETICS, edited by Edmund K. Miller

Preface

The technology of making time-domain electromagnetic measurements continues to evolve as better sources, sensors, and instrumentation become available. Progressing from the radars of the 1940s with bandwidths of a few percent; to early TDR and EMP test systems of the 1960s with bandwidths of 2:1, or so; to the short-pulse hardware used today with bandwidths that can exceed 10:1—throughout all this development, there has occurred a steady improvement in time-domain technology. This improvement has accompanied an increasing variety of applications that benefit from or require wide-bandwidth systems. These include radar-target identification from short-pulse signatures, lightning and EMP analysis and protection, inverse and nondestructive evaluation, radiation and interference in high-speed digital systems, and various measurement applications ranging from basic studies to meeting routine data needs.

Although the technical literature exhibits a growing attention to time-domain technology, there does not yet exist a dedicated reference source that provides a comprehensive overview of the area. Were this source to be developed, it could contribute significantly to accelerating the development and exploitation of this important technology. The primary purpose of this book is to provide such a source by collecting together in one place an overview of the state-of-the-art in EM time-domain measurements. It is intended as a resource for those working in the area and as an introductory reference for the larger community of potential users. We are especially interested in seeing time-domain techniques, both experimental and computational, become more widely taught at the college level than is now the case. Thus, we stress the physical insight such techniques provide, as well as the relatively small expense of setting up a basic time-domain laboratory.

The specific idea for this book came about from my being invited to organize a special session of Commission A (Metrology) of the International Scientific Radio Union (URSI) at the 1982 United States National Meeting in Boulder, CO. The invitation was extended by Dr. Helmut Hellwig, then Commission A chairman, who felt that holding topical special sessions of the Commission dealing with various topics in measurement technology could contribute to the Commission's focus on metrology. Since my interest in computational and experimental time-domain electromagnetics extends back to the late 60s, I readily accepted, as the time seemed especially appropriate for a review of this important and fascinating area.

Special sessions at professional-society meetings are typically organized in one of two general ways, or some combination thereof. Perhaps the more common, simpler approach is to announce in the meeting's call for papers that certain topics will receive special attention, with the sessions then organized to accord with the papers submitted. As the other alternative, a specific outline is developed and presentations are identified so that some desired theme can be developed or a set of viewpoints presented. The latter approach was used for the Commission A session organized for the 1982 meeting, whose final title was "Time-Domain Measurements."

After discussion of these matters with various colleagues at Lawrence Livermore National Laboratory (H. C. Cabayan, K. F. Casey, R. J. Lytle, and A. J. Poggio), it was concluded that the overall topic of the special session could best be treated by subdividing it into the four basic areas of (1) background, benefits, and opportunities; (2) measurement technology; (3) applications; and (4) signal processing. These subjects form the main sections of this book, with the addition of an appendix dealing with design considerations and the hardware selection involved in setting up a time-domain range.

Although the session was developed for the sole purpose of providing a review of electromagnetics time-domain technology at the URSI meeting, response to the presentations was so positive that a book organized along similar lines seemed worthwhile considering. Consequently, I prepared a prospectus and outline, which Van Nostrand soon agreed to publish as an edited book, the results of which appear here.

The first section on background, benefits, and opportunities contains three chapters, which address, in turn, (1) historical development and motivations, (2) current status and future directions, and (3) instructional opportunities offered by time-domain measurements in electromagnetics. Chapter 1 by Dr. G. F. Ross, one of the pioneers of time-domain electromagnetics, sets the tone for the rest of the book. It does this by establishing the rationale for looking at EM phenomena in the time domain from the viewpoints of physical insight and efficiency of characterization. Dr. Ross demonstrates how representing equivalent information in the time domain differs from its frequency-domain counterpart and illustrates the benefits of time-domain analysis through the use of numerous examples. In Chap. 2 by Dr. N. S. Nahman, another long-time contributor to time-domain technology, the present status and future directions of time-domain measurements are reviewed. He summarizes information collected from numerous sources in text descriptions and tables and provides an extensive bibliography. This section concludes with a discussion of the instructional opportunities afforded by time-domain measurements by Dr. S. M. Riad in Chap. 3. Professor Riad describes three approaches that might be used to obtain transient information for linear problems and illustrates their attributes for the specific problem of deriving the impulse response of an RC network.

The second section on measurement technology describes the basic hardware needed to implement a time-domain range and contains four chapters that deal with (4) fast pulsers, (5) sensors and radiators, (6) instrumentation, and (7) the validation and calibration of a representative time-domain range. Dr. J. R. Andrews surveys subnanosecond, fast-pulse generators in Chap. 4, ranging from the low-energy extreme of milli-Joule transistorized pulsers to the high-energy extreme of mega-Joule linear accelerators. Dr. Andrews provides a figure of merit for comparing pulsers in terms of the time-rate-of-change of the output voltage, with values, in V/nsec, ranging from ~ 1 for transistor pulse generators to several thousand for spark gaps and linear accelerators. In Chap. 5, Dr. M. Kanda covers the kinds of radiators and sensors that are needed to produce and measure impulsive fields. Dr. Kanda provides both mathematical and physical explanations for how impulsively excited antennas behave when used as transmitters or receivers and presents extensive results from analysis and measurement. The instrumentation required to complete the measurement system is discussed in Chap. 6 by Dr. G. D. Sower and Mr. J. R. Pressley. They describe the various kinds of hardware that might be used for this purpose, ranging from film-based recording systems to electronic transient digitizers. To conclude the section on measurement technology, Drs. R. M. Bevensee and E. K. Miller demonstrate how all of the various components are brought together in developing a working time-domain range. In particular, they describe the Electromagnetic Transient Facility at Lawrence Livermore National Laboratory, the procedures employed for its calibration and validation, and various applications for which it has been used.

In the third section are a variety of examples of time-domain applications in electromagnetics. The four chapters included here deal with (8) EMP measurement techniques, (9) applications of a subsurface, transient radar, (10) the time-domain measurement of components and materials, and (11) the measurement of lightning waveforms. Drs. J. P. Castillo and L. Marin present an introduction to making EMP measurements in Chap. 8. They discuss the simulators (radiators), instrumentation, and data-acquisition and processing systems employed in EMP applications; they also explain how EMP simulation needs provided the impetus for much of the development that has occurred in short-pulse technology in electromagnetics. The intriguing area of near subsurface exploration using a transient radar is covered in Chap. 9 by Drs. J. D. Young and L. Peters. They consider requirements unique to this kind of application, including pulse propagation in a lossy, dispersive medium that typifies most real grounds and comment on the data-processing needs for this particular application. In Chap. 10, the use of time-domain measurement techniques to characterize both components and materials is explored by Dr. H. M. Cronson. He demonstrates the advantages and limitations of using short-pulse data to obtain wide-band information in closed systems. The concluding applications discussion in Chapter 11 by Drs.

E. F. Vance and J. Nanovitz covers the real-world problem of measuring lightning-induced phenomena. They examine the almost unique requirements of measuring lightning waveforms and present current capabilities and representative results in this particular area.

The text proper concludes with two chapters on some signal-processing issues. The general problem of processing transient signals is discussed in Chap. 12 by Drs. J. V. Candy, D. B. Harris, and D. M. Goodman, who provide an overview of the techniques presently being used. The closely related, but more specialized, topic of electromagnetic parameter estimation from transient waveforms is treated in Chap. 13 by Drs. D. G. Dudley and D. M. Goodman. The message conveyed by both chapters is that for fuller realization of the benefits of time-domain measurements, it is mandatory to give appropriate attention to processing the waveforms involved. Failure to do so may result in available information not being extracted from the data, or, even worse, incorrect answers being obtained.

Closing out the book, an appendix by Dr. J. A. Landt summarizes the issues involved in selecting hardware and designing a time-domain range. This discussion reviews the various tradeoffs and design choices to be kept in mind in the design process. Dr. Landt includes performance specifications and price information and provides a nominal design for a time-domain range.

Other features of the book include comprehensive references at the end of every chapter (more than 400 entries in all), which provide an excellent resource base for more detailed information about the subjects treated. Finally, a keyword and abbreviation list should help the reader decipher the specialized and sometimes cryptic terminology employed in the text for the sake of conciseness.

I would like to acknowledge the cooperation of the chapter authors cited here for their hard work and persistence in seeing this project through to its conclusion. It has been a pleasure to work with them. I also want to express my appreciation to Rose O'Brien and Jan Grimm of LLNL, who provided expert and enthusiastic secretarial help during the course of this endeavor. Finally, the contributions of Barbara Sokolowski, also of LLNL, were above and beyond the call of duty in her dedicated typing of the entire manuscript.

EDMUND K. MILLER

Contents

TIME-DOMAIN MEASUREMENTS IN ELECTROMAGNETICS

1
Early Developments and Motivations for Time-Domain Analysis and Application

Gerald F. Ross, Ph.D

Anro Engineering Consultants, Inc.

1.0 INTRODUCTION

Although the use of the time domain for analysis and synthesis became prominent with the corresponding advances in the development of short pulse hardware and instrumentation, one may ask what the motivation is for using the time domain or, for that matter, any other domain for the analysis and synthesis of lumped and distributed networks. In classical analysis, we know that a linear, time-invariant system can be described uniquely by its impulse response, $h(t)$. And, from the impulse response, we can find the amplitude spectrum, $A(\omega)$, and the phase function, $\phi(\omega)$, by a Fourier transformation as follows:

$$F(\omega) = A(\omega)e^{-j\phi(\omega)} \overset{\mathbf{F}}{\leftrightarrow} h(t) \tag{1-1}$$

But we may also describe a network in the complex domain, $p = \sigma + j\omega$, by a Laplace transform, or in the domain, $Z = e^{+pT,}$ where poles and zeros in the complex plane fall within the unit circle (T is a sampling constant). Other representations, of course, are also possible and useful.

As engineers, we seek simply the domain that presents a mathematical solution most compactly: a domain where the response of a network in time, or its real or complex spectrum, can readily be defined. The use of such a domain, almost as a by-product, provides insight for system behavior.

As an example, consider a high Q series or parallel resonant circuit. The duration of the impulse response of such a circuit may exist over most of the time domain, but in the frequency domain, it is well contained over a narrow bandwidth. The time domain would be the wrong domain in which to describe the behavior of the network or to measure its properties; the network can be readily described, however, in the frequency domain.

In a stable, sample-data system where a closed-form solution for the system function exists, the Laplace transform consists of an infinitely denumerable number of poles and zeros that are periodic in the left hand plane. By describing the system in the $Z = e^{+pT}$ plane, the periodic poles and zeros all coalesce and map as a finite number of points within the unit circle.[1] Clearly, the complex Z plane provides the most convenient domain to represent the properties of the networks.

In certain very wideband networks (for example, a length of TEM mode line), the time-domain representation is the obvious choice. The impulse response is simply a delayed impulse. We cannot generate or realize a perfect impulse, but we can generate a short pulse, one whose duration is much shorter than the duration of the impulse response. Recent advances in pulse technology using tunnel diodes permit the generation of pulses as short as 20 picoseconds for diagnostic purposes, and recent developments in sampling oscilloscope technology permit one to display these pulses.[2] The capability of generating and displaying pulses of such short duration leads to many interesting system applications, as will be described in later sections of this chapter.

In Sec. 1.1, direct time-domain techniques, both theoretical and experimental, are described. In Sec. 1.2, these techniques are applied to the analysis of microwave networks, antennas, target scattering, and material diagnostics. Some system applications that developed directly from the analytical tools are described in Sec. 1.3, and a brief summary is presented in Sec. 1.4.

1.1 ANALYSIS IN THE TIME DOMAIN

Theoretical Formulation

As indicated in the introduction, when the response of a network or system is well contained in the time domain (i.e., time-limited), analysis in the time domain is most appropriate.

The convolution integral is used in linear systems analysis to find the response, $r(t)$, of a system to an excitation, $f(t)$, knowing its impulse response, $h(t)$. The most general form of the convolution integral is given by the following[3]:

$$\begin{aligned} r(t) &= \int_{-\infty}^{+\infty} h(x) f(t - x) dx \\ &= \int_{-\infty}^{+\infty} f(x) h(t - x) dx \end{aligned} \tag{1-2}$$

where x is a dummy variable. Here, the upper and lower limits on the integral x are $+\infty$ and $-\infty$, respectively. Often, however, neither the function $h(x)$ nor

$f(x)$ exists over the entire space, $-\infty < x < \infty$, and it is appropriate to modify the limits on Eq. 1-2. The choice of the integration interval (i.e., the choice of upper and lower limits) often is not obvious. In the paragraphs that follow, the limits on the integral will be given for the most functions $h(t)$ and $f(t)$. The method involves the novel use of the step function.[4]

The step function method. Assume it is required to convolve the two functions, $h(t)$ and $f(t)$, shown in Fig. 1-1, where $h(t)$ exists for all positive time, t (abbreviated $\forall_{t+}$), and $f(t)$ exists for negative time, $\forall_{t-}$, but is time limited for t positive (abbreviated tl_{t+}). The existence of these signals in the time domain can be conveniently described by the use of step functions. For example, the functions $h(t)$ and $f(t)$ shown in Fig. 1-1 may be described as

$$\begin{aligned} h(t) &= 0 \qquad t < 0 \\ h(t) & \qquad\quad t > 0 \end{aligned} \tag{1-3a}$$

$$\begin{aligned} f(t) &= 0 \qquad t > T \\ f(t) & \qquad\quad -\infty < t < T \end{aligned} \tag{1-3b}$$

or more compactly as

$$\begin{aligned} h &= h(t) \times U(t) \\ f &= f(t) \times U(T - t) \end{aligned} \tag{1-4}$$

where

$$\begin{aligned} U(t) &= 0,\ t < 0 \\ U(t) &= 1,\ t > 0 \end{aligned}$$

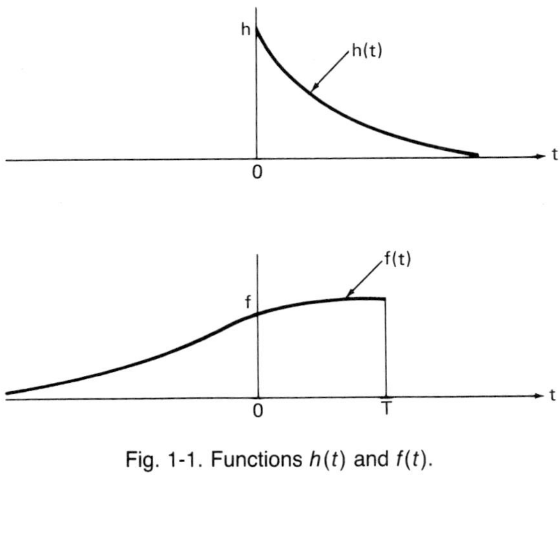

Fig. 1-1. Functions $h(t)$ and $f(t)$.

To find $h(t) * f(t)$, or the convolution of $h(t)$ with $f(t)$, one substitutes Eq. 1-4 into Eq. 1-2 and obtains

$$\begin{aligned} r(t) &= \int_{-\infty}^{+\infty} [h(x)U(x)][f(t - x) \cdot U(x + T - t)]dx \\ &= \int_{-\infty}^{+\infty} h(x)f(t - x)\frac{U(x)}{2} \cdot 2U(x + T - t)dx \end{aligned} \tag{1-5}$$

where x and $(t - x)$ have been substituted for t in the expressions for $h(t)$ and $f(t)$, respectively.

The bracketed term in the integrand of Eq. 1-5 can be used to determine the specific limits on the integral. This term is shown in Fig. 1-2. When the constant, $t - T$, is greater than zero (i.e., with respect to the variable of integration x), the value of the integrand (determined by the product of the two displaced step functions) is zero $\forall x < t - T$, as shown in Fig. 1-2.

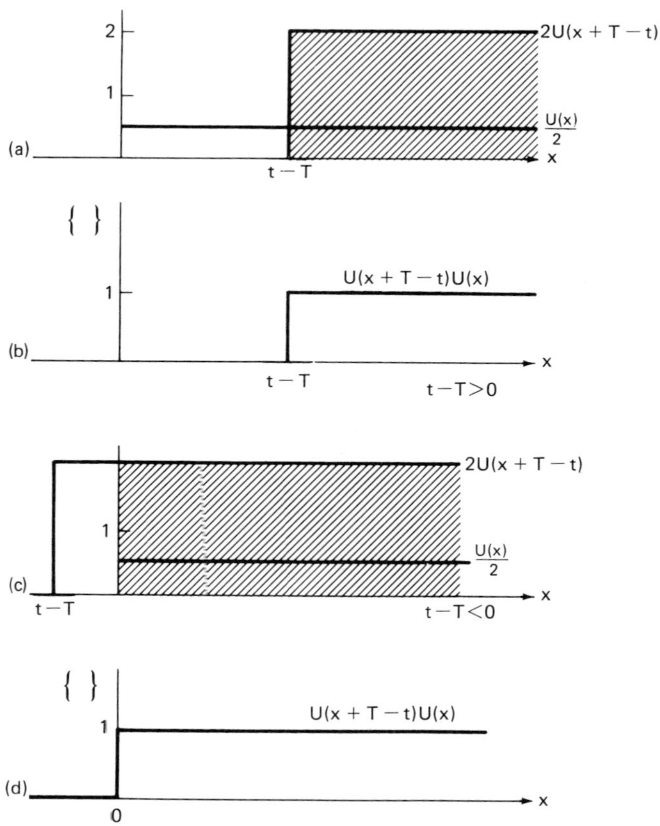

Fig. 1-2. The product of step functions.

When $t - T < 0$, the product of the two displaced step functions is zero $\forall x < 0$, as shown in Figs. 1-2(c) and 1-2(d). Thus,

$$r(t = \int_0^{\infty} h(x)f(t - x)dx \qquad \text{for } t - T < 0, \text{ or } \forall_{t < T} \qquad (1\text{-}6)$$

$$r(t) = \int_{t - T}^{\infty} h(x)f(t - x)dx \qquad \text{for } t - T > 0, \text{ or } \forall_{t > T} \qquad (1\text{-}7)$$

The integration indicated by Eq. 1-5 is with respect to the dummy variable x, and t and T are simply constants, as illustrated in Fig. 1-2. The limits on the integral are determined by assuming, alternately, that the constant is greater than or less than zero and then solving the inequality for t. The solution of t determines the region over which the answer, $r(t)$, is valid. A chart showing various combinations of possible functions, $h(t)$ and $f(t)$, is given in Fig. 1-3. The limits on the convolution integral of these combinations is presented in Table 1-1. In Fig. 1-3, some functions that are tl were selected, for convenience, to originate

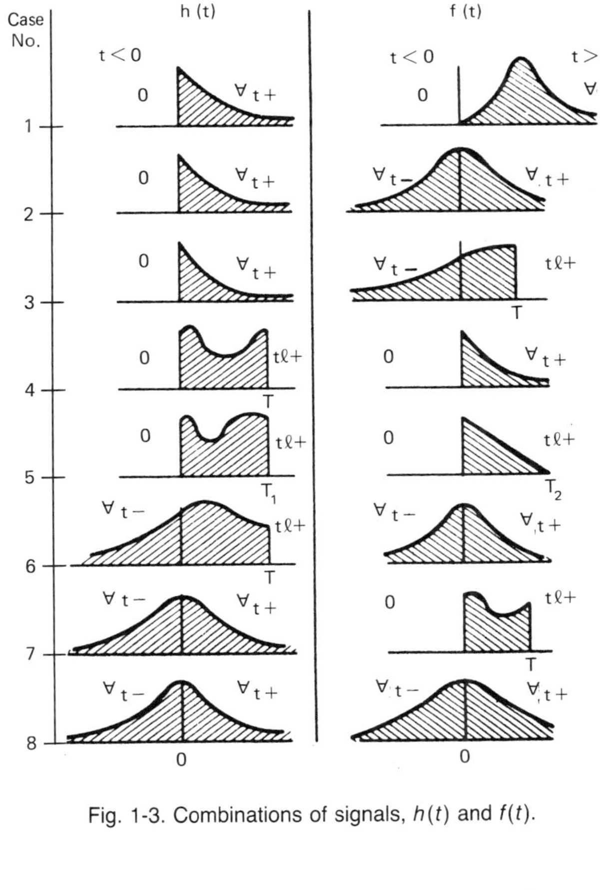

Fig. 1-3. Combinations of signals, $h(t)$ and $f(t)$.

Table 1-1. Summary Table for Convolution Operations.

CASE NO.	$h(t)$		$f(t)$		RESPONSE, $r(t)$
	$t < 0$	$t < 0$	$t < 0$	$t < 0$	
1	0	$\forall_t +$	0	$\forall_t +$	$r(t) = \int_0^t h(t - x)f(x)dx;\ t > 0 \qquad r(t) = 0;\ t < 0$
2	0	$\forall_t +$	$\forall_t -$	$\forall_t +$	$r(t) = \int_0^\infty f(t - x)h(x)dx \qquad r(t) = 0;\ t < 0$
3	0	$\forall_t +$	$\forall_t -$	$t1_t +$	$r(t) = \int_{t-T}^\infty [h(x)f(t - x)]dx;\ t > T \qquad r(t) = \int_0^\infty [\ ]dx;\ t > T$
4	0	$tl_t +$	0	$\forall_t +$	$r(t) = 0;\ t < 0 \qquad r = \int_0^t [h(t - x)f(x)]dx;\ 0 < t < T$ $r = \int_{t-T}^t [\ ]dx;\ t > T$

					Case (a): $T_2 > T_1$	Case (b): $T_1 > T_2$
5	0	tl_t+	0	tl_t+	$r = \int[f(x)h(t-x)]dx = 0;\ t < 0$ $r = \int_0^t [\ ]dx;\ 0 < t < T_1$ $r = \int_0^{T_1} [\ ]dx;\ T_1 < t < T_2$ $r = \int_{t-T_2}^{T_1} [\ ]dx;\ T_2 < t < T_1 + T_2$ $r = 0;\ t > T_1 + T_2$	$r = 0;\ t < 0$ $r = \int_0^t [\ ]dx;\ 0 < t < T_2$ $r = \int_{t-T_2}^{t} [\ ]dx;\ T;_2 < t < T_1$ $r = \int_{t-T_2}^{T_1} [\ ]dx;$ $T_1 < t < T_1 + T_2$ $r = 0;\ t > T_1 + T_2$
6	$\forall_t-$	tl_t+	$\forall_t-$	$\forall_t+$	$r(t) = \int_{-\infty}^{T} h(x)f(t-x)dx$	
7	$\forall_t-$	$\forall_t+$	0	tl_t+	$r(t) = \int_0^{T} f(x)h(t-x)dx$	
8	$\forall_t-$	$\forall_t+$	$\forall_t-$	$\forall_t+$	$r(t) = \int_{-\infty}^{+\infty} h(t-x)f(x)dx$	

at $t = 0$. The results of the table can be applied to functions that are *tl* but do not start at $t = 0$ simply by delaying or advancing $h(t)$ and $f(t)$ to the origin and appropriately delaying or advancing $r(t)$ by an equal amount of time.

It should be noted that the step function method can conveniently be used to find the limits on other integral relationships, e.g., evaluating the cross-correlation function of two given signals. The method, although somewhat protracted, will result in the choice of the narrowest limit on the integral with a minimum chance of error.

Many times an analysis in one domain provides insight into behavior in another domain. For example, a review of Hallen's early work (1938) on the driving-point impedance of a dipole (monopole) antenna[5] shows that

$$Z(\omega) \approx \sum_{-\infty}^{+\infty} Z\left(\omega + \frac{2\pi n}{T}\right) \tag{1-8}$$

i.e., the impedance is approximately periodic with frequency. This, of course, implies that the inverse Fourier transform of $Z(\omega) \overset{\mathbf{F}}{\leftrightarrow} z(t)$ is a train of weighted impulses, that is,

$$\sum_{-\infty}^{+\infty} Z\left(\omega + \frac{2\pi n}{T}\right) \overset{\mathbf{F}}{\leftrightarrow} \sum_{-\infty}^{+\infty} a_k\, \delta(t - kT) \tag{1-9}$$

where $\delta(t)$ is the Dirac delta function. Later (1972), Schmidt and King[6] showed that this result was indeed true, and DeLorenzo[7] based development of the time-domain, ground-plane scattering range on this result. DeLorenzo increased the time between impulses by using a long monopole wire antenna placed perpendicular to a large ground plane. Then, to a target placed on the ground plane, the impulsive radiation from the base or feed of a simple monopole produces the impulse response of the target directly, provided that this response decays below the noise level before the arrival of the next impulse.

In the preceding example, the results in the frequency domain implied a response in the time domain and helped bring about important invention in the field of target-scattering analysis and synthesis.

Another interesting interpretation of behavior in the time domain provides a better understanding of resonance in the frequency domain. For example, consider the case where equal-length TEM mode lines are interconnected to form microwave hybrid junctions, for example, the branch line and ring couplers. Here, the impulse response between any two ports of the network is given by

$$h(t) = \sum_{k=1}^{\infty} A_k\, \delta(t - k\tau) \tag{1-10}$$

where A_k is a function of the reflection coefficients at the junctions and τ is a time delay depending on the geometry of the network and the propagation constant of the medium.

When the impulse response of a linear, time-invariant network consists of a train of impulses of varied weighting, the response to an input, $f(t)$, may be found conveniently by appropriately weighting and delaying replicas of the input signal, $f(t)$. In particular, if the impulse response is uniformly spaced $\lambda_o/2c$ seconds apart, the response to a step-modulated signal of the form

$$F(t) = \sin \omega_o t \times U(t) \tag{1-11}$$

where $\omega_o = 2\pi f_o = 2\pi c/\lambda_o$ can be found in any given half period by a simple addition. This follows since the phase difference between weighted replicas of the input signal is either zero or 180 degrees (i.e., either plus or minus). Viewing the step-modulated response of the coupler at any of the output ports provides a revealing picture of the constructive and destructive resonant buildup in a microwave network. These buildup and settling times can be shown to be related to the energy stored in the network. More will be said about these networks in Chap. 10.

Experimental Techniques

As previously indicated, analysis in the time domain often provides intuition that, although present in the mathematical formulation for other domains, is not immediately obvious. For most engineers, a necessary part of any total analysis is an experimental capability. It is this capability that allows the investigator to iterate between theory and experiment and converge on a solution.

The tools available for experimental analysis in the early 1960s were limited. A major breakthrough was the development of the sampling oscilloscope by Hewlett-Packard.[8] This display device provided short duration time measurement by an "equivalent" method. Here a single sample from each successive repetition of a pulse train was painted on the screen with the next sample being taken at a slightly different time relative to the starting pulse. The measurement time required to display a nanosecond waveform could thus take milliseconds, corresponding to multiples of the pulse repetition frequency. Hewlett-Packard developed high-speed sampling gates that used tunnel diodes in precisely engineered microwave cavities. They also marketed, separately, tunnel diode step generators having an amplitude of 0.5 V and a risetime of about 100 picoseconds (ps). About this time, manufacturers were also developing avalanche transistors for high-voltage (but slower risetime) applications. By 1970, the step-recovery diode also became commercially available. Mercury switches were used for higher voltage target-identification applications, but at a much slower PRF. A summary of the characteristics of these pulse sources is presented in Table 1-2.

For direct evaluation of the buildup time of a network or system to a step-modulated signal at resonance, ω_o, it is most convenient to generate a step-modulated signal fo the form in Eq. 1-11, namely,

$$f(t) = sin(\omega_o t) \times U(t) \tag{1-12}$$

Early system analysis involved phased-array networks at L band where a single RF cycle corresponds to about 1 nanosecond (ns).[10]

It is difficult to generate a signal described by Eq. 1-12. For example, gating an oscillator introduces synchronization difficulties as well as its own transient degradation caused by bandwidth limitations. The problem was solved by generating the required waveform synthetically.[11] That is, as shown in Fig. 1-4, we first generate a step function using a tunnel diode source (risetime $<$ 100 ps) and feed this into a short-circuited stub junction of length L that has one-half the characteristic impedance of the main line. The result is a pulse of $2L/c$ s duration and an amplitude of 0.5 V. By feeding this signal to an identical short-circuited stub, the resulting waveform becomes ± 0.25 V in amplitude and resembles one square cycle of an RF carrier of period $4L/c$ s. Next, feeding this output to a series of open-circuited stubs of lengths $2L$, $4L$, $8L$, . . . , we produce 2, 4, 8, . . . square cycles of microwave energy, provided the stub separation is greater than the longest stub.

The measuring instrument (sampling oscilloscope) behaves as a low-pass filter,

Table 1-2. Typical Characteristics of Pulse Sources.[9]

TYPE	STEP/PULSE	BEST AVAILABLE RISETIME AT AMPLITUDE		NOTES
Mercury switch	Step	70 ps	300 V	Max. PRF = 200 Hz
Avalanche transistor	Pulse	150 ps	12 V	Device selection necessary
Tunnel diode	Step	25 ps	0.25 V	Fastest transistion time
		100 ps	1.0 V	
Step recovery	Step	60 ps	20 V	Commercially available
		100 ps	50 V	
		200 ps	200 V	Specially ordered four-stack
Hertzian	Impulse; also pulse modualted	100 ps	1000 V	Limited lifetime, spark gap
		1 ns	1000 V	
Avalanche diode	Impulse	400 ps	125 V	MHz rep. rate

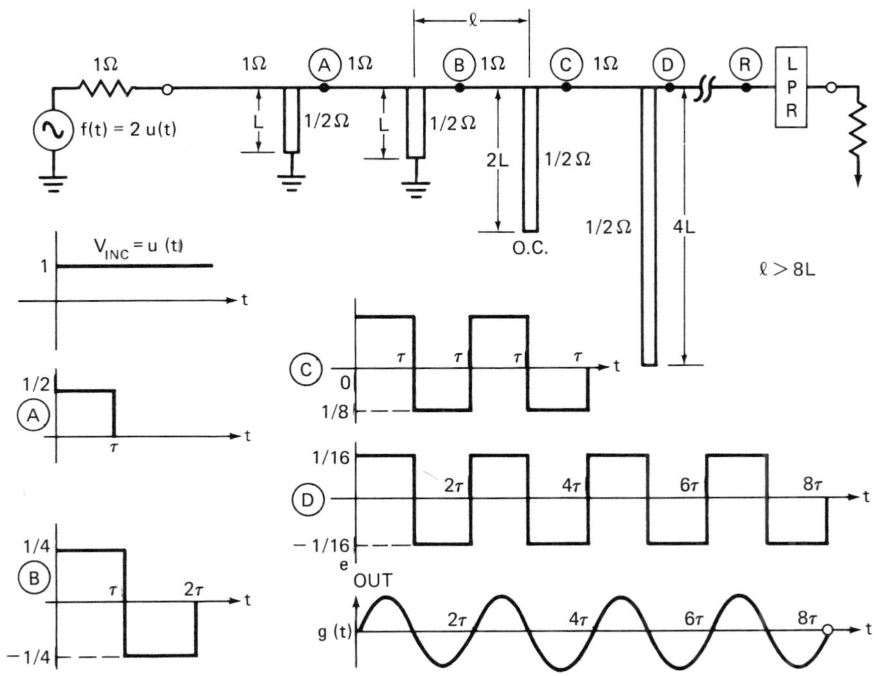

Fig. 1-4. Synthetic generation of wave energy.

rounding off the sharp corners of the wave and making the resulting waveform indistinguishable from the desired test waveform of Eq. 1-12. The disadvantage of this technique is that the output voltage, V_o, is reduced by the factor

$$V_o = 2^{-k} \tag{1-13}$$

where k is equal to the number of stubs employed. This limits to 8 or 16 the number of cycles that can be practically generated. Later inventions by Proud and Cronson[12] circumvented the problems of loss by the use of line lengths separated by spark-gap devices or step-recovery diodes, respectively. These experiments were important because they showed that virtually any frequency-coded pulse could be generated synthetically from a simple unit-step function; a unipolar baseband pulse was simply the addition of two opposite-polarity step functions shifted in time. These simple observations stimulated the work that followed in the antenna and metrology areas of investigation.

Improvements in solid-state devices since these initial experiments have resulted in significant improvements in both the risetime and amplitude of generated step and pulse functions.[13] In the sections that follow, the use of these waveform sources will be demonstrated.

1.2 AREAS OF TIME-DOMAIN ANALYSIS

Networks

Flow graph techniques[14] are chosen as the primary analysis tool in time domain studies because (1) they afford a clear physical representation of the interaction of reflections between junctions, and (2) they permit by application of a simple formula an evaluation of the overall response or system function of the network in closed form, thus revealing the pole-zero (resonance) pattern of the system. This approach is not limited to the analysis of symmetrical networks but can also be applied to solving more general network configurations. The more conventional approach involves tedious conversion of the scattering matrix of each junction to a transmission matrix and then the multiplication of a number of such matrices and the inversion of the result to form a new scattering matrix. The process is repeated until the desired terminals are reached in a manner analogous to finding the driving-point impedance or admittance of a ladder network with complicated elements.

The flow graph representation of a given network is not unique; that is, a network may be represented by several flow graphs each having equal validity. Conceptually, some graphs may be easier to visualize or construct. As an illustration of the technique, consider the flow graph representing the interconnection of four three-port TEM mode junctions forming a four-port network, shown in Fig. 1-5.[15] The output ports of the network are assumed to be matched to the impedance of the driving source, which, for convenience, has been normalized to 1 Ω. In the discussion that follows, the reference to "ports" refers to the physical network, whereas the reference to "nodes" refers to the topological representation of the network, namely, the flow graph. Each port in the network is represented by two nodes in the flow graph, for example, A and A′. This distinction is required since waves traveling in a clockwise (cw) direction in the network may experience a different reflection coefficient at a port than waves entering the same port from a counterclockwise (ccw) direction. The signal flow is described in the following discussion.

An incident (impulse) voltage entering port A is partially reflected (as a result of mismatch) whereas the two transmitted impulses travel in a cw direction toward port B and simultaneously in a ccw direction toward port D. In the graph, cw signals travel along the outer square entering the unprimed nodes; ccw signals travel along the inner square entering the primed nodes. When the transmitted impulse reaches port B, for example, a fraction of the energy is reflected back

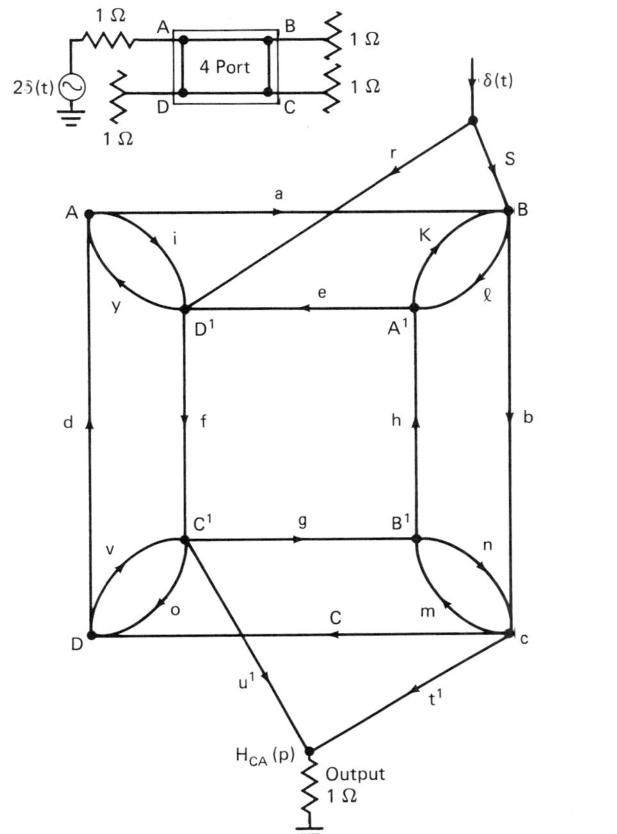

Where

$a = a(p) = a^1 e^{-p\tau_a}$

$b = b(p) = b^1 e^{-p\tau_b}$

⋮

Subject To The Constraint

$d + i = A$

$a + \ell = b$

⋮

Fig. 1-5. Flow graph for the four-port network.

to port A, while the transmitted impulse continues on to port C. The impulse reflected toward port A now travels in a ccw direction and, in the graph, enters node A′. Part of this signal is re-reflected toward port B (in a cw direction), again entering node B in the graph, forming 2 closed loop (BlA′k). The portion of the signal transmitted beyond node A′ travels in a ccw direction toward node D′, resulting in similar reflections between nodes A and D′. This process is continued until the transmitted signals from nodes B′ and D enter nodes A′ and A, respectively, and the process repeats. The values of the branch transmittances (for example, a, b, c, . . .) in Fig. 1-5 are chosen to be the Laplace transforms of the voltage transfer coefficients between nodes. Since the impulse response of each branch is a delayed impulse, the Laplace transforms of the voltage transfer coefficients are of the form

$$a = a(p) = a'e^{-p\tau} \tag{1-14}$$

where a' is real and a function of the characteristic impedances of the interconnecting lines at a junction; τ depends on the length and propagation constant of the line; and $p = \sigma + j\omega$, the complex frequency variable. The equation at each node is simple

$$E_{\text{incident}} + E_{\text{reflected}} = E_{\text{transmitted}} \tag{1-15}$$

At node A, for example, Eq. 1-15 becomes

$$d(p) + y(p) + i(p) = a(p) \tag{1-16}$$

Since primed and unprimed nodes having the same letter designations physically represent the same port or junction in the network, special care must be exercised in introducing the excitation and taking the output from the graph. For example, if the excitation were introduced directly into nodes A and A′, representing the input port, two false signals would appear at output port C. This result follows since injection at these nodes would produce an initial voltage at the output port of (ifu) + (abt) + (kbt) + (efu) V, as indicated in Fig. 1-5. From the actual network, it is clear that the initial voltage must be (abt) + (efu) V. To prevent this from occurring, the input excitation weighted by the appropriate branch transmittances is introduced into the nodes that follow, A, A′ for both the cw and ccw waves. To obtain the output signal at port C, the outputs from both node C and C′ must then be combined. Mason's formula or reduction theorem, can be used to evaluate the graph.[15] The formula states that the transmission, H, between any two nodes in the graph is given by

$$H = \frac{\sum^{m} H_m \Delta_m}{\Delta} \tag{1-17}$$

where

H_m = the product of the branch transmittances of the mth forward path

$\Delta = 1 - \sum^{n1} L_1 + \sum^{n2} L_2 - \sum^{n3} L_3 + \ldots (-1)^r \sum^{nr} L_r$

L_r = the product of the branch transmittances of nontouching loops taken r at a time

r = an integer
n_r = the number of closed loops
Δ_m = the value of Δ for that part of the graph not touching the mth forward path.

The use of this formula may be found in other references.[16] In particular, when the line lengths between junctions are integral multiples of a given length, the system function representing the transmission from ports A to C is given by

$$H_{CA}(p) = \frac{N_{CA}\,(e^{-p\tau})}{D(e^{-p\tau})} \tag{1-18}$$

and for the class of TEM mode components, this is a meromorphic function.[17] A meromorphic function has in every finite region in the complex p plane at most a finite number of poles. Thus, $H_{CA}(p)$ is the ratio of two finite polynomials in p, or

$$H_{CA}(p) = \frac{\prod_{L=1}^{\infty} (p - p_i)}{\prod_{J=1}^{\infty} (p - p_j)} \tag{1-19}$$

This function can also be expressed as an infinite train of exponentials by dividing the numerator of Eq. 1-18 by its denominator. The same result can also be obtained by taking the Laplace transform of a train of weighted impulses, as follows:

$$\begin{aligned} H_{CA}(p) &= L\,h(t) = \sum_{k=1}^{\infty} A_k\,\delta(t - k\tau) \\ &= \sum_{k=1}^{\infty} A_k\,e^{-pk\tau} \end{aligned} \tag{1-20}$$

Equating Eqs. 1-19 and 1-20, one obtains

$$\frac{\prod_{i=1}^{\infty} (p - p_i)}{\prod_{j=1}^{\infty} (p - p_j)} = \sum_{k=1}^{\infty} A_k e^{-pk\tau} \tag{1-21}$$

It is clear from Eq. 1-19 that if $p = p_i$ or p_j is a root of either the numerator or denominator of Eq. 1-17, then

$$p = p_i \text{ or } p_j + j\frac{2\pi}{\tau}n \tag{1-22}$$

is also a root, where n is an integer. Therefore, the poles and zeros of the system function are periodic in the complex p plane and also infinitely denumerable. The introduction of the complex transmission

$$z = e^{-p\tau} \tag{1-23}$$

results in the periodic poles in the p plane coalescing into a finite number of poles in the z domain. The system function then becomes the ratio of two finite polynomials in z.

Examples[18,19] It will be instructive to apply the theory presented in this section to the analysis of several frequently used microwave networks. These networks are used in feed networks for phased-array systems, monopulse radar-tracking-schemes, and filter design.

1. *Ring hybrid.* Consider the familiar ring hybrid and 3-db branch-line couplers shown in Fig. 1-6 together with their respective flow graphs. For this ring:

$$\begin{aligned} x &= x(p) = x'\,e^{-p\tau} = x'z = \frac{2\sqrt{2}}{1+\sqrt{2}}z \\ \Gamma_1 &= \Gamma'_1\,z = -\frac{2-\sqrt{2}}{2+\sqrt{2}}z \\ y &= y'_1\,z = \frac{\sqrt{2}}{1+\sqrt{2}}z \\ \Gamma_2 &= \Gamma_2'z = \frac{-1}{1+\sqrt{2}}z \end{aligned} \tag{1-24}$$

with τ the time required for an impulse to travel $\frac{\lambda_o}{4}$ meters. Setting $Z = z^{-2}$ and solving the graph, we find that $H_{DA}(Z) = N_{DA}(Z)/D(Z)$ (sum port) and $H_{BA}(Z) = N_{BA}(Z)/D(Z)$ (delta port). Here,

$$\begin{aligned} N_{DA}(Z) &= 0.485Z(Z + 0.1338 \pm j0.1738)(Z - 0.04744 \\ &\quad \pm j0.8140) \\ N_{BA}(Z) &= 0.485Z(Z - 0.06495 \pm j0.1905)(Z + 0.2719 \\ &\quad \pm j0.8014) \\ D(Z) &= (Z - 0.04932)(Z + 0.06286 \pm j0.2034)(Z + \\ &\quad 0.2042 \pm j0.5859) \end{aligned} \tag{1-25}$$

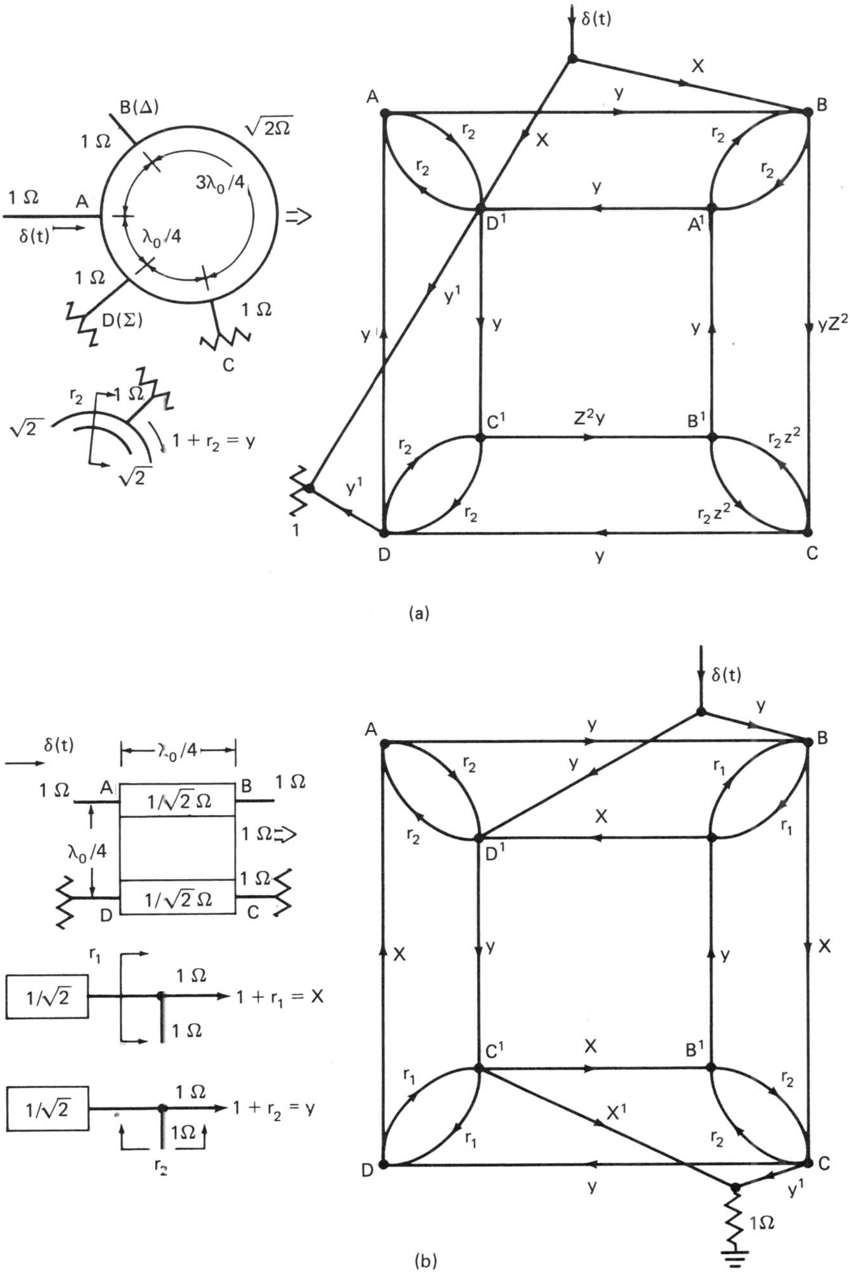

Fig. 1-6. Flow graph for (a) the ring hybrid, and (b) the 3-dB branch-line coupler.

The amplitude spectrum is found by setting $p = j\omega$ and determining the magnitude $|H(Z)|$. The envelope of the magnitude of the step-modulated response at resonance of each network can be found by inverting Eq. 1-25 and convolving the resulting impulse trains with a step-modulated signal.

2. *The 3-dB branch-line coupler.* In similar manner, it can be shown that

$$\begin{aligned} H_{BA}(Z) &= \frac{N_{BA}(Z)}{D(Z)} \\ H_{CA}(Z) &= \frac{N_{CA}(Z)}{D(Z)} \end{aligned} \tag{1-26}$$

where

$$\begin{aligned} N_{BA}(Z) &= 0.485Z(Z + 0.2425 \pm j0.3357) \\ N_{CA}(Z) &= 0.5682Z^2(Z + 0.4140) \\ D(Z) &= (Z - 0.1716)(Z + 0.3846 \pm j0.1531) \end{aligned}$$

3. *A cascade connection of equal length lines.* As a final example of the efficacy of the flow graph method, consider the important physical problem of a cascade connection of equal length TEM mode lines. Many physical situations in network theory, propagation, and optics can be represented by a tandem connection of commensurate-length TEM mode lines.

   The general flow graph is shown in Fig. 1-7. It can be shown that the general form for the forward transmission or scattering coefficient, $S_{21}(z)$, can be expressed in closed form. The inverse Laplace transform of this function, of course, represents a train of impulses. In the graph, the branch transmittances are defined in terms of the driving-point reflection coefficient, Γ_ℓ, at each junction. Thus, in Fig. 1-7,

$$\begin{aligned} \Gamma_1 &= \frac{Z_1 - 1}{Z_1 + 1} \\ \Gamma_2 &= \frac{Z_2 - Z_1}{Z_2 + Z_1} \\ &\vdots \\ \Gamma_\ell &= \frac{Z_\ell - Z_{\ell - 1}}{Z_\ell - Z_{\ell + 1}} \end{aligned} \tag{1-27}$$

where

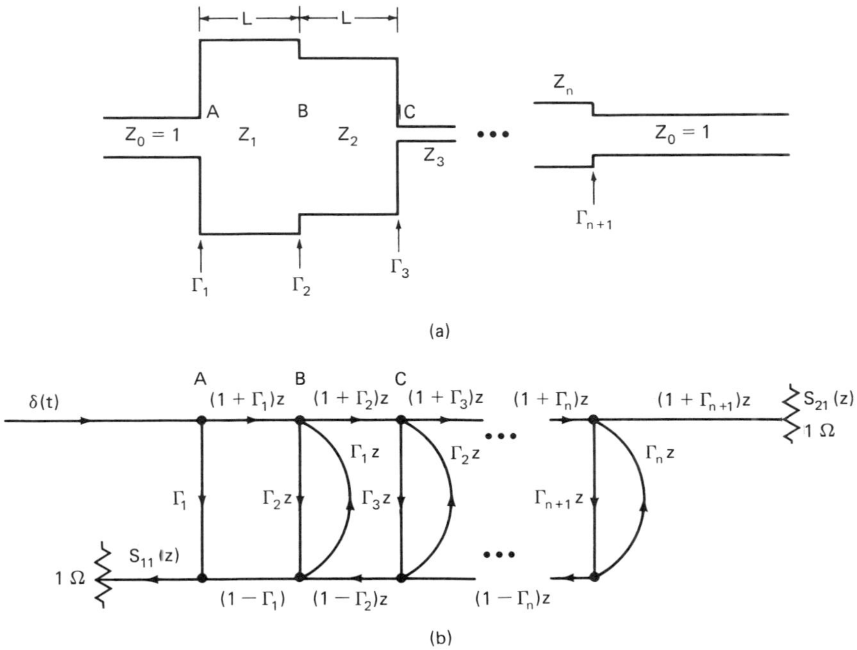

Fig. 1-7. Cascade connection of TEM lines: (a) normalized filter configuration, and (b) flow graph.

$z = e^{-p\tau}$

Z_1 = the normalized characteristic impedance of the ℓ'th line (lines are normalized with respect to generator and load impedances of 1 ohm)

$p = \sigma + j\omega$, the complex frequency variable

$\tau = L/c$

L = the length of each line

c = the speed of light in the medium

$S_{21}(z)$ is derived from the graph in Fig. 1-7 and is given by the equation:

$$S_{21}(z) = \frac{\prod_{\ell=1}^{n+1} (1 + \Gamma_l)z^n}{1 + C_1 z^2 + C_2 z^4 + \ldots C_k z^{2k} + \ldots C_n z^{2n}} \tag{1-28}$$

where n is the number of line sections. The C_k coefficients are also derived from graph theory; these coefficients are shown in Table 1-3 as a function

Table 1-3. Relationship Between C_k and Γ_l.

C_k COEFFICIENT	ONE LINE	TWO LINES	THREE LINES	FOUR LINES	FIVE LINES
C_1	$\Gamma_1\Gamma_2 = -\Gamma_1^2$	$\Gamma_1\Gamma_2 + \Gamma_2\Gamma_3$	$\Gamma_1\Gamma_2 + \Gamma_2\Gamma_3 + \Gamma_3\Gamma_4$	$\Gamma_1\Gamma_2 + \Gamma_2\Gamma_3 + \Gamma_3\Gamma_4 + \Gamma_4\Gamma_5$	$\Gamma_1\Gamma_2 + \Gamma_2\Gamma_3 + \Gamma_3\Gamma_4 + \Gamma_4\Gamma_5 + \Gamma_5\Gamma_6$
C_2		$\Gamma_1\Gamma_3$	$\Gamma_1\Gamma_3 + \Gamma_2\Gamma_4$ $+ \Gamma_1\Gamma_2\Gamma_3\Gamma_4$	$\Gamma_1\Gamma_3 + \Gamma_2\Gamma_4 + \Gamma_3\Gamma_5$ $+ \Gamma_1\Gamma_2\Gamma_3\Gamma_4 + \Gamma_1\Gamma_2\Gamma_4\Gamma_5$ $+ \Gamma_2\Gamma_3\Gamma_4\Gamma_5$	$\Gamma_1\Gamma_3 + \Gamma_2\Gamma_4 + \Gamma_3\Gamma_5 + \Gamma_4\Gamma_6$ $+ \Gamma_1\Gamma_2\Gamma_3\Gamma_4 + \Gamma_1\Gamma_2\Gamma_4\Gamma_5$ $+ \Gamma_1\Gamma_2\Gamma_5\Gamma_6 + \Gamma_2\Gamma_3\Gamma_4\Gamma_5$ $\Gamma_2\Gamma_3\Gamma_5\Gamma_6 + \Gamma_3\Gamma_4\Gamma_5\Gamma_6$
C_3			$\Gamma_1\Gamma_4$	$\Gamma_1\Gamma_4 + \Gamma_2\Gamma_5$ $+ \Gamma_1\Gamma_2\Gamma_3\Gamma_5 + \Gamma_1\Gamma_3\Gamma_4\Gamma_5$	$\Gamma_1\Gamma_4 + \Gamma_2\Gamma_5 + \Gamma_3\Gamma_6$ $+ \Gamma_1\Gamma_3\Gamma_4\Gamma_5 + \Gamma_1\Gamma_3\Gamma_5\Gamma_6$ $\Gamma_2\Gamma_4\Gamma_5\Gamma_6 + \Gamma_1\Gamma_2\Gamma_3\Gamma_5$ $+ \Gamma_1\Gamma_2\Gamma_4\Gamma_6 + \Gamma_2\Gamma_3\Gamma_4\Gamma_6$ $+ \Gamma_1\Gamma_2\Gamma_3\Gamma_4\Gamma_5\Gamma_6$
C_4				$\Gamma_1\Gamma_5$	$\Gamma_1\Gamma_5 + \Gamma_2\Gamma_6 + \Gamma_1\Gamma_2\Gamma_3\Gamma_6$ $+ \Gamma_1\Gamma_4\Gamma_5\Gamma_6 + \Gamma_1\Gamma_3\Gamma_4\Gamma_6$
C_5					$\Gamma_1\Gamma_6$

of the number of cascade line sections. A careful examination of the table will enable one to induce the rule determining the interrelationship between Γ_ℓ and C_k. As indicated, n constants need be specified for an n-section filter. The first and n'th coefficients are the easiest to determine. They are as follows:

$$C_1 = \Gamma_1\Gamma_2 + \Gamma_2\Gamma_3 + \ldots + \Gamma_\ell\Gamma_{\ell+1} + \ldots \Gamma_{n-1}\Gamma_n + \Gamma_n\Gamma_{n+1} \tag{1-29}$$

and

$$C_n = \Gamma_1\Gamma_{n+1} \tag{1-30}$$

Graph theory permits a general formula for determining the C_k coefficients. It is a recursive relationship involving derivations of previous terms and may be found in complete form in the work by Cohn, et al.[20]

Concluding Remarks on Analysis. The details for finding the impulse and step-modulated response of many familiar microwave networks together with the amplitude spectrum and phase function of these networks derived from time-domain analysis using graph techniques may be found in the literature. The flow-graph approach described here has also led to the development of numerous and novel experimental techniques.

Material Diagnostics

In the previous section, it was shown how flow graph theory was applied to the solution of network problems in microwaves. The advent of the sampling oscilloscope and the development of baseband pulse generators with subnanosecond risetimes together with graph theory and more recently developed fast Fourier transform techniques contributed significantly, in the early 1970s, to the understanding and measurement of the intrinsic properties of materials.

In 1969, Fellner-Feldegg[21] sparked interest among physical chemists by reporting on a reflection method for determining the dielectric constant and relaxation time of certain liquids. Independently, in 1969, Bagozzi[22] developed the theory for polar liquid dielectric coaxial lines, including skin-effect losses. Concurrently, Bagozzi et al. applied the theory to the dielectric relaxation time in Debye binary liquids using a time-domain method. At the 1970 Conference on Precision Electromagnetic Measurements, Nicholson and Ross[23] presented a new method utilizing reflected and transmitted pulses from a material located in a coaxial fixture to obtain its complex permittivity and permeability over a broad range of microwave frequencies. The breakthrough in material metrology was made possible by the convenience and accuracy of time-to-frequency translations (e.g., the Fast Fourier Transform technique) and through the use of computer-controlled scanning procedures. The computer was used not only for averaging the data for improved accuracy and for inverse transform calculations, but also to monitor and dynamically correct for errors caused by additive voltage drift and timing jitter in the sampling oscilloscope circuitry. The results provide, via a single impulsive but repetitive time-domain measurement, accurate complex permeability and permitivity data for a wide range of linear materials over three decades in frequency.

The configuration used for the measurements is shown in Fig. 1-8. The pulse from the pulse generator typically has a risetime of less than 50 ps, propagates down the coaxial delay line through the feed-through sampling head of the oscilloscope, and is incident on the sample of material in the test-sample holder. The pulse is partially transmitted through the sample. The transmitted pulse reflects from a short circuit, returns up the line, and passes again through the sample on its way to the oscilloscope sampling head. It can be shown that these pulses yield $S_{11}(\omega)$ and $S_{21}^2(\omega)$, where $S_{21}(\omega)$ and $S_{11}(\omega)$ are the scattering coefficients of the sample under test.

The basic principle of the measurement can be defined with the aid of Fig. 1-8. Consider an annular disk of material with permeability μ ($\mu = \mu_o \times \mu_r$) and thickness d, installed in a coaxial air-filled line with characteristic impedance Z_o, as shown in Fig. 1-8 (a).

Within the region $0 \leqslant x \leqq d$, the line will assume a new characteristic impedance, $Z = \sqrt{\mu_r/\varepsilon_r} \times Z_c$, where μ_r and ε_r may be complex. If d were infinite,

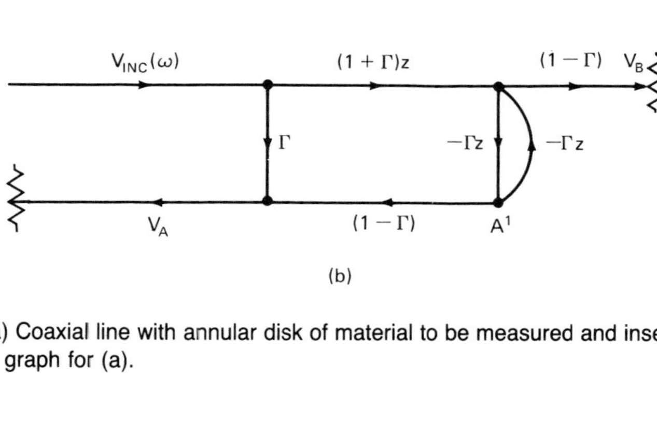

Fig. 1-8. (a) Coaxial line with annular disk of material to be measured and inserted, and (b) signal-flow graph for (a).

then the reflection coefficient of a wave incident on the interface from the air-filled line would be given simply by

$$\Gamma = \frac{Z - Z_o}{Z + Z_o} = \frac{\sqrt{\mu_r/\varepsilon_r} - 1}{\sqrt{\mu_r/\varepsilon_r} + 1} \tag{1-31}$$

For finite d, the transmission coefficient between faces A and B of the slab may be written

$$z = \exp(-j\omega\sqrt{\mu\varepsilon d}) = \exp[-j\frac{\omega}{c}\sqrt{\mu_r\varepsilon_r} \times d] \tag{1-32}$$

The scattering coefficients S_{21} and S_{11} of the slab may be obtained from the flow graph in Fig. 1-8(b) so that

$$S_{21}(\omega) = \frac{V_B}{V_{\text{inc}}} = \frac{(1 - \Gamma^2)z}{1 - \Gamma^2 z^2} \tag{1-33}$$

$$S_{11}(\omega) = \frac{V_A}{V_{\text{inc}}} = \frac{(1 - z^2)\Gamma}{1 - \Gamma^2 z^2} \tag{1-34}$$

Writing the sums and differences of the scattering coefficients as

$$\begin{aligned} V_1 &= S_{21} + S_{11} \\ V_2 &= S_{21} - S_{11} \end{aligned} \tag{1-35}$$

and letting

$$X = \frac{1 - V_1 V_2}{V_1 - V_2}$$

it can then be shown that Γ is given by

$$\Gamma = X \pm \sqrt{X^2 - 1} \tag{1-36}$$

with the sign chosen so that $|\Gamma| < 1$. Also,

$$z = \frac{V_1 - \Gamma}{1 - \Gamma V_1} \tag{1-37}$$

Now from Eq. 1-31, define

$$\frac{\mu_r}{\varepsilon_r} = \left(\frac{1 + \Gamma}{1 - \Gamma}\right)^2 = C_1 \qquad (1\text{-}38)$$

and from Eq. 1-32, define

$$\mu_r\varepsilon_r = -\left[\frac{c}{\omega d}\,\ell n\left(\frac{1}{z}\right)\right]^2 = C_2 \qquad (1\text{-}39)$$

so that

$$\begin{aligned}\mu_r &= \sqrt{C_1 C_2} \\ \varepsilon_r &= \sqrt{\frac{C_2}{C_1}}\end{aligned} \qquad (1\text{-}40)$$

Complex permeability and permitivity are thus obtained from measurements of the transmission and reflection scattering coefficients of a slab of the material. The delay-line lengths are chosen sufficiently long so that the incident, the reflected, and the doubly transmitted transient-voltage waveforms do not overlap at the sampling head. Consideration of the lengths shown in Fig. 1-8 shows that the reflected and transmitted waveforms each have a "time window" over which they may be measured without contamination by other reflections of $2L/c$ s. Typical results of this technique for the measurement of Teflon are shown in Fig. 1-9. The use of time-domain techniques to further evaluate the properties of microwave components is covered in more detail in Chap. 10.

Antenna Analysis

An understanding of the short-pulse behavior of antennas is important because processes in nature are usually not sinusoidal but rather impulsive in time. Also, in modern radar applications, very broadband signals (i.e., signals having an effective short-pulse duration) are being used to aid in target identification. The effect of short-pulse signals in antenna systems becomes observable wherever the propagation delay of the signal on the structure is comparable to the risetime or duration of the pulse. When this situation occurs, only certain portions of the antenna are involved in the radiation process at any one time.

For example, consider the radiation field of a dipole of total length $2h$, radius a, and driven by a source of internal impedance Z_g. If $h >> a$ so that the circumferential variations can be neglected, then it can be shown that the broadside-direction impulse response of such a structure is well approximated by a

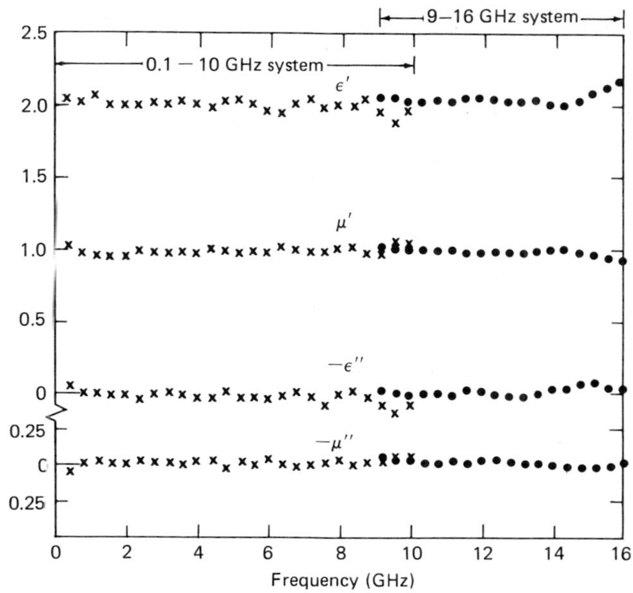

Fig. 1-9. Relative permittivity, ε, and permeability, μ, for 5.0-mm-thick Teflon.

train of impulses of varying weight and of spacing *h/c* seconds, where *c* is the speed of light.[24] In any time window less than *h/c* seconds in duration, the impulse response can be represented by a single impulse. This implies that any time-limited source waveform of duration less than *h/c* seconds can be transmitted and replicated during this time window without distortion.

It is interesting to note that once the radiation response of a given element is known, the response of that same antenna as a receiver can be found from the Rayleigh-Carson relationship derived by R.W.P. King.[25] Using this relationship, Schmitt showed that the receive response is approximately the integral of the impulse response of the element used in the transmit mode.[26]

The time-domain approach to antenna theory provides a significant degree of insight that is not obvious from the conventional frequency-domain approach. For example, the time-domain approach suggests new criteria for the design and synthesis of wide-band radiating elements.[27] Conventionally, to minimize distortion requires that the system function (i.e., Fourier transform of the impulse response) of an antenna have a flat amplitude spectrum and a linear phase function over a given band, with minimum specifications outside the band. The difficulty here is that the in-band data can be used to predict the gross response of the system to a step or pulse-modulated input but cannot accurately be used to predict the settling time or time-domain residues, respectively. The settling time is defined as the time required for the response to settle to within 5 percent of its steady-state value. The fine structure of the response or the time-domain sidelobes (or residues) depends on the out-of-band characteristics.

Specifications in the time domain can be written that conveniently describe and constrain the impulse response of an antenna. In particular, the duration of the impulse response can be constrained to be precisely τ seconds. Devices of this type can be described as having a time-limited impulse response (TLIR). In general, systems having a TLIR have interesting mathematical properties. For example, their Laplace transform is an "entire function," having no poles in the entire complex p plane. Practically speaking, the duration of the transient for a step-modulated excitation of the form

$$f(t) = 2\pi f_o t \times U(t) \tag{1-41}$$

cannot exceed the duration of the impulse response. For a pulse-modulated input of the form

$$p(t) = sin\ 2\pi f_o t\ [U(t) - U(t - n\mathrm{T})] \tag{1-42}$$

where $T = 1/f_o$ and n is an integer, the total dispersion is limited to twice the duration of the impulse response. There are no time-domain residues or distortion introduced by an antenna after this time. If the duration of the impulse response $h(t)$ is less than $T/2$, then signal distortion is limited to less than half of an RF cycle at resonant frequency f_o.

The shape of $h(t)$ determines the distortion introduced during the duration of the transient, and the magnitude of $h(t)$ determines the steady-state sinusoidal response for $t > \tau$ seconds. As an example, consider a causal system having a TLIR of duration τ given by

$$h(t) = \delta(t) + \delta(t - \tau) \tag{1-43}$$

where $\delta(t)$ is a delta function. The response of a step-modulated input described by Eq. 1-42, where $f_o = 1/2\tau$, is given by

$$\begin{aligned} r(t) &= sin\ \frac{\pi}{\tau}\ t && \text{when } 0 \leqslant t \leqslant \tau \\ r(t) &= 0 && \text{when } t > \tau \end{aligned} \tag{1-44}$$

Steady state is reached after τ seconds. If the sign of the second impulse is reversed, then the response is given by

$$r(t) = sin \frac{\pi t}{\tau} \qquad \text{when } 0 < t < \tau$$
$$r(t) = 2\ sin \frac{\pi t}{\tau} \qquad \text{when } t > \tau \tag{1-45}$$

The system "settles" to its steady-state response after τ seconds, but $r(t)$ becomes sinusoidal for $t > \tau$.

As a second and more pertinent example, consider a thin vertical monopole mounted above a ground plane, as shown in Fig. 1-10. Here the driving-point surge impedance of the monopole has been approximately matched at the base by using a long, tapered, conical TEM-mode section beneath the ground plane. A detailed description of this element appears in the literature.[28] The impulse response in the receive mode for such an antenna is shown at broadside and for an angle of incidence of 30 degrees in Figs. 1-10(b) and (c), respectively. The response of this antenna to a step-modulated signal described by Eq. 1-42, where $f_o = 1/2\tau$, can be found by convolution. The experimental result obtained with the waveform generator described earlier is shown in Fig. 1-11 and indicates that the buildup time of the response is less than one RF cycle, as predicted.

In addition to the matched dipole, a so-called wave-trapping type (WTT) has

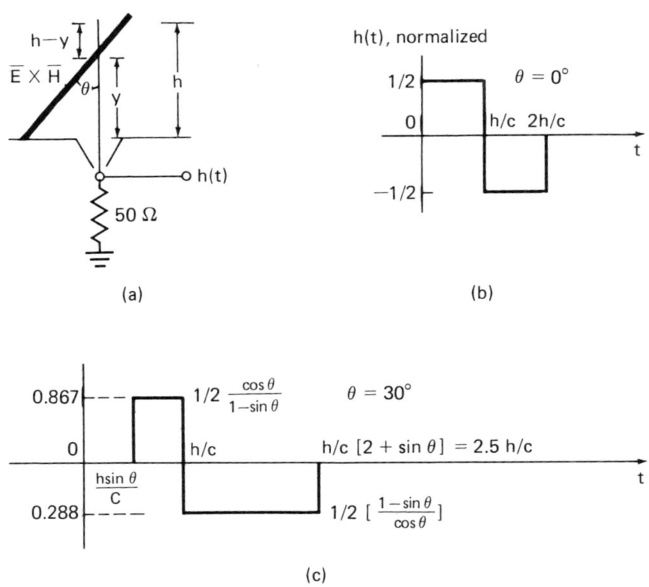

Fig. 1-10. Impulse response of a matched dipole: (a) configuration, (b) θ = 0 degrees, and (c) θ = 30 degrees.

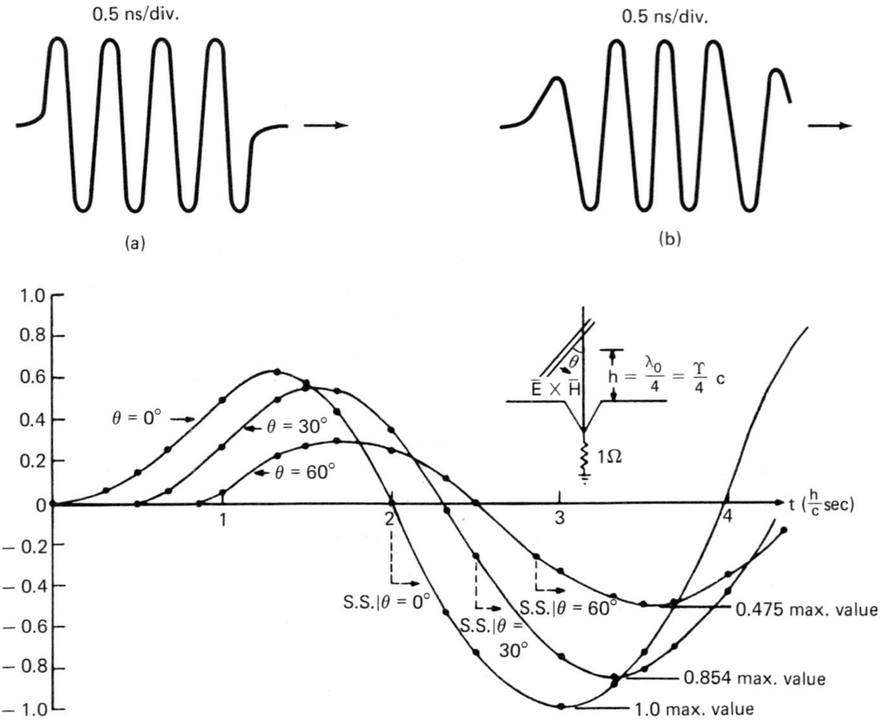

Note: S.S. = steady state, h/c seconds = T/4 seconds or ¼ rf cycle.
Step modulated response, calculated.

Fig. 1-11. Pulse-modulated response, actual (θ = 0 degrees): (a) transmitted pulse, and (b) received pulse.

been built that is characterized by a short wire of length h and a thin metal top hat.[29] There is also a matched-monopole type (MMT) where h is relatively long and the wave is matched at the tip by a metallic film of submicrometer thickness deposited on a thin glass plate. Both antennas have a TLIR. The WTT has also been used very effectively as a wide-band probe in a cavity.[30] Other members of the class of antennas that have essentially a TLIR are used in the many baseband radar applications described in Sec. 1.3.

The development and perfection of the technology necessary to obtain the transient solution to electromagnetic-scattering problems directly in the time domain has coincided with a growing interest in wide-band radar design for higher resolution and target-classification problems. There has also been great interest in reducing the radar cross section of targets, such as aircraft, over a wide frequency band by placing radar-absorbing materials in the vicinity of target scattering centers. The electromagnetic-pulse (EMP) problem associated with nuclear explosions, which has been the object of much attention recently, is also an area that is, essentially, a transient problem.

Having assessed the conception, advantages, and applications of the time-domain approach to obtaining radar target signatures, this section will now address some fundamental concepts of the time-domain signature analysis. The configuration and measurement characteristics of the time-domain scattering range will be described as well as a generalized approach for deriving the space-time integro-differential equations that describe the scattered field at any point in space.

Target Scattering Analysis

The conventional experimental approach to determining target signatures is to illuminate the target with a single-frequency waveform and measure the amplitude and phase of the return.[31] Although this approach provides narrow beamwidths and high signal-to-noise (SNR), extraneous returns from objects in the vicinity of the target can be eliminated only be placing the target in an expensive anechoic chamber. To obtain the broadband or transient response, the measurement must be repeated over the entire band of frequencies contained in the spectrum of the transient excitation.

The cost of an anechoic chamber and the difficulty in making coherent measurements over a broad frequency band led to the development in 1967 of a scattering range that measures the transient response directly in the time domain by illuminating the target with a smoothed-impulse waveform.[32] The resulting smoothed-impulse response is useful for several reasons. First, it contains all the information about the electromagnetic scattering properties of the target over the frequency band defined by the incident-smoothed impulse. Since the impulse response is closely related to the actual target geometry, it is particularly useful for target classification. Second, the radar cross section (or, equivalently, the frequency response) can be obtained from the impulse response by a Fourier transform. Third, the response of the target as a result of any waveform can be obtained from the impulse response by a simple convolution procedure.

For the purpose of target classification, the impulse response is a particularly useful characterization since it is closely related to the actual target geometry. Finally, the impulse response can provide a better understanding of electromagnetic scattering phenomena. This section describes the basic operating principle of the original time-domain scattering range developed by DeLorenzo[7] and later improved by Bennett.[9]

A functional block diagram of a representative time-domain scattering range is shown in Fig. 1-12. The system signal source is a high-voltage switch that generates a 300-V step function with a risetime less than 100 ps. The signal is

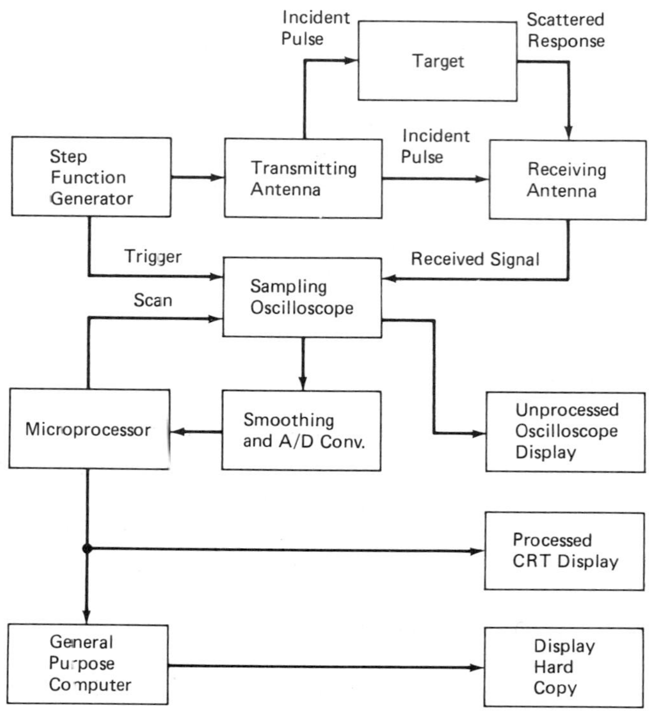

Fig. 1-12. Video time-domain scattering range.

radiated from a vertical monopole antenna protruding through a circular ground plane 20 feet in diameter. The ground plane provides isolation from equipment and cabling that is necessary for its operation. Since the ground plane acts as an electromagnetic mirror, this system is limited to the measurement of targets with mirror symmetry. The radiated wave scatters from the target and is received on a flush-mounted coaxial horn antenna; the latter essentially smooths and differentiates the signal and thus provides the smoothed-impulse response of the target. The received waveform is sampled by a sampling oscilloscope that has been triggered by the initial pulse and whose sampling-gate deflection is under the control of a microprocessor.

Unprocessed data is displayed on the oscilloscope CRT, while the sampled-and-held waveform is passed through a low-pass filter, digitized, read into the microprocessor, and stored on magnetic tape automatically. This system has been designed to correct for long-term timing drift and/or amplifier drift. In addition, the waveforms are stored in such a way that they are ready for the subsequent operations of averaging (to remove short-term noise) and baseline jitter. For example, the effects of a time-varying baseline are subtracted from measured waveforms to improve system accuracy. Finally, the microprocessor is connected to a general-purpose computer for more sophisticated and/or more rapid signal processing.

The salient characteristics of such a scattering range are the speed and simplicity with which transient responses (or, equivalently, broad-band frequency-domain data; e.g., over almost three decades in bandwidth) can be obtained. These advantages accrue because the time-domain scattering range yields an "uncontaminated" interval of time between the arrival of the direct wave and the arrival of unwanted reflections. Figure 1-13 shows the relative location of the elements on the ground plane, and Fig. 1-14 shows the range response as it appears at the oscilloscope (no data processing has been used at this point). The transmitted signal travels outward from the base of the wire antenna and is received at R at time $t_o = d/c$ (where c is the speed of light). This time is marked by the pulse at the left end of the trace in Fig. 1-14. The outgoing wave reaches the target at $t = r/c$, is reflected, and arrives at the receiver at $t_1 = (2r + d)/c = t_o + (2r/c)$.

The targets are usually located anywhere from two to five feet from the transmitting antenna; therefore, target returns lie in the region marked by the doublets at t_1 and t_2 in the lower photograph. The response at the right edge of the trace that occurs after t_3 marks the arrival of the pulse reflected by the table edge and the effects of the pulse radiated from the tip of the transmitting antenna. The entire region between the direct transmission and the table-edge response forms a convenient time "window" to view the target response and allows one to "gate out" (in time) unwanted reflections. In this way, undistorted transient target responses can be viewed without resorting to anechoic chambers. In addition, a single time-domain measurement obviates the requirement for measurement of the amplitude and phase response at many frequencies.

The accuracy of the measurement system can be estimated by measuring the

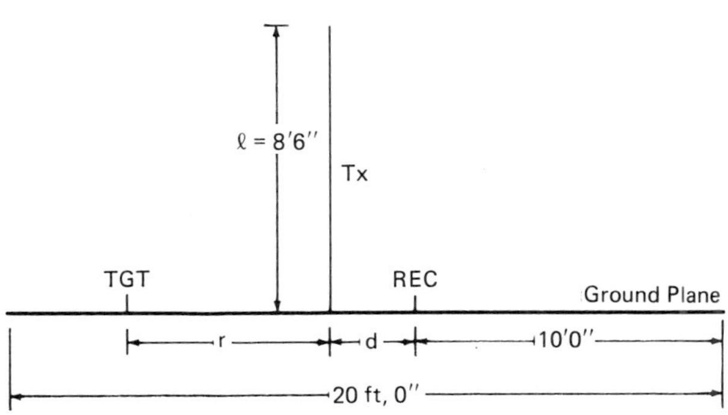

Fig. 1-13. Geometrical configuration of video time-domain scattering range (d = 18 in.; r = 36 in.).

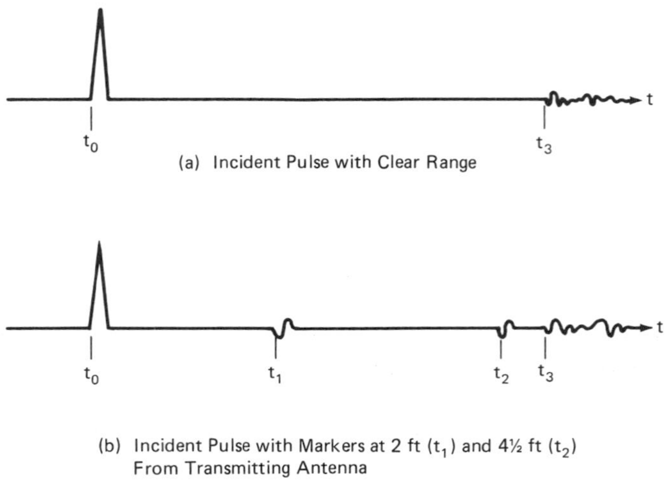

Fig. 1-14. Response of video time-domain scattering range showing incident pulse and time window (horizontal scale: 2 ns/div; vertical scale: 200 mV/div.): (a) incident pulse with clear range, and (b) incident pulse with markers at 2 ft (t_1) and 5 ft (t_2) from the transmitting antenna.

standard deviation of the measured voltage. For the case where the peak of the incident pulse is approximately 500 mV and the 10-mV scale on the sampling oscilloscope is used, the standard deviation of the measured response has been measured as 0.06 mV, or approximately 0.6 percent of the peak value of the measured return.[33] In addition, the measured responses can be further processed on a general purpose computer by means of a convolution procedure to obtain the response resulting from a Gaussian-shaped incident pulse (or any other incident waveform whose spectrum is contained within that of the measure incident pulse) rather than from the approximate smoothed impulse used in the actual measurements.

Such a scattering range produces a radiated signal having an effective pulse width of approximately 200 psec at the 3-dB points. The measurement system can be used to determine the target response over a frequency band that extends from 100 MHz to approximately 4.5 GHz. The high-frequency limit is approximately measured by the reciprocal of the pulse width; the low-frequency cutoff is a function of the finite length of the wire radiating structure. This facility has been used to measure, directly, the smoothed impulse response of targets and to verify the results obtained by calculations. It has also been used in its own right to evaluate the response of multi-scattering center targets, to identify the scattering centers, and to measure their amplitude. The target response caused by an arbitrary radar waveform can then be calculated by a convolution calculation. Proper scaling must be used for larger targets.

Until now, we have addressed a time-domain measurement approach to the scattering problem. Analytic solutions to the scattering problem have been, for the most part, solved in the frequency domain. Application of the classic boundary-value solution technique to problems where the target surface coincides with the coordinate surfaces of a separable coordinate system is limited to a few target classes. Approximate techniques such as geometric optics, physical optics, and the Rayleigh approximation have been applied to other problems to obtain estimates of the scattered response. With the development of the digital computer, however, frequency-domain integral-equation techniques extended the solution of the scattering problem to simple bodies of arbitrary shape from near zero frequency well into the resonance region. These integral-equation solutions have provided many interesting and useful results; however, certain difficulties can exist near the resonances of the internal-scattering problem.[34]

It has been observed more recently that an integro-differential equation and its solution can also be obtained directly in the time domain. This alternative approach of solving Maxwell's equation directly in the time domain was developed by Bennett et al.[9] and applied to a number of problems. Since the integro-differential-equation approach requires a solution only on the surface of the scatterer, it provides advantages in both computer storage and running time over a finite-difference approach for scattering problems involving perfect conductors.

1.3 SYSTEMS APPLICATION OF TIME-DOMAIN TECHNOLOGY

Introduction

In the above discussion, we mentioned that time-domain techniques were motivated by the need to make certain measurements on microwave networks and later scattering targets. As the measurement technology developed, numerous new system concepts became apparent.

The development of the sampling oscilloscope provided a new and convenient method of viewing very short pulses and transients and also resulted in new step-function (and pulse-) generating schemes. Introduction of the avalanche transistor provided higher voltage pulses for application areas other than diagnostics. Design of new dispersionless probes and radiating elements contained many of the elements necessary to build a time-domain reflectometer or radar system.

The final and key component required to complete the requirement was a sensitive, range-gated receiver. Such a receiver, developed in 1972 by Nicolson, involved the concept of maintaining a constant false alarm rate by having a tunnel diode held in a closed loop by triggering on system noise.[35] Using this receiver and the other elements, a simple radar system was developed for use as a precollision sensor for an airbag on an automobile; this sensor had a range of 10 feet.[36] Later, as improvements in receiver sensitivity and transmitter peak power were realized, the radar range was extended to 25 feet for use in NASA spaceship docking, to 150 feet for radar braking for cars,[37] to 250 feet for airport runway and taxiway traffic control, and to 5000 feet for tanker ship docking. The range resolution and accuracy (a function of the signal-to-noise ratio) of these systems were less than 1 foot.

In addition to free-space reflectometry, the time-domain technique was used as a single-wire reflectometer for liquid-level sensing and collision avoidance.[38] Here the transmitted pulse is coupled to a single-wire dielectric coated transmission line (i.e., a Goubau line), and any obstacle near the wire (e.g., an air-water interface) produces a major reflection. A properly designed Goubau line has been experimentally determined to be dispersionless.

Another area that received early attention using time-domain network techniques was multiplexing.[39] Gigabit logic design first appeared in 1973 with the development of a 1-gigabit data system using simple transmission lines and comparative circuitry. Some of this circuitry was also tested in Atlanta, Georgia, by the Sperry Systems Management Division for use as a baseband communications link for traffic control application (e.g., between the control box at an intersection and a central processing unit).

Free-Space Reflectometry

In this section, we will describe in more detail the design of the CFAR tunnel-diode receiver and its application to free-space reflectometry.

The Constant False-Alarm-Rate Receiver. By employing processing techniques, the sensitivity of a tunnel-diode receiver can be increased significantly at the expense of circuit complexity. The scheme used to accomplish the increase in sensitivity employs a constant false-alarm-rate (CFAR) receiver.[40] As the threshold level of the single-hit receiver is reduced, the sensitivity is obviously increased, but at the expense of the false-alarm rate. If the hits obtained from noise alone are fed to a register that assigns a + 1 each time a noise spike exceeds the threshold and a zero when it does not, then the outputs from, say, 32 stages of this register can conveniently be summed. By appropriately feeding the register, a sliding average of detector hits in any group of 32 consecutive periods can be obtained. The threshold can be arbitrarily lowered until, on the average, three or four hits caused by noise alone are contained in the register; the transmitter is disabled during this time. A dc voltage derived from the sum port of the register can be used via a feedback loop to maintain automatically, on the average, an arbitrary number of pulses in the register caused by noise. This loop sets the constant-bias current through the tunnel-diode threshold detector in the absence of signal (i.e., on alternate pulse-repetition periods when the transmitter is disabled). When the transmitter is enabled, the threshold for establishing the presence of a target is set into a separate target register-and-sum network. For example, a target is said to be present when 25 or more of the 32 stages of the second register are filled. A block diagram of the CFAR receiver is shown in Fig. 1-15.

A Low-Cost Radar for Free-Space Time-Domain Reflectometry. The perfection of a waveform source, efficient radiating and receiving antenna elements, and inexpensive threshold-sensing devices leads naturally to the development of a short-range free-space time-domain reflectometry or *BA*seband *R*adar (BAR) system.[41] There are many useful applications for devices of this type, especially if the cost can be kept well below $1000 for a transceiver pair. The range of such a system, naturally, is limited because very little energy is transmitted and detection is primarily a function of energy on target. A unique feature

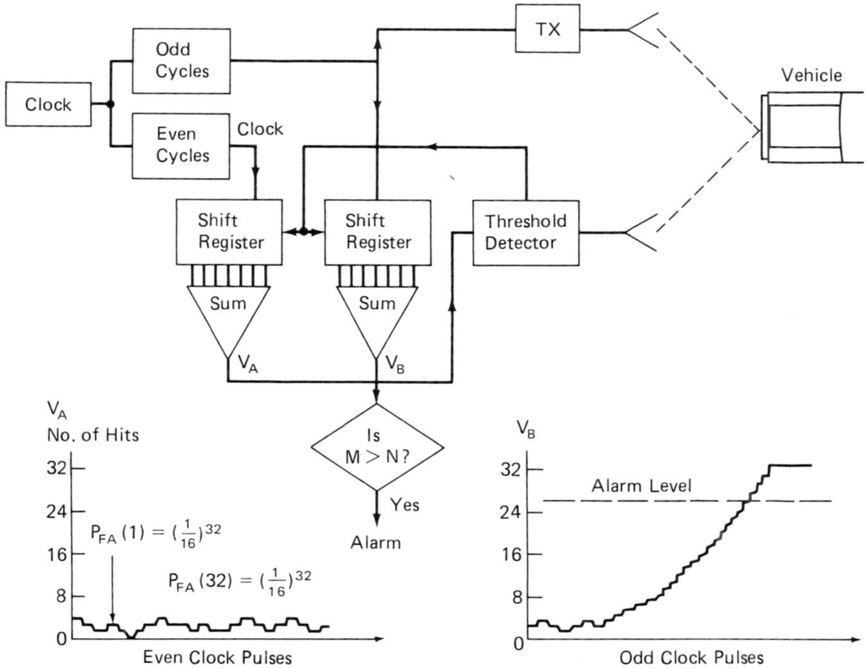

Fig. 1-15. Constant false-alarm rate (CFAR) receiver.

of a BAR system is its high-resolution capability and virtually interference-free operation because of range gating and the processing action of the CFAR receiver.

Applications

Some primary application areas for this technology appear to be the following:

1. Automobile electronics, where a BAR sensor has been successfully demonstrated for both precollision sensing and braking
2. Airport ground-traffic control, where intersection control at high-density terminals and independent altitude warning is desired, especially under Category 3 landing conditions
3. Ship docking for both spaceship and large oil tanker use.

Automobile Electronic Sensing. At the Fourth European Microwave Symposium, progress was reported in automobile precollision sensing for air bag application (e.g., the BARBI system) for a range of 10 feet or less.[42] Work on a BAR braking sensor was conducted later. In an audited test by the U.S. Government in Boulder, CO, in May 1975 and later in June 1977, the use of the BAR sensor in a binaural configuration was demonstrated for sensing objects within a lane width at a range of about 150 feet.[43] Beam narrowing is accomplished by signal processing using the transmitter receiver configuration shown in Fig. 1-16 in conjunction with a microwave tapped delay line. By preselecting the ports at which the signals from both receive channels coalesce, it is not difficult to show that the effective beamwidth θ can be narrowed in accordance with the relationship,

$$\theta = 2\, sin^{-1} \frac{\Delta}{2a} \tag{1-47}$$

where Δ is the difference between the tap locations at which the pulses first coalesce and a is the separation between the transmitter and receiving units. This technique makes it possible to achieve an effective beamwidth of several degrees. It was demonstrated at these tests that the resolution of the system is fine enough to distinguish between a human being, a bicyclist, a motorcyclist, and an automobile only a few feet apart.

Airport Ground-Control Sensors. There are several situations in an airport where a need exists for a short-range high-resolution sensor, among them the following:

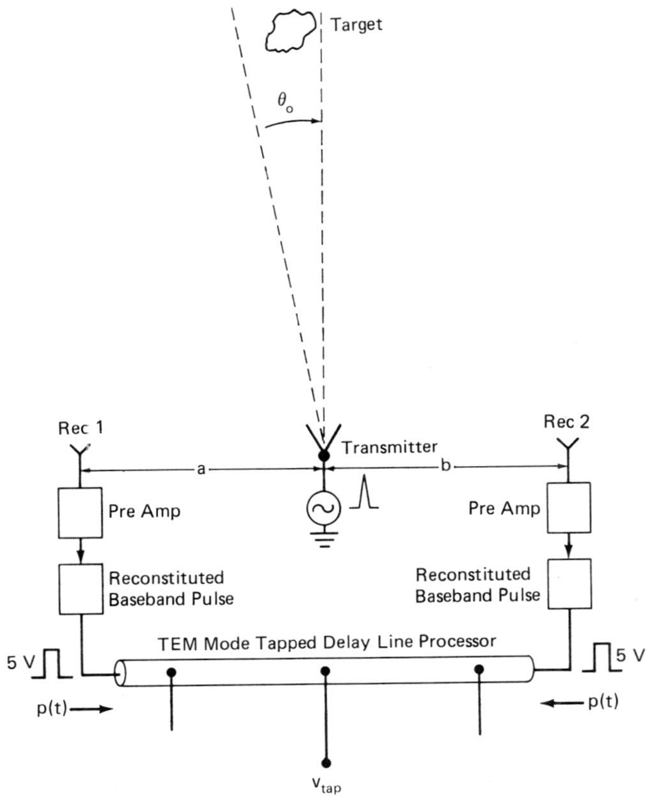

Fig. 1-16. Beam shaping by time-domain coincidence techniques.

1. Runway clear: for determining in bad visibility whether the complete runway is clear of traffic; if not, for determining where the obstacle is located
2. Taxiway intersection control: for determining the presence, direction of travel, and speed of an aircraft approaching an intersection
3. Taxiway guidance: for giving an indication of aircraft position relative to a taxiway centerline under very low-visibility conditions
4. Ground-based altimeters at the approach end of a runway at the far, mid, and near markers: for providing an indication of aircraft height from the ground when it is at altitudes up to 1000 feet.

Tests have been made on all of these applications, one of them being illustrated in Fig. 1-17. The transmitter and receiver units were set up adjacent to an active taxiway with two receivers flanking the transmitter. As the aircraft taxied past

Fig. 1-17. Airport intersection-control set.

the equipment, the difference in times of acquisition by the two receivers was used to gate the clock to a counter whose output, in turn, fed a small calculator programmed to display velocity directly. Tests on a variety of aircraft from small private planes to large jets gave a range of velocity readings from 10 to 30 mph, which correlated well with speed measurements made using a stop watch. A similar system has been used to measure automobile velocity from the side of a road (instead of looking along the road as in Doppler methods), giving results of 20 mph, which agreed within ± 1 mph with a pair of mechanical switch pads on the roadway. For runway/clear applications, tests have shown that complete coverage requires a several-hundred-foot spacing between successive units alongside the runway.

Ship-Docking Sensor. The capability to display directly the range to a large broadside target in feet has been used to demonstrate use of a baseband radar for tanker docking at distances as great as 1600 feet. It is significant that a return is obtained only when some metallic part of the target is normal to the line of the radar; hence, although the system has broad angular coverage, it can detect the tanker only when it is normal to the beam-pointing angle. The ultimate range of the present receiver against the side of a large tanker is about 5000 feet. With a 1-s integration interval, range uncertainty during tests was about 1 foot rms out to 300 feet and 3 foot rms to 1000 feet. For a slowly docking ship, range rates can be measured to about 1 ft/min by digital smoothing and differentiation.

Guided-Wave Reflectometry

The previous subsections have dealt with the free-space radiation of basebandlike pulses in a radar mode. There are also applications of baseband techniques in guided-wave modes where there is little external radiation, and the system performs strictly as a reflectometer. Examples of these applications include liquid-level sensing and a new collision-avoidance system for rapid-transit vehicles.

Sensing Liquid Level. There are many applications where the accurate sensing of liquid level is of importance.[44,45] Some of these applications include the sensing of oil or liquid-natural-gas levels in tanker ships and the measurement of fuel supply in aircraft. Present techniques for accomplishing liquid-level measurement primarily use float-type devices and suffer from a lack of accuracy and reliability. Especially in those areas where highly viscous or caustic fluids are involved, floats seem to fail after brief use. Ultrasonic as well as EM sensors have also been employed. Two EM types that have been employed with limited success (i.e., depending on the corrupting effects of the fluid) have been the straight (air dielectric) coaxially supported cable and the segmented coaxial capacitor. In both devices, the presence of the fluid changes the capacity per unit length of the line and the accompanying resonant frequency and/or the reflection coefficient at the air-fluid interface, thereby becoming a measure of the liquid level. Eventually, the space between the inner and outer conductor of the line or coaxial capacitor becomes contaminated.

The time-domain reflectometer method described here has been experimentally verified and is similar to the coaxial-cable approach inasmuch as the air-liquid interface results in a mismatch and a measurable change in the reflection coefficient. The major difference is that there is no need for an outer conductor; with the air of a single-wire Goubau-line system one is able to measure the liquid level in a tank, and, with the aid of a processor, one can also identify the liquid contained in the tank. The scheme makes use of a baseband pulse source (e.g., 200 ps in duration) feeding a single Teflon-coated wire that extends to the bottom of perhaps a 100-ft tank (for a 100-ft tank, the diameter of the wire should be greater than 1/8 in.). Processing equipment can be minimal if liquid level is the only concern. Somewhat more sophistication is required for the quantitative analysis of the liquid in the tank.

The scheme can be best explained with the aid of Fig. 1-18. A rectangular pulse $p(t)$ is fed past a parallel line coupler to a coaxial horn and a dielectric coated line. The leakage signal fed to the processor via the coupler establishes

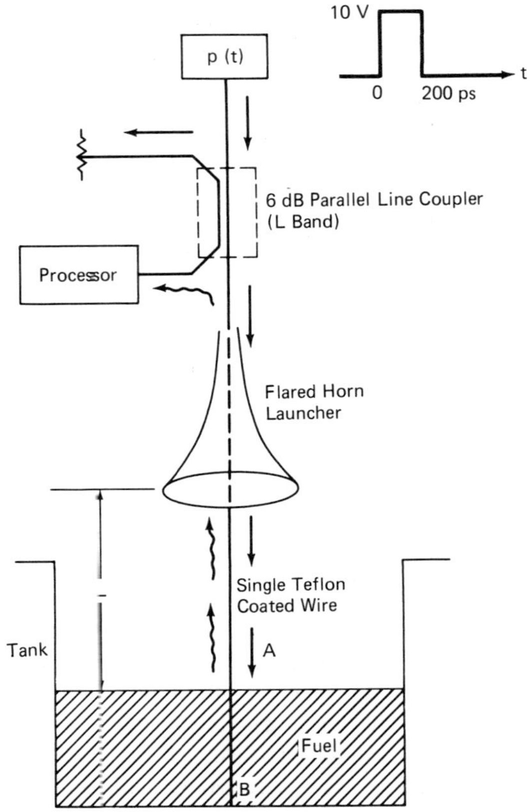

Fig. 1-18. Time-domain reflectometer scheme.

the reference signal, $t = 0$. The signal reflecting off the edge of the horn produces an unwanted signal in the processor at a known distance after $p(t)$ and is gated out. The signals of importance are the following:

1. The amplitude and time delay of the signal reflecting off the surface at A, which uniquely determine the height of the liquid in the tank and the properties of the liquid
2. The signal reflecting off the bottom of the tank at B for calibration purposes, especially if the tank also contains a second liquid (e.g., water)

This technique results in a liquid-level accuracy in the order of an inch or better for the first 10 feet into the liquid. Early experiments conducted indoors used oil and water (e.g., the oil had a dielectric constant of 5) in a large laboratory sink. Other experiments conducted in an outdoor water tower further verified the level-determining theory. Additional work is currently proceeding to optimize the design of the horn, coupler, and processor. In cases where only the liquid level is of concern, the processor design is considerably simplified. It was determined that the energy in the incident pulse is at least three orders of magnitude below the danger level for the explosion of aviation fuel.

Collision Avoidance System Using Baseband Reflectometry (CASBAR). A guided-wave approach for avoiding collisions between rapid-transit vehicles is particularly attractive when the vehicles are committed to a track or guideway. This approach permits one to see around curves as well as eliminating anomalous responses from overhead structures and tunnels. In addition, there is virtually no radiation in a guided-wave system.

One such proposed scheme, known as CASBAR,[46] employs baseband techniques and is shown in Fig. 1-19. A surface-wave guiding structure similar to that used in the liquid-level sensor lies parallel to the track. Each vehicle carries a self-contained radar system with logic circuits connected to the train control. The operational sequence starts with a short pulse launched on the main guide via a surface wave directional coupler. The wave travels along the guide until it is partially reflected by one or more passive reflectors adjacent to the guide mounted on the forward vehicle. This reflected signal is coupled into the tunnel diode receiver by another directional coupler. (For redundancy, another system may be installed on the other side of the guideway.) A target may be detected at a range of several hundred feet or more; the range rate is also computed by the receiver processor. A breadboard model of this system has recently been

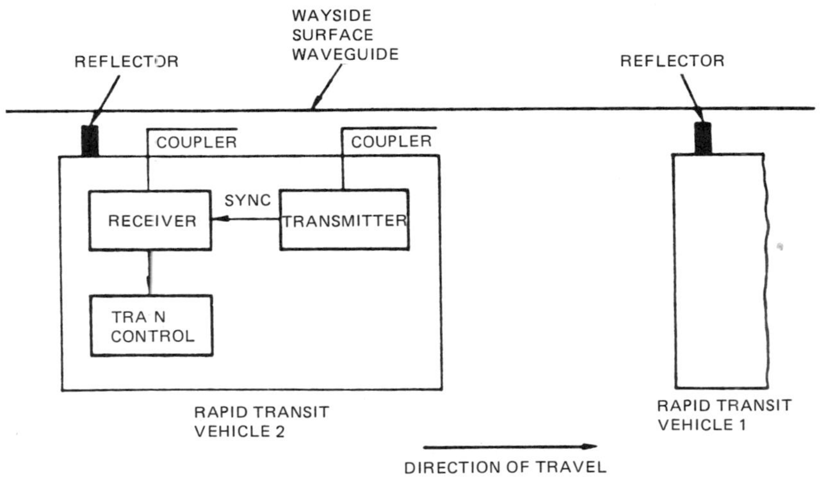

Fig. 1-19. CASBAR scheme.

completed and tested in cooperation with the Boeing Aerospace Company and the Urban Mass Transit Administration, Department of Transportation.

1.4 SUMMARY

The advances in the state of the art in subnanosecond hardware and software technology have brought the engineer much more powerful tools for the solution of time-domain electromagnetic problems. The development of subnanosecond-pulse techniques, which began with waveform generation, has now progressed beyond the fields of metrology and is being applied to computer applications, radar, time-domain reflectometry, target-signature analysis, and communications theory. The time-domain scattering range represents a valuable aid for verification of calculations, for determination of target responses, and for better understanding of the transient scattering phenomena. The space-time integral equation technique provides a powerful tool for calculation of the transient scattered response of targets and provides valuable insight into the solution of target-classification problems. The BAR approach to sensing shows that a fresh look at the radar problem, at least for short-range application, can result in a high-resolution and inexpensive sensor that requires only a few microwave components. Previous radar approaches to the short-range problem have involved scaled-down versions of the conventional and more complex long-range systems; the BAR unit more closely resembles a free-space time-domain reflectometer.

In the chapters that follow, the work by other investigators in the field of time-domain measurements on materials and components, instrumentation for transient measurement, and the application to subsurface radar (e.g., for detecting buried pipes and tunnels) will be presented. In addition, the usefulness of target scaling will be compared to actual measurement.

REFERENCES

1. Truxal, J. 1955. *Control system synthesis*. New York: McGraw Hill, p. 509.
2. Nahman, N. 1983. Picosecond-domain waveform measurement, status and future directions. *IEEE Trans. on Instrumentation and Measurement* IM-32, No. 1, 117–123.
3. Papoulis, A. 1962. *The Fourier integral and its applications*. New York: McGraw Hill, pp. 25–27.
4. Ross, G. F. 1967. Choosing the limits on the convolution integral, *IEEE Trans. on Education* E-10, No. 1, 45–47.
5. Hallen, E. 1948. Admittance diagrams for antennas and the relation between antenna theories. *Technical Report 46*, Cruft Laboratory, Harvard University.
6. Schmitt, H. J. 1959. Transients in cylindrical antennas. *Technical Report 296*. Cruft Laboratory, Harvard University.
7. DeLorenzo, J. D. 1967. A range for measuring the impulse response of scattering objects. *IEEE NEREM Record* 9: 80–81.
8. Hewlett Packard Co. 1964. Time domain reflectometry. *Application Note 62*. Palo Alto, CA.

9. Bennett, C. L., and Ross, G. F. 1978. Time domain electromagnetics and its applications, *Proc. IEEE* 66, No. 3: 299–318.
10. Ross, G. F. 1963. The transient analysis of multiple beam feed networks for array systems. Ph.D. dissertation, Polytech Institute of New York.
11. Ross, G. F. 1958. High frequency phase-synchronized signal synthesizer. U.S. Patent No. 3,418,604.
12. Cronson, H. 1975. Picosecond pulse sequential waveform generation. *Proc. IEEE, 1975 MTS International Symposium:* 185–186.
13. Nicolson, A. M. 1972. Advances in subnanosecond pulse technology. Physics and Electronics Dept., Royal Radar Establishment Seminar, Malvern, England.
14. Mason, S. 1955. Feedback theory—further properties of signal flow graphs, *Proc. IRE* 44: 920–926.
15. Ross, G. F. 1966. The transient analysis of certain TEM mode four-port networks. *IEEE Trans.* MTT-14, No. 11: 528–542.
16. Robichand, L.; Boisvert, M.; and Robert, Jean. 1962. Signal Flow Graphs and Applications. New York: Prentice Hall.
17. Knopp, K. 1947. *Theory of Functions,* Part II. Dover Publications, pp. 35–39.
18. Getsinger, W. 1960. Analysis of certain transmission-line networks in the time domain. *IRE Trans.* MTT-8 301–309.
19. Ross, G. F. 1970. A simple method for obtaining the system function of a cascade connection of transmission lines. *IEEE Trans.* MTT-18, No. 10: 738–740.
20. Cohen, M., and Ott, E. 1966. The impulse response of cascaded transmission lines. Sperry Research Center Research Report.
21. Fellner-Feldegg, H. 1969. The measurement of dielectrics in the time domain. *J. Phys. Chem.* 73: 616–623.
22. Bagozzi, R. P.; Ives, W.; and Nahman, N. 1969. Determination of the dielectric relaxation time in a Debye binary liquid by pulse measurements. Int. URSI 16th Gen. Assembly, Ottawa, Canada.
23. Nicolson, A. M., and Ross, G. F. 1970. Measurement of the intrinsic properties of materials by time domain techniques. *IEEE Trans.* IM-19: 377–383.
24. Ross, G. F. 1970. The system function of a microwave dipole aerial derived from time domain considerations. 1st European Microwave Symposium, London, England, pp. 393–396.
25. King, R. W. P. 1956. *Theory of the linear anteea*, Cambridge, MA: Harvard Press.
26. Schmidt, H. 1967. A survey of time domain techniques in antenna theory. 1967 IEEE NEREM paper, Plenary Talk, Boston, MA.
27. Ross, G. F. 1968. A time domain criterion for the design of wideband radiating elements. *IEEE Trans. Antennas and Propagation* 16, No. 3: 355.
28. King, R. W. P., and Schmidt, H. 1962. The transient response of linear antennas and loops. *IRE Trans.* AP-10: 222–228.
29. Ross, G. F. 1968. Members of the class of TEM mode wire antennas with a TLIR. IEEE International Symposium on Antennas and Propagation, Symposium Record, pp. 249–252.
30. Fenster, P. 1970. Wideband Wave Trapping Antenna Having a TLIR. U.S. Patent No. 3,545,002.
31. Andreasen, M. G., "Scattering from Bodies of Revolution," IEEE Trans. Antennas and Propagation, Vol. AP-13, p. 303, March 1965.
32. DeLorenzo, J. D. 1970. Radar target identifying apparatus. U.S. Patent No. 3,523,292.
33. Bennett, C L.; DeLorenzo, J.; Auckenthaler, A; and Smith, R. 1973. Space-time integral equation to the large body scattering problem. Contract Report USAF, RADC-CR-73-70.
34. Mitzner, K. M. 1968. Numerical solution of the exterior scattering problem at the eigenfrequencies of the interior problem. Fall URSI Meeting, Boston, MA.

35. Nicolson, A. M., and Brophy, R. 1973. Detector having a constant false alarm rate and method for providing same. U.S. Patent No. 3,755,696.
36. Ross, G. F. 1974. BARBI, a new radar concept for precollision sensing. Soc. of Auto. Engineers, 3rd International Conference on Occupant Protection, Paper 740574, pp. 141–152.
37. Ross, G. F. 1978. A baseband radar system for auto braking application. Soc. Automotive International Meeting, Paper 780262.
38. Ross, G. F., 1974. Apparatus and method for measuring the level of a contained liquid. U.S. Patent No. 3,832,900.
39. Ross, G. F.; Susman, L.; and Eves, E. 1973. Time division multiplexor–demultiplexor for digital transmission at gigaHertz rates. U.S. Patent No. 3,763,318.
40. Skolnik, M. 1962. *Introduction to radar systems*. New York: McGraw-Hill, pp. 566–567.
41. Nicolson, A. M., and Ross, G. F. 1975. A new radar concept for short range application. IEEE International Radar Conference Symposium Record, Washington, D.C., pp. 146–151.
42. Ross, G. F. 1964. BARBI. Fourth European Microwave Conference, Montreaux, Switzerland, pp. 146–149.
43. Ross, G. F., and Robbins, K. 1977. Baseband radar braking system for automobile application. Final Report, U.S. Dept. of Commerce, Contract #OT-0142.
44. Ross, G. F. 1976. Apparatus and method for sensing a liquid with a single wire transmission line. U.S. Patent No. 3,995,212.
45. Rao, R., and Frew, J. 1981. Surface wave reflectometer for liquid level measurements. Internal Research Report #SRC-RR-81-42, Sperry Research Center, Sudbury, MA.
46. Cronson, H,; Ross, G. F ; and Rao, R. 1981. Collision avoidance system using short pulse reflectometry. U.S. Patent No. 4,254,418.

2
Status and Future Directions of Picosecond Domain Waveform Measurements

Norris S. Nahman

National Bureau of Standards

2.0 INTRODUCTION

The main purpose of this chapter is to (1) summarize the present-day capability in time-domain measurements for pulses or transient signals whose shape (evolution with time) occurs on a time scale of the order of picoseconds, and (2) offer some opinions as to the future directions of electrical and optical picosecond-domain pulse waveform measurements.* Consequently, the techniques and other data presented here are limited to those suitable for such fast signals. For example, all of the methods cited recognize that the signal-circuit interconnections or transmission paths in devices and in instrumentation are distributed or waveguiding in nature.

The secondary purposes of the chapter are (1) to point out succinctly the significance of IEEE Pulse Standards 181 and 194, and (2) to provide clarification of the classification of measurement methods defined in References 1 and 2 by presenting some illustrative instrumentation block diagrams.

The review of the state of the art of picosecond time-domain waveform measurements presented here includes measurements in both the electrical (dc through mm maves, etc.) and optical regions (infrared through visible, etc.) of the electromagnetic spectrum. This review is the latest edition of a series of reviews[1–3] on high-speed pulse measurements compiled by the author commencing in 1967; specifically, this review updates the 1983 review.[3] The significance of the Institute of Electrical and Electronic Engineers (IEEE) Pulse Standards 181 and

194 and of the identical International Electrotechnical Commission (IEC) Standards 469-1 and 469-2 is discussed briefly. The classification of time-domain measurements from the 1978 review[2] is summarized and augmented with basic instrumentation block diagrams. The present-day capabilities are presented via temporal resolution state-of-the-art charts using the 1978 format. For the reader's convenience, the 1978 charts are reproduced here, along with charts that show only those changes that have occurred since 1978. The references cited in all of the chart entries are contained in the list of references.

2.1 PULSE STANDARDS

For measurement purposes, fundamental time-domain definitions for pulse and wave form terms along with definitions and descriptions of time-domain pulse measurements are required. With such definitions and descriptions in hand, it is then possible to determine the characteristics of practical and hypothetical pulses (waveforms)—regardless of accuracy or precision—that may occur in a wide range of technologies. IEEE Pulse Standards 194 and 181[4,5] provide the definitions and description; the same definitions and descriptions are to be found in IEC Publications 469-1 and 469-2,[6,7] respectively.

During the past 15 years, two technological developments have had a major impact on picosecond-domain measurements: (1) the minicomputer, and (2) the laser. The former has led to automatic time-domain measurement systems that utilize computer control and data reduction, while the latter has spawned the new field of optoelectronics. The transformation of laser pulse sources from laboratory devices into commercially available optical pulse sources has resulted in a confluence of picosecond-domain measurement methods from the electrical and optical regions of the electromagnetic spectrum, respectively, with the interface occurring in the new field of optoelectronics. Accordingly, electrical and optical time-domain measurement standards are inherently drawn together because the ensuing measurement problems require correlation between electrical and optical time-domain properties and pulse-waveform parameters. IEEE Pulse Standards 194 and 181 definitions are valid for all pulses, waveforms, etc., across the entire electromagnetic spectrum, ranging from baseband through millimeter wavelengths and on through the ultraviolet. Thus, the words *baseband, electrical,* or *optical* are not defined in these standards, but pulse and pulse parameters or pulse features are defined and can be applied to time-varying electromagnetic fields in any region of the spectrum.

Unfortunately, many engineers and scientists involved in time-domain measurements of pulses (waveforms) do not use the terms, procedures, etc., that appear in the IEEE Standards and, as a result, at times "reinvent the wheel," i.e., develop definitions already embodied in the standards. The short paragraphs comprising each Section 1 in Standards 194 and 181 are required reading for those involved in pulse measurements and pulse technology in general; a careful reading of those sections will lead to further study of the Standards.

As was mentioned earlier, the minicomputer has made a major impact on time-domain measurements leading to automated time-domain measurements that utilize computer control and data reduction. IEEE Standards 194 and 181 were developed with computer analysis and measurement implementation in mind. In fact, some of the measurement methods are impractical without computer-controlled measurements and on-line data analysis.

Some of the terms in the IEEE Standards have a profoundly particular significance even though they are conceptually simple. For example, the term *pulse* is used to describe only the physical entity involved, i.e., the pulse that travels along a transmission line or appears at the terminals of some element. On the other hand, the term *waveform* is used to describe only the manifestation of the pulse, i.e., the measured quantity or shape corresponding to the physical pulse. These definitions fundamentally recognize that a pulse measurement system can never be perfect; "the observed (measured) waveform is always different from the pulse (being measured)." The degree of such difference can be very important in some situations.

2.2 DEFINITION OF SOME PULSE TERMS

A brief discussion of some of the IEEE/IEC Standard's pulse terms will now be given. The terms themselves will be presented in upper-case letters. The term PULSE is defined as the physical modifications of a state, such as a voltage, current, or the like, whereas the term WAVEFORM is reserved for the observed or measured shape of the pulse. Clearly, the PULSE and its WAVEFORM can differ; put another way, the measurement process does, in fact, yield a WAVEFORM that is not an identical replica of the pulsed physical state, i.e., the PULSE. Accordingly, the terms PULSE and WAVEFORM must also be used as adjectives when discussing features of the two entities, e.g., PULSE ENERGY as contrasted to that computed using the WAVEFORM, i.e., WAVEFORM ENERGY. Also acceptable is such usage as PULSE WAVEFORM, even though the adjective PULSE is redundant because WAVEFORM is the observed measurement system entity corresponding to the physical pulse.

Frequently, the signals under consideration are generally comprised of a sequence of pulses, each pulse being limited in DURATION because each pulse in the sequence exists only over a finite time interval. Note that the term DURATION is used in contrast to width, the latter being an English language term pertaining to a length measure rather than a time measure. On the other hand, an electromagnetic pulse propagating in a transmission medium does have a (spatial) width, i.e., the pulse occupies a certian spatial region; consequently, the term WIDTH should only be used for spatial description.

In the past, for a simple pulse in which the modifications of state occur in a smooth quasi-increasing and/or decreasing manner, the terms "risetime" and "falltime" were used to denote the amounts of time required for the transitions between the off-to-on and on-to-off states, respectively. The term "rise" was used for the first change of state even when the waveform fell to its on-state. In the IEEE 194 and 181 Standards terminology, the terms FIRST TRANSITION DURATION and LAST or SECOND TRANSITION DURATION are used. Note that for a waveform that does not turn on and off in a monotomic manner, several transition durations may be needed to characterize the waveform's leading/trailing epochs. Also, the term TIME denotes an instant, whereas DURATION is the difference between two TIMES or instants.

It is common practice to describe a WAVEFORM TRANSITION in terms of three TIMES, e.g., the times at which the waveform crosses the 10-percent, 50-percent, and 90-percent levels. Now note that it was stated, "e.g. . . . 10-percent, 50-percent, and 90-percent levels." These three levels are commonly used, but, in general, other levels may be much more appropriate. Accordingly, these three levels are defined as the PROXIMAL, MESIAL, and DISTAL levels, these terms being English language words denoting "nearer," "toward the middle," and "farther," respectively. They are used in the present context to denote nearest to, toward the middle of, or farthest from *the waveform baseline*. In a given application, if the levels are not otherwise defined, they are to be assumed to be the 10-percent, 50-percent, and 90-percent levels, respectively. The TRANSITION DURATION is defined as the difference between the PROXIMAL and DISTAL POINTS (in time). A WAVEFORM DURATION is defined as the difference between the MESIAL POINTS (in time) of the FIRST and LAST TRANSITIONS. Figure 2-1 illustrates these definitions for a single pulse.[4,6] One final comment: The determination of the TOPLINE/BASELINE requires the use of some specified algorithm. Typical algorithms are given in the IEEE/IEC document,[5,7] which includes a magnitude histograph method.

If a given pulse is duration-limited, it may be repeated to yield a sequence of pulses, i.e., a PULSE TRAIN. Each pulse in the sequence is said to be COHERENT if the functional time dependence over the pulse duration is the same in each pulse. For example, consider a sequence of pulses in which each duration-limited pulse is generated by a phase-locked (synchronized) combination of pulsed amplitude and phase modulation applied to, and phase-locked with, a sinusoidal carrier; assume the AM modulation to be 100 percent. The resulting individual pulses in the train (sequence) of pulses all have the same functional time dependence but occupy different time intervals, i.e., the carrier oscillations vary in amplitude and frequency in exactly the same way. In other words, if the pulses occur at an aperiodic rate, an oscilloscope operating with a signal-triggered time base would display each successive waveform (which triggers the time base)

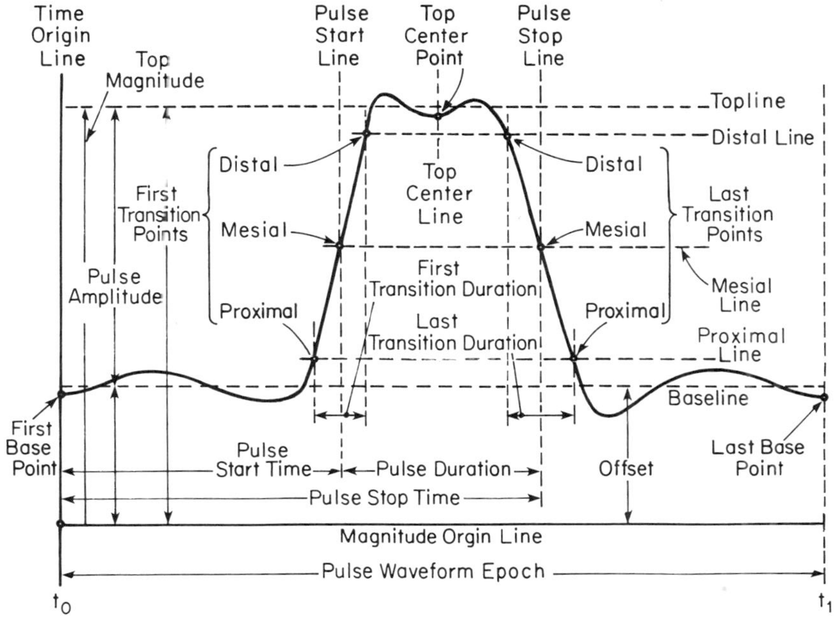

Fig. 2-1. Parameters for describing a single pulse (refer to IEEE Standards 194 and 181[4,5] or IEC Standards 469-1 and 2[6,7]).

as exactly overlapping with previous one, thereby producing a waveform display having an unchanging (stationary) shape or time dependence.

On the other hand, a NONCOHERENT pulse would not provide a stationary time dependence in all of the waveform features. For example, consider a hypothetical case in which a sinusoidal carrier having a random time-varying phase angle (a time-jittering sinusoid) is 100-percent amplitude-modulated by a train of stationary pulses. With the oscilloscope time base being triggered by the amplitude-modulation signal, the resultant waveform could possess an almost stationary ENVELOPE, but inside the envelope the display could be incoherent and blurred into a possibly uniform intensity, depending upon the carrier frequency and the modulation pulse duration.

The manner of occurrence for a pulse train must also be subject to definition. At one extreme is the SINGLE EVENT or single-shot case of a single pulse. If the pulse recurs, it is repetitive and does so in either a PERIODIC or APERIODIC manner. PERIODIC pulses recur with a fixed period of T seconds; APERIODIC pulses recur in some nonperiodic manner, the intervals between pulses varying in either some deterministic way or in a stochastic manner.

2.3 BASIC TIME-DOMAIN MEASUREMENT SYSTEM

As considered here, the basic time-domain measurement system is a transmission system suitable for measuring the output of pulse generators and the effects of transmission through an inserted element (see Fig. 2-2). In the latter case, the before and after insertion waveforms can be related to the frequency-domain scattering parameter, $S_{21}(s)$. Other arrangements of the depicted components plus others as necessary can be used to obtain reflection and/or transmission waveforms that can be related to a full set of network scattering parameters (or other parameters).[9] For incoherent optical pulses,[10,11] the parameters would be intensity (irradiance) parameters such as modulation transfer functions.[12,13,14]

In Fig. 2-2, note that the time-domain quantities on the input side of the time-domain instrumentation are denoted as pulses, whereas that on the output side is denoted as the waveform. Furthermore, at the actual point of a time-domain measurement, the signal channel input Pulse-2 differs from the generator output Pulse-1 as a result of passage through the cascade of connecting elements.

The interconnections shown are wave propagation paths, typically coaxial 50-ohm transmission lines for guided electrical pulses and optical fibers for guided light pulses. The system can also be implemented using unguided propagation paths, e.g., time-domain antenna/scattering measurements[15] and laser pulse measurements.[16,17] Care must be exercised to insure that the overall measurement signal channel possesses an impulse response adequate for the measurement. The signal channel limitations include linear-system transmission/delay path distortions, connector distortion/reflection, and the distortion caused by the network

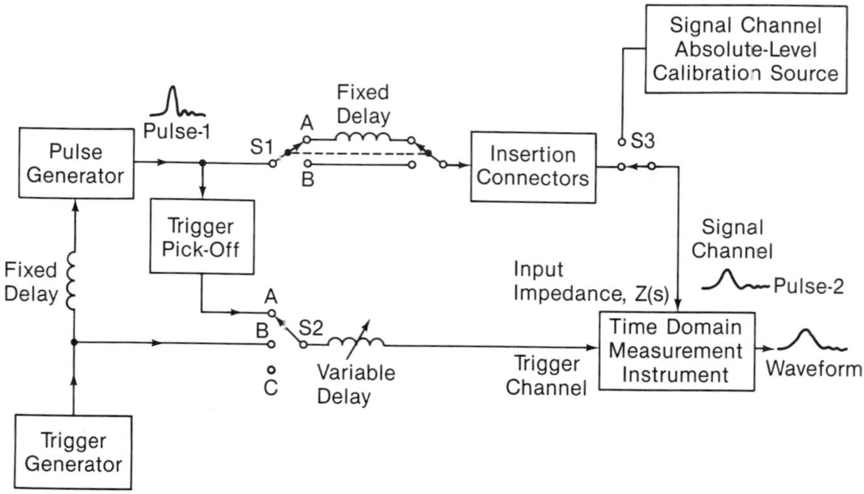

Fig. 2-2. Basic time-domain measurement system.

driving point/transfer properties of the time-domain instrumentation. Moreover, timing errors such as jitter and drift can change the shape of the instrumentation output wave form, thus further affecting the overall measurement system impulse response.

The triggering circuits or the switch settings for S1 and S2 are dependent on the type of pulse generator and type of time-domain instrumentation system used. Typically, positions S1A and S2A would provide triggering circuitry for non-triggerable pulse generators; positions S1B and S2B, for triggerable pulse generators; and positions S1B and S2C, for equivalent-time random-sampling measurements. Equivalent-time measurement systems require repetitive signals (not necessarily periodic), whereas real-time systems do not.

Various methods are used to calibrate the time-domain instrumentation system signal channel and time base. Typical methods include known dc voltages and known amplitude, low-frequency square waves for the signal channel[18] and pulsed CW sources,[19] pulse-excited transmission-line structures,[20] and calibrated delay lines[1] for the time base. In optical systems, variable path lengths using pairs of turning prisms (roof prisms) with precision mechanical positioners[11] are used to provide a known delay-shift on an optical pulse. The amplitude calibration of an optical signal channel is accomplished using calorimetry, calibrated attenuators, and calibrated photodiodes to establish the power response, because an optical pulse is detected or measured in terms of its power/cm^2 (irradiance).[2,16,17]

2.4 SOME INSTRUMENTATION METHODS

There are two general classes of time-domain measurements: (1) real time, and (2) equivalent time. A real-time measurement acquires the signal data from a single occurrence of the signal, whereas an equivalent-time measurement requires a number of reoccurrences to acquire the signal data. Historically, a real-time measurement employed a conventional signal-deflection-structure oscilloscope of sufficient writing rate that it could photographically record a single-occurrence (shot) waveform corresponding to the applied pulse. On the other hand, an equivalent-time measurement employed a sampling oscilloscope that obtained a single sample from each successive repetition of the applied pulse, with each sample being taken at a different time relative to the start of the applied pulse. Consequently, the pulse-measurement process was carried out on an "equivalent-time" basis where the actual measurement times were, say, in milliseconds, corresponding to multiples of the pulse repetition rate, while the corresponding physical pulse duration was, say, in nanoseconds. Thus, common usage relegated real-time measurements to conventional oscilloscopes and sampled data measurements to equivalent-time sampling oscilloscopes. Today, such is not the case as an inspection of the headings in Tables 2-1 through 2-4 shows.

Table 2-1. 1978 State-of-the-Art Temporal Resolution Using Real-Time Methods.
This is a corrected version of the chart given in Reference 2. The "see text" citations refer to the text of Reference 2. All numbers in brackets indicate references at end of chapter.

	REAL TIME METHODS													
	CONTINUOUS MEASUREMENT						SAMPLED MEASUREMENT							
	OSCILLOGRAPHIC		CORRELATORS				MULTIPLE SAMPLING			MULT. CORRELATIONS				MULT. AMPL. DISCR.
	T. WAVE CRT	STREAK CAMERA	AUTO		CROSS		SINGLE GATE		MULT GATES	AUTO		CROSS		
			L	N	L	N	L RR/SR	HRR		L	N	L	N	
ELECTRICAL FIELD	50ps [28] [1, 24]						See text in [2]	See text in [2]	100ps [77] [1, 78, 26]					See text in [2]
OPTICAL FIELD AND IRRAD.	100ps [79] [80–82]	0.5ps [83] [84–88, 89] [90,91]		0.2ps [92] [93] [94]		0.2ps [92, 95] [96]	3.6ps [97] [98–103] [104]	2.0ps [100]					5ps [133]	

KEY:

IRRAD—Irradiance, in watts/m^2 (intensity)
T. WAVE CRT—Traveling wave cathode ray tube
L—Linear; N—Nonlinear

LRR—Low repetition rate (sampling gate)
HRR—High repetition rate (sampling gate)
SR—Signal replication

MULT—Multiple
AMP. DISCR—Amplitude discriminators

Table 2-2. 1978 State-of-the-Art Temporal Resolution Using Equivalent-Time Methods as Given in Reference 2.

	EQUIVALENT TIME METHODS						
	CONTINUOUS MEASUREMENT				SAMPLED MEASUREMENT		
	CORRELATION				SEQUENT. SAMPLING	RANDOM SAMPLING	AMPL. DISCR.
	AUTO		CROSS				
	L	N	L	N			
ELECTRICAL FIELD	8ps [25] [134]				20ps [114] [115–117] 18ps [118] [21, 119–125]	25ps [116] 20ps [126]	165ps [127]
OPTICAL FIELD AND IRRAD.	1ps [105] [135, 136]	0.5ps [106, 107] [108, 109]	1ps [105, 110] [135]	0.5ps [106] [109, 111–113]	50ps [22] [128–130] [131, 132]		

KEY:

IRRAD.—Irradiance (Intensity)
MEAS.—Measurement
AMPL. DISCR.—Amplitude Discriminator

L—Linear; N—Nonlinear
SEQUENT.—Sequential
SAMPL.—Sampling

There are two models for samplers (sampling devices), one being a product model, whereas the other is an integrated product model. The product device instantaneously forms the product of the high-repetition-rate periodic-sampling pulse train and a single-occurrence signal. The resultant sampled signal consists of a train of nearly identical pulses amplitude-modulated by the signal variations. In principle, if the signal is band-limited, then the sampling rate need only be equal to twice the highest frequency, or Nyquist Frequency, in the signal (see Fig. 2-3).

The integrated product model is fundamentally the cross-correlation integral

$$g(\tau) = \int_{-\infty}^{\infty} s(t - \tau) f(t)\, dt$$

where $s(t)$, $ft)$, and $g(\tau)$ are the sampling pulse, the signal, and the resultant correlation function, respectively.[2] Such functions may be encountered in real-time and in equivalent-time measurements.

Even though sampling and correlation methods have the same integral form,

Table 2-3. Present State-of-the-Art Temporal Resolution Using Real-Time Methods. Only Changes Since 1978 Are Cited Here. (See Table 2-1.)

	REAL TIME METHODS													
	CONTINUOUS MEASUREMENT						SAMPLED MEASUREMENT							
	OSCILLOGRAPHIC		CORRELATORS				MULTIPLE SAMPLING			MULT. CORRELATIONS				MULT.
	T. WAVE	STREAK	AUTO		CROSS		SINGLE GATE		MULT	AUTO		CROSS		AMPL.
	C R T	CAMERA	L	N	L	N	L RR/SR	HRR	GATES	L	N	L	N	DISCR.
ELECTRICAL FIELD	[29]													
OPTICAL FIELD AND IRRAD.		[30–35]									10ps [36]			

Table 2-4. Present State-of-the-Art Temporal Resolution Using Equivalent-Time Methods. Only Changes Since 1978 Are Cited Here. (See Table 2-2.)

	EQUIVALENT-TIME METHODS						
	CONTINUOUS MEASUREMENT				SAMPLED MEASUREMENT		
	CORRELATION				SEQUENT SAMPLING	RANDOM SAMPLING	AMPL. DISCR.
	AUTO		CROSS				
	L	N	L	N			
ELECTRICAL FIELD					500fs [37, 38] [39–42]		2.1ps [43] [44–48]
OPTICAL FIELD AND IRRAD.		30fs [49]			190fs [51] 7ps [52] [53]		

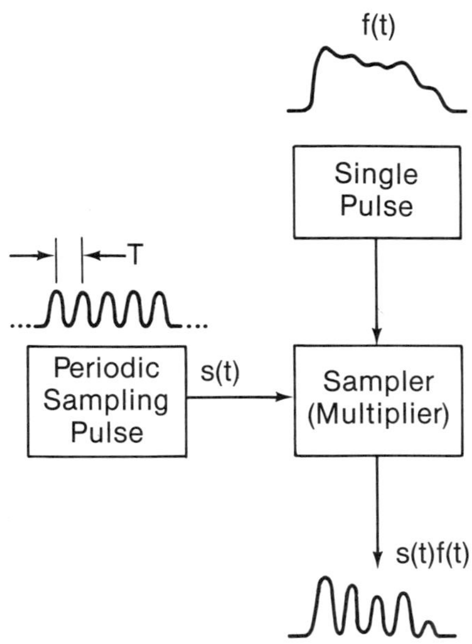

Fig. 2-3. Real-time sampling with a high-repetition-rate (product) sampler where $s(t)$ is the periodic sampling pulse of period T and $f(t)$ is a single-occurrence signal.

they are usually implemented by entirely different techniques. To form the product of the two functions, the sampling method uses switching (or gating) operations, whereas the correlation method employs a mixing process in a non-linear element. Electrical sampling includes gating, switching, and amplitude discrimination, whereas optical sampling includes shuttering, polarization switching, and beam deflection in space or waveguiding structures. By using photoconducting processes in solids, electrical fields (voltages) and optical irradiances (intensities) may be sampled in the same device.[21,22]

Figures 2-4 and 2-5 show the ideal equivalent-time sampling systems: sequential and random. Here the sampling is assumed to be ideal, that is, $s(t)$ is an ideal impulse and the signal path distortionless. In the sequential method, the signal is passed through a delay line to allow time for the sampling pulse to be generated; on successive signal occurrences, the delay generator shifts the sampling time in a known a priori way, usually uniformly. The successive samples are stored in the signal (vertical) channel memory. No signal channel delay line may be required for triggerable signal sources if satisfactory electronic delay is available.

In the random method, the time value of each sample is not known a priori but is determined by measuring the relative time position between the start of the signal and the sampling pulse; the value so determined is stored in the time-base memory (see Fig. 2-5). Note that (1) the sampling pulse is not synchronized to the signal, (2) no signal delay line is thus required, and (3) non zero samples are obtained whenever the signal and sampling pulse occur simultaneously.[1] In Fig. 2-5, the time ramp could just as well be started by the signal and stopped by the sampling pulse.

If a means exists for implementing multiple sampling (see Fig. 2-6) or pulse replication (see Fig. 2-7), it is then possible to apply equivalent-time sampling measurement methods to the measurement of a single-occurrence pulse. Effective signal branching or splitting is relatively easy to implement for optical signals but not for electrical signals where pulse distortion occurs as a result of parasitic

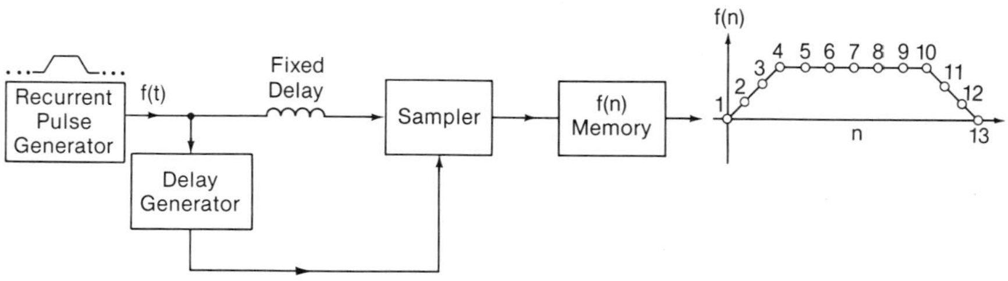

Fig. 2-4. Ideal equivalent-time sequential sampling: The samples, $f(n)$, are taken in a sequential or known order as indicated by the numbers at the points. Note that $f(t)$ is a repetitive signal, not necessarily periodic.

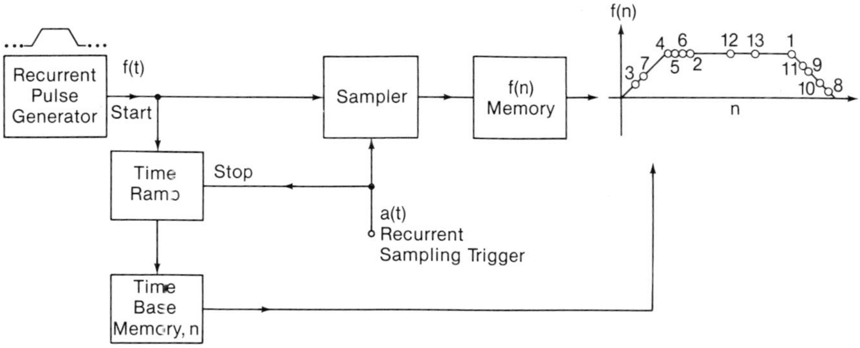

Fig. 2-5. Ideal equivalent-time random sampling: The samples, $f(n)$, are taken in a random order as indicated by the numbers at the points. Note that $f(t)$ is a repetitive signal, not necessarily periodic, and is not synchronized to the sampling trigger, $a(t)$.

capacitances and the like at the branching points. Furthermore, small electrical transmission lines (coaxial, strip, etc.) are lossy.

Real-time continuous (as contrasted to sampled) correlation measurements can also be made on single optical pulses using fluorescence and film or electronic recording (see Fig. 2-8).

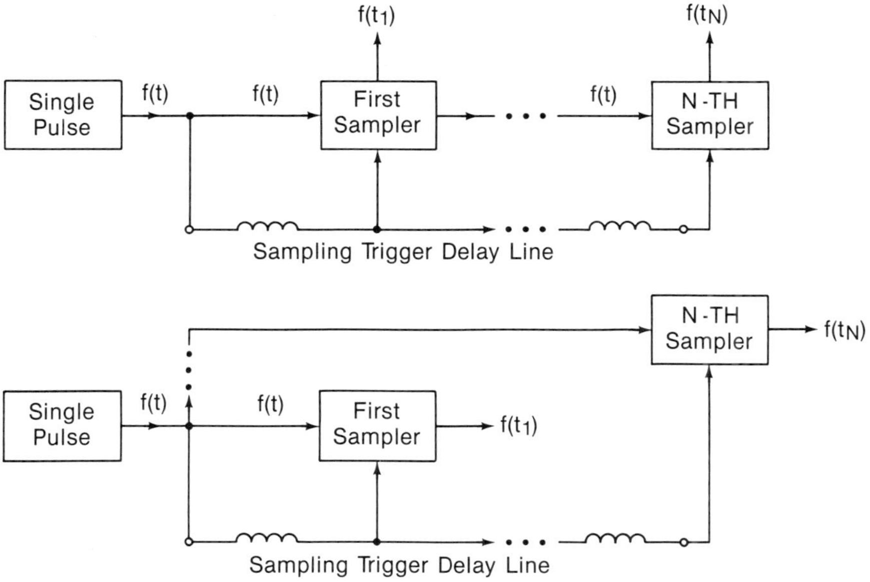

Fig. 2-6. Ideal real-time sampling with multiple low-repetition-rate (equivalent-time) samplers: In the upper figure, the samplers are feed-through units, and $f(t)$ is a nonrecurrent pulse.

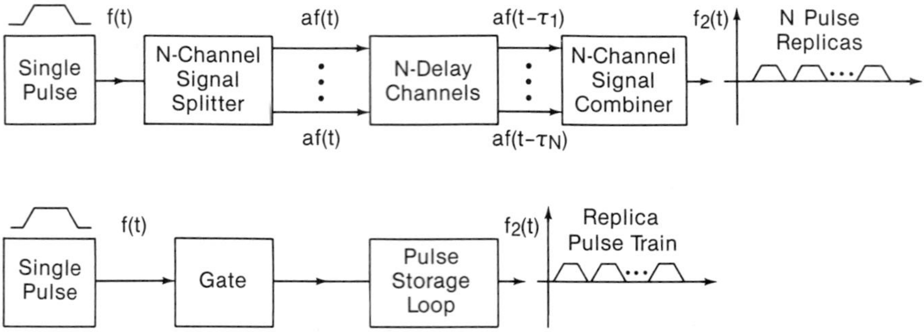

Fig. 2-7. Real-time sampling by signal replication and a low-repetition-rate (equivalent-time) sampler: The upper figure shows a passive system and the lower, an active one; $f(t)$ is a nonrecurrent pulse.

Sampling devices themselves fall into two categories: (1) gates, and (2) amplitude discriminators (see Fig. 2-9). The former gates or transmits a relatively short duration slice (between two time values) of the signal. The amplitude-discriminating device provides a small slice between two amplitude levels. The resultant time resolution of either device is dependent upon the complete sampling

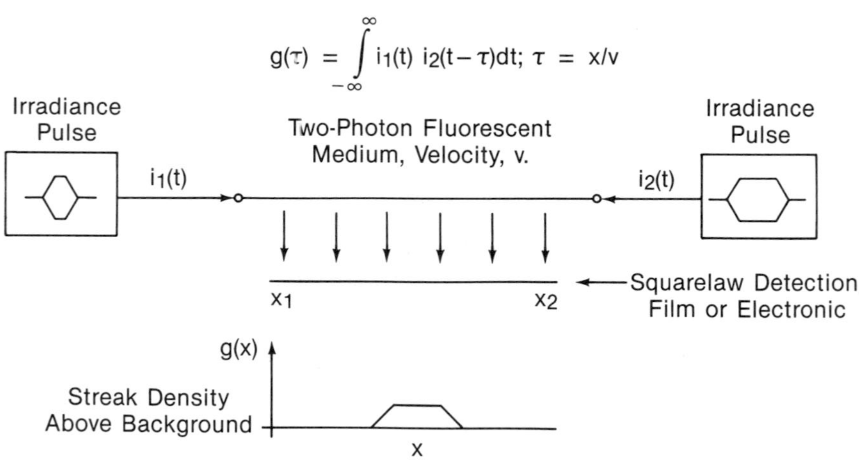

Fig. 2-8. Real-time nonlinear correlation: Both $i_1(t)$ and $i_2(t)$ are nonrecurrent, optical-carrier pulses of the same wavelength. As irradiances, they are proportional to the square of the electromagnetic-field quantities. Second-harmonic radiation proportional to the sum of the instantaneous powers is produced as the pulses pass each other in the fluorescent medium, yielding a streak of light.

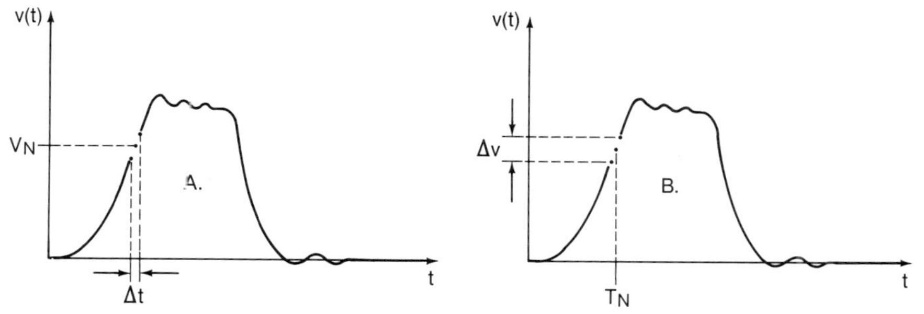

Fig. 2-9. The two types of sampling devices: (a) gate, and (b) amplitude discriminator.

network's inherent speed of response, which, in turn, determines the slice thickness. For the gate, the response can be visualized as an effective gating function, i.e., the result of all of the circuit elements and driving signals that make the sampler operate. Similarly, for the amplitude discriminator, the composite result can be envisioned as a switching or transition function representing the response of the discriminator from its rest state to the state where its action becomes autonomous, i.e., the state in which the input signal no longer affects the ongoing switching action. Notice that in both cases the slice thickness can never be exactly zero, which also says that all sampling devices require a time interval greater than zero in which to respond. In turn, this says that all sampling devices provide a filtered or smoothed replica of the signal being sampled.

Amplitude-discriminator pulse-sampling measurements can be implemented by two general approaches: (1) the sampling position (see Fig. 2-10) selected primarily by the time position of a sampling pulse relative to the signal,[1] and (2) the sampling position (see Fig. 2-11) determined by a specified amplitude level.[2] For both cases, effective sampling results when the signal amplitude is much less than that of an auxiliary pulse.

In the first case, a (relatively) narrow sampling pulse is added to the signal, and the sum trips the amplitude discriminator; the sampling-pulse time position is shifted through the epoch of the signal and the amplitude-discrimination level for each position recorded. Comparing the discrimination levels with and without the signal present provides a waveform corresponding to the signal (pulse).

In the second case, a ramp of sufficient amplitude is added to the signal to yield a monotonic (always increasing or decreasing) sum. The amplitude-discrimination level is scanned to obtain a waveform corresponding to the sum of the signal and the ramp. When the ramp level vs. time is known, the signal level vs. time can be determined.

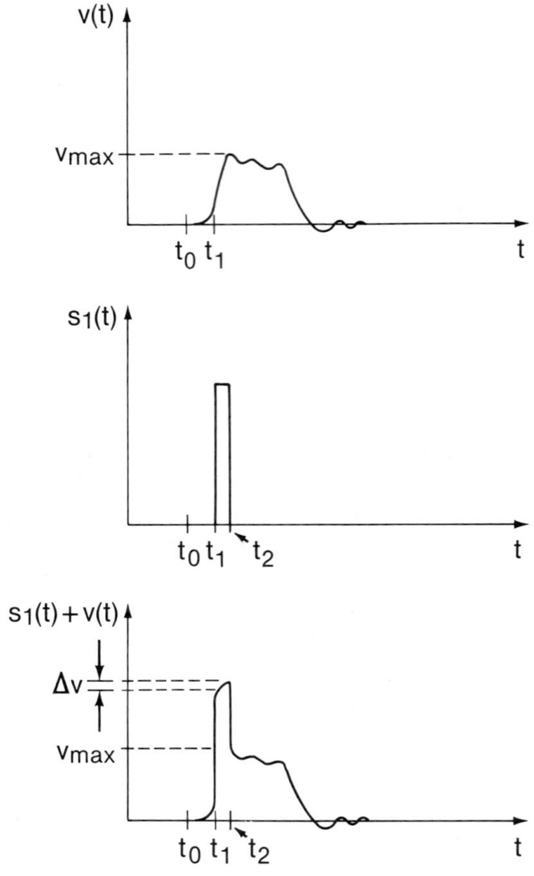

Fig. 2-10. An amplitude-discriminator sampling method employing a sampling pulse, $s_1(t)$, whose nominal time position selects the sampling position: The sum, $s_1(t) + v(t)$, trips the amplitude discriminator.

2.5 STATE OF THE ART

State-of-the-art reports are in themselves time-domain data on the latest results in the development and application of techniques. As such, they suffer from observer errors and are always behind the times because of the noninfinitesimal time interval required to write, print, etc.

As was pointed out in the introduction, the state-of-the-art charts consist of those from the 1978 review[2] and a present set containing only those entries that have changed since 1978 (at the time when the present work was written). Also, the present set is different from the 1982 set published in 1983.[3] Furthermore,

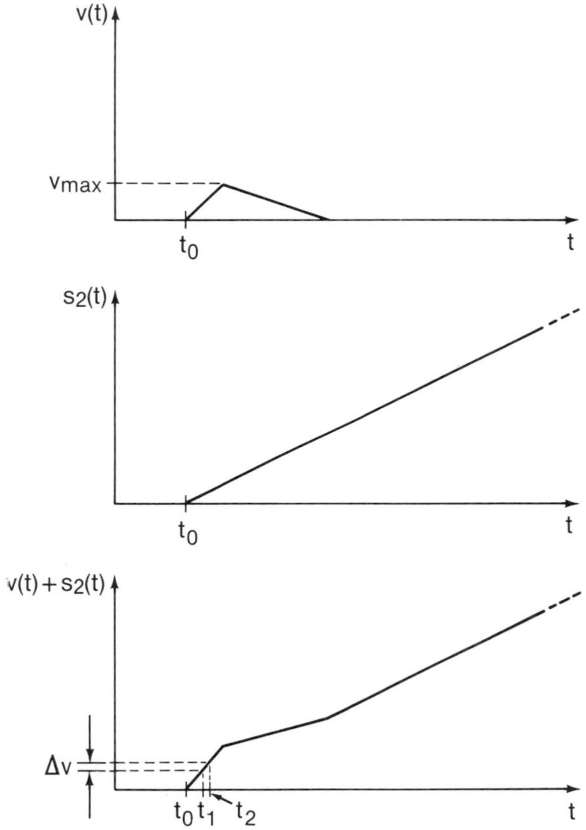

Fig. 2-11. An amplitude-discriminator sampling method whose sampling position is determined by the nominal discrimination level of the amplitude discriminator: The ramp, $s_2(t)$, possesses a large enough slope for the sum, $s_2(t) + v(t)$, to be monotonic, and thus the amplitude discriminator is driven by a single-valued function of time.

those readers with a continuing interest in high speed time domain measurements should read the earlier reviews.[1–3] Make no mistake about it, what is written in Refs. 1 through 3 and the present review represent one person's opinions and insights; they are not absolute and continue to evolve with time. Useful ideas however, can be found in the earlier work cited in Refs. 1 and 2.

Furthermore, because there is no universal usage of some set of terms and definitions, all values for temporal resolution listed in the state-of-the-art tables are those stated by the cited investigator. It would be very difficult, if not impossible, for the author to reduce others' results to a common definition. The term *temporal resolution* as used here (and in Ref. 2) is an all-inclusive one that takes on the appropriate definition for the measurement under discussion; generally speaking, it relates to the steplike 10-to-90-percent transition duration (rise-time time)* or the impulse like response duration at the 50-percent response level, full duration at half maximum (FDHM), of the observed waveform produced by the oscilloscope, streak-camera, or correlation-system. Moreover, the terms *linear* and *nonlinear,* with respect to correlation measurements, are called *first order* and *higher order,* respectively.

From a mathematical point of view, linear and nonlinear correlation have integrands of the form $f_1(t)f_2(t - \tau)$ and $f_1^n(t)f_2^n(t - \tau)$, respectively. When $f_1(t$ and $f_2(t)$ are proportional to the electric or magnetic field, then $f_1^2(t)$ and $f_2^2(t)$ are proportional to their powers. Optical detectors, streak cameras, etc., respond to the irradiance (W/m^2), or square of the electromagnetic field quantities. A general nonlinear correlation formula is given in Ref. 2.

Before discussing the entries in the state-of-the-art charts, mention will be made of a state-of-the-art device that doesn't appear in them but is very important for certain electrical pulse measurements and should be mentioned. It is the superconducting (coaxial) delay line.[23] For some electrical pulse measurements, a fixed-signal delay is required, as indicated in Sec. 2.4 and Figs. 2-2 and 2-4. As reported, such delay lines[24] can provide a nominal delay of 70 ns with a system step response transition duration of 18 ps.

In the following discussion, reference to a given entry in a state-of-the-art chart will be given as the temporal resolution followed by a reference citation, e.g., 8 ps.[25] Where reference citations are given without a temporal resolution, the citation relates (1) to some improvement not associated with speed, and/or (2) to some similar work of slower speed.

Real-Time Measurements

Referring to the real-time methods of the 1978 state-of-the-art chart in Table 2-1, the listing under multiple sampling with multiple gates is a corrected version of that in the 1978 review,[2] the correction being the addition of Ref. 26. This system was listed under continuous oscillographic measurements in the 1978 paper (Ref. 42 in the 1978 paper) because, within the device, the signal continuously acts on an electron beam; hence, the fundamental device limitation is contained in a continuous process. As a measurement instrument however, it provides a sampled data output. In Ref. 26, the real-time sampling is accomplished by deflecting a ribbon-electron beam across an apertured target that is in front of a semiconductor array. The portions of the beam passing through the apertures and striking the semiconductor array provide the digitized samples.

*The terms *transition duration* and *rise time* are commonly used in the literature at the discretion of the author, as is also true of this book.

The apertures are arranged so that the beam position is read out as a digital (Grey-coded) nine-bit number; the beam deflection is proportional to the time dependence of the signal being measured. For a 1-gigahertz signal-channel bandwidth, the system step response would consist of 512 samples over a 5-ns time window with a transition duration of about 350 ps. Some limitations of signal-channel bandwidth, sampling rate, and timing for real-time sampling are discussed by Linnenbrink.[27]

Although traveling-wave cathode-ray tubes have undergone some refinements, Ref. 28 remains the state of the art: a step response transition duration of 50 ps with a sensitivity of 0.14 V/cm. Recent developments include a 50-ps, 5-V/cm unit and others of slower speeds and sensitivities in the range of 0.15 to 16 V/cm.[29]

The speed of streak cameras for state-of-the-art optical-pulse measurements has remained mainly unchanged at about 0.5 ps. A new approach for a streak camera based upon interference patterns is reported in which an electro-optic crystal deflector is used.[30] Such crystal deflectors have also been used in other streak cameras that do not use interference phenomena.[31,32] Furthermore, a new method for triggering the electrical deflection circuits is described by Mourou and Knox.[33] Improvements to conventional streak tubes have been made.[34,35] In the former case,[34] the improvements include increasing the stability of an S1 photocathode. Such cathodes have been known to degrade, with a shelf life of only a year or so. In the latter case,[35] a tube designed for circular sweep operation coupled to a diode array is described.

A multiple nonlinear autocorrelator for optical pulses has been built using an echelon and thin photoconductive detectors[36]; in use, the experimental arrangement is the same as that for a two-photon fluorescence pulse autocorrelation experiment but differs in that the echelon/photoconductor assembly replaces the two-photon fluorescent medium.

Equivalent-Time Measurements

An equivalent time-sampling system has been developed for the measurement of subpicosecond electrical pulses and has produced an observed transition duration (10 to 90 percent) of about 500 fs with a voltage sensitivity of about 50 microvolts.[37,38] This device uses a very short-duration periodic optical pulse to sample a periodic electrical pulse. The sampler is a traveling-wave Pockels cell whose electric field is proportional to the electrical pulse being measured. The optical pulse duration is about 120 fs and is produced by a mode-locked colliding-pulse dye laser operating with a 100-MHz repetition rate. The electrical pulse observed with the system was a step like pulse produced by a GaAs Cr-doped photoconductive detector excited by a separate optical beam of the 120-fs pulse train.

Another equivalent-time sampling system that uses a very short-duration optical pulse from a mode-locked dye laser to sample an electrical pulse employs radiation-damaged (short-lifetime) photoconductors as switching elements. With identical photoconductors for electrical pulse generation and sampling, this system has achieved a resolution of 3.6 ps.[39] Also, photoconductive systems have been applied to measure the impulse response of a GaAs field-effect transistor having an unpackaged bandwidth to about 16 GHz.[41,42]

Josephson-junction (JJ) amplitude discriminators have been employed in equivalent-time sampling systems yielding resolutions of 2.1 ps[43] and 8.5 ps[44] for measuring on-chip signals in superconducitng pulse IC's. Another system demonstrated a time-domain reflectometer for measuring the superconducting transmission-line elements employed in superconducting integrated circuits, with a resolution of 10 ps.[45] The incident current step was produced by a JJ pulse generator, and all of the components were fabricated on a single substrate. In these cases,[43,44] the amplitude-discriminator sampler was one in which a narrow pulse was added to the signal.[46] Other such sampling circuits using JJ's have been built.[47,48] All of the measurements reported here [43–48] were applied to signals generated on chip in the cryogenic environment of 4.2 K.

Optical nonlinear ($n = 2$) autocorrelation measurements have been made that gave an autocorrelation waveform of 30 femtoseconds full duration at half maximum (FDHM).[49] These measurements were done using a zero-background correlation measurement method.[50] Cross-correlation measurements between pulses of the same wavelength have been made giving a cross-correlation waveform with 190 fs FDHM.[51]

Optical nonlinear ($n = 2$) cross-correlation measurements between two optical pulses of different wavelengths have also been made, yielding resolutions of 7 ps (1060 nm and 840 nm)[52] and 7.76 ps (597.5 nm and 630 nm),[53] respectively. In the former, 1060-nm pumping pulses were cross-correlated with 840-nm pulse emission from a bulk GaAs and a synchronously mode-locked bulk GaAs laser. In the latter, the pulses were synchronously generated by a double mode-locked dye laser using Rhodamine 6G and cresyl violet dyes as the simultaneous lasing mediums. For completely mode-locked pulses, the cross-correlation waveform had a value of 7.76 ps FDHM. These measurements[52,53] were made using the zero-background correlation method described by Ippen and Shank[50]; however, in this case, the correlation optical ray was phase-matched in the crystal at the sum frequency of the two laser frequencies.

2.6 REMARKS AND FUTURE DIRECTIONS

The main objective of this chapter has been to summarize the present-day capability in picosecond-domain waveform measurements. In doing so, electrical and optical pulse measurement methods have been drawn together in the same manner as in the 1978 review[2] in order to exhibit the commonality between electrical and optical methods and to encourage the exchange of ideas between electrical and optical researchers. The importance of bringing together electrical and optical picosecond-domain measurements has been emphasized by the selection, translation, and subsequent publication of the 1978 review in Japanese.[54]

In the author's view, there are three areas in present-day activities that provide precursors to future developments.

First, there is electron device research that is leading to faster or more versatile devices. The superconducting devices referred to in Table 2-4[43–48] will be improved, and additional elements such as A/D converters, memories, etc., will become available.[55,56] Another interesting class of devices is the proposed ballistic electron semiconductor devices[57–60] that are envisioned to possess submicron geometries in which electron velocities will be larger and responses faster. The realization of submicron geometries is central to the development of faster devices[61]; in particular, 2-ps switching delays are projected for the permeable-base transistor.[62] Photoconductive detectors[62–66] and metal tunnel-barrier photodetectors[67] are other devices that are projected to be faster as a result of submicron techniques. Furthermore, submicron techniques among others will probably be applied to speed up charge-coupled devices such as those used in real-time sampling.[68,69] Finally, electron-beam techniques may be used to evaluate submicron devices by making direct measurements on the devices themselves; such a measurement method has been used to study the transient currents in a semiconductor deivce.[70]

Second, there is the application of optical pulse technology to electrical pulse measurements. In addition to those optically driven electrical devices already cited,[37,38,40–42] there are the electronic correlation,[71] the microwave pulse generator,[72] and the high-speed switch.[73] More ways will evolve to use optical devices in electrical pulse measurements.

Third, there is optical pulse gating that falls into two categories: optically driven[74] and electrically driven.[75,76] Such methods will lead to improved and more versatile optical pulse measurement systems.

In summary, these three activities will strongly stimulate and directly contribute to the further development of waveform measurements into the fractional picosecond (femtosecond) domain. As shown in Table 2-4, present state-of-the-art equivalent-time measurements have reached well into the femtosecond range (30 fs).[49] The frontiers of femtosecond domain research are rapidly moving toward measurement of dye laser pulse durations of less than 10 fs.[137]

REFERENCES

1. Nahman, N. S. 1967. The measurement of baseband-pulse risetimes of less than 10^{-9}s. *Proc. IEEE* 55:855–864.
2. Nahman, N. S. 1978. Picosecond-domain waveform measurements. *Proc. IEEE* 66:441–454.

3. Nahman, N. S. 1983. Picosecond-domain measurements: Status and future directions. *IEEE Trans. I&M* IM 32, No. 1.
4. *IEEE glossary of pulse terms and definitions.* IEEE Standard 194 Revised, July 26, 1977.
5. *IEEE pulse measurement and analysis, general considerations.* IEEE Standards 181 Revised, July 22, 1977.
6. IEC. 1974. *Pulse techniques and apparatus, Part 1: Pulse terms and definitions.* Publication 469-1, 1st ed. International Electrotechnical Commission, Geneva, Switzerland.
7. IEC. 1974. *Pulse techniques and apparatus, Part 2: Pulse measurement and analysis, general considerations.* Publication 469-2, 1st ed. International Electrotechnical Commission, Geneva, Switzerland.
8. Nicolson, A. M., and Ross, G. F. 1970. Measurement of intrinsic properties of materials by time-domain techniques. *IEEE Trans. Instrum. Meas.* IM-19:377–382.
9. Andrews, J. R. 1978. Automatic network measurements in the time domain. *Proc. IEEE* 66:414–423.
10. Personick, S. D. 1977. Photon probe—An optical-fiber time-domain reflectormeter. *Bell Syst. Tech. J.* 56:355–366.
11. Fontaine, J. J.; Diels, J.; Wang, C.; and Sallaba, H. 1981. Subpicosecond time domain reflectometry. *Optics Lett.* 6:405–407.
12. Born, M., and Wolf, E. 1970. *Principles of Optics.* 4th ed. London, Eng.: Pergamon Press, p. 483.
13. Young, M., 1977. *Optics and Lasers.* New York: Springer-Verlag, p. 138.
14. Wolfe, W. L., ed. *Handbook of Military Infrared Technology.* Office of Naval Research, U.S. Government Printing Office, Washington, DC, p. 621.
15. Lawton, R. A.; and Ondrejka, A. R. 1978. Antennas and the associated time domain range for the measurement of impulsive fields. *NBS Technical Note 1008,* National Bureau of Standards, Boulder, CO.
16. Honda, T., and Endo, M. 1976. Measurements of picosecond laser pulses from mode-locked Nd:YAG laser. *IEEE Trans. Instru. Meas.* IM-25:397–400.
17. Young, M., and Lawton, R. A. 1979. Measurement of pulsed-laser power. *NBS Technical Note 1019.* National Bureau of Standards, Boulder, CO.
18. Andrews, J. R., and Baldwin, E. E. 1981. Amplitude calibrator for oscilloscopes. *NBSIR 81-1646,* National Bureau of Standards, Boulder, CO.
19. Andrews, J. R., and Gans, W. L. 1975. Pulsed wavemeter timing reference for sampling oscilloscope calibration. *IEEE Trans. Instrum. Meas., Vol. IM-24,* March 1975, p. 82.
20. Cronson, H. M., and Mitchell, P. G. 1974. A 2.5-ns timing standard. *IEEE Trans. Instrum. Meas.* IM-23:97–99.
21. Lawton, R. A., and Andrews, J. R., 1976. Optically strobed sampling oscilloscope. *IEEE Trans. Instru. and Meas.* 25, No. 1:56–60.
22. Andrews, J. R., and Lawton, R. A. 1976. Electrically strobed optical sampling oscilloscope. *Rev. Sci. Instrum. 47,* No. 3:311–313.
23. Nahman, N. S. 1973. Miniature superconductive coaxial transmission lines. *Proc. IEEE* 61, No. 1:76–79.
24. Andrews, J. R. 1974. Precision picosecond-pulse measurements using a high-quality superconducting delay line. *IEEE Trans. Instru. Meas.* IM-23, No. 4:468–472.
25. Lawton, R. A. 1972. The application of pyroelectric detectors to baseband pulse measurements. Ph.D. dissertation, Electrical Engineering Dept., University of Colorado, Boulder, CO.
26. Robin, N. A., and Ramirez, R. 1980. Capture Fast Waveforms Accurately With a 2-Channel Programmable Digitizer, *Electronic Design* 3:50–55.

27. Linnenbrink, T. E. 1981. Waveform acquisition methods for single transient measurements. *Proceedings of the ATE/Seminar,* Pasadena, CA. Boston: Benwill Publications pp. IV-3 to IV-41.
28. Clement, G., and Loty, C. 1973. Channel plate electron multipliers in cathode ray tubes. *Acta Electronica* 16, No. 1:102–111.
29. Campbell, D. E. 1983. A review of single transient oscillographic recorders with gigahertz bandwidth. *IEEE Trans. Nuclear Science* NS-30, No. 1:267–271.
30. Honda, T. 1978. Picosecond time-resolved interferometric streak unit. *Proc. of 13th International Congress of High Speed Photography and Photonics,* Tokyo, Japan, pp. 513–516.
31. Thomas, S. W 1975. Electro-optic deflector design considerations for use in the crystal streak camera. *Proc. of 13th International Congress of High Speed Photography and Photonics,* Tokyo, Japan, pp. 499–503.
32. Ireland, C. L. M. 1979. A ~20 ps resolution crystal streak camera, *Opt. Commun.* 30, No. 1:99–103.
33. Mourou, G., and Knox, W. 1980. A picosecond jitter streak camera. *Appl. Phys. Lett.* 36:623–626.
34. Roddie, A. G.; Edwards, J. G.; Murphy, T.; and Maisey, S. 1981. Development of a miniaturized picosecond streak camera. Part 1: Cathode stability and initial electrode design *NPL Report DES 72*. National Physical Laboratory, Teddington, Eng.
35. Johnson, C. B.; Nevin, S.; Bebris, J; and Abshire, J. B. 1980. Circular-scan streak tube with solid-state readout. *Appl. Opt.* 19:3491–3495.
36. Lee, C. H., and Mak, P. S. 1980. Measurements of ultrashort light pulses with a multichannel two-photon photoconductivity detector. *The Optical Society of America Digest of Technical Papers, Topical Meeting on Picosecond Phenomena*. Cape Cod, MA, pp. THC15-1 to THC15-6.
37. Valdmanis J. A.; Mourou, G.; and Gabel, C. W. 1983. Subpicosecond electrical sampling. *Proc. of SPIE—International Society of Optical Engineering* 439:142–148.
38. Valdmanis, J. A.; Mourou, G.; and Gabel, C. W. 1983. Subpicosecond electrical sampling. *IEEE J. Quantum Electron.* QE-19, No. 4:664–667.
39. Auston, D. H. 1983. Impulse response of photoconductors in transmission lines. *IEEE J. Quantum Electron.* QE-19, No. 4:639–648.
40. Low, A. J., and Carroll, J. E. 1978. 10 ps optoelectronic sampling system. *IEEE Solid State and Electron Devices* 2:185–190.
41. Smith, P. R.; Auston, D. H., and Augustyniak, W. A. 1981. Measurement of GaAs field-effect transistor electronic impulse response by picosecond optical electronics. *Appl Phys. Lett.* 39:739–741.
42. Auston, D. H., and Smith, P. R., 1983. Picosecond Optical Electronics. *Bell Labs. Record* 61, No. 1:6–11.
43. Wolf, P.; Van Zeghbroeck, B. J.; and Deutsch, A. 1984. A Josephson sampler with a 2.1 ps resolution. *Proceed. of the 1984 Applied Superconductivity Conference* San Diego, Calif.
44. Harris, R. E.; Wolf, P.; and Moore, D. F. 1982. Electronically adjustable delay technology. *IEEE Electron Device Letters* EDL-3, No. 9:261–263.
45. Tuckerman, David B. 1980. A Josephson ultrahigh-resolution sampling system. *Applied Phys. Lett.* 36:1008–1010.
46. Hamilton, C. 1981. Sampling circuit and method therefor. U.S. Patent 4,245,169. (Original patent application dated Nov. 21, 1977.)
47. Faris, Sadeg M. 1980. Generation and measurement of ultrashort current pulses with Josephson devices. *Appl. Phys. Lett.* 36:1005–1007.
48. Hamilton, C. A.; Lloyd, F. L.; Peterson, R. L.; and Andrews, J. R. 1979. A superconducting sampler for Josephson logic circuits. *Appl. Phys. Lett.* 35.

49. Shank, C. V.; Fork, R. L.; Yen, R.; Stolen, R. H.; and Tomlinson, W. J. 1982. Compression of femtosecond optical pulses. *Appl. Phys. Lett.* 40:761–763.
50. Ippen, E. P., and Shank, C. V. 1979. Ultrashort light pulses. In *Topics in Applied Physics*. Vol. 18, edited by S. L. Shapiro. New York: Springer, pp. 90–91.
51. Halbout, J. M., and Tsang, C. L. 1982. Femtosecond interferometry for nonlinear optics. *Appl. Physics Lett.* 40, No. 9:765–767.
52. Vaucher, A. M.; Cao, Wei-Lou; Ling, Jun-Da; and Lee, Chi H. 1982. Generation of tunable picosecond pulses from a bulk GaAs laser. *IEEE J. Quantum Electron.* QE-18, No. 2:187–192.
53. Bourkoff, E.; Dienes, A.; and Whinnery, J. R. 1979. Measurements of picosecond pulse shapes for the highly synchronous CW double mode-locked dye laser. *Appl. Phys. Lett.* 34:455–457.
54. Japanese translation of Ref. 2. 1979. In *Nikkei Electronics*. Tokyo, Japan: Nikkei-McGraw-Hill, pp. 68–94.
55. Hamilton, C. A.; Lloyd, Frances L.; and Kautz, R. L. 1981. Analog Measurement Applications for High Speed Josephson Switches. *IEEE Trans. on Magnetics* MAG-17:577–582.
56. Hamilton, C. A.; Lloyd, F. L. 1981. Design limitation for superconducting A/D converters. *IEEE Trans. on Magnetics* 17:3414–3419.
57. Shur, M. S. 1980. Ballistic and collision dominated transport in a short semiconductor diode. *Proceedings of the Electron Device Meeting*. IEEE Publication 0. CH1616-2/80/0000-0618. Washington, DC, pp. 618–621.
58. Johnson, Lee 1981. The probability for ballistic electron motion in n-GaAs. *IEEE Electron Device Letters* EDL-2:167–169.
59. Ferry, D. K.; Zimmermann, J.; Lugli, P.; and Brubin, H. 1981. Limitations to ballistic transport in semiconductors. *IEEE Electron Device Letters* EDL-2:228–230.
60. Shank, C. V.; Fork, R. L.; Greene, B. I.; Reinhard, F. K.; and Logan, R. A. 1981. Picosecond Nonequilibrium Carrier Transport in GaAs, *Appl. Phys. Lett., Vol. 38,* January 15, pp. 104–105.
61. Special Issue on Very Fast Solid-State Technology, *Proc. IEEE* 70, No. 1.
62. Bozler, C. O., and Alley, G. D. The Permeable Base Transistor and Its Application to Logic Circuits. *Proc. IEEE* 70, No. 1:46–52.
63. Degani, J.; Nahory, R. F.; Pollock, M. A.; Heritage, J. P.; and DeWinter, J. C. 1982. Fast photoconductive detector using ρ-$In_{0.53}Ga_{0.47}$ As with response to 1.7 μm. *Appl. Phys. Lett.*, pp. 27–29.
64. Defonso, A. P. 1981. Picosecond photoconductivity in germanium films. *Appl. Phys. Lett.* 39:480–482.
65. Foyt, A. G.; Leonberger, F. J.; and Williamson, R. C. 1982. Picosecond InP optoelectronic switches. *Appl. Phys. Lett.* 40:447–449.
66. Bokor, J.; Bucksbaum, P. H.; and Auston, D.H. 1983. Measurement of picosecond ultraviolet laser pulse widths using an electrical autocorrelator. *Appl. Phys. Lett. 42,* No. 4:342–344.
67. Thaniyavarn, S., and Gustafson, T. K. 1962. Metal/tunnel-barrier/semi-conductor/tunnel-barrier/metal fast photodetector. *Appl. Phys. Lett.* 40:255–257.
68. Deyhimy, I.; Hill, W. A.; and Anderson, R. J. 1981. Continuously Clocked 1 GHz GaAs CCD. *IEEE Electron Device Letters* EDL-2:70–72.
69. Linnenbrink, T. E.; Gradl, D. A.; Ritt, D. M., Peterson, G. A.; DeWitte, G. J.; and Hutton, J. D. 1983. One gigasample per second transient recorder: A performance demonstration. *IEEE Trans. Nuclear Science* NS-30, No. 1:278–291.
70. Fukioka, Hiromu, and Ura, Katsumi: Waveform measurements on gigahertz semiconductor devices by scanning electron microscope stroboscopy. *Appl. Phys. Lett.* 39:81–82.

71. Yamado, S.; Urisu, T.; Sugeta, T.; and Mizuchima, Y. 1983. First Order Electronic Correlation Utilizing Metal-Semiconductor-Metal Photodetectors. *Appl. Phys. Lett.* 42, No. 6:482–484.
72. Mourou, G.; Stancampiano, C. V.; Antonetti, A.; and Orszag, A. 1981. Picosecond microwave pulses generated with a subpicosecond laser-driven semiconductor switch. *Appl. Phys. Lett.* 39:295–596.
73. Williamson, S.; Letzring, S.; and Mourou, G. 1981. Picosecond switching with photoelectrons. *IEEE J. Quantum Electron.* QE-17, Part II, 70.
74. Jamison, S. A., and Nurmikko, A. V. 1978. Generation of picosecond pulses of variable duration at 10.6 μm. *Appl. Phys. Lett.* 33:598–600.
75. Marcatili, E. A. J. 1980. Optical subpicosecond date, *Appl. Optics* 19:1468–1475.
76. Haus, Hermann A.; Kirsch, Steven T.; Mathyssek, Konrad; and Leonberger, Frederick J. 1980. Picosecond optical sampling. *IEEE J. Quantum Electron.* QE-16:870–873.
77. Schwarte, R. 1972. New results of an experimental sampling system for recording single events. *Electronics Letters* 8, No. 4:95–96.
78. Davies, T. J., and Nelson, M. A. 1976. Optical Sampling Transient Analyzer System. *Appl. Optics* 15, No. 6:1404–1410.
79. Deloron, M. A 1972. Photodiode ultra-rapide (Ultra-Fast photodiode). *Acta Electronica* 15, No. 4.
80. McLellan, E. J., and Lunsford, J. S. 1977. A single-sweep 5 GHz oscilloscope detector combination for CO_2 laser pulse measurements. Paper 8.10, Conference on Laser Engineering & Applications, Washington DC.
81. Beck, G. 1976. Photodiode and holder with a 60 ps response time. *Rev. Sci. Instru.* 47, No. 7.
82. Green, S. I. 1976. 50 ps Detector Laser Pulse Monitor. *Rev. Sci. Instru.* 47, No. 9.
83. Bird, P. R.; Bradley, D. J.; and Sibbett, W. 1975. *Proc. 11th International Congress on High Speed Photography*. London, England: Chapman and Hall, pp. 112–117.
84. Adams, M. C.; Ruddock, I. S.; Sibbett, W.; and Bradley, D. J. 1976. *12th International Congress on High Speed Photography*. Toronto, Canada.
85. Bradley, D. J.; Liddy, B.; and Sleat, W. E. 1971. Direct measurement of ultra-short light pulses with a picosecond streak camera. *Opt. Commun.* 2, No. 8:391–395.
86. Johnson, C. B. 1976. Photoelectronic Streak-Tube Technology Review. Proc. of SPIE, High Speed Optical Techniques 94:13–18.
87. Lieber, A. J.: Sutphin, H. D.; and Webb, C. B. 1976. Subpicosecond proximity-focused streak camera for X-ray and visibile light. Proc. of SPIE, High Speed Optical Techniques 94:7–12.
88. Zaviosky, E. K., and Fanchenko, S. D. 1965. Image converter high-speed photography with 10^{-9} to 10^{-14} sec. time resolution. *Appl. Optics* 4, No. 9:1155–1167.
89. Walker, A. C., and Alcock, A. J. 1976. Picosecond resolution, real-time linear detection system for 10 micrometer radiation. *Rev. Sci. Instrum.* 47, No. 8:915–920.
90. Prosser, R. D. 1972. An electron-dispersion tube for the observation of sub-nanosecond transient signal. *Proc. IEEE* 60, No. 5:645–646.
91. Prosser, R. D. 1976. Electron-dispersion techniques for observation of fast transient signals. *J. of Physics E: Scientific Instruments* 9, No. 1:57–59.
92. Caughey, S. J. 1970. A study of mode-locked giant pulse lasers. Ph.D. dissertation, Queen's University, Belfast, Northern Ireland.
93. Von Der Linde, D. 1972. Experimental study of single picosecond light pulses. *IEEE, J. Quantum Electron.* QE-8, No. 3:328–338.
94. Auston, D. H. 1971. Higher Order Intensity Correlations of Optical Pulses. *IEEE, J. Quantum Electron.* QE-7, No. 9:465–467.
95. Bradley, D. J.; New, G. H. C.; and Caughey, S. J. 1970. Amplitude and phase structure of picosecond pulses from Nd:glass laser. *Phys. Lett.* 32A:313–314.

96. Auston, D. H. 1971. Nonlinear spectroscopy of picosecond pulses. *Opt. Commun.* 3:272–276.
97. Topp, M. R.; Rentzepis, P. M.; and Jones, R. P. 1971. Time resolved picosecond emission spectroscopy of organic dye lasers. *Chem. Phys. Lett.* 9, No. 1:1–5.
98. Duguay, M. A. 1976. The ultrafast optical Kerr shutter. In *Progress in Optics*. Vol. XIV, edited by E. Wolf. North-Holland Co., pp. 161–193.
99. Mourou, G.; Drouin, B.; and Denariez-Roberge, M.M. 1972. Variable ultrafast photographic shutter. *Appl. Phys. Lett.* 20:453–455.
100. Topp, M. R.; Rentzepis, P. M.; and Jones, R. P. 1971. Time resolved absorption spectroscopy in the 10^{-12}-sec range. *J. Appl. Phys.* 42, No. 9:3415–3419.
101. Netzel, T.; Rentzepis, P. M.; and Leigh, J. 1973. Picosecond kinetics of reaction enters containing bacteriochlorophyll. *Science* 182:238–241.
102. Duguay, M. A., and Savage, A. 1973. Picosecond Optical Oscilloscope. *Opt. Commun.* 9, No. 2:212–215.
103. Vogel, G. C.; Savage, A.; and Duguay, M. A. 1974. Picosecond optical sampling. *IEEE J. of Quantum Electron.* QE-10, No. 9:642–646.
104. Duguay, M. A., and Hansen, J. W., 1969. An ultrafast light gate. *Appl. Phys. Lett.* 15, No. 6:192–194.
105. Shah, R. R.; Rabson, T. A.; and Kim, D. M. 1975. Interferometric studies of mode-locked Nd^{+3}: Glass lasers. *IEEE J. Quantum Electron.* QE-11, No. 9:783–788.
106. Ippen, E. P., and Shank, C. V. 1975. Dynamic spectroscopy and subpicosecond pulse compression. *Appl. Phys. Lett.* 27, No. 9:488–490.
107. Ruddock, I. S., and Bradley, D. J. 1976. Bandwidth limited subpicosecond pulse generation in mode-locked cw dye lasers. *Appl. Phys. Lett.* 29, No. 5: 296–297.
108. Yassa, Z. A.; Dienes, A.; and Whinnery, J. R. 1977. Subpicosecond pulses from cw double mode-locked dye laser. *Appl. Phys. Lett.* 30, No. 1:24–26.
109. Ippen, E. P.; Shank, C. V.; and Bergman, A. 1976. Picosecond recovery dynamics of malachite green, *Chem. Phys. Lett.* 30, No. 3:611–614.
110. Uhlhorn, R. W., and Holshouser, D. F. 1970. Cross-correlation detection of sub-nanosecond optical pulses. *IEEE J. Quantum Electron.* QE-6, No. 12:775–782.
111. Shank, C. V., and Ippen, E. P. July 1977. Subpicosecond pulses. *Laser Focus,* pp. 44–49.
112. Shank, C. V., and Ippen, E. P. 1975. Anisotropic absorption saturation with picosecond pulses. *Appl. Phys. Lett.* 26, No. 2:62–63.
113. Shank, C. V.; Ippen, E. P.; and Teschke, O. 1977. Sub-picosecond relaxation of large organic molecules in solution. *Chem. Phys. Lett.* 45, No. 2:291–294.
114. Grove, W. 1966. Sampling for oscilloscopes and other rf systems: DC through X-band. *IEEE Trans. MTT* MTT-14, No. 12:629–635.
115. Frye, G. 1968. A new approach to fast gate design. *Tektronix Service Scope* 52:8–9.
116. Zimmerman, A. 1968. The state of the art in sampling. *Tektronix Service Scope* 52:2–7.
117. Merkelo, J. 1973. A dc-to-20 GHz thin-film signal sampler for microwave instrumentation. *Hewlett-Packard Journal* 24, No. 8:10–13.
118. Meyer, F. 1970. Pulse measurements with resolution time of 18 ps. *Electronics Letters,* 6, No. 22:703–704.
119. Auston, D. H. 1974. Picosecond optoelectronic switching and gating in silicon. *Appl. Phys. Letters* 26, No. 3:101–103.
120. Lawton, R. A., and Andrews, J. R., 1975. Pulsed-laser application to sampling oscilloscope. *Electronics Letters* 11, No. 7:138.
121. Johnson, A. M., and Auston, D. H., 1975. Microwave switching by picosecond photoconductivity. *IEEE J. Quantum Electron.* QE-11, No. 6:283–286.
122. LeFur, P., and Auston, D. H. 1976. Picosecond Optoelectronic switch and Pockel's cell. *Appl. Phys. Lett.* 28, No. 1:221–23.

123. Antonetti, A.; Migus, A.; Malley, M. M.; and Mourou, G. 1977. Opto-electronic sampling in the picosecond range. *Opt. Commun.* 21, No. 2:211.
124. Lee, C. H.; Antonetti, A.; and Mourou, G. 1977. Measurements of the photoconductive lifetime of carriers in GaAs by optoelectronic gating technique. *Opt. Commun.* 21, No. 1:158–161.
125. Meyer, F. 1973. Analysis of voltage steps with a time resolution of 12 picoseconds. *Arch. Elektron and Ubertragungstech* 27, No. 1:19–24.
126. Andrews, J. R. 1971. An interfacing unit for a 28 picosecond feed-through sampling head and random sampling oscilloscope. *Rev. Sci. Instrum.* 42, No. 12:1882–1885.
127. Henkels, W. H. 1974. An elementary logic circuit employing superconducting Josephson tunneling gates. *IEEE Trans. Magn.* MAG-10:860–863.
128. Lawton, R. A., and Andrews, J. R. Nov. 1976. Electrical strobing of photoconductor cuts sampling oscilloscopes' risetime. *Laser Focus,* pp. 62–65.
129. Applied Optics 1976. Technical development note. *Appl. Opt.* 15, No. 19:2098.
130. Lawton, R. A., and Andrews, J. R. 1977. Waveform sampler. *U.S. Patent 4,030* [sic].
131. Merkelo, H., and Wiczer, J. J. 1973. A high speed sampling photo multiplier. *Technical Digest of the International Electron Devices Meeting, Session 11.2.* New York: IEEE, pp. 217–219.
132. Wiczer, J. J., and Merkelo, H. 1975. Picosecond optical detection by high speed sampling of photoelectrons. *Appl. Phys. Lett., Vol. 27,* No. 7, pp. 397–399.
133. Richardson, M. C., and Sala, K. 1973. Picosecond framing photography of a laser-produced plasma. *Appl. Phys. Lett.* 23, No. 8.
134. Crisp, M. D. 1972. Comment on a linear technique for the measurement of ultrashort optical pulse widths. *Opt. Commun.* 3, No. 2:111–112.
135. Ruiz, H. J.; Turner, J. J.; and Rabson, T. A. 1972. Interpulse coherence of picosecond pulses. *Opt. Commun.* 6, No. 4:435–436.
136. Rabson, T. A. 1975. Phase characteristics of mode-locked lasers. *Proceedings of the National Science Foundation Users Conference, Opt. Commun.,* Boulder, CO.
137. Holmes, L. 1985. Frontiers of femtosecond research. *Laser Focus/Electro-Optics,* April 1985, pp. 65–76.

3
Instructional Opportunities Offered by the Time-Domain Measurement Technology

Sedki M. Riad, Ph. D., P. E.

Virginia Polytechnic Institute and State University

3.0 INTRODUCTION

The term time-domain technology is used to refer to methods of performing measurements and solving analysis and design problems while dealing with physical quantities in their time-dependent form. An alternative to this term would be *transient techniques or methods*. The latter arises from the fact that time-domain or transient methods are used in most cases only when dealing with nonsteady-state cases to resolve problems in the transition durations between two steady states.

This chapter aims at emphasizing the advantages of introducing the subject of time-domain measurement techniques to an electrical engineering education and demonstrates how electrical engineering students should expect to benefit from it. For those of us who have been long involved in using time-domain (TD) techniques, their advantages seem to be self-evident. This is particularly the case when it is realized that the time domain is the real physical domain where problems originate and where the desired solutions should eventually be found. The conveniences offered by the frequency-domain (FD) techniques that have been used over the past decades for dealing with electrical engineering problems have made these techniques the dominant mode of operation and caused time-domain techniques not only to be secondary but sometimes also forgotten.

The case of time-domain versus frequency-domain techniques should not be surprising if we realize the nature of how the human mind operates. When one is faced with a difficult problem, there is a tendency to escape the real world to an imaginary one—a "dream" world if you wish. In this imaginary world, an attempt is made to resolve a transformed image of the difficulty. If the image is resolved, a return to the real world and an attempt to implement the obtained solution may follow. The process of transforming the real-world problem to the imaginary world as well as the process of implementing the imaginary world solution in the real world (transforming it back) is very important and decides the degree of success of the entire operation.

In this light, the frequency domain is that imaginary, hypothetical domain that we have invented to escape from the complexity of solving electrical engineering problems in the time domain, i.e., in the real world. It is the Fourier transformation that has made that possible to happen.

There is no doubt that solving problems entirely in the real world (the time domain) requires a high level of mathematical abilities. A definite advantage of such solutions is that one keeps in touch with physical quantities and phenomena during the solution process and thus has an improved conceptual understanding of problem detail. This fact does not lower the value of the transformation approach (the frequency domain) by any means. The transformation approach is an indispensable one, especially for those problems that cannot be solved entirely in the real world. Even for some of the problems that can be, some of the tools used in the solution process are more naturally applicable to the transformation approach.

Conventionally, there exist three alternatives for attempting a solution of electrical engineering problems, as follow:

1. *Working the solution formulation entirely in the time domain:* In this method, the problem equations are formulated in the time domain. These equations may include nonlinearities and involve differential equations and/or integral equations. Solutions of these equations are attempted entirely in the time-domain with no transformations.
2. *Working the solution formulation "almost entirely" in the frequency domain:* In this method, the problem equations are formulated in the frequency domain (the conventional frequency, Laplace, or z domains), and the problem statement must be transformed to the frequency domain at the outset. The obtained equations are then solved in the transformation domain, and, finally, the achieved solution is transformed back to the time domain.
3. *Working the solution formulation partially in both domains:* In this method, the problem equations are formulated in the time domain. The equations are then transformed to the frequency domain, where a solution is achieved. The obtained solution is then transformed back to the time domain.

It is relevant to remind the reader here that methods (2) and (3) are not well-suited for nonlinear or time-varying systems. For such systems, method (1) must generally be used.

To demonstrate the basic conceptual differences between these alternatives of problem solving, we will consider the case of deriving the impulse response, $h(t,)$ of the RC network shown in Fig. 3-1. This problem will be solved using each of the three conventional approaches—(1), (2), and (3).

Solution using approach (1)—entirely in the time domain

The desired $h(t)$ is the network's output when its input is a Dirac-delta function $\delta(t)$. Thus, the network equations in the time domain are as follows:

$$v_{\text{in}}(t) = \delta(t) \tag{3-1}$$

$$v_{\text{out}}(t) = h(t) \tag{3-2}$$

Writing the Kirchhoff's voltage law around the loop, we obtain

$$v_{\text{in}}(t) = Ri(t) + v_{\text{out}}(t) \tag{3-3}$$

where

$$i(t) = C\left(\frac{d}{dt}\right) v_{\text{out}}(t) \tag{3-4}$$

Combining all four equations, defining the time constant $\tau = RC$, and rearranging, we obtain

$$\left(\frac{d}{dt}\right) h(t) + \left(\frac{1}{\tau}\right) h(t) = \left(\frac{1}{\tau}\right) \delta(t) \tag{3-5}$$

To solve this first-order linear differential equation, we define a function $g(t)$ such that

$$\left[\left(\frac{d}{dt}\right) g(t)\right] \Big/ g(t) = \frac{1}{\tau} \tag{3-6}$$

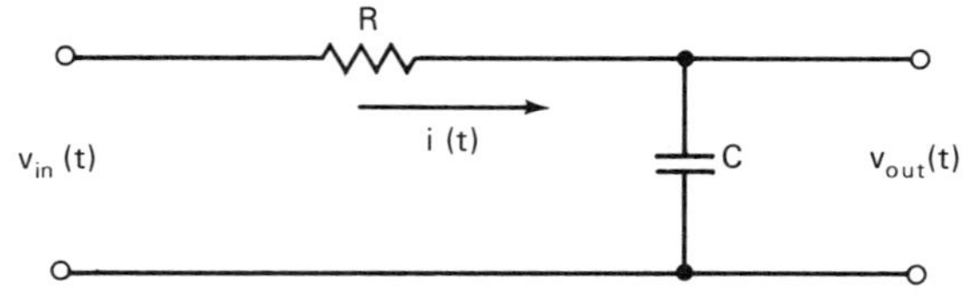

Fig. 3-1. RC network whose impluse response is sought.

This enables us to rewrite Eq. 3-5 in the form

$$\left(\frac{d}{dt}\right) h(t) + \left[\left(\frac{d}{dt}\right) g(t) \Big/ g(t)\right] h(t) = \left(\frac{1}{\tau}\right) \delta(t) \qquad (3\text{-}7)$$

or

$$g(t)\left[\left(\frac{d}{dt}\right) h(t)\right] + \left[\left(\frac{d}{dt}\right) g(t)\right] h(t) = \left(\frac{1}{\tau}\right) \delta(t)\, g(t) \qquad (3\text{-}8)$$

$$\left(\frac{d}{dt}\right)\left[g(t)\, h(t)\right] = \left(\frac{1}{\tau}\right) \delta(t)\, g(t) \qquad (3\text{-}9)$$

Integrating Eq. 3-9, we can obtain the desired solution, as follows:

$$\begin{aligned} h(t) &= \frac{1}{\tau g(t)} \int \delta(t)\, g(t)\, dt \\ &= \left(\frac{g(0)}{\tau g(t)}\right) u(t) \end{aligned} \qquad (3\text{-}10)$$

Using Eq. 3-7, the time function $g(t)$ is found to be

$$\ln g(t) = \left(\frac{1}{\tau}\right) \int dt = \frac{t}{\tau}$$

or

$$g(t) = \exp\left(\frac{t}{\tau}\right) \qquad (3\text{-}11)$$

Combining Eqs. 3-10 and 3-11, we obtain the final form for $h(t)$ as follows

$$\begin{aligned} h(t) &= \left(\frac{1}{\tau}\right)\left[1 \Big/ \exp\left(\frac{t}{\tau}\right)\right] u(t) \\ &= \left(\frac{1}{\tau}\right) \exp\left(\frac{-t}{\tau}\right) u(t) \end{aligned} \qquad (3\text{-}12)$$

Solution using approach (2)—"almost entirely" in the frequency domain

To obtain the impulse response, we first find the network's transfer function in the frequency domain (or more conveniently in the Laplace domain) and then perform an inverse transformation to obtain the desired $h(t)$. As discussed earlier, the problem statement needs to be presented in the frequency domain at the very beginning. We do so in Fig. 3-2 by redrawing the Laplace domain representation of the network. (Note that zero initial capacitor voltage is assumed in this Laplace transformation.)

The transfer function, $H(s)$, is defined as the ratio between the output and the input voltage transforms as follows:

$$
\begin{aligned}
H(s) &= V_{\text{out}}(s)/V_{\text{in}}(s) \\
&= (1/sC)/[R + (1/sC)] \\
&= \left(\frac{1}{\tau}\right) \Big/ \left[s + \left(\frac{1}{\tau}\right)\right]
\end{aligned}
\tag{3-13}
$$

and the impulse response, $h(t)$, can then be obtained by applying the inverse Laplace transform to $H(s)$, as follows:

$$
h(t) = L^{-1}[H(s)] = \left(\frac{1}{\tau}\right) \exp\left(\frac{-t}{\tau}\right) u(t) \tag{3-14}
$$

Solution using approach (3)—partially in both domains

In this case, we first derive the network's time-domain equations leading to differential Eq. 3-5, a step similar to that of approach (1). To solve the differential equation, we then transform it to the Laplace domain as follows:

$$
s\,H(s) - h(0^-) + \left(\frac{1}{\tau}\right)H(s) = \left(\frac{1}{\tau}\right) \tag{3-15}
$$

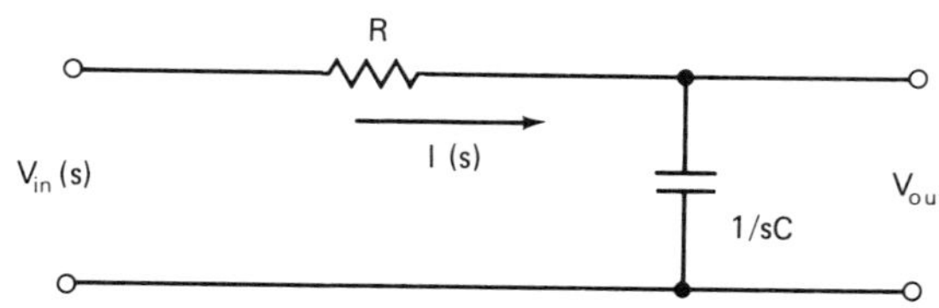

Fig. 3-2. Laplace-domain representation of the network in Fig. 3-1 (zero initial capacitor voltage is assumed in this transformation).

Substituting $h(0^-) = 0$ for zero initial condition and rearranging, we get

$$H(s) = \left(\frac{1}{\tau}\right) \bigg/ \left[s + \left(\frac{1}{\tau}\right)\right] \tag{3-16}$$

and the inverse Laplace transform yields the desired impulse response:

$$h(t) = \left(\frac{1}{\tau}\right) \exp\left(\frac{-t}{\tau}\right) u(t) \tag{3-17}$$

Examining these three solutions, it appears that the frequency-domain approach (2) is the shortest, and the time-domain approach (1), the longest. This may be true in general, but the fact of the matter is that one has to be familiar with transformation theorems and techniques to be able to do the fewer steps shown in approach (2). This requires a certain degree of training that electrical engineering students typically acquire in their networks courses.

The attractiveness of the frequency-domain technique lies in its having fewer steps as well being free of complicated operations such as integrations or differential equations. Such advantages make the study of transformation theorems and techniques a worthwhile one. The main drawback of using the transformation approach however, is its failure to provide physical insights during the various solution steps. In particular, consider the fact that after the first few courses in networks, students being trained in the transformation approach are seldom requested to present final solutions to problems in the time-domain. Consequently, students forget how to do inverse transformations to the time domain and lose physical insight of the problem completely. In some applications, this situation may be tolerated, but, in general, it is definitely a dangerous attitude to ignore returning to the real world, i.e., the time domain.

Keeping in touch with what physically happens within a network or system components at all times is the main advantage of performing solutions entirely in the time domain. The principal drawback lies in the mathematical complications that may arise in the solution process.

In the following section, we will proceed with this discussion of the relative advantages of time-domain measurements over their counterparts in the frequency-domain.

3.1 TIME-DOMAIN MEASUREMENTS

As just discussed, current electrical engineering curricula emphasize how to deal with electrical engineering analysis, design, and measurement problems using frequency-domain techniques. Accordingly, students develop a certain degree of expertise in working with frequency-domain quantities, such as impedance, transfer function, phase, and propagation constant. In accordance with these studies, laboratory courses are designed to train students in the various kinds of frequency-domain measurement techniques. Most laboratory experiments are designed to use a harmonic generator and a real-time oscilloscope. By observing the amplitudes and relative phases of the sinusoidal waveforms, complex frequency-domain quantities such as impedance, gain, and bandwidth can be determined. The main reason for using such simple equipment is that most frequency-domain measurement equipment is costly and requires a high level of expertise for proper use and operation.

Performing measurements at a single frequency permits frequency-domain quantities to be determined only for that particular frequency (assuming a linear, time-invariant system). To study the frequency-domain dependence of the measured parameters, the generator frequency must be swept over the band of interest. At each frequency, the amplitudes and phases of the various waveforms must be recorded.

It is important to realize that it may not be convenient to determine some physical quantities with frequency-domain measurements, in particular when a system's excitation is not a simple harmonic one. Examples of such quantities include the maximum amplitude of a system's response waveform, the speed of the response, and its duration. Such quantities can be conveniently measured only with time-domain measurements and may not require any equipment other than a function generator and an oscilloscope.

To demonstrate some of these concepts, let us consider the case of measuring the impulse response/transfer function of a linear time-invariant system, e.g., the RC network of Fig. 3-1. The block-diagram representation for the system's excitation and response signals can take the form given in Fig. 3-3, in which both the time- and frequency-domain representations are presented. The notations x and y are used instead of v_{in} and v_{out} for the sake of convenience.

The system's response, $y(t)$, is related to both its excitation, $x(t)$, and its impulse response, $h(t)$, by the convolution integral,

$$y(t) = \int_{-\infty}^{+\infty} x(t - \theta)h(\theta)d\theta \tag{3-18a}$$

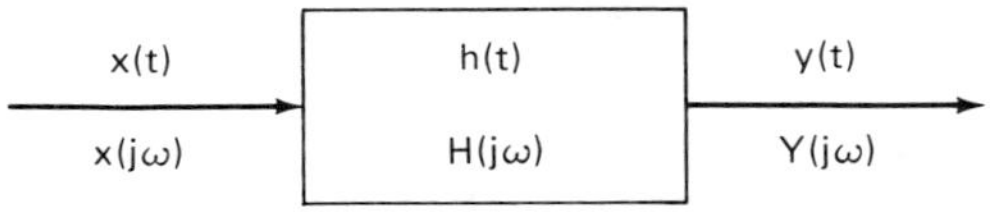

Fig. 3-3. Representation for a general network.

or

$$y(t) = x(t) * h(t) \tag{3-18b}$$

where "$*$" denotes the convolution operation. In the frequency-domain representation, the corresponding relation is given by the Fourier transformation of Eq. (3-18), or

$$Y(j\omega) = X(j\omega) \cdot H(j\omega) \tag{3-19}$$

Time-Domain Measurement of *h*(*t*)

According to Eq. 3-18, if the system is excited by a Dirac delta function, then the system yields $h(t)$ as its response, or

$$x_d(t) = \delta(t) \tag{3-20a}$$

and

$$\begin{aligned} y_d(t) &= \delta(t) * h(t) \\ &= h(t) \end{aligned} \tag{3-20b}$$

Also, for a step excitation, the system yields its step response as follows:

$$x_s(t) = u(t) \tag{3-21a}$$

$$\begin{aligned} y_s(t) &= u(t) * h(t) \\ &= \int_{\infty}^{\infty} h(t)\, dt \end{aligned} \tag{3-21b}$$

in which case

$$h(t) = (d/dt)y_s(t) \tag{3-21c}$$

An experimental setup based on these concepts would consist of a pulse generator and a waveform recorder (simply, an oscilloscope). The purpose of the generator is to generate the excitation pulse (delta or step), which has to be as fast as possible to simulate the ideal. The waveform recorder detects the response waveform.

If high accuracy is needed in these measurements, deconvolution must be applied to remove the effect of the nonideal excitation waveform, $x(t)$, from the detected response, $y(t)$. The deconvolution operation and its limitations will be discussed in a separate section of this chapter. However, if $x(t)$ is much faster than $h(t)$ (its spectral content covers a much wider bandwidth), then fairly adequate measurement accuracy can be obtained without deconvolution.

Frequency-Domain Measurement of $H(j\omega)$

According to Eq. 3-19, the system's transfer function, $H(j\omega)$, can be determined as the complex ratio of the response and excitation frequency-domain transforms. If the linear system's excitation waveform is a simple harmonic one, its response waveform will be a simple harmonic one as well. Consequently, the frequency-domain transforms of both the excitation and response waveforms take the forms of delta functions and are completely characterized by the amplitudes and phases of the harmonic waveforms. At the excitation frequency, the magnitude of the transfer function is equal to the ratio of waveform amplitudes, and its phase is equal to the difference between the waveform phases. Performing this measurement is fairly straightforward. The main problem lies in the fact that one obtains only one frequency data point at a time and that to cover a frequency band, the excitation frequency has to sweep this band.

Using a simple harmonic generator with variable frequency (sweep generator) and a display oscilloscope, one can perform the transfer function measurement of interest. The measurement can be conveniently performed using a dedicated network analyzer, however. This is an instrument with sweep frequency capability that measures and computes the complex ratio and finally displays the results versus frequency.

Discussion

In comparing the two methods here discussed, we should realize that the two quantities of interest—the impulse response and the transfer function—are not only equivalent but ideally related to each other by the Fourier transform pair,

$$H(j\omega) = \int_{-\infty}^{\infty} h(t) \exp(-j\omega t)\, dt \tag{3-22a}$$

and

$$h(t) = (1/2\pi) \int_{-\infty}^{\infty} H(j\omega) \exp(j\omega t) \, d\omega \tag{3-22b}$$

As the transformation equations indicate, information covering the entire time duration from $-\infty$ to $+\infty$ is needed to compute any frequency-domain component in the transformation. Likewise, all frequency-domain components covering the frequency band from $-\infty$ to $+\infty$ are needed to determine any time-domain value of the corresponding waveform. Neither of these two requirements can be satisfied in experimental measurements, and, consequently, only approximate transformations are possible. Moreover, to perform transformations, experimental data need to be acquired, digitized, and supplied to a processor to execute the necessary computations. Because of these processes, further errors accumulate in the computed transforms.

In light of these considerations, it is preferable, if at all possible, to avoid performing domain transformation in measurement procedures, particularly when using unsophisticated (or low-precision) equipment. Consequently, it is better to use time-domain techniques for measuring time-domain quantities and frequency-domain techniques for measuring frequency-domain quantities. In cases where errors are not excessive, a single-domain measurement can adequately determine quantities in both domains. Some preference may be given to one domain or the other depending on the nature of the system in use, its typical excitation, and the quantities to be determined. In this regard, some approximate formulae are available to determine quantities in one domain in terms of those measured in the other domain. Some of these formulae will be discussed in the following section.

3.2 EQUIVALENCE BETWEEN TIME- AND FREQUENCY-DOMAIN QUANTITIES

It should have become evident by now from these discussions that time-domain techniques in general and time-domain measurements in particular form a very essential part of any electrical engineering study and should be given proper attention when designing electrical engineering curricula.

This section will demonstrate how to integrate time-domain techniques and measurements into current electrical engineering curricula. It is proposed that the physical correspondences between time- and frequency-domain quantities be used to serve this purpose, a method that should be effective since it introduces time-domain concepts by relating them to the familiar frequency-domain ones. Demonstrating these correspondences should have a great impact on the student's physical insight into the subject.

Correspondences Between Time- and Frequency-Domain Quantities

We will now present three demonstrations of how various correspondences between time- and frequency-domain phenomena can be used to integrate time-domain techniques and concepts into electrical engineering curricula.

Correspondence Between Bandwidth and Dispersion. Consider the case of measuring the transfer characteristics of a small-signal, low-frequency (audio) amplifier. Using the frequency-domain techniques discussed in Sec. 3.1, the magnitude of the amplifier's transfer function can be measured. Figure 3-4 displays a typical response.

The physical insight gained in performing this frequency-domain measurement lies in observing how the amplifier behaves at various frequencies—the low-, mid-, and high-frequency bands—and in seeing how the various amplifier network elements affect the amplifier's performance parameters, such as the midband gain, A; the low-frequency cutoff, f_L; and the high-frequency cutoff, f_H. Such parameters are important to characterize the amplifier's performance and appear directly in the frequency-domain measurement results. These parameters can also be determined using time-domain techniques. If the amplifier is excited using a step waveform, a response $w(t)$ similar to that of Fig. 3-5 is obtained.

The amplitude of the response, W_{max}, is equal to the midband gain; the cutoff frequencies can be determined from the transition duration, τ_h; and the sag slope is established using the approximate formulae,[1]

$$f_H - f_L = 0.35/\tau_h \tag{3-23}$$

$$\text{Sag slope} = -2\pi A \cdot f_L \tag{3-24}$$

It is clear that either a time-domain or frequency-domain measurement can be used to determine the network's midband gain and cutoff frequencies. Since they are frequency-domain quantities, however, they are more directly obtained from a frequency-domain measurement. On the other hand, the use of time-

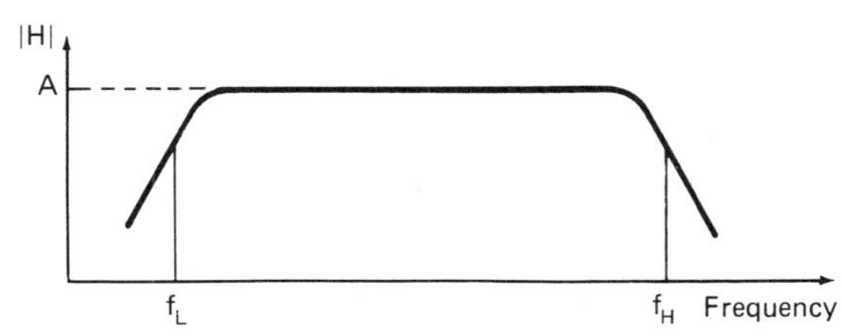

Fig. 3-4. Typical response or transfer function of a band-pass amplifier.

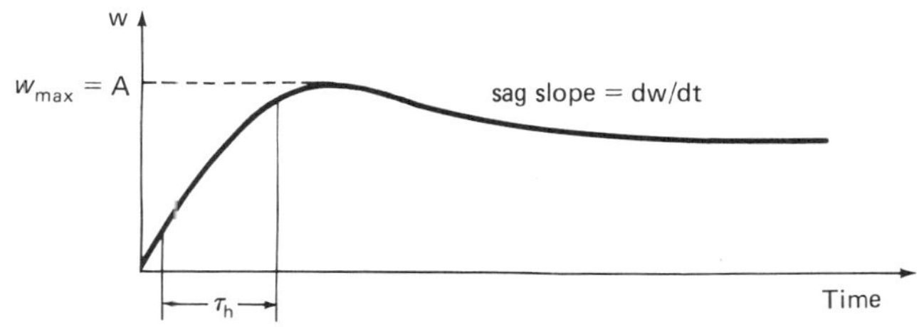

Fig. 3-5. Step response of the band-pass amplifier of Fig. 3-4.

domain techniques saves on experimental steps since only one measurement is needed, whereas several measurements are needed to sweep an adequate frequency range in the frequency-domain. Another advantage of using the time domain for this problem is that the amplifier's effect on waveform dispersion becomes more physically evident to the viewer.

It is important to note here that the experimental setup used to perform a measurement can be the same for both domains if a waveform generator capable of producing both harmonic and pulse waveforms is used. Such a setup allows for more experimentation with the amplifier response. The student can test the amplifier's response to a variety of waveforms at various repetition rates, in addition to the harmonic and step excitations already mentioned.

Moreover, if the generator chosen is capable of changing the transition duration of the generated step, the need for deconvolution in time-domain measurement can be demonstrated. If the generated step has a fixed transition duration, however, the same need can be demonstrated by applying an externally connected filter to slow its transition down. In both cases, the exact deconvolution of waveforms will not be possible using the simple equipment designated here. Instead, approximate formulae relating the various signal bandwidths or transition durations should be used.[1] If the slow excitation waveform is labeled $x(t)$, the amplifier's response, $y(t)$, will be given by the convolution in Eq. 3-18. If the corresponding transition durations and bandwidths are labeled τ_x, τ_y, BW_x, and BW_y, respectively, then one can use the following approximate relations to estimate the amplifier's transition duration, τ_h, and bandwidth, BW_h:

$$\tau_y^2 = \tau_x^2 + \tau_h^2 \tag{3-25}$$

$$BW_y^{-2} = BW_x^{-2} + BW_h^{-2} \tag{3-26}$$

Correspondence Between Phase and Time Delay. Consider the measurement setup of Fig. 3-3, and let a time delay, T_d, occur to either $x(t)$ or $h(t)$ by introducing a delaying network (simply, a delay line) between the pulse generator, $x(t,)$ and the system. If the delay is considerd as part of the pulse generator, the network's excitation can be written as $x(t - T_d)$, and the system's impulse response is still $h(t)$. On the other hand, if the delay is considered as part of the network, the exciting signal is x(t), and the new system's impulse response becomes the convolution of those of the delayng network, $\delta(t - T_d)$, and of the old system's response, $h(t)$, thus becoming $\delta(t - T_d) * h(t)$, or $h(t - T_d)$. In either case, the response of the new setup takes the form

$$x(t - T_d) * h(t) = x(t) * h(t - T_d) = y(t - T_d) \tag{3-27}$$

which is the same as the response of the old setup except for the delay T_d. In a frequency-domain representation, the response takes the form,

$$X(j\omega) \cdot \exp(j\omega T_d) \cdot H(j\omega) = Y(j\omega) \cdot \exp(j\omega T_d) \tag{3-28}$$

which has exactly the same magnitude function as $Y(j\omega)$, the response of the old system. The only difference between the two responses is the additional linear phase component (frequency function), ωT_d.

This demonstration clarifies the correspondence between the phase in the frequency domain and the delay in the time domain. It also demonstrates how time-domain measurements can directly reveal time delays, and hence phase changes.

To introduce a bit of sophistication into this demonstration, we will discuss dispersion as a time-delay phenomenon. It is known that dispersion in transmission networks is accounted for by the nonuniform delays that occur in the various modes of propagation. In other words, dispersion is associated with nonlinear phase changes (versus frequency). It is also known that, for minimum phase networks, a relationship exists between this phase nonlinearity and bandwidth limitation.[2] Consequently, one can see the correspondence between dispersion and bandwidth limitation in terms of that between phase and delay.

In the light of this discussion, we return to the case of the RC network, where the slow accumulation of charges on the capacitor results in nonuniform delays of the output voltage components. These delays form the slow rise of the voltage waveform and hence produce dispersion. The explanation of such events in the frequency-domain is based on the correspondence between the nonuniform delay and nonlinear phase component, which results in bandwidth limitations and consequently exhibits the correspondence between bandwidth limitations and dispersion.

Correspondence Between Resonances and Multiple Reflections (Standing Waves). Consider the case of the transmission network shown in Fig. 3-6. This network is composed of a lossless transmission section with two

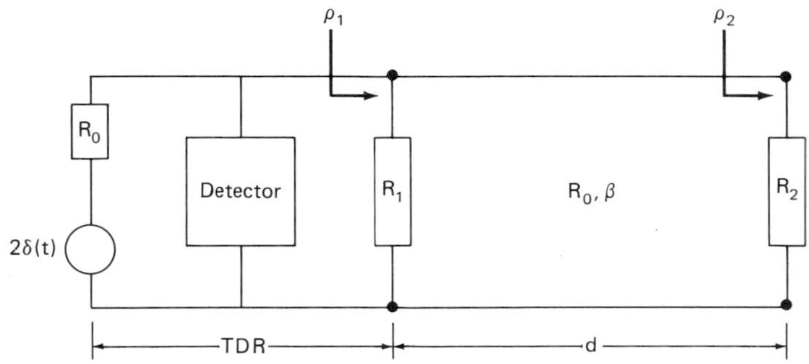

Fig. 3-6. Transmission network to illustrate the correspondence between resonances and multiple reflections.

resistive discontinuities, R_1 and R_2, separated by a distance, d. Let the section's characteristic impedance be R_0 and its phase propagation constant β. The section is then excited by a voltage pulse through a time-domain reflectometer (TDR). Without any significant loss of generality, let the voltage of the TDR generator be a Dirac-delta function of time with an intensity of 2, i.e., $2\ \delta(t)$.

In Fig. 3-6, the local reflection coefficients ρ_1 and ρ_2 are given by

$$\begin{aligned}\rho_1 &= (R_1//R_0 - R_0)/(R_1//R_0 + R_0)\\ &= -\ R_0/(2R_1 + R_0) \end{aligned} \tag{3-29}$$

$$\rho_2 = (R_2 - R_0)/(R_2 + R_0) \tag{3-30}$$

Using the space-time (bounce) diagram,[3] it can be shown that the TDR detected voltage, $v_{\text{TDR}}(t)$, has the form,

$$\begin{aligned} v_{\text{TDR}}(t) &= (1 + \rho_1)^2\ \delta(t) + \rho_2\ (1 + \rho_1)^3\ \delta(t - 2T)\\ &+ \rho_1\rho_2^2\ (1 + \rho_1)^3\ \delta(t - 4T)\\ &+ \rho_1^2\ \rho_2^3\ (1 + \rho_1)^3\ \delta(t - 6T) + \ldots \end{aligned} \tag{3-31}$$

where T is the one-way travel time of the distance d, which can be expressed in terms of the speed of propagation c as follows:

$$T = d/c \tag{3-32}$$

Equation 3-31 shows the detected TDR voltage as a sequence of delta functions corresponding to the multiple reflections between the line discontinuities. If the excitation were not a delta function, the TDR voltage would still take the sequence form except that it would be a sequence of the same waveform as that of the excitation and weighted in the same manner as given in Eq. 3-31.

Using frequency-domain techniques to express the TDR voltage, the result will be the Fourier transformation of Eq. 3-31. Denoting the Fourier representation of $v_{TDR}(t)$ by $V_{TDR}(j\omega)$, we get

$$\begin{aligned}(3\text{-}33)V_{TDR}(j\omega) &= (1+\rho_1)^2 + \rho_2(1+\rho_1)^3\exp(j2\beta d)\\ &+ \rho_1\rho_2^2(1+\rho_1)^3\exp(j4\beta d)\\ &+ \rho_1^2\rho_2^3(1+\rho_1)^3\exp(j6\beta d) + \ldots \qquad (3\text{-}33)\end{aligned}$$

which contains a summable geometric series that yields

$$V_{TDR}(j\omega) = (1+\rho_1)^2\{1+\rho_2\exp(j2\beta d)\}/\{1-\rho_1\rho_2\exp(j2\beta d)\} \qquad (3\text{-}34)$$

This frequency-domain form has a periodic structure. The peaks occur when $\exp(j2\beta d)$ is a positive real quantity and the valleys when it is a negative real quantity, that is, peaks occur at $2\beta d = 2n\pi$, or

$$f_{np} = (2n)c/(4d) \qquad (3\text{-}35a)$$

and valleys occur at $2\beta d = (2n+1)\pi$, or

$$f_{nv} = (2n+1)c/(4d) \qquad (3\text{-}35b)$$

where n is an integer and f_{np} and f_{nv} are the frequency coordinates of the peaks and valleys, respectively.

The multiple peaks of the frequency-domain form in Eq. 3-34 correspond to multiple resonances in the network under consideration. The fundamental resonance frequency is the frequency of the first peak, or

$$f_{1p} = c/(2d) = 1/(2T) \qquad (3\text{-}36)$$

and all the higher frequency resonances occur at the harmonics of this fundamental.

As revealed by Eq. 3-36, the frequency of the fundamental resonance, as well as that of the other resonances, is related to $2T$, the two-way travel time around the transmission section. According to Eq. 3-31, $2T$ is also the time interval between two consecutive delta functions in the time-domain waveform.

This demonstration clearly shows the correspondence between the multiple reflections in the transmission network and the observed resonances of the network. With a bit of imagination, this correspondence can be discussed for nontransmission networks as well. Considering the simple LC resonant network, one can view the energy exchange between the inductor and capacitor as a form of multiple reflections, with the time interval of the two-way transfer as $2T$.

The demonstrations given here are but a few of the many that can be used to help electrical engineering students relate to time-domain techniques and concepts. Laboratory courses can benefit from them by designing experiments to introduce time-domain techniques and comparing results obtained from these techniques with frequency-domain measurements.

The important thing to keep in mind is that measurement precision may be sacrificed if one tries to avoid performing Fourier transformations in general and deconvolution in particular. In addition to requiring acquisition and signal-processing capabilities, such operations are not as simple and as straightforward as their analytical forms might seem to indicate. In the following section, we will discuss the deconvolution operation in order to reveal a problem associated with its practical applications, a problem that handicapped time-domain measurements for a long time.

3.3. TIME-DOMAIN TECHNIQUE'S MAIN HANDICAP—DECONVOLUTION

This section provides a tutorial review of the deconvolution problem. It reviews the definitions of convolution and deconvolution and discusses both the time-domain and frequency-domain methods of performing the latter. It also identifies the true nature of the practical deconvolution problem as an approximation or estimation problem, and it discusses the concepts of achieving an "acceptable" solution. Deconvolution is also discussed in Chaps. 12 and 13.

Definition of Deconvolution

Unlike most operations, deconvolution has no direct mathematical definition. It is defined merely as the inverse of the convolution operation. Consequently, convolution must first be reviewed before discussing deconvolution.

In reference to Fig. 3-3, if an input signal, $x(t)$, is applied to excite a linear, time-invariant system with an impulse response, $h(t)$, the system's response, $y(t)$, due to the imposed excitation is given by the convolution[4] defined in Eq. 3-18, restated here for convenience,

$$y(t) = \int_{-\infty}^{+\infty} x(t - \theta) \cdot h(\theta) d\theta \tag{3-18a}$$

or

$$y(t) = x(t) * h(t) \tag{3-18b}$$

with the corresponding frequency-domain form given by Eq. 3-19,

$$Y(j\omega) = X(j\omega) \cdot H(j\omega) \tag{3-19}$$

As the inverse process of the convolution operation, deconvolution involves the determination of either $x(t)$ or $h(t)$, assuming prior knowledge of the other two time functions. Without any loss of generality, only the deconvolution for determining $h(t)$ will be referred to henceforth.

Ideally, with the exact and complete knowledge of $x(t)$ and $y(t)$, deconvolution can be performed to compute $h(t)$ exactly. The computation can be performed in either the time- or the frequency-domains. In the frequency-domain method, the first step is to use the Fourier transformation to compute frequency-domain forms $X(j\omega)$ and $Y(j\omega)$. Next, Eq. 3-19 is used to compute $H(j\omega)$, to which the inverse Fourier transform can be applied to yield $h(t)$. In the time-domain methods, the first step is to form a set of linear simultaneous equations by using the discrete sequence representations of time functions $x(t)$, $y(t)$, and $H(t)$ in the integral of Eq. 3-18. This set of equations is then solved to yield the time sequence of $h(t)$.[5,6] Such a solution is typically achieved by using either matrix-inversion or difference-equation techniques.

In the case of ideal or analytical deconvolution such as that just considered, the procedure for performing the deconvolution operation has been shown to be fairly straightforward. In a fashion analogous to numerical differentiation, however, the practical implementation is troublesome. It will now be shown that the deconvolution operation is very sensitive to data errors. Any small error in the processed data multiplies significantly and will be exaggerated in the final result. The latter can be accurate only if the processed data exhibit absolute accuracy.

Practical Deconvolution

In practical measurements, exact knowledge of the time signals is not possible. As a result of bandwidth limitation of the time-signal detector and the digitization and acquisition processes, the acquired waveform differs to a certain extent from the true time signal. Denoting the acquired waveforms by $x_w(t)$ and $y_w(t)$, they can be related to their corresponding true time signals $x(t)$ and $y(t)$, respectively, as follows:

$$x_w(t) = x(t) + x_e(t) \tag{3-37a}$$

and

$$y_w(t) = y(t) + y_e(t) \tag{3-37b}$$

where $x_e(t)$ and $y_e(t)$ are unknown error components (time functions). Using Eq. 3-37, the convolution of Eq. 3-18b takes the form,

$$[y_w(t) - y_e(t)] = h(t) * [x_w(t) - x_e(t)] \tag{3-38}$$

and the corresponding frequency domain form is

$$[Y_w(j\omega) - Y_e(j\omega)] = H(j\omega) \cdot [X_w(j\omega) - X_e(j\omega)] \tag{3-39}$$

Since the error quantities, $x_e(t)$ and $y_e(t)$ [or their transforms, $X_e(j\omega)$ and $Y_e(j\omega)$] are not known, it is not possible to solve Eqs. 3–38 or 3–39 to obtain an exact solution for $h(t)$ [or $H(j\omega)$]. Instead, these equations can yield exact solutions for $h_w(t)$ and $H_w(j\omega)$, which are defined as follows:

$$y_w(t) = h_w(t) * x_w(t) \tag{3-40}$$

$$Y_w(j\omega) = H_w(j\omega) \cdot X_w(j\omega) \tag{3-41}$$

where $h_w(t)$ is the inverse Fourier transform or $H_w(j\omega)$. Consequently, using Eqs. 3-39 and 3-41, $H_w\,(j\omega)$ can be written as

$$H_w(j\omega) = H(j\omega) + H_e(j\omega) \tag{3-42a}$$

where

$$H_e(j\omega) = [Y_e(j\omega) - H\,(j\omega)\cdot X_e(j\omega)(j\omega)]/X_w(j\omega) \tag{3-42b}$$

Assuming that error components $h_e(t)$ and $H_e(j\omega)$ are small, the solutions $h_w(t)$ and $H_w(j\omega)$ may provide acceptable approximations to the desired $h(t)$ and $H(j\omega)$, respectively, or

$$h(t) \approx h_w(t) \tag{3-43a}$$

and

$$H(j\omega) \approx H_w(j\omega) \tag{3-43b}$$

On the other hand, significant errors may be encountered in this approximation because of the neglected error components, a situation that can be explained by examining the nature of the approximation in Eq. 3-43. From Eqs. 3-19, 3-39, 3-41, and 3-43, the frequency-domain form of this approximation can be written as

$$H(j\omega) = \frac{Y(j\omega)}{X(j\omega)} = \frac{Y_w(j\omega) - Y_e(j\omega)}{X_w(j\omega) - X_e(j\omega)} \approx \frac{Y_w(j\omega)}{X_w(j\omega)} \approx H_w(j\omega) \tag{3-44}$$

In practical situations, $x(t)$ is a band-limited signal, i.e., $X(j\omega)$ has a finite bandwidth, and, consequently, there exists a region (or regions) of frequencies for which $X(j\omega)$ is fairly small and may even be zero. In these regions, the approximation, $X(j\omega) \approx X_w(j\omega)$, is not only not a good one, it can be orders of magnitudes off. In the same regions, $Y(j\omega)$ is expected to be small, too, and, consequently, the approximation $Y(j\omega) \approx Y_w(j\omega)$ is not a good one either. As a result, the computed ratio, $Y_w(j\omega)/X_w(j\omega) = H_w(j\omega)$, can be expected to differ greatly from the correct ratio, $Y(j\omega)/X(j\omega) = H(j\omega)$. In fact, the deconvolution result $H_w(j\omega)$ will contain some fairly high values (some may be infinite).[7–10] These high values can be viewed as a sequence of spikes or $\delta(j\omega)$ functions. Upon the application of the inverse Fourier transformation to $H_w(j\omega)$ in order to compute its time-domain form, $h_w(t)$, the $\delta(j\omega)$ error spikes that were concentrated in the frequency regions of small $X(j\omega)$ produce error contributions that are spread over the entire time-domain transform epoch (or most of it). And if the error components are high, they may dominate the time-domain transformation, thereby hiding all or most of the details of $h_w(t)$.[7–10]

At this point, it is important to mention that although we have been considering the frequency-domain deconvolution method, the conclusions of this discussion are valid for the time-domain deconvolution as well because, assuming no computational errors, both the time-domain and frequency-domain deconvolution procedures are mathematically equivalent and should yield identical answers. They are just two different methods for solving the same equation. Obviously, computational errors will add to the magnitude of the problem and the errors encountered. Such errors can be minimized, however, and are typically fairly small compared to the approximation errors; they can thus be overlooked in this presentation.

As should be clear by now, the problem at hand is an estimation problem and does not have an exact solution because of the approximations assumed in the equations to be solved and the lack of information about the measurement errors. Such problems are often referred to as "ill-conditioned" or "ill-posed" problems.[11–13] For this class of problems, more than one approximate solution may be achieved, some of which can be considered "acceptable" estimates of the exact solution, and some of these are just "better" than the others.

Achieving Acceptable Estimates

The basic concept for achieving an acceptable solution to practical deconvolution is to accept $H_w(j\omega)$ as an available approximation (or preliminary estimate) of $H(j\omega)$ and try to minimize the contribution of the $\delta(j\omega)$ error spikes in a way that will permit a meaningful inverse Fourier transformation. This can be achieved by applying a properly designed filter to $H_w(j\omega)$, one whose stop-band characteristics will eliminate or reduce the $\delta(j\omega)$ error spikes. Denoting the filter's transfer function by $F(j\omega)$, an acceptable estimate $H_{acc}(j\omega)$ of $H(j\omega)$ can be written as

$$\begin{aligned} H_{acc}(j\omega) &= H_w(j\omega) \cdot F(j\omega) \\ &= [Y_w(j\omega) \cdot F(j\omega)]/X_w(j\omega) \\ &= Y_w(j\omega)/[X_w(j\omega)/F(j\omega)] \end{aligned} \tag{3-45}$$

Depending on the filter characteristics, $F(j\omega)$ can be applied to either $H_w(j\omega)$, $X_w(j\omega)$, or $H_w(j\omega)$, as shown above in Eq. 3-45.

The main concern in designing $F(j\omega)$ is how to achieve the "best" $\delta(j\omega)$ error-spike reduction without introducing any serious errors in the $H_w(j\omega)$ information. Denoting the transfer function of the optimum filter by $F_{opt}(j\omega)$, the best (or optimum) estimate for the deconvolution result, $H_{opt}(j\omega)$, can be written as

$$H_{opt}(j\omega) = H_w(j\omega) \cdot F_{opt}(j\omega) \tag{3-46}$$

Typically, the optimization process is performed iteratively while monitoring some performance indicator(s).[6–10,14,16] Performance monitoring can be done either manually via operator intervention[7–9,16] or automatically, as in a computer algorithm.[6-10,14,16] The method is usually based on the type of filter used and the nature of the performance perameters to be monitored.

Some examples of frequency-domain filters having different degrees of sophistication will now be demonstrated. The first example presents the simple ideal filter with abrupt cutoff.[6,7,15] In this filter, two parameters are to be optimized: the cutoff frequency and the attenuation in the stop-band. The optimization process is expected to be fairly lengthy since there are two variables (parameters) to adjust. Typically, the performance of such a filter is only marginally acceptable even when optimized because of the abrupt discontinuity(ies) introduced at the cutoff frequency(ies). In order to improve its performance, a bit of sophistication can be introducted by changing the abrupt cutoff to a linear (or, in general, a gradual) one. In doing so, however, the number of optimization parameters increases to three (or more), thus complicating the optimization process.

A fairly sophisticated class of filters that have demonstrated quality performance is characterized by the form[6,8,10,14,17]

$$F(j\omega) = \frac{1}{[1 + \phi(\omega)/|X_w(j\omega)|^2]} \tag{3-47}$$

where $\phi(w)$ is a positive function of frequency. Basically, this filter is designed so that the quantity $\phi(\omega)/|X_w(j\omega)|^2$ is fairly small, and, consequently, $F(jw)$ is almost unity in the regions where $X_w(j\omega)$ is not very small. When $X_w(j\omega)$ is fairly small, however, the quantity $\phi(\omega)/|X_w(j\omega)|^2$ should become significant and cause adequate filter attenuation.

Two cases of this class of filters have appeared in the literature; they are those in which $\phi(\omega) = \lambda$ (see References 8, 10, and 14) or $\phi(w) = \gamma\omega^4$ (see Reference 6), where both λ and γ are positive quantities. In both these cases, there is only one parameter to optimize, and this optimization is typically carried out through iteration while monitoring some performance indicator(s).

Filtering can be performed completely in the time-domain as well. An acceptable estimate, $h_{acc}(t)$ for the system's impulse response is related to the preliminary approximation $h_w(t)$ and to the filter's impulse response $f(t)$ by the relationship,

$$h_{acc}(t) = h_w(t) * f(t) \tag{3-48}$$

which is the inverse Fourier transformation of Eq. 3-45. Similarly, Eq. 3-46 transforms to yield the corresponding relationship between the optimized estimate, $h_{opt}(t)$, and the optimum filter, $f_{opt}(t)$, as follows:

$$h_{opt}(t) = h_w(t) * f_{opt}(t) \tag{3-49}$$

Some examples of time-domain filteration techniques include averaging,[6] least squares,[6,17–19] and conjugate gradient methods.[20] Two other examples are worth mentioning here, the first because of its simplicity, and the second because of its correspondence to a previously mentioned frequency-domain filter. The first[21] is a simple smoothing filter to be applied to the computed system's step response [obtained by integrating $h_w(t)$]. The smoothed step response is then differentiated to obtain $h_{acc}(t)$. The second filter of interest[16] is obtained by considering the time-domain form of filter Eq. 3-47; the case of filters in which $\phi(\omega) = \lambda$ has been discussed in the literature.

REFERENCES

1. Millman, J. 1979. *Microelectronics: digital and analog circuits and systems*. New York: McGraw-Hill.
2. Oppenheim, A. V., and Schafer, R. W. 1975. *Digital Signal Processing*. Englewood Cliffs, NJ: Prentice-Hall.
3. Nahman, N. S., and Jickling, R. M. 1973. Frequency domain measurements of baseband instrumentation, *NBS IR 73-330,* National Bureau of Standards, Boulder, CO.
4. McGillem, C. D., and Cooper, G. R. 1974. *Continuous and discrete signal and system analysis*. New York: Holt, Reinhard, and Winston.
5. Gans, W. L., and Andrews, J. R. 1975. Time domain automatic network analyzer for measurement of RF and microwave components. *NBS Tech. Note 672,* National Bureau of Standards, Boulder, CO.
6. Nahman, N. S., and Guillaume, M. E. 1981. Deconvolution of time domain waveforms in the presence of noise. *NBS Techn. Note 1047,* National Bureau of Standards, Boulder, CO.
7. Riad, S. M., and Riad, A. A. 1979. Optical fiber dispersion characterization study, *NASA Contract Report 165624,* Kennedy Space Center, NASA, FL.
8. Riad, S. M. 1980. Optical fiber impulse response evaluation using frequency domain optimal compensation deconvolution, *Proceedings of FOC '80,* International Fiber Optic and Communication Exposition, San Francisco, CA, pp. 210–213.
9. Riad, S. M., and Stafford, R. B. 1980. Impulse response evaluation using frequency domain optimal compensation deconvolution, *Proceedings of 23rd Midwest Symposium on Circuits and Systems,* Toledo, OH, pp. 521–525.
10. Parruck, B., and Riad, S. M. 1983. An optimization criterion for iterative deconvolution. *IEEE Trans. on Insrum. & Meas.* IM-32, No. 1: 137–140.
11. Tikhonov, A. N., and Arsenin, V. Y. 1977. *Solution of Ill Posed Problems*. New York: Halsted Press.
12. Ekstrom, M. P. 1973. A spectral characterization of the ill-conditioning in numerical deconvolution. *IEEE Trans. on Audio and Electroacoustics* AU-21, No. 4: 344–348.
13. Sarkar, T. K.; Weiner, D. D.; and Jain, V. K. 1981. Some mathetmatical considerations in dealing with the inverse problem. *IEEE Trans. Antennas & Propagat.* AP-29, No. 2: 373–379.
14. Parruck, B., and Riad, S. M. 1984. Study and performance evaluation of two iterative frequency domain deconvolution techniques. *IEEE Trans. on Insrum. & Meas.* IM–39, No. 4:281–287.
15. Parruck, B. 1983. Study and performance evaluation of some iterative frequency domain deconvolution techniques. M. S. thesis, Virginia Polytechnic Institute and State University, Blacksburg, VA.
16. Ekstrom, M. P. 1969. An iterative time domain method of system response correction. *WESCON Tech. Papers,* Pt. 14.
17. Wing, G. M. 1984. A primer on integral equations of the first kind: The problem of deconvolution and unfolding. *Los Alamos National Laboratory Report* LA-UR-84-1234, Los Alamos, NM.
18. Sarkar, T. K.; Weiner, D. D.; Jain, V. K.; and Dianat, S. A. 1982. Impulse response determination in the time domain. *IEEE Trans. Antennas & Propagat.* AP-30, No. 4: 657–663.

19. Preis, D. 1977. Least-square time-domain deconvolution for transversal-filter equalizers. *Electronics Letters* 13, No. 12:356–357.
20. Sarkar, T. K.; Tseng, F. I.; Dianat, S. A.; and Hollmann, B. Z. 1983. L_2-approximation of impulse response from time limited input and output: Theory and experiment. *URSI*, US National Radio Science Meeting, Boulder, CO.
21. Okoshi, T., and Sasaki, K. 1978. Precise measurement of the impulse response of an optical fiber, *The Trans. of the IECE of Japan* E 61, No. 12: 964–965.

4
Fast-Pulse Generator Survey

James R. Andrews, Ph.D.

Picosecond Pulse Labs, Inc.

4.0 INTRODUCTION

The purpose of this chapter is to survey the state of the art in pulse generators.* Emphasis is given to commercially available pulse generators with transition durations (rise times) of 1 nsec or faster. The survey papers by Runyon[1] and Nicolson et. al[2] are also recommended reading.

Each recent decade has seen a continuing shrinking of the time scales of interest in electronic instrumentation. The forties were concerned with 100-nsec time periods. In the fifties, instrumentation entered the 10-nsec range, and by the early sixties, it had reached time scales of 1 nsec. The introduction of avalanche transistors, step-recovery diodes, tunnel diodes, and Schottky diodes provided early entry into the picosecond domain. For the most part, this progress did not continue during the seventies. Rather, this decade was a time for refinement and commercial application of many of the advances of the sixties. The major exception was the work in superconductivity, mode-locked lasers, and photoconductivity, which yielded durations as short as 1 psec.

The development of the sixties culminated in the Hewlett-Packard 1106**, a 20-psec tunnel-diode pulse generator, and now the well-known H-P 1430A** sampling oscilloscope, eventually reaching a resolution of 20 psec in the H-P 1430C. Unfortunately, the Hewlett-Packard 1106 and 1430 are no longer manufactured. As a consequence, the Tektronix S-6 30-psec instrument is the only wide-band sampling oscilloscope that has been available since the late seventies.

*This chapter was originally presented as a talk at the National Radio Science Meeting, 13 January 1982, Boulder, CO, URSI-USNC, Comm. A, Session A1, Time-Domain Measurements. The talk was entitled "Fast Pulsers—Small and Large."

**Items mentioned in the text and identified with a double asterisk (**) are, to the best of the author's knowledge, no longer commercially available, except perhaps on the used equipment market.

The pulse generators to be discussed in this chapter range from the conventional bench models that approach the microscopic at one extreme to the very large pulsed-power machines that can fill an entire building at the other extreme. In like manner, the power levels range from microwatts to megawatts. Risetimes or impulse durations also range from femtoseconds (0.001 psec) to microseconds. A useful way to compare various generators is to introduce a figure of merit based on the time-rate-of-change of the output voltage, i.e., output amplitude/risetime, in units of V/nsec.

4.1 CONVENTIONAL TRANSISTOR PULSE GENERATORS

Many manufacturers throughout the world make conventional transistor-pulse generators. Most of these generators provide a repetition rate of less than 100 MHz, with pulse amplitudes of 10 V into 50 ohms and transition durations (risetimes) on the order of 3 nsec. For rep. rates in this range, quite a few manufacturers offer the added feature of adjustable risetimes. The fastest generator of this type is the HP-8161A, with a 100-MHz rep. rate, 5-V amplitude, and 1.3-nsec (or slower) variable risetime. Quite a few generators in this category are also programmable via the IEEE 488 Buss.

Above 100 MHz, the number of generators offered drops dramatically. The fastest is Colby Instruments Model PG1000A, which operates at a maximum

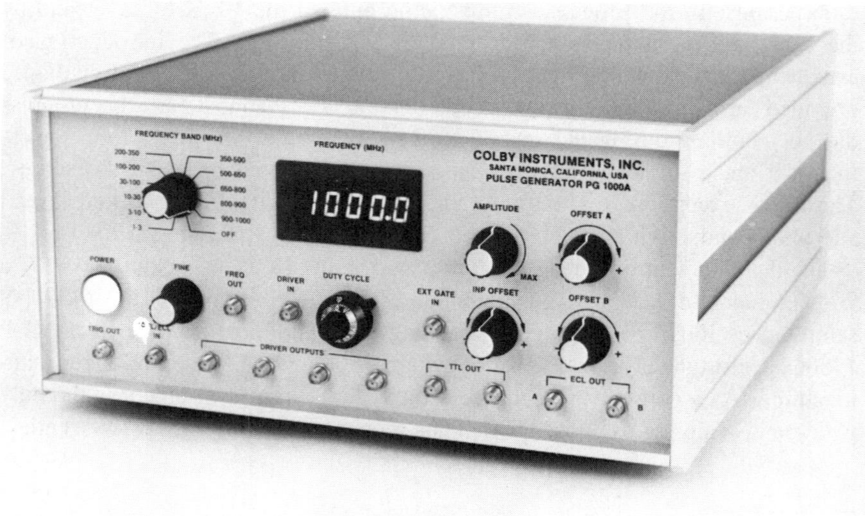

(a)

Fig. 4-1. (a) Colby 1-GHz PG-1000A pulse generator; (b) 200 mV/div. and 200 ps/div. (leading edge); and (c) 500 mV/div. and 500 ps/div. (1-GHz rep. rate).

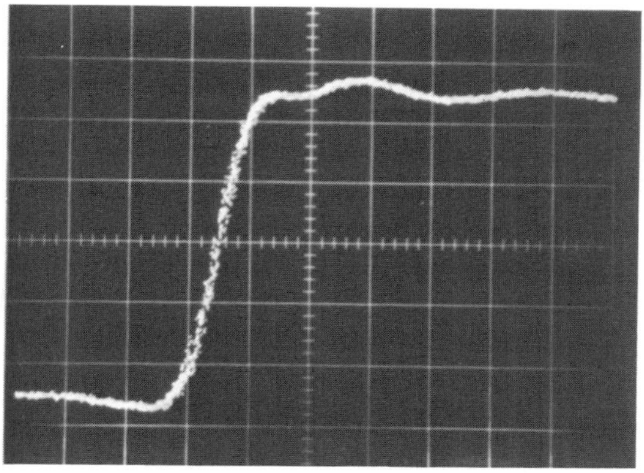

(b)

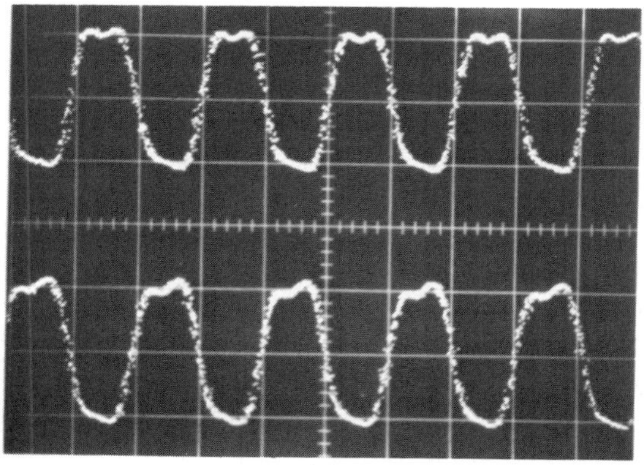

(c)

Fig. 4-1. (continued)

rep. rate of 1 GHz and provides 1 V into 50 ohms and 200-psec risetimes. Figure 4-1 shows this generator along with some measured waveforms. Other generators in this category are listed in Table 4-1. The figure of merit for this class of pulse generators ranges from 2 to 8 V/nsec.

Pulse generators typically use standard transistor-transistor-logic (TTL) or emitter-coupled-logic (ECL) integrated circuits to perform the functions of repetition-rate generation, trigger output, pulse delay, and pulse duration. Earlier designs used discrete components. The output stage of conventional pulse generators usually uses moderate power, RF, bipolar transistors.

Adjustable risetime generators use constant-current sources to charge and discharge a capacitor to obtain linear ramps for the leading and trailing edges,

Table 4-1. Conventional High-Speed Pulse Generators.

MANUFACTURER	MODEL	REP. RATE	AMPLITUDE, V	TRANSITION DURATION (RISETIME)
Colby Instruments	PG1000A	1 GHz	1	200 ps
Takeda-Riken**	4200	1 GHz	2	250 ps
Hewlett-Packard	8080	1 GHz	1.2	300 ps
Tau-Tron	MN-1001	1 GHz	1	500 ps
E-H Int.	129	500 MHz	2	500 ps
E-H Int.	122	250 MHz	5	1 ns
Systron-Donner	113	250 MHz	5	1.3 ns
Tektronix	PG502	250 MHz	5	1 ns
Tekelec-Airtronic	TE 20	200 MHz	5	1 ns
Hewlett-Packard	8008A	200 MHz	4	1.2 ns
E-H Int.	1516	100 MHz	2.5	600 ps

as shown in Fig. 4-2. Diode limiters set the pulse baseline and topline. The output stage following this capacitor is a dc-coupled, linear, video-power amplifier.

Designs for fixed-risetime generators usually use a switching-circuit output stage rather than a linear amplifier. Figure 4-3 shows a typical output stage based upon ECL design. The balanced transistor pair, Q_1–Q_2, is driven with differential logic pulses. Initially, transistor Q_2 is nonconducting while Q_1 is on and conducts all of the emitter current, I_2. DC current I_1 flows through resistor R_g, establishing output voltage V_1 : $V_1 = I_1 \times R_g$. When the generator logic-drive pulse changes states, Q_1 ceases conducting while Q_2 rapidly turns on, pulling current I_2 away

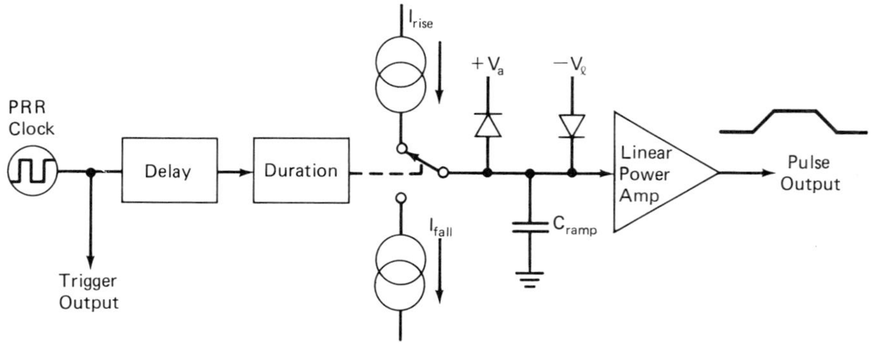

Fig. 4-2. Adjustable transition-duration pulse generator.

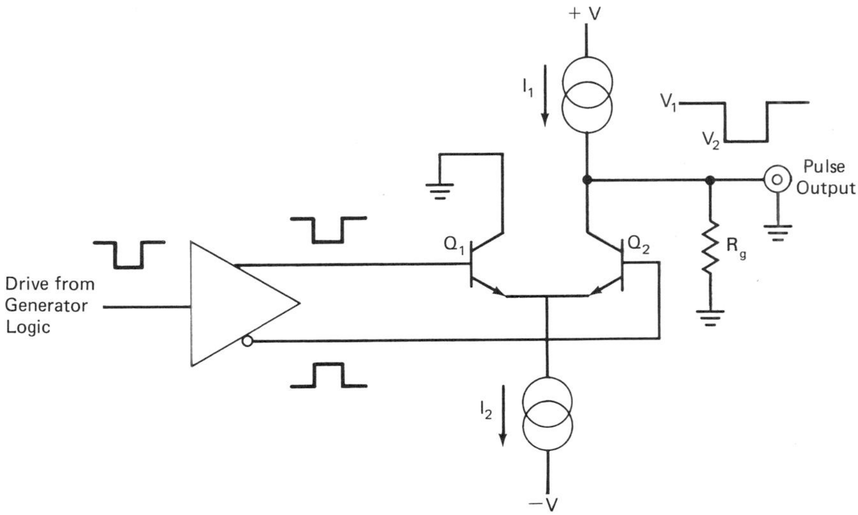

Fig. 4-3. ECL type of output stage.

from the output. This forces the output voltage to drop to $V_2 : V_2 = (I_1 - I_2) \times R_g$. The generator output impedance is set by resistor R_g.

The Colby PG1000A uses a different technique, as illustrated in Fig. 4-4. Instead of switching circuits, it employs overdriven linear amplifiers. The generator has a 1- to 1000-HMz sine-wave generator but no circuits to establish

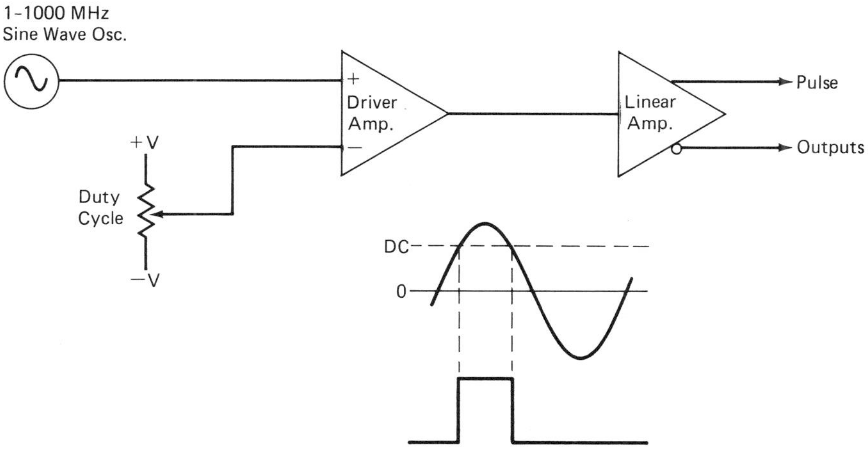

Fig. 4-4. Simplified block diagram of Colby PG1000A.

pulse delay. The pulse duration is varied by changing the duty cycle of the driver-amplifier output. The input of the driver amplifier is strongly overdriven by the rep.-rate-generator sine wave. Because of the overdrive, the clipped sine-wave output appears as a pulse. The input also receives a dc voltage from the front-panel duty-cycle pot. This dc voltage establishes the portion of the sine wave that is in the linear region of the amplifier.

4.2. AVALANCHE-TRANSISTOR PULSE GENERATORS

A considerably higher figure of merit of 200 V/nsec can be obtained if avalanche transistors are used as high-speed switching elements rather than as linear amplifiers as in the conventional transistor amplifiers. Figure 4-5 shows a simple avalanche circuit. In the avalanche mode, an extremely high voltage is applied to the collector of the transistor through a large current-limiting resistor. The base has no bias on it and is either shorted to the emitter or connected through a small-value resistor of less than 100 ohms. It can also be reverse-biased. The collector–emitter voltage is thus at the breakdown voltage, $BV(ces)$. In this condition, an extremely high electric field (E) exists across the collector-base junction. If a small amount of base current is injected into the junction, the charge carriers are subjected to the extremely high E field and are very quickly accelerated to a high velocity. When a charge carrier finally strikes the lattice, it has enough momentum to break loose several additional charge carriers. These additional charge carriers are likewise quickly accelerated, and the process is then repeated. This process, called *avalanche multiplication,* is extremely rapid. When the transistor starts to conduct, the collector voltage switches along the 50-ohm ac load line to the lower breakdown voltage of $BV(ceo)$. In some transistors, the switching time is in the subnanosecond region. Currents up to a

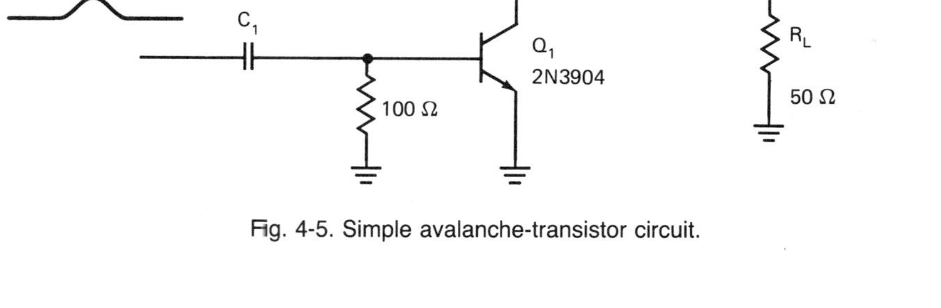

Fig. 4-5. Simple avalanche-transistor circuit.

few amps can be obtained. A paper by Mitchell[3] is highly recommended reading for further details.

Various pulse amplitudes can be obtained by selecting transistors with appropriate breakdown voltage specs. The switching time, however, is somewhat proportional to the voltage, with a 200 V/nsec figure of merit, i.e., 200 V in 1 nsec, 100 V in 1/2 nsec, 50 V in 1/4 nsec, etc. Pfeiffer[4] and Rein and Zahn[5] have reported obtaining 10 V in 120 psec and 100-psec pulses, respectively. It is also possible to stack several avalanche transistors in series to obtain higher pulse amplitudes. For example, Metzger[6] used 36 transistors to obtain 3.6 kV with a 10-nsec risetime. Another configuration to obtain higher voltages is the Marx generator circuit of Fig. 4-6. In this circuit, the capacitors are charged in parallel and discharged in series. With this particular circuit, Optoelectronics, Ltd., has achieved a 4.5-kV pulse with a rise of 23 nsec.

The major limitation on using avalanche transistors is the maximum rep. rate achievable. The duty cycle must be kept low to avoid destroying the transistor by overheating. The peak power dissipated in the transistor is quite large in comparison to its cw power rating.

Figure 4-7 shows examples of waveforms from commercially available avalanche-transistor pulse generators. The Picosecond Pulse Labs Model 1000B produces 40-V, 400-psec-wide impulses at a maximum rep. rate of 1 MHz. The Picosecond Pulse Labs Model 2500 produces a 60-V step function pulse lasting for 100 nsec. The leading-edge risetime is 300 psec, and the maximum rep. rate is 100 kHz.

4.3 STEP-RECOVERY DIODE (SRD) PULSE GENERATORS

With a step-recovery diode, it is possible to obtain a figure of merit on the order of 400 V/nsec with risetimes as fast as 50 psec (Fig. 4-8). The typical SRD pulse-generator configuration[7–9] uses an avalanche pulse generator as a driver (Fig. 4-9). The SRD is mounted in shunt across a coaxial line and forward-biased with a dc current. A voltage step is applied to the diode from the avalanche pulser of such polarity as to reverse-bias the SRD. The SRD is initially conducting and appears as a short circuit across the coaxial line, thus reflecting the incident wave back towards the driver. The diode impedance remains low while the stored charge caused by the forward-biased dc current is being removed. When all the stored charge has been removed, the SRD rapidly switches to a nonconducting, a high-impedance state that allows the incident-voltage step from the driver to pass on to the output load. The net effect is a sharpening of the leading edge of the pulse from the driver. Multiple SRDs can be cascaded to obtain faster transitions. The transmission lines are used for time-window isolation of various reflections.

SRDs with breakdown voltages ranging from 10 to 100 V are available from

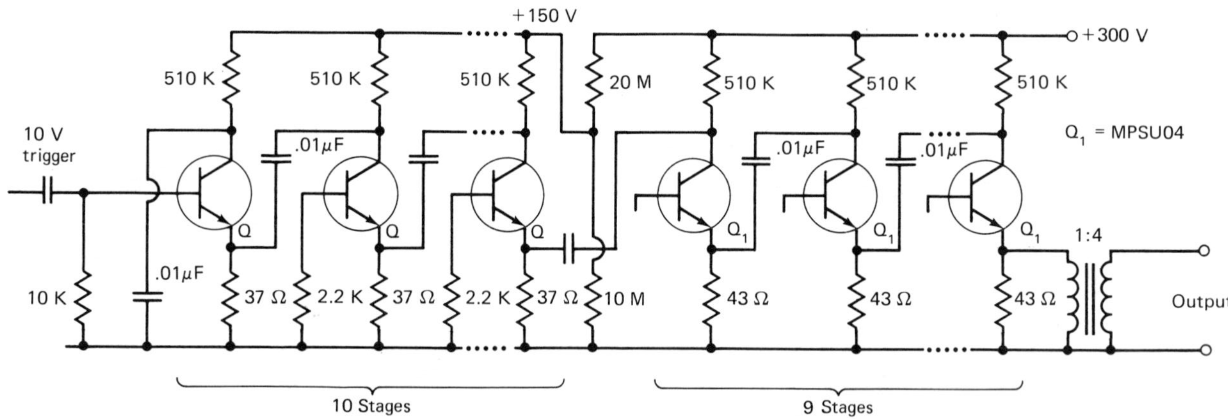

Fig. 4-6. Avalanche transistor in Marx generator circuit: The capacitors are charged in parallel and discharged in series.

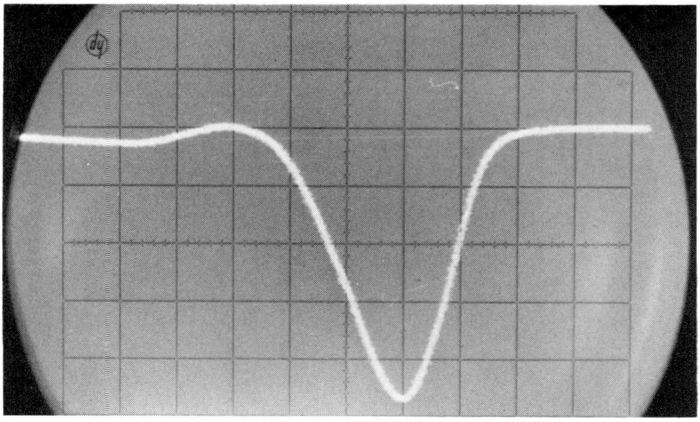

(a)

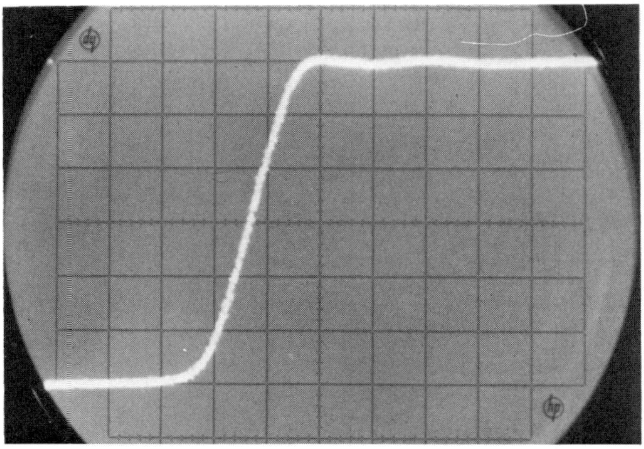

(b)

Fig. 4-7. Picosecond Pulse Labs avalanche-transistor pulse generators (vert. = 10 V/div.; horiz. = 200 ps/div.): (a) Model 1000B, and (b) Model 2000.

several different semiconductor manufacturers. In addition, ALPHA offers SRDs in which multiple identical chips are mounted in series within the same package to obtain breakdown voltages as high as 225 V at 500 psec.

Figure 4-10 shows examples of commercially available SRD pulse generators. The Picosecond Pulse Labs Model 3000A produces 5-V, 65-ps-wide impulses at 1 MHz, whereas the Model 4500 produces 25-V, 100-ps risetime step functions at 100 kHz.

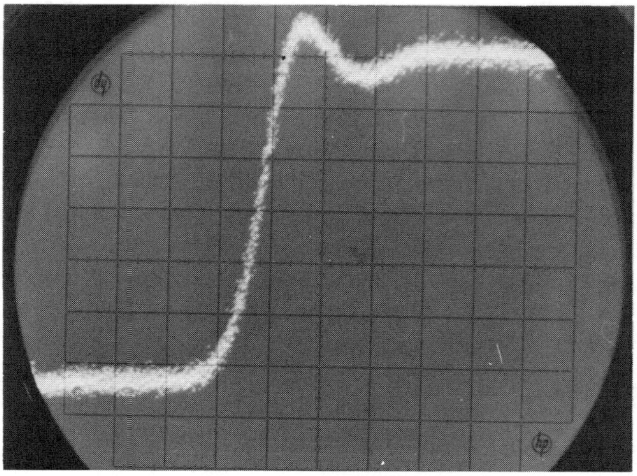

Fig. 4-8. Ultra-fast, step-recovery diode pulse (vert. = 2 V/div.; horiz. = 50 ps/div.).

4.4 TUNNEL-DIODE PULSE GENERATORS

The fastest conventional electronic-pulse generator uses a germanium tunnel diode (TD). The typical pulse amplitude is 250 mV, with a maximum figure of merit of 15 V/nsec. The manufacturers of TD pulse generators are listed in Table 4-2. Figure 4-11(b) shows the typical waveform obtained from the fastest TD pulser, the HP-1106**. This was measured by the HP-1430A (28-psec) sampling

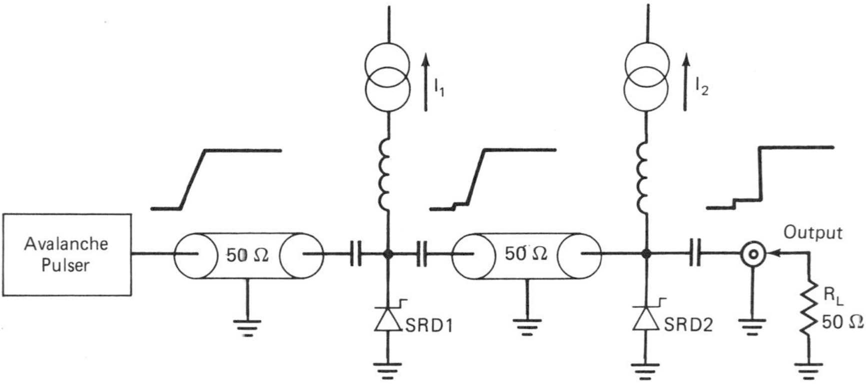

Fig. 4-9. Typical SRD pulse generator.

oscilloscope. Unfortunately, Hewlett-Packard has discontinued making their tunnel-diode pulsers.

Figure 4-12 shows a typical TD pulse-generator circuit (a) and the characteristic curve of a TD(b). The TD is initially biased by V_{bb} at point A, slightly below peak point B. A positive trigger pulse is coupled through R_1 into the TD, thereby forcing the TD current beyond the peak point current, I_p, and into an unstable,

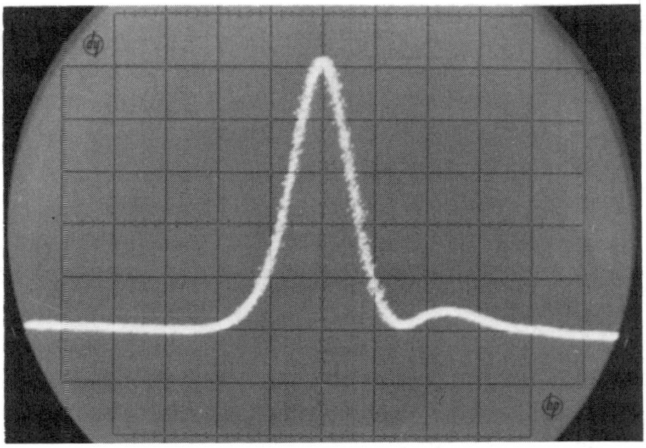

(a)

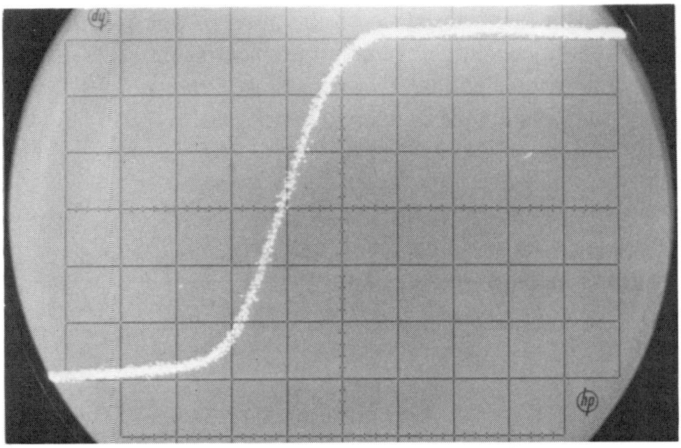

(b)

Fig. 4-10. Picosecond Pulse Labs SRD pulse generators: (a) Model 3000A (1 V/div. and 50 ps/div.); (b) and (c) Model 4500 (5 V/div. and 50 ps/div.).

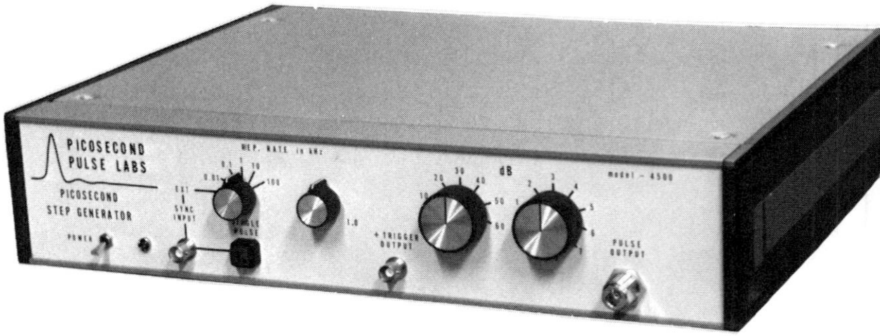

(c)

Fig. 4-10. (continued)

negative-resistance region. The only stable point is at point C. With inductor L appearing as a constant current source of approximately I_p, the TD rapidly switches to point C. This is the fast transition shown in Fig. 4-11(b). Eventually, inductor L can no longer supply current I_p, and the operating point slowly moves down the characteristic curve towards point D. When point D is reached, the TD again enters an unstable, negative-resistance region. It thus rapidly switches to point E, with L again acting as a constant current source of I_v. The operating point then slowly climbs up the curve and stops at its initial quiescent point, point A. The TD is a low impedance on the order of a few ohms. R_2 is added to provide a 50-ohm generator, reverse-impedance match.

For fast-switching applications, TDs with peak currents of 10 to 100 mA are used. For Ge TDs, typical values are as follows: $I_p/I_v = 10:1$; $V_p = 100$ mV; $V_f = 600$ mV; and $V_v = 400$ mV.

Table 4-2. Tunnel diode pulse generators.

MANUFACTURER	MODEL	TRANSITION DURATION, ps
Hewlett-Packard	1106**	20
	1108**	60
Tektronix	S-52	25
	284	$<$ 70 (Ref. 9)
Picosecond Pulse Labs	TD-1107	45
Colby Instruments	TD-50PA	50
	P5A	250
E-H Int.	902A	100

(a)

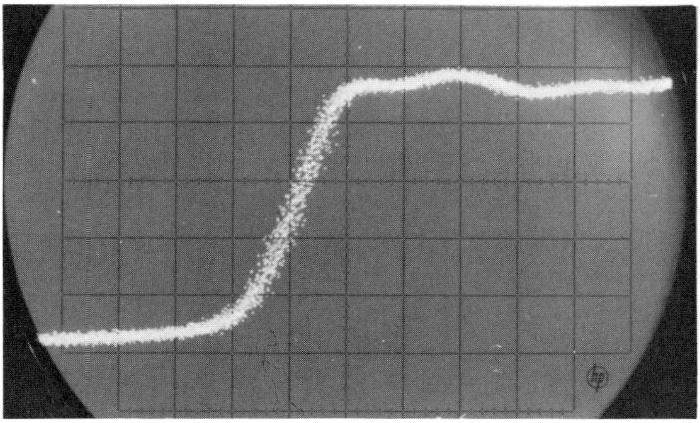

(b)

Fig. 4-11. Hewlett-Packard Model 1106A tunnel-diode pulse generator (vert. = 50 mV/div.; horiz. = 20 ps/div.).

To a first-order approximation, the switching speed, T_s, of a TD is given by

$$T_s = C\hat{v} \times [V\hat{f} - V\hat{p}] \,/\, [I\hat{p} - I\hat{v}] \tag{4-1}$$

where C_v is the valley point—terminal capacitance. When TDS are made of GaAs, the voltage swing is twice as large as it is for Ge TDs. In general, GaAs TDs are slower than Ge TDs.

At the present time, to the best of the author's knowledge, there are only two manufacturers producing fast-switching Ge TDs: General Electric and Custom Components. The several other manufacturers that make TDs specialize in low-current TDs for use as microwave amplifiers and detectors.

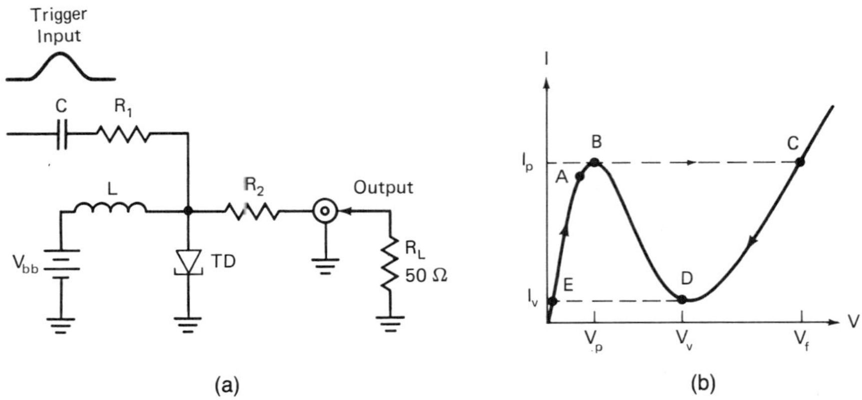

Fig. 4-12. (a) Typical tunnel-diode pulse-generator circuit, and (b) tunnel-diode characteristic curve.

4.5 VACUUM-TUBE PULSE GENERATORS

Vacuum tubes are also used in pulse generators, primarily for high-voltage, high-current, high-average-power applications for which there are not suitable semiconductors available. High-power transmitting tubes are typically used for power levels from kilowatts to megawatts. At these power levels, the risetimes range from tens of nsec to the microsecond range. Table 4-3 lists some of the commercially available pulsers in this category. The Cober 606 and the Velonex 360 are typical examples, providing 2.5-kV, 31-kW pulses with 20-nsec transitions. Most of these manufacturers also build transformers to provide either higher voltage or higher current outputs.

The Alpha pulse generator[10] of Power Spectra Inc. has very impressive per-

Table 4-3. Typical Vacuum Tube Pulsers.

MANUFACTURER	MODEL	PEAK POWER	PEAK VOLTAGE	TRANSITION DURATION
E-H Int.	1253	2 W	−10 V	200 ps
Power Spectra	Alpha	500 kW	5 kV	1.5 ns
Velonex	380	5 kW	500 V	7 ns
Velonex	360	31 kW	2.5 kV	20 ns
Cober	606	31 kW	2.5 kV	20 ns
Cober	605	24 kW	2.2 kV	20 ns
Cober	604	9 kW	1.5 kV	25 ns
Velonex	350	26 kW	2.1 kV	30 ns
Velonex	345	6 kW	1 kV	35 ns
Cober	1674	16 MW	150 kV	>500 ns

formance specifications. With an output of 5 kV into 50 ohms and a 1.5-nsec risetime, this generator has a figure of merit of 3.3 kV/nsec. It employs seven linear amplifier stages to amplify low-level TTL or ECL pulses, the first four of which use bipolar and field-effect transistor (FET) power transistors and the last three, planar triodes. The final stage uses three planar triodes in parallel, with a 10-kV plate voltage.

The fastest conventional vacuum-tube pulse generator is the E-H 125B, which uses a small microwave triode to produce −10-V pulses with 200-psec risetimes.

4.6 HIGH-CURRENT PULSERS

In addition to vacuum tubes, thyratrons are used for very-high-current pulsers such as might be needed in high-power radar transmitters. Risetimes are typically 50 nsec or slower. Because of the statistical nature of the gaseous breakdown mechanism in a thyratron, several nanoseconds of timing jitter is common. Krytrons are often used in applications requiring less current and faster switching. A krytron is basically a cold-cathode thyratron with an internal radioactive source to provide the necessary pre-ionization of the gas. Typical results obtained with a krytron are 4-kV, 10-nsec pulses with 3-nsec jitter.[11]

The solid-state equivalent of the thyratron, the silicon-controlled rectifier (SCR), is also often used in high-power, high-current applications. The fastest, low-power SCR switches 1 A in about 10 nsec, whereas the high-power ones require hundreds of nsec to microseconds to switch.

Power Technology, Inc., builds a line of solid-state, high-current, laser-diode pulsers that produce currents of 10 to 100 A with rise times of 2 to 10 nsec, respectively.

4.7 MERCURY-WETTED MECHANICAL SWITCHES

Mercury-wetted, mechanical-switch relays have been used for many years as subnanosecond pulse generators. With conventional relays, pulses up to 1 kV and 1/2-nsec risetimes are easily obtained. The usual configuration consists of a charged transmission line, as in Fig. 4-13, or a capacitor bank that is connected to the output load using a relay-operated switch. Initially, with the switch open, the coaxial charge line (R_O, TD) is charged up to the supply voltage, V_{bb}. At this time, the transmission line functions as a charge-storage capacitor. When the switch connecting the charge line to the output load is closed, the energy stored in the line starts to flow into the load resistor, R_L. The current flow is limited by the "surge," or characteristic impedance, R_O, of the charge line. The voltage developed across the load is given by

$$V_{out} = V_{bb} \times R_L/(R_L + R_O)$$

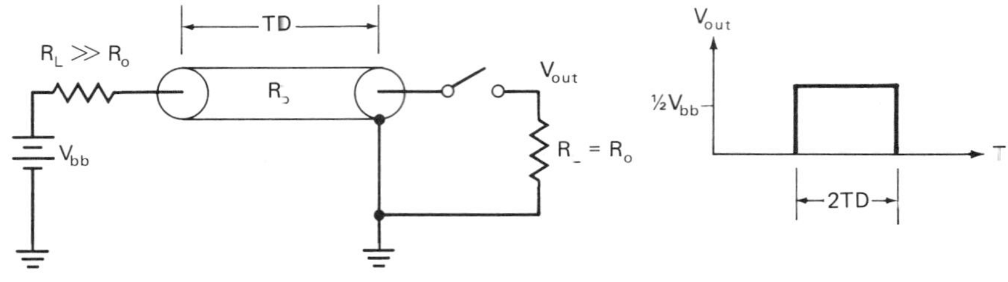

Fig. 4-13. Mercury-switch pulse generator.

If $R_L = R_O$, then

$$V_{\text{out}} = V_{bb}/2$$

Because of the distributed properties of the line, energy initially flows from the right side nearest the switch. A discharge current wave propagates to the left, progressively discharging the line. When this wave reaches the far end of the line, it encounters an open circuit, $R_C >> R_O$. This open circuit reflects the current wave back toward the right. As the reflected wave travels toward the right, it cancels out the leftward-traveling wave, and, upon reaching the right end of the charge line, terminates the output pulse. The output-pulse duration is thus twice the propagation delay time (TD) of the charge line. If the output load, R_L, is not matched to the charge-line impedance, R_O, the output waveform will have multiple reflections following the main pulse. Each reflection will be a pulse of 2TD duration, and the amplitudes will be dependent upon the output, mismatch reflection coefficient.

The fastest mercury switch in existence is a very small unit called the "Log-Cell," made by 5th Dimension. Meyer[12] reported obtaining a 2.5-V pulse with a 12-psec risetime using a Log-Cell. NBS has built a 50-V, 40-psec risetime pulser using the Log-Cell.[13,14] Workers at Los Alamos Scientific Labs and also at EG&G have obtained similar results. Risetimes of 1/2 ns are more typical of conventional mercury-switch pulsers.

The mercury-switch pulser has several advantages. It is easy to build, and its output waveform can be easily predicted. The duration is fixed by the length of charge-line cable used. The amplitude is independently adjustable by varying the supply voltage, V_{bb}. For these reasons, it has been used for many years as a pulse-amplitude calibration source. It has also been used as a spectrum-amplitude calibration source for electromagnetic interference (EMI) measurements. However, a word of caution is in order. The preceding equations are valid only to within a few percent accuracy. Actual waveforms and amplitudes will deviate from the ideal rectangular pulse shown in Fig. 4-13. The causes for this discrepancy are primarily skin-effect losses in the charge line and parasitic reactances within the switch itself.

The mercury switch also has several disadvantages. It will operate only at low rep. rates—up to 100 Hz, typically—although some switches have been operated at a few kHz by driving them at a mechanical resonant frequency. Being a mechanical device, the mercury switch cannot be precisely triggered with an external electrical trigger. Trigger jitter of a few ms is typical. Moreover, because of statistical variations in the contact surface at the instant of closure, the output pulse amplitude sometimes exhibits amplitude instabilities of 1 to 2 percent. Mercury switches also have a limited lifetime and degrade with use.

At the present time, only one manufacturer, the SPIRE Corp., is making a general-purpose mercury-switch pulse generator, the SPI-PULSE 25 Transmission Line Pulser. It can produce up to a 1000-V pulse into 50 ohms (20 kW) with a risetime of 1/2 nsec. The pulse duration can be varied from 1 to 500 nsec by changing the length of an external charge cable. The rep. rate is adjustable up to 275 Hz. The switch lifetime is approximately 10^8 pulses.

Singer and Electro-Metrics also make mercury-switch pulsers. Their units have a fixed duration impulse of either 1/2 or 10 nsec and were designed for use as a calibration source of broadband impulsive noise. As such, their output is calibrated in "spectrum amplitude" (in dB microvolts/MHz) and not in pulse amplitude (in volts).

4.8 SPARK GAPS

For higher-voltage applications than can be handled by a mercury switch, a spark gap is often used as the switching element that connects the charged transmission line or capacitor bank to the output load (as in Fig. 4-13). A free-running, spark-gap pulser simply uses a supply voltage, V_{bb}, larger than the breakdown voltage of the gap. The charge line slowly charges up through the charging resistor, R_C. When the charge-line voltage reaches the breakdown voltage, the gap arcs over and allows the charge line to discharge into the load resistor, R_L, in much the same fashion as the mercury-switch pulser. The typical switching time of a spark gap can be well under 1 nsec. The actual gap-breakdown voltage is a function of the gap geometry, spacing, and dielectric materials. Breakdown voltages of a few hundred volts or a few kV are easily obtained with simple geometrics and air dielectric. For higher breakdown voltages, other dielectrics such as pressurized gases, oil, water, etc., are used.

The leading-edge risetime of a spark-gap pulse can be further sharpened by passing it through successive spark gaps, as shown in Fig. 4-14. The author has built a small, dual-gap pulser that produced 2-kV pulses into 50 ohms with a risetime of less than 200 ps. McDonald et al.[15] have built a dual-gap pulser and obtained steplike pulses up to 60 kV with risetimes of 300 ps.

The spark gap has the advantage of being reliably triggered. Triggering can be accomplished either by introducing an electrical trigger into the gap via a third electrode or by illuminating the gap with an intense laser pulse. Timing

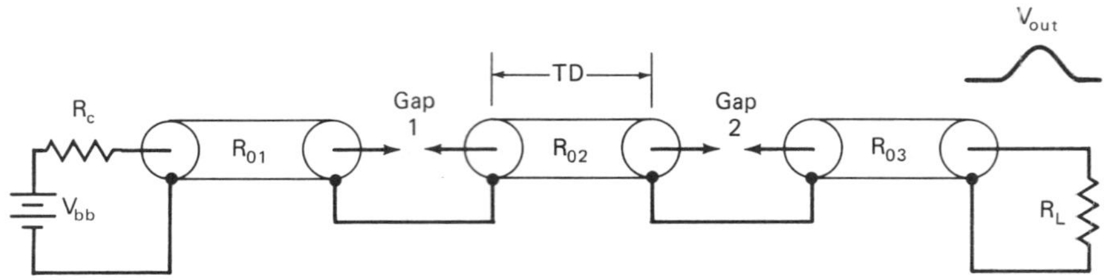

Fig. 4-14. Dual spark-gap impulse generator.

jitters on the order of 1 nsec are typical. Spark gaps, like mercury switches, have the disadvantages of limited lifetime and degradation with use.

The fastest commercial spark-gap pulse generator was the IKOR Model R100,** which produced a 1-kV, 250-ps-wide impulse, as illustrated in Fig. 4-15, and free-ran at approximately 250 Hz. It could not be triggered externally or be single-pulsed. The R100 is now built by Omni-Wave.

Spark gaps are used in the largest pulse generators, both in terms of output and physical size. The huge machines built by Maxwell Labs, Sandia Labs, Naval Research Lab, EG&G Bedford, Cornell, and Physics International, to mention a few, produce pulses of kV to MV and kA to MA with rise times of tens of nanoseconds to microseconds. Among some of the applications of these generators are to drive E-beam and ion-beam accelerators and for lightning simulation.

Figure 4-16 shows one of the larger machines, the Maxwell Labs Blackjack 5.[16] It produces 10-TW, 50- to 100-nsec pulses into very low impedance loads. Figure 4-17(a) shows the internal construction of the Blackjack 5 pulse line: a

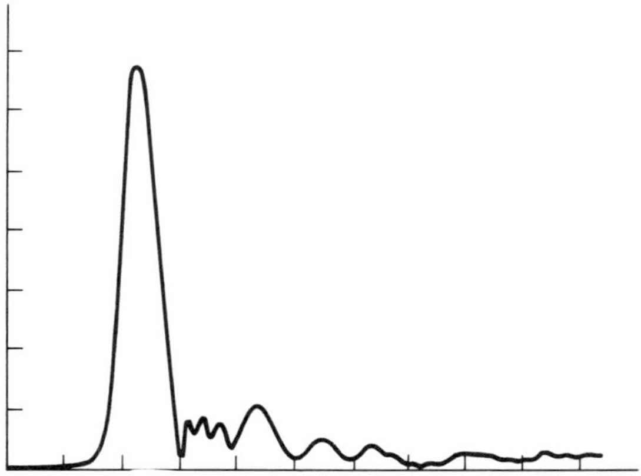

Fig. 4-15. IKOR R100 spark-gap impulse (vert. = 160 V/div.; horiz. = 500 ps/div.).

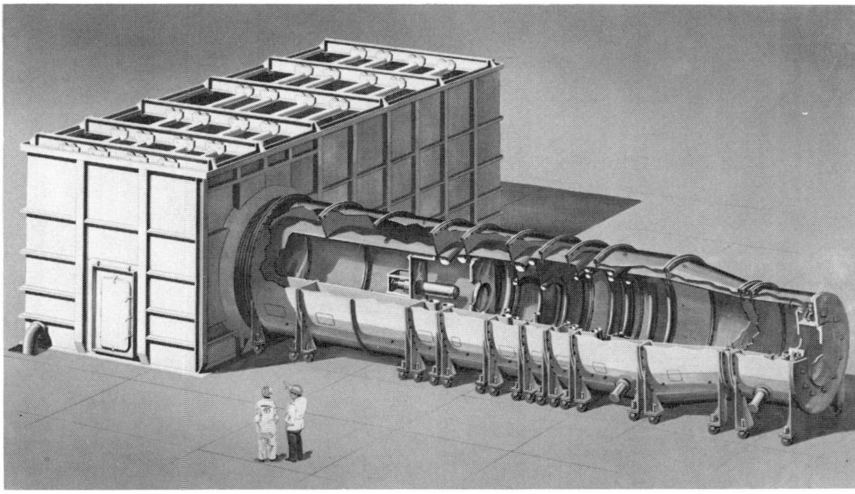

Fig. 4-16. Maxwell Labs 10-TW pulse machine, BLACKJACK 5.

cascade of various transmission lines separated by spark gaps. Figure 4-17(b) shows the waveforms obtained at various locations. The maximum output current is 5 MA with a *di/dt* of 100 kA/nsec.

Physics International has built its EAGLE machine for the Defense Nuclear Agency for above-ground, nuclear-weapon-effects testing. The EAGLE is a high-power, Marx-generator-driven, water-dielectric, low-impedance pulse generator.

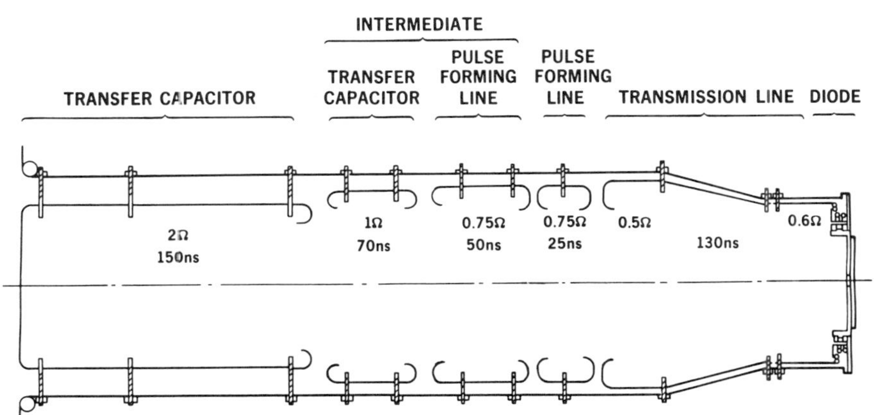

(a)

Fig. 4-17. BLACKJACK 5: (a) internal construction; and (b) waveforms.

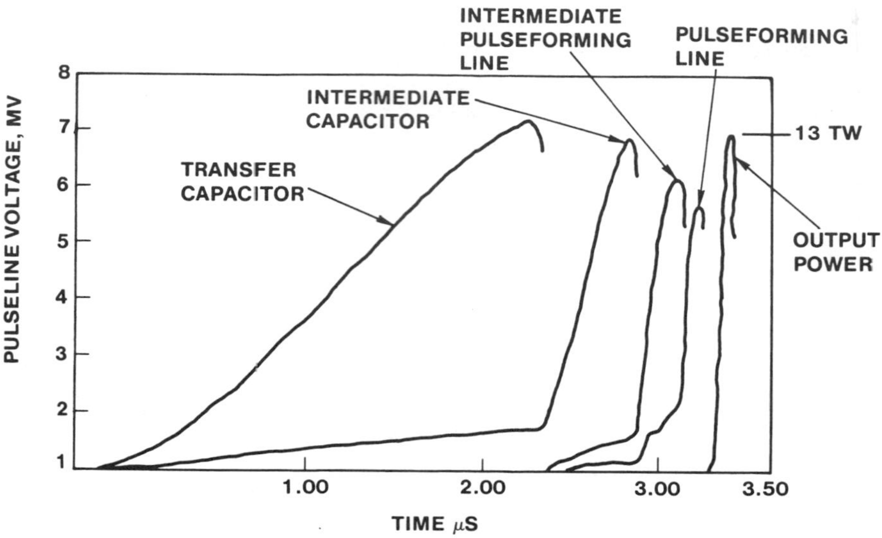

(b)

Fig. 4-17. (continued)

The measured peak-output pulse parameters are as follows: 4.6 ± 0.5 TW, 1.6 MA, 275 kJ, and 75-nsec (FWHM) pulse width.[17–20] Figure 4-18(a) shows the wedge-shaped EAGLE with a man at the back of it for size comparison.

The EAGLE is eventually to be used as a module for a future giant pulse generator, the ROULETTE-X. Figure 4-18(b) is an artist's drawing of the 20-stage ROULETTE-X, along with future intermediate level generators, the DOUBLE-EAGLE and ROULETTE-1. Extrapolation of EAGLE data suggests that the ROULETTE-X will produce 70 TW, 5 MJ, and 75-nsec pulses.

4.9 LINEAR ACCELERATOR

The linear accelerator (LINAC) is essentially an extremely large vacuum-tube pulse generator. Although extremely expensive and immobile, it is unsurpassed for generating narrow, repetitive, high-voltage electrical pulses. Most LINACS are owned by the Department of Energy and used for nuclear research. Although their primary purpose is to generate short, intense bursts of radiation such as electrons, X-rays, gamma rays, and neutrons, they have also been used as fast-pulse generators. Most LINACs generate a continuous train of pulses at a high RF frequency, but some have been modified to produce single impulses.[21,22]

Electrons accelerated in multiple-RF-cavity structures, either standing-wave

(a)

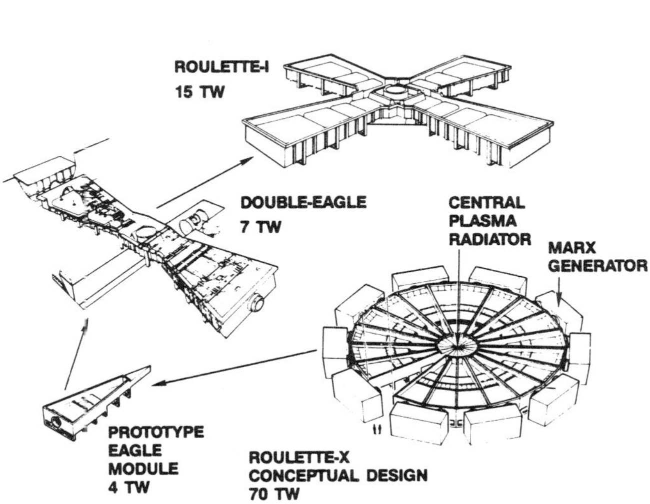

(b)

Fig. 4-18. (a) Physics International 4.6-TW pulse generator, EAGLE, and (b) artist's conception of future ROULETTE-X using 20 EAGLE modules.

or traveling-wave, are inherently bunched into a small interval, typically 15 degrees, of the RF period. In most such accelerators, which typically operate at either S-band (2856 MHz) or L-band (1300 MHz), electrons are injected from the electron gun for several microseconds. The accelerated output beam of electrons consists of a train of very short pulses separated by the period of the accelerating wave and adjusted to "travel" (advance in phase) at essentially the velocity of light, which is essentially also the velocity of the relativistic electrons.

A single one of the microstructure pulses can be produced by injecting electrons into only one RF cycle of the fundamental accelerating frequency. Initially, the injection was done by modulating the gridded electron gun with a triangular voltage pulse and a superimposed bias voltage. The gun conduction could be limited to about 1 nsec, or approximately 1 RF cycle of the L-band frequency. More recent developments have employed velocity-modulation bunching in several types of structures at subharmonics of the fundamental frequency to inject large quantities of electrons (40 nC in an Argonne Nat. Lab. experiment) into about 25 psec of a single RF cycle. If all this charge were captured in the center conductor of a coaxial 50-ohm Faraday cup, 1.6 kA could produce 80 kV!

A stable c.w. master oscillator operating on the twenty-fourth (typically) subharmonic of the accelerator frequency serves as the master clock for the bunchers and gun pulser. Phase-synchronous RF amplifiers are easily built to drive the subharmonic bunchers in phase with the accelerator. A trigger signal is derived from one of the subharmonic frequencies that is phase synchronous with all the RF frequencies and can be timed to occur at the desired part of the accelerator's microsecond-long macropulse. This trigger signal, sharpened and amplified, is used to drive the grid-cathode electrodes of the electron gun. To achieve reasonable pulse-to-pulse stability, the jitter in the gun pulse must be a small fraction (1 to 2 percent) of the RF period, a prerequisite that can be achieved by modern techniques, particularly by using avalanche transistors and triode amplifier stages to produce the kilovolt pulse voltages required by many electron guns. An important feature of this trigger system is that experimental apparatus and sampling oscilloscopes can be triggered synchronously with the beam pulse, allowing high-bandwidth measurements with low jitter.

The bunch of accelerated electrons can be converted into an electrical pulse by intercepting them in the center conductor of coaxial Faraday cup.[23] The time response is limited by the capacity of the open end of the coax, the resonances in the volume of the coaxial structure, and the transit time of the electrons across the gap. With relativistic electrons at energies greater than 2 MeV, the transit time is manageable, and Faraday cups with small dimensions can have a frequency response to 7 to 8 GHz or better. Using the fast Faraday cup, four feet of good coaxial cable, and a sampling oscilloscope, a typical LINAC pulse will be like that shown in Fig. 4-19.

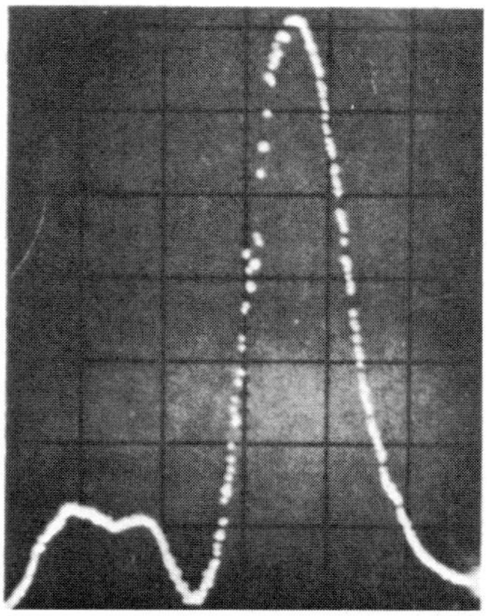

Fig. 4-19. LINAC Farady-cup output pulse (vert. = 100 V/div.; horiz. = 50 ps/div.).

4.10 JOSEPHSON JUNCTIONS

At the opposite extreme from the huge spark-gap machines and LINACs lies the field of superconducting cryoelectronics. The discovery of the superconducting Josephson junction has opened up a whole new field of extremely fast, picosecond electronics. To date, all of the work in this area has been done in sophisticated research labs. Hamilton at the National Bureau of Standards in Boulder, Colorado has demonstrated that a Josephson junction can be switched in less than 10 psec.[24] Theory predicts switching transition times on the order of 2 psec. IBM has attempted to develop a superconducting computer using Josephson junctions.

The Josephson junction, however, has several major disadvantages for widespread commercial applications. First and foremost is its requirement for a cryogenic, liquid-helium environment, and second, the extremely low, millivolt-level signals involved.

4.11 OPTICAL PULSE GENERATORS—LASERS

The shortest pulses ever reported have been obtained from mode-locked lasers. The December 1981 issue of *Laser Focus*[25] reported that G. Mourou and T. Sizer of the University of Rochester had obtained dye laser pulses of 70 fsec (0.070 psec). A pulse this brief is also extremely short physically, the wave packet being only 21 microns long. To obtain these extremely short optical pulses, however, requires a large optical-bench setup with high-powered pump lasers and experienced personnel to assemble and operate it.

Using injection laser diodes, it is now possible to obtain subnanosecond optical pulses in a conventionally packaged, simply operated pulse generator. Andrews at CNET in France showed in 1972 how pulses as short at 110 psec could be easily obtained from an injection laser diode with an avalanche-transistor, 10-A pulser.[26,27] Several companies now offer these optical pulse generators, including Hamamatsu of Japan, Optel of Germany, Opto-Electronics of Canada (see Fig. 4-20), and Lasertron of the USA (see Table 4-4 for the characteristics of the latter). The electrical pulse drivers for these laser diodes are typically avalanche-transistor pulsers, whereas the faster units use SRDs. In 1983, Lasertron introduced the fastest commercially available optical pulse generator with a 30-psec impulse. Still briefer optical pulses have been obtained in research labs by mode-locking, CW-injection laser diodes. The best results to date are 20-psec impulses at a 3-GHz rep. rate.[28]

4.12 PHOTOCONDUCTIVITY

Considerable work is presently being done in several research labs in the U.S. and Europe in an attempt to convert the extremely brief optical pulses into electrical pulses. At present, investigations into fast photoconductive effects seem to be most promising. The major materials being studied at present are Si, GaAs, and InP. The fastest results reported to date are 100-V, 10-psec pulses by Austin of Bell Labs.[29] Antonetti et al.[30] have also reported producing 10-kV pulses with

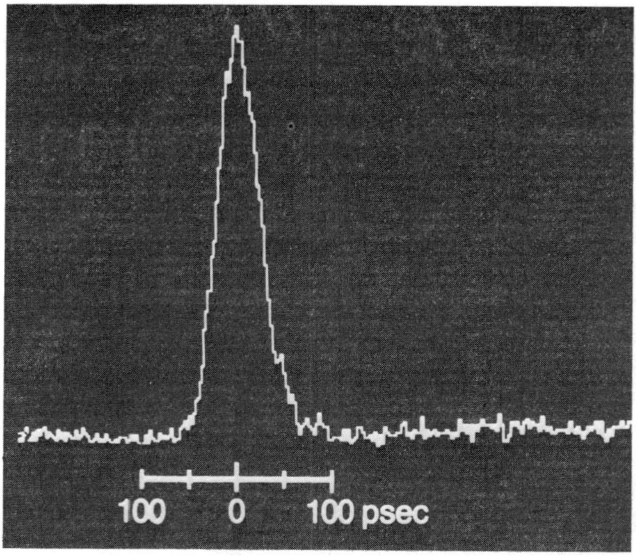

Fig. 4-20. Output of an Opto-Electronics PLS10 laser-diode pulse generator (900 nm) as monitored by a 10-ps-resolution streak camera.

Table 4-4. Laser-Diode, Optical Pulse Generators.

MANUFACTURER	MODEL	IMPULSE DURATION ps	PEAK POWER	WAVELENGTH, nm
Lasertron	QLPX	35	> 40 mW	1300
			> 30 mW	1530
Opto-Electronics	PLS10	< 75	100–600 mW	810–870
				900
				1100–1600
Hamamatsu	C1308	> 100	> 750 mW	820
Optel	CL1003	100	1 W	820
				904

risetimes of less than 50 psec. These results give a figure of merit of 200 kV/nsec.

4.13 CONCLUSIONS AND FORECAST

In conclusion, it can be observed that a wide variety of subnanosecond pulse generators are available. As always, the best choice is never completely clear since it depends strongly upon the needs of the end user.

In the future, it can be expected that research labs will continue to push for shorter rise times and pulse widths with lasers, photoconductive switches, and cryoelectronics. The most significant impact on conventional bench-type pulse generators will come from the large effort in the late seventies to develop microwave and millimeter-wave power FETs. As manufacturers start to design around these new FETs, 5-GHz rep. rate, 100-psec bench-style pulse generators should soon appear.

Acknowledgements. I want to extend a special "thank you" to Dr. N. S. Nahman, who has been my guiding light throughout my professional career, as my thesis and dissertation advisor at the University of Kansas, as my supervisor at NBS, and as a personal friend.

I would also like to acknowledge the assistance of a number of people who have contributed material for this chapter. They include: Saul Haffner, Cober Electronics, Stamford, Conn.; Dr. Siegfried Knorr, Colby Instruments, Inc., Santa Monica, Calif.; Richard Miller, Maxwell Labs, Inc., San Diego, Calif.; Dr. Brian Garside, Opto-Electronics, Inc., Oakville, Ontario; Dr. Marco DiCapua, Physics International, San Leandro, Calif.; Kenneth Santoro, Power Spectra, Inc., San Leandro, Calif.; Ward Halverson, Spire Corp., Bedford, Mass.; Douglas Greenwood, Velonex, Santa Clara, Calif.; as well as several others who chose to remain anonymous.

REFERENCES

1. Runyon, S. 1973. Focus on pulse & word generators. *Electronic Design,* May 24, pp. 128–137.
2. Nicolson, A. M.; Cronson, H.; and Mitchell, P. 1974. Subnanosecond risetime pulse generators. Research paper, Sperry Research Center, Sudberry, MA.
3. Mitchell, W. P. 1968. Avalanche transistors give fast pulses. *Electronic Design* 6:202–209.
4. Pfeiffer, W. 1971. Ein einfacher impulsgenerator for reflexionsfaktorund sprung uber tragungsmessungen. *Inter. Elek Rundschau* 25:268–272.
5. Rein, H. M., and Zahn, M. 1975. Subnanosecond pulse generator with variable pulsewidth using avalanche transistors. *Electronics Letters* 11, No. 1:21–23.
6. Metzger, G.; Schmitt, F.; Riedinger, M.; Gresser, J.; and Sutter, G. Circuits rapides utilisant des transistors en régime d'avalanche. *Revue de Physique Appliques* 4:223–224.
7. Pulse & waveform generation with step-recovery diodes. Hewlett-Packard Application Note 918.
8. Tielert, R. 1976. Subnanosecond pulse generator employing 2 stage pulse step sharpener. *Electronics Letters* 12, No. 3:84–85.
9. Frye, G. 1968. A predictable subnanosecond step generator. *IEEE Trans. I&M* IM17, No. 4:324–329.
10. Davis, S. J. 1982. Fast, high-powered triode pulse amplifiers. *IEEE Conf. Record, 15th Power Modulator Symposium,* pp. 227–230.
11. Vandre, R. and Molen, G. 1976. High-voltage pulse generator offers variable delay and only 3 ns jitter. *Electronic Design,* April 26, p. 86.
12. Meyer, F. 1970. Impulsmessplatz mis einer zeitauflosung von 18 ps. *Electronics Letters* 6:703–704.
13. Andrews, J. 1976. Picosecond pulse generators using microminiature mercury switches. *NBSIR 75-819,* NBS, Boulder, CO.
14. Andrews, J. 1976. Precision picosecond pulse measurements using a high quality superconducting delay line. *IEEE Trans. Inst. & Meas.* IM-23:468–472.
15. McDonald, D. F., and Benning, C. J. 1965. Subnanosecond risetime multikilovolt pulse generator. *Rev. Sci. Inst.* 36, No. 4:504–506.
16. Miller, A. R., 1981. Sub-ohm coaxial pulse generators, Blackjack 3, 4 & 5. Third IEEE Int. Pulsed Power Conf., Albuquerque, NM, pp. 200–205.
17. Frazier, G. B.; Ashby, S. R.; Barrett, D. M.; DiCapua, M. S.; Demeter, L. J.; Osios, D. E.; Ryan, R.; Spence, P.; Strachan, D. F.; and Sullivan, T. S. The design of a 2 TW pulsed power research facility. 1981 Particle Accelerator Conf., Washington, DC.
18. Frazier, G. B.; Ashby, S. R.; Barrett, D. M.; DiCapua, M. S.; Demeter, L. J.; Huff, R.; Osias, D. E.; Ryan, R.; Spence, P.; Strachen, D. F.; and Sullivan, T. S. 1981. Eagle, a 2 TW pulsed power research facility. Third IEEE Int. Pulsed Power Conf., Albuquerque, NM.
19. Frazier, G. B., et al. (see Ref. 18). 1981. Pulse power research on Eagle. 4th Int. topical conf. on high-power electron & ion-beam research & tech. Ecole Polytechnique, Palaiseau, France.
20. Frazier, G. B.; Ashby, S. R.; Demeter, L. J.; DiCapua, M. S.; Douglas, J.; Glidden, S. C.; Hammon, H. G. III; Huff, R.; Lan, S. K.; Rutherford, A.; Ryan, R.; Sincerny, P.; and Strachan, D. F. 1983. Eagle & Double-Eagle. Fourth IEEE Pulsed Power Conf., Albuquerque, NM.
21. Hanst, R. K., and Norris, N. J. 1965. Injector Pulser for Linac Picosecond Operation. *IEEE Trans. Nuc. Sci.* NS-12:830–835.
22. Norris, N. J., and Hanst, R. K. 1969. Velocity modulation system for enhancement of 50 Picosecond radiation pulse. *IEEE Trans. Nuc. Sci.* NS-16.
23. Norris, N., and Hanst, R. 1969. Picosecond beam monitors & data-acquisition system. *IEEE Trans. Nuc. Sci.* NS-16, 3:927–931.
24. Hamilton, C.; Lloyd, F. L.; Peterson, R. L.; and Andrews, J. R. 1979. A superconducting sampler for Josephson logic circuits. *Appl. Phys. Lett.* 35(9):718–719.

25. Saturable absorber helps Rochester team produce 70 fs pulses. 1981. *Laser Focus,* December, p. 14.
26. Andrews, J. 1972. Light impulse from GaAs injection laser. Talk presented at the 1st European Conf. on Fibre Optical Communications. Imperial College, London, England.
27. Andrews, J. 1974. Inexpensive laser diode pulse generator for optical waveguide studies. *Rev. Scien. Insts.* 45, No. 1:22–24.
28. Ho, P.; Glasser, L. A.; Ippen, E. P.; and Hans, H. A. 1978. Picosecond pulse generation with a CW GaAlAs laser diode. *Appl. Phys. Lett.* 33, No. 3:241–242.
29. Auston, D. 1975. Picosecond optoelectronic switching & gating in silicon. *Appl. Phys. Lett.* 26, No. 3:101–103.
30. Antonetti, A., et al. 1977. High power switching with picosecond precision: Applications to high speed Kerr cell & Pockets cell. *Optics Comm.* 23, No. 3:435–439.

5
Time-Domain Sensors and Radiators

Motohisa Kanda

National Bureau of Standards

5.0 INTRODUCTION

Recent study related to electromagnetic-pulse (EMP) phenomena has focused strong attention on the subject of transient electromagnetic (EM) fields. Antenna structures able to preserve the time-domain waveform of EMP must be inherently broadband and nondispersive.[1,2]

The purpose of this chapter is to discuss various sensors and radiators commonly used for time-domain antenna measurements. The sensors and radiators discussed here are passive analog devices that convert the electromagnetic quantity of interest to a voltage or current at their terminal ports. Moreover, they are primary standards in the sense that their transfer functions can be calculated from their geometries and are flat (frequency-independent) across a wide frequency range. One of the major requirements of these sensors and radiators is that the electromagnetic far field, for transmission or reception, be a replica or high-fidelity derivative of the original pulse.

Because of their usefulness in electric-field strength measurements, linear antennas nonuniformly and continuously loaded with resistance, or both resistance and capacitance, are discussed. Also, a conical antenna and an asymptotic conical antenna are examined from the standpoint of improved antenna characteristics. Various types of transverse electromagnetic (TEM) horns are considered for improved directivity, e.g., a conducting TEM horn and a resistively loaded TEM horn. For magnetic-field strength measurements, a loop antenna with uniform resistive loading is discussed.

Section 5.1 discusses one of the important properties governing the transmitting and receiving transfer functions of an antenna. The frequency-domain and time-domain characteristics of the various electric- and magnetic-field sensors mentioned above are then covered in Secs. 5.2 and 5.3, respectively.

5.1 ANTENNA RECEIVING AND TRANSMITTING TRANSIENT RESPONSES[3,4]

It is rather well known that the transmitting transient response of an antenna is proportional to the time derivative of the receiving transient response of the same antenna. For example, consider an antenna designed to receive an impulse with minimum distortion. If this antenna is driven with an impulse input, the radiated waveform is a differentiated impulse, i.e., a doublet.

To explain this properly, let us consider a specific linear antenna. The radiated field of a center-driven electric dipole of length $2h$ and wire radius a, as shown in Fig. 5-1, is given by the formula,[5]

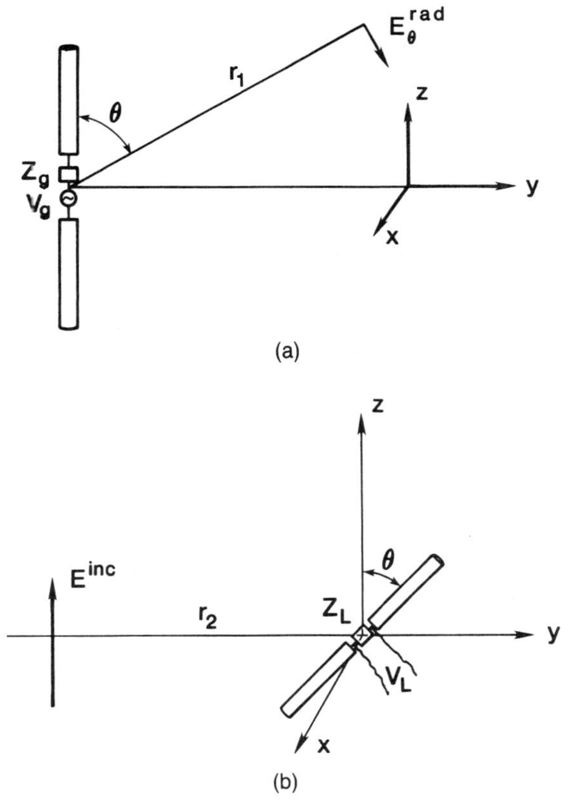

Fig. 5-1. Configuration of a linear antenna: (a) transmitting antenna, and (b) receiving antenna.

$$E_\theta^{\text{rad}} = \frac{60jV_0}{Z_{\text{in}} + Z_g} F\left(\theta, kh, \frac{a}{h}\right) \frac{e^{j\omega\left(t - \frac{r_1}{c}\right)}}{r_1} \tag{5-1}$$

where E_θ^{rad} is the radiated electric field at a distance r_1 from a linear antenna; $V_0 e^{j\omega t}$ is the driving voltage of a generator with impedance Z_g; $F\left(\theta, kh, \frac{a}{h}\right)$ is the antenna field factor of an antenna with impedance Z_{in}; (r, θ, ϕ) represents a spherical coordinate system as shown in Fig. 5-1(a); k is the free-space wave number; and c is the velocity of light.

When the same antenna, used for receiving, is placed at the origin of the coordinate system and inclined in the direction of angles θ and ϕ, as shown in Fig. 5-1(b), the current I_L induced in the load impedance Z_L becomes

$$I_L = \frac{-2h_e\left(\theta, kh, \frac{a}{h}\right)}{Z_{\text{in}} + Z_L} E^{\text{inc}} e^{j\omega\left(t - \frac{r_2}{c}\right)} \cos\phi \tag{5-2}$$

where $h_e\left(\theta, kh, \frac{a}{h}\right)$ is half the effective length of the antenna, $E^{\text{inc}}e^{j\omega t}$ is the incident electric field, and the term $e^{-j\omega r_2/c}$ gives the phase measured at a point in space at a distance r_2 from the center of the dipole.

Applying the Rayleigh-Carson reciprocity theorem,[5] we can show that the field factor, F, and half the effective length, h_e, satisfy the following reciprocity relation:

$$F\left(\theta, kh, \frac{a}{h}\right) = kh_e\left(\theta, kh, \frac{a}{h}\right) \tag{5-3}$$

Though perhaps not well known, the reciprocity theorem given in Eq. 5-3 is analogous to the more familiar formula,

$$G_e = \frac{4\pi}{\lambda^2} A_e \tag{5-4}$$

where G_e is an antenna power gain that is a transmitting characteristic, A_e is an antenna effective area that is a receiving characteristic, and λ is the wavelength.

By differentiating Eq. 5-2 with respect to time and using the reciprocity theorem given in Eq. 5-3, one can obtain, when $Z_L = Z_g$, the relation,

$$\frac{\partial I_L}{\partial t} = - \frac{cr_1 \cos\phi \, E^{\text{inc}} e^{j\omega(r_1 - r_2)/c}}{30 \, V_0} E_\theta^{\text{rad}} \tag{5-5}$$

Equation 5-5 indicates that the response of a receiving antenna differentiated with respect to time is proportional to the time dependence of the radiated field. The proportionality constant depends neither on the particular antenna size nor on the frequency.

Although this equation is derived specifically for a linear antenna, the same holds for any antenna. This property can be explained elegantly by the use of plane-wave scattering theory.[6] The transmitting transfer function, $T_t(f)$, of an antenna is defined as the ratio of the radiated electric field, $E^{\text{rad}}(f,r)$, of frequency f, to the phasor wave amplitude, $a_0(f)$, of one plane wave incident at the antenna terminal, and is given by the formula,

$$T_t(f) = \frac{E^{\text{rad}}(f,r)}{a_0(f)} = jkS_{10}(K=0)\,\frac{e^{-jkr}}{r} \tag{5-6}$$

where $S_{10}(K=0)$ is the transmitting characteristic of an antenna scattering matrix with K being the transverse part of k and r the distance from the antenna to the field point. The receiving transfer function $T_r(f)$ is defined as the ratio of the phasor amplitude, $b_0(f)$, of the emergent wave at the antenna terminal to the electric field, $E^{\text{inc}}(f)$, incident on the antenna, and is given by

$$T_r(f) = \frac{b_0(f)}{E^{inc}(f)} = 2\pi S_{01}(K=0) \tag{5-7}$$

where $S_{01}(K=0)$ is the receiving characteristic of an antenna scattering matrix.

The scattering matrix elements, $S_{10}(K=0)$ and $S_{01}(K=0)$, of an antenna are related through the reciprocity theorem,

$$\eta_L S_{01}(K=0) = -\eta_0 S_{10}(K=0) \tag{5-8}$$

where η_L is the antenna load admittance and η_0 is the admittance of free space. Thus,

$$\frac{T_t(f)}{T_r(f)} = - \frac{jk\eta_L}{2\pi\eta_0}\,\frac{e^{-jkr}}{r} \tag{5-9}$$

which indicates that the transmitting transient response is, in general, proportional to the time derivative of the receiving response when η_L is real and frequency independent because of the jk (or $j\omega$) factor in this expression.

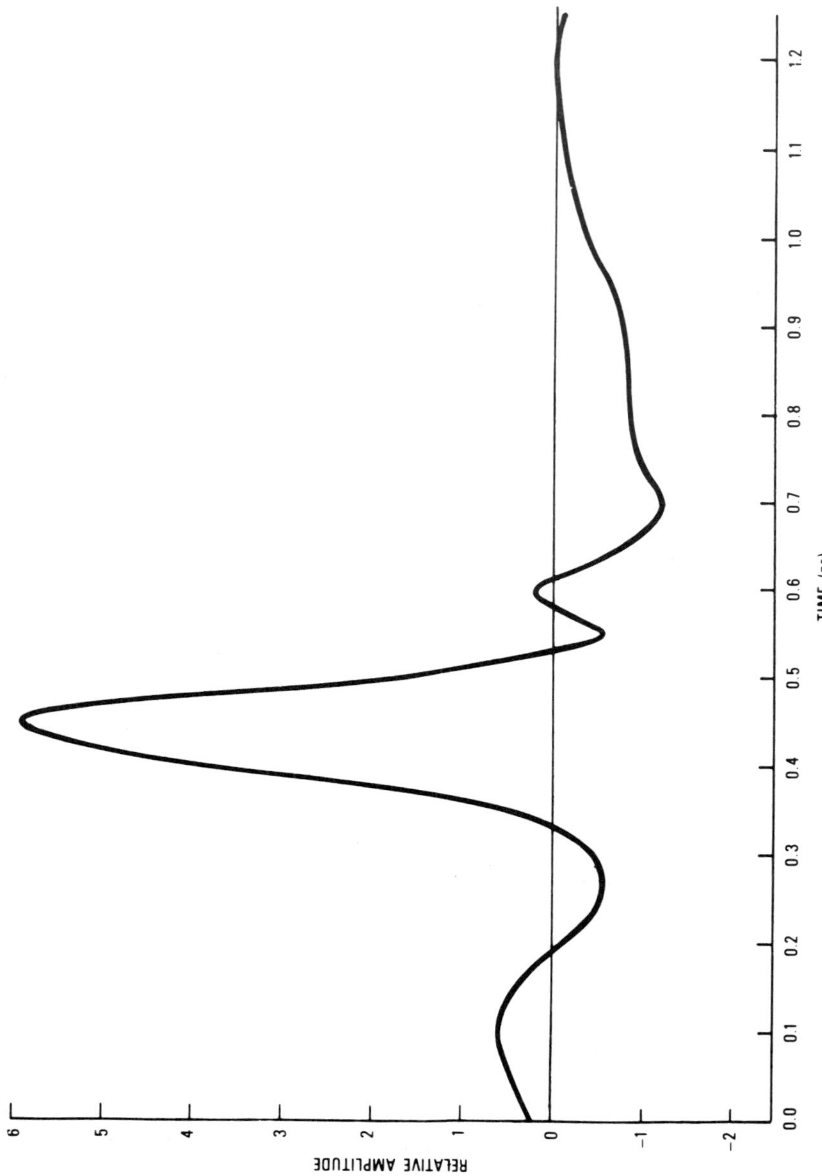

(a)

Fig. 5-2. Time-domain measurements using picosecond impulses: (a) time-domain waveform, and (b) spectrum amplitude.

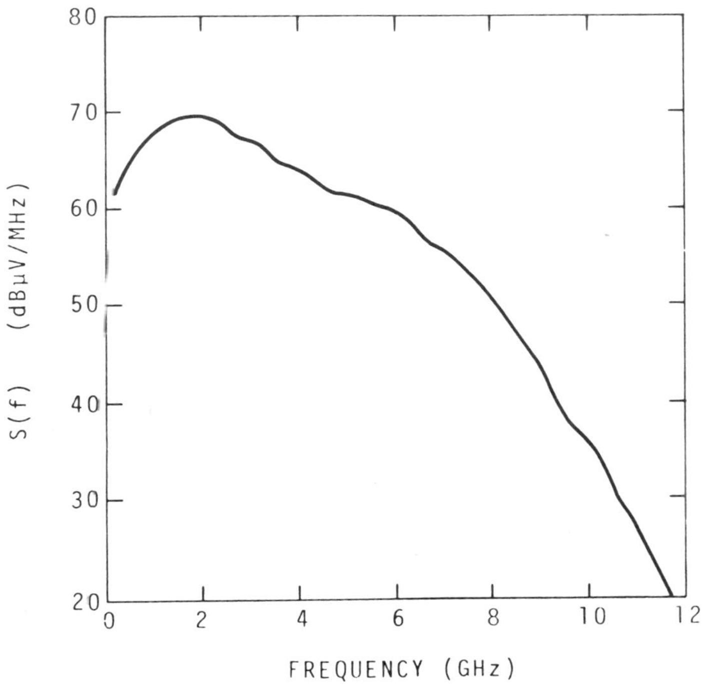

(b)

Fig. 5-2. Time-domain measurements (*continued*)

If the input excitation to an antenna is a nonsinusoidal transient it can be represented by a Fourier spectrum. Since Eqs. 5-5 and 5-9 hold for all single frequency components, they hold for a nonsinusoidal transient as well. Hence, when the current, I_L, induced in the load impedance of a receiving antenna is known, the radiated field of the same antenna when used for transmitting can be calculated directly by differentiating the induced current. On the other hand, when the radiated transient field of a driven antenna is known, the induced current of the same antenna when used for receiving can be computed by integration.

Theoretical and experimental investigations of the time-domain characteristics of various sensors and radiators discussed in this chapter have been performed using extremely short 70-psec impulses as a reference. The time-domain waveform of the pulse shown in Fig. 5-2(a) has a relatively flat spectrum amplitude $s(f)$ greater than 60 dB μV/MHz up to 5 GHz, as shown in Fig. 5-2(b). The 70-psec impulse is used both as a driving voltage for the investigation of transmitting transient response and as an incident impulse field for the investigation of receiving transient response. The experiments are performed using a time-domain antenna range in conjunction with a time-domain automatic network analyzer.[8]

5.2 ELECTRIC-FIELD SENSORS

This section considers the receiving and transmitting transient characteristics of the following electric-field sensors and radiators specially designed for time-domain measurement use: (1) a linear antenna with resistive loading, (2) a linear antenna with resistive and capacitive loading, (3) a conical antenna, (4) an asymptotic conical antenna, (5) a conducting TEM horn, and (6) a resistively loaded TEM horn.

Linear Antenna with Resistive Loading[4,9,10,11,12]

This antenna consists of a cylinder of length $2h$ and radius a that is aligned along the z axis and centered at the origin. It has an internal impedance $Z^i(z)$, due to continuous complex impedance loading, and carries an axial current $I_z(z)$, which is assumed to be uniformly distributed around the periphery of the cylinder since the radius is much less than the wavelength, λ. When the antenna is driven by a delta-function voltage, V_0, at $z = 0$, the axial component of the vector potential, $A_z(z)$, on the surface of the antenna satisfies the one-dimensional wave equation,

$$\left(\frac{\partial^2}{\partial z^2} + k^2\right)A_z(z) = \frac{jk^2}{\omega}\,[Z^i(z)I_z(z) - V_0\delta(z)] \tag{5-10}$$

where

$$A_z(z) = \frac{\mu_0}{4\pi}\int_{-h}^{h} I_z(z')K(z,z')dz' \tag{5-11}$$

and

$$K(z,z') \cong \frac{e^{-jk\sqrt{(z-z')^2+a^2}}}{\sqrt{(z-z')^2+a^2}} \tag{5-12}$$

When the cylindrical antenna has a continuous complex impedance loading such that the internal impedance per unit length $Z^i(z)$ is a particular function of the position along the antenna, Wu and King[9] found that a pure outward traveling wave can exist on an antenna of finite length; that is, if

$$Z^i(z) = \frac{60\psi}{h - |z|} \tag{5-13}$$

where

$$\psi = \left. \frac{\int_{-h}^{h} I(z')K(z,z')dz'}{I(z)} \right|_{z=0}$$
$$\cong 2\left[\sinh^{-1}\frac{h}{a} - C(2ka, 2kh) - jS(2ka, 2kh)\right] + \frac{j}{kh}(1 - e^{-j2kh}) \tag{5-14}$$

and $C(a,x)$ and $S(a,x)$ are the generalized cosine and sine integrals, then the approximate current distribution for the one-dimensional wave equation given in Eq. 5-10 has the nature of a traveling wave, i.e.,

$$I_z(z) = \frac{V_0}{60\psi(1 - j/kh)}\left(1 - \frac{|z|}{h}\right) e^{-jk|z|} \tag{5-15}$$

Since the solution of $e^{+jk|z|}$ no longer satisfies the original wave equation, the antenna does not support a standing wave. Since the frequency dependence of $Z^i(z)$ appears only in ψ in the form of a logarithm for small kh, the antenna has much broader frequency characteristics than an antenna with a lumped resistor located at a quarter wavelength from the ends.[13]

It should be noted, however, that the loading given by Eq. 5-13 is a complex impedance since the constant ψ is a complex number. As will be shown later, the real part of ψ is weakly dependent on frequency; its imaginary part, negative and proportional to frequency. For this reason, it is not possible to realize the loading given by Eq. 5-13 using passive elements. It will be shown in Fig. 5-6 that, at frequencies where $kh < \pi/2$, the imaginary part of ψ is usually smaller than its real part.[11] For this reason, a short linear antenna with tapered resistive loading has been examined.

The current distribution on a resistively loaded antenna can be calculated by numerically solving the one-dimensional wave equation given in Eq. 5-10 using the method of moments.[14] To simplify the integral equation, it is assumed that wire radius a is much smaller than wavelength λ ($a \ll \lambda$) so that the current can be considered to flow in a filament along the z axis rather than on the surface of a cylinder. To permit a solution for the current distribution, suitable testing functions for basis functions for expansion are needed. We have chosen piecewise linear testing functions and piecewise constant basis functions. Other choices of testing functions and basis functions for expansion are possible—e.g., piecewise constant, piecewise linear, piecewise sinusoidal, etc. Although combinations other than piecewise linear testing functions and piecewise constant basis functions for expansion of the current might lead to better convergence, it is beyond the scope of this section to discuss all the possible series representations. A detailed discussion of this subject can be found elsewhere.[15]

Figure 5-3 shows the coordinate system of a typical one-dimensional wire antenna of radius a. The incident electric field, E_{inc}, associated with the current, I, and the charge density, ρ, on a wire antenna is given in terms of the usual vector and scalar potentials, $\vec{A}$ and Φ, as follows:

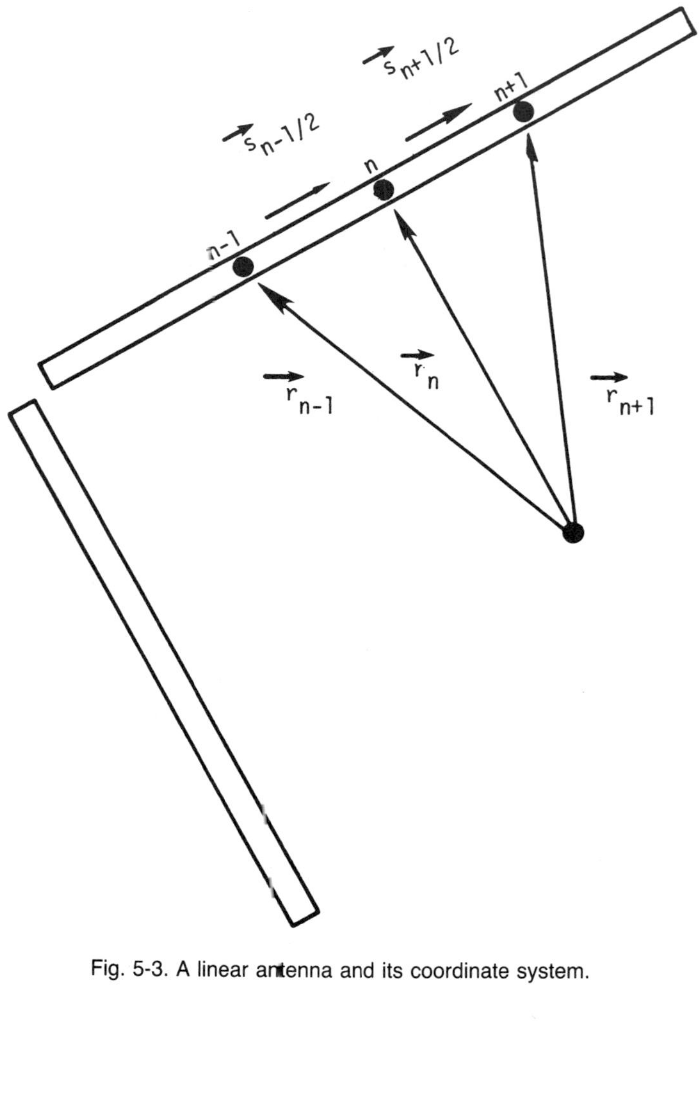

Fig. 5-3. A linear antenna and its coordinate system.

$$\vec{E}_{\text{inc}} = -j\omega\vec{A} - \nabla\Phi \tag{5-16}$$

where

$$\vec{A} = \frac{\mu}{4\pi}\int_c I(s')\vec{s}(s')\,\frac{e^{-jkR}}{R}\,ds' \tag{5-17}$$

and

$$\Phi = \frac{1}{4\pi\varepsilon}\int_c \rho(s')\,\frac{e^{-jkR}}{R}\,ds' \tag{5-18}$$

and where R is defined to be

$$R = \sqrt{|\vec{r} - \vec{r}'(s')|^2 + a^2} \tag{5-19}$$

The one-dimensional charge density, $\rho(s)$, is related to the current, $I(s)$, through the continuity equation,

$$\rho(s) = \frac{j}{\omega}\,\frac{dI(s)}{ds} \tag{5-20}$$

The unit direction vector, $\vec{s}$, which is parallel to the axis of each segment, is defined as

$$\vec{s}_{n+1/2} = \frac{\vec{r}_{n+1} - \vec{r}_n}{|\vec{r}_{n+1} - \vec{r}_n|} \tag{5-21}$$

Using the pulse testing function, expanding the current in pulses, and applying a finite-difference approximation to compute charge density ρ, we can rearrange Eq. 5-16 in terms of the usual impedance matrix elements, Z_{mn}, associated with the current and voltage matrix elements, I_n and V_m, i.e.,

$$\sum_{n=1}^{N} I_n Z_{mn} = V_m \tag{5-22}$$

The matrix element, Z_{mn}, associated with the nth current at the observation point, s_m, is given by[16]

$$Z_{mn} = \frac{j}{4\pi\omega\varepsilon}\left[k^2(\vec{r}_{m+1/2} - \vec{r}_{m-1/2}) \cdot (\vec{s}_{n+1/2}\, g_{m,n,n+1/2} + \vec{s}_{n-1/2}\, g_{m,n-1/2,n}) - \frac{g_{m+1/2,n,n+1}}{s_{n+1} - s_n} + \frac{g_{m+1/2,n-1,n}}{s_n - s_{n-1}} + \frac{g_{m-1/2,n,n+1}}{s_{n+1} - s_n} - \frac{g_{m-1/2,n-1,n}}{s_n - s_{n-1}}\right] + Z(s_m)(s_{m+1/2} - s_{m-1/2}) \tag{5-23}$$

where $Z(s_m)$ is the internal resistance per unit length due to resistive loading. Both the vector and scalar potential terms involve an integral of the form

$$g_{m,\mu,\nu} = \int_{s_\mu}^{s_\nu} \frac{e^{-jkR_m}}{R_m}\, ds' \tag{5-24}$$

where

$$R_m = \sqrt{|\vec{r}_m - \vec{r}\,'(s')|^2 + a^2} \tag{5-25}$$

The matrix element V_m in Eq. 5-22 is simply given by

$$V_m = \vec{E}^{\text{inc}}(s_m) \cdot (\vec{r}_{m+1/2} - \vec{r}_{m-1/2}) \tag{5-26}$$

The elements of the vector I_n, which give the current distribution, can then be obtained by solving a matrix equation.[14]

Numerical solutions using the method of moments and the analysis based on the approximations by Wu and King have been applied to the resistively loaded antenna. In the method of moments, we calculate the current distributions by subdividing the dipole into 61 subsections. In order to observe the convergence of the series representation for current on the dipole, a numerical solution using 121 subsections was also examined. The results indicate that the convergence for a numerical solution with 61 subsections is very good.

Resistive antenna elements are made by depositing a linearly tapered thin-film alloy on a glass rod. A glass rod, 7.5 cm long and 0.254 cm in diameter, was chosen as the substrate. These dimensions yield a quarter-wave monopole, $kh = \pi/2$, at 1 GHz. To achieve a dipole configuration, a glass rod 15 cm long is used as the substrate. Once the antenna dimensions have been determined, the desired resistive loading profile is calculated using Eq. 5-13. This resistive loading profile, which turns out to be a relatively weak function of frequency and is calculated at 1 GHz, is shown in Fig. 5-4. The constant ψ in Eq. 5-14, which is the ratio of the vector potential to the current along the antenna and which is, in general, complex, is assumed to be real and is represented by the absolute value of the complex constant ψ. Typically, the resistive loading required

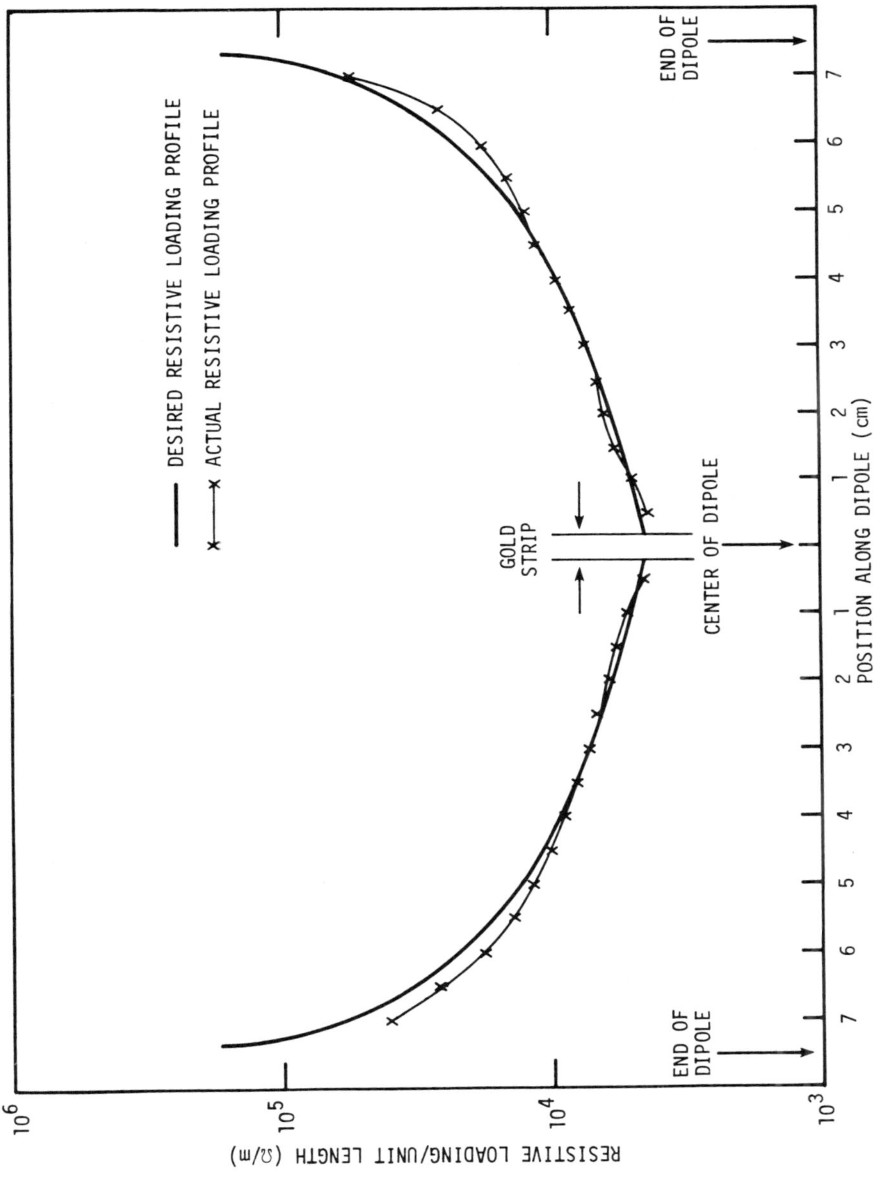

Fig. 5-4. Resistive loading profile of the linear antenna.

is about 5 kΩ/m at the driving point, about 10 kΩ/m at the midpoint, and infinite at the end of the antenna.

The magnitude and the phase of the current distributions on the resistively loaded antenna are calculated using both the method of moments and the Wu–King approximation; they are shown in Figs. 5-5(a) and (b) for different frequencies. The experimental results for the current distribution at 1 GHz are also shown in Fig. 5-5(b).

The ratio, ψ, of the vector potential to current at the antenna driving point is also calculated by using the current distributions from both the method of moments and the Wu–King approximation. The real and imaginary parts of ψ are shown in Figs. 5-6(a) and (b), respectively, where the real part of ψ is seen to be usually larger than the imaginary part. As mentioned previously, the real part of ψ is a very slowly varying function of frequency, and, therefore, the resistive loading profile $Z^i(z)$ given in Eq. 5-13 and the current distribution $I_z(z)$ given in Eq. 5-15 have only a weak frequency dependence, which results in the desired broadband characteristics.

The driving-point admittances (conductance and susceptance) are obtained from the method of moments and the Wu–King approximation for the current distribution. For example, the Wu–King approximation yields

$$Y_0 = \frac{1}{60\psi(1 - j/kh)} \tag{5-27}$$

Results for the driving-point conductance and susceptance of the antenna are shown in Figs. 5-7(a) and (b) compared with experimental results.

The effective length, $h_e(f)$, of the resistively loaded linear antenna is then evaluated according to the formula,

$$h_e(f) = \frac{1}{I_z(0)} \int_{-h}^{h} I_z(z)dz \tag{5-28}$$

Using the Wu–King approximation for $I_z(z)$ in Eq. 5-15, the approximate effective length, $\tilde{h}_e(f)$, is found to be

$$\tilde{h}_e(f) = \frac{2}{k^2h}(1 - jkh - e^{-jkh}) \tag{5-29}$$

The results for $\tilde{h}_e(f)$ using the current distributions obtained from the method of moments and the Wu-King approximation are shown in Fig. 5-8, where the moment-method solution involves numerically integrating the current distribution as given in Eq. 5-28.

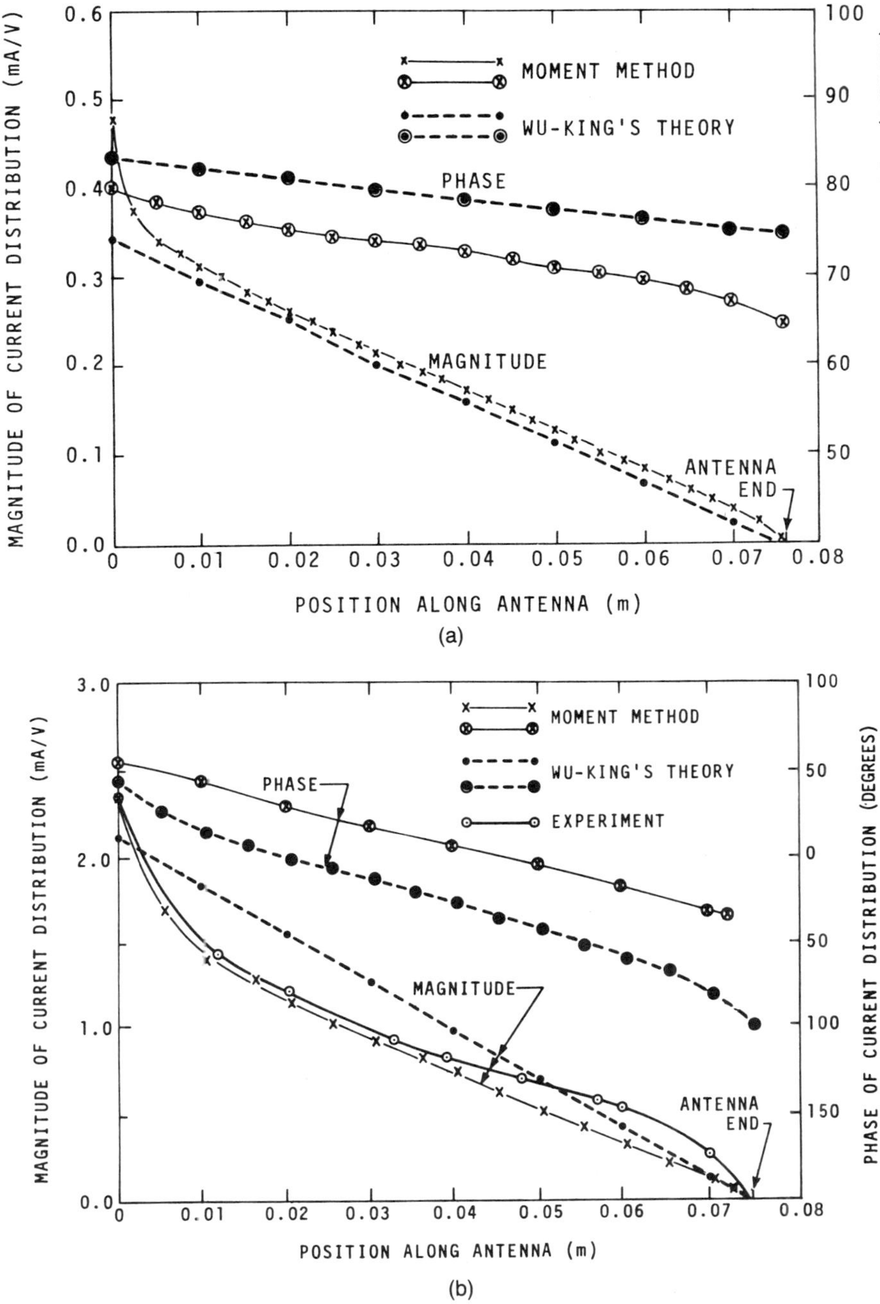

Fig. 5-5. Current distributions of a resistively loaded linear antenna: (a) 100 MHz, and (b) 1 GHz.

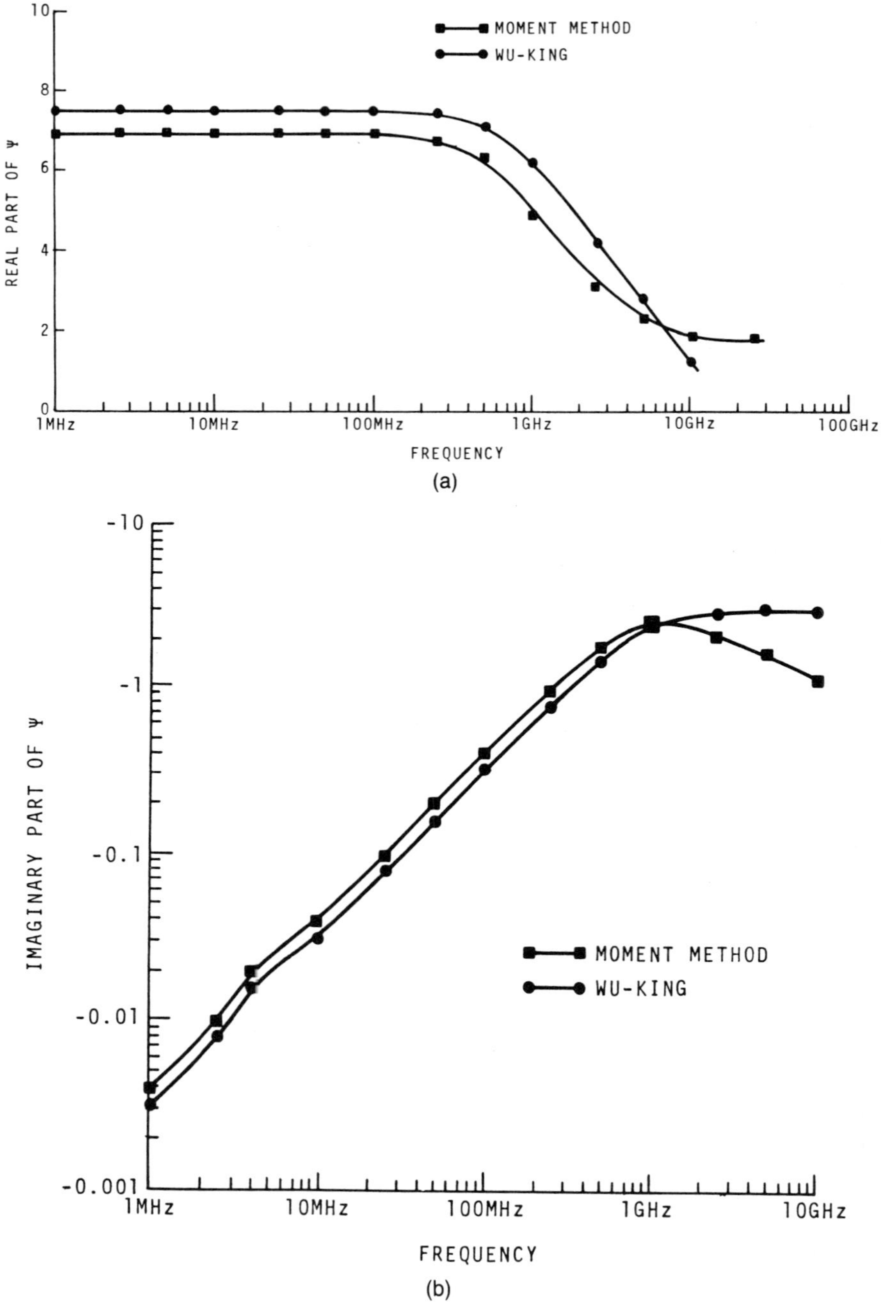

Fig. 5-6. (a) Real part of Ψ, and (b) imaginary part of Ψ.

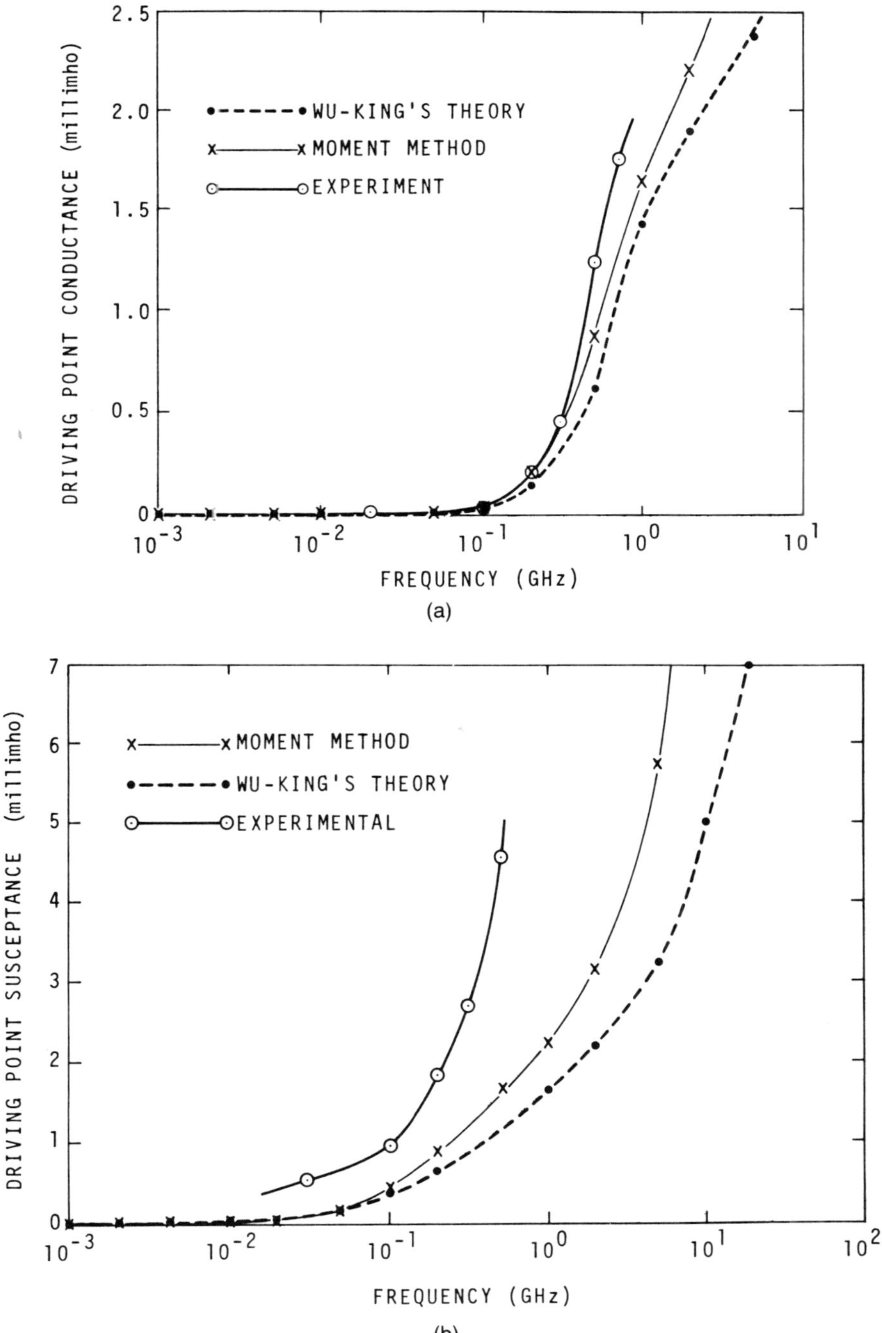

Fig. 5-7. Driving point admittance of the resistively loaded linear antenna: (a) conductance, and (b) susceptance.

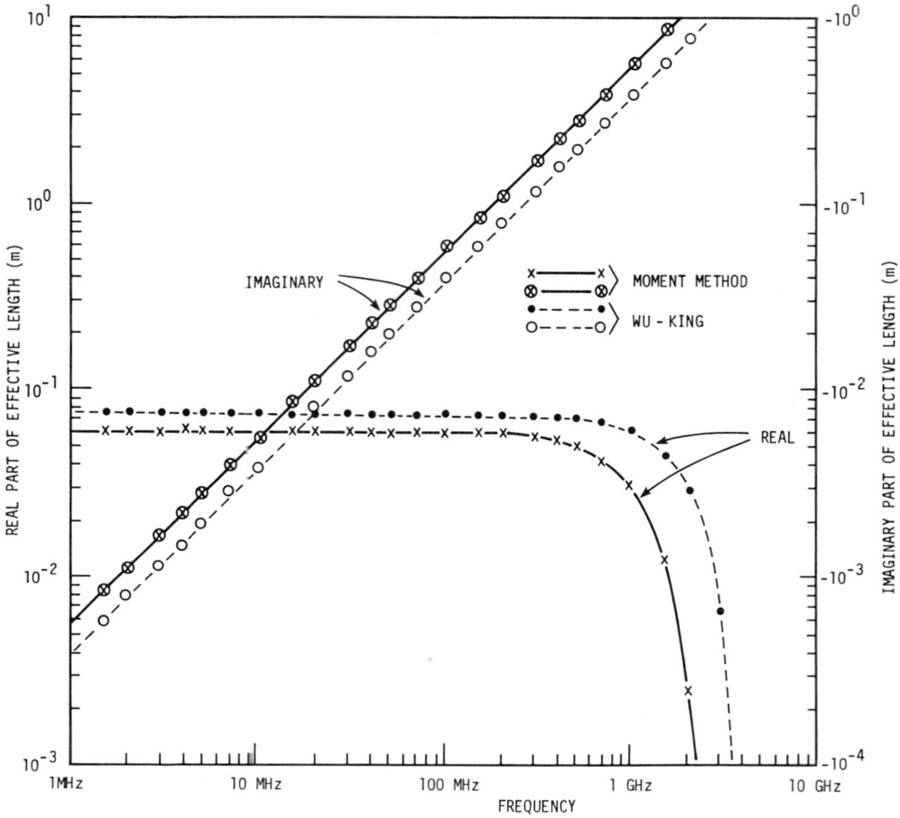

Fig. 5-8. Effective length of the resistively loaded linear antenna.

The effective lengths and the driving-point impedances obtained from both the Wu–King current distribution and the moment-method current distribution are used in Eq. 5-1 and 5-2 to calculate the receiving and transmitting transient responses in the frequency domain. The use of a fast Fourier transform (FFT) then allows the determination of transient fields for a known input waveform.

The calculated receiving and transmitting transient responses of a resistively loaded linear antenna are shown in Fig. 5-9, along with experimental results. As indicated there, the original 70-psec pulse is broadened as a result of high-frequency dispersion above 100 MHz. Figure 5-9 also demonstrates clearly that the transmitting characteristic of the transient response is proportional to the receiving characteristic differentiated with respect to time, as discussed in Sec. 5.1.

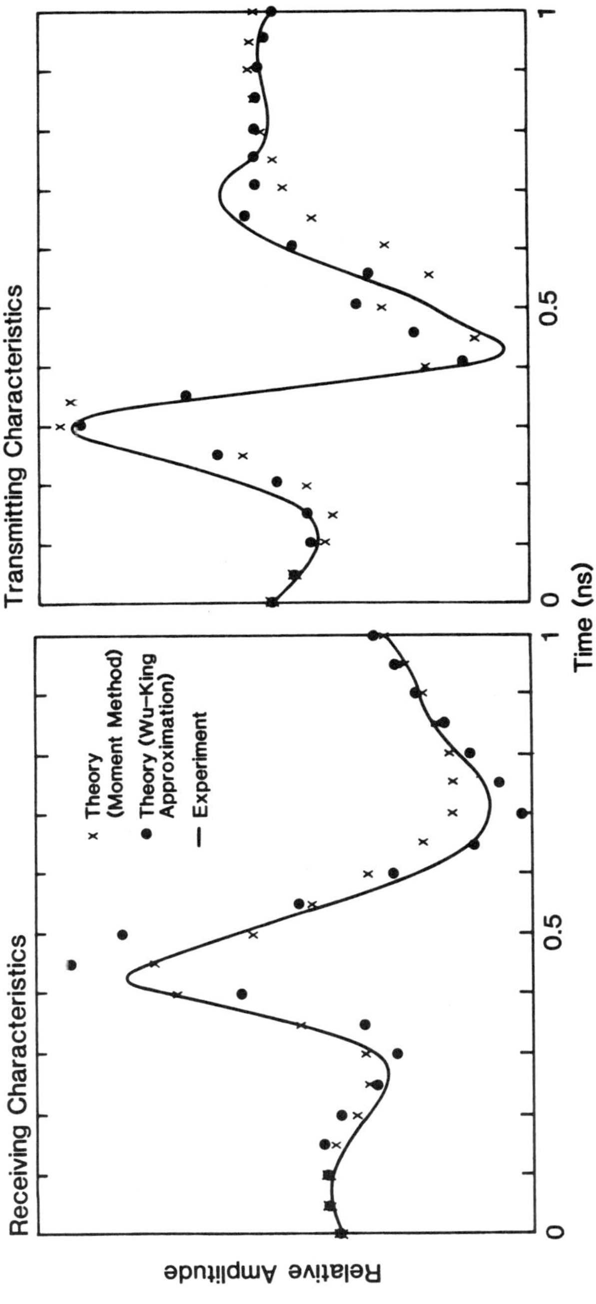

Fig. 5-9. Receiving and transmitting characteristics of the resistively loaded linear antenna.

Linear Antenna with Tapered Resistive and Capacitive Loading[17,18]

The impedance loading given by Eq. 5-13 is required in order to obtain a traveling-wave current distribution on a linear antenna. It was noted in Fig. 5-6 that the imaginary part of ψ is usually less than its real part. However, at frequencies where $kh > 10(\pi/2)$, the imaginary part of ψ, which turns out to be negative, becomes comparable to its real part.[11] Therefore, to create an outward traveling-wave current distribution on a linear antenna at frequencies where $kh \gg \pi/2$, the required impedance loading along the antenna must be complex, with both resistance and capacitive reactance. For this reason, the characteristics of a linear antenna with resistive and capacitive loading have been investigated.

For a linear antenna with a length $2h$ of 7.5 cm and a radius a of 0.3 cm, the ψ term as given by Eq. 5-14 becomes $3.06 - j2.39$ at 2 GHz, where $kh = \pi/2$. To implement the real part of the impedance loading given by Eq. 5-13, the antenna element was fabricated by depositing a tapered thin-film alloy on a glass rod.[11] To achieve the imaginary part of the impedance loading given by Eq. 5-13, narrow rings on the deposited thin-film alloy on the glass rod were burned away by an argon laser to form a segmented antenna. The neighboring edges of each gap on this segmented antenna act as a lumped capacitance inserted in series with the antenna. When the number of segments per wavelength becomes large, the antenna can be considered as being loaded with continuously distributed capacitance. The gap width t between segments is chosen to provide the properly tapered capacitive reactance loading as given by Eq. 5-13,

$$\mathcal{I}m[Z^i(z)] = -\frac{60 \times 2.39}{h - |z|} = -\frac{1}{\omega d C_t} \tag{5-30}$$

where the distance d between two adjacent gaps is chosen to be 5 mm. The lumped capacitance of each gap, C_t, is given by $\pi a^2 \varepsilon_g / t$, where ε_g is the effective

Table 5-1. Capacitive Loading Profile of the Linear Antenna.

POSITION ALONG ANTENNA, z(cm)	GAP WIDTH, t(mm)	LUMPED CAPACITANCE, C_t(pF)
0.5	6.1×10^{-2}	7.1
1.0	7.2×10^{-2}	6.0
1.5	8.8×10^{-2}	4.9
2.0	1.13×10^{-1}	3.8
2.5	1.58×10^{-1}	2.7
3.0	2.64×10^{-1}	1.6

permittivity of the glass in the gap. The required lumped capacitance of each gap and corresponding gap width are given in Table 5-1.

The integral equation outlined in the previous section was solved using the method of moments. For this study, as for the case of resistive loading only, piecewise linear testing functions and piecewise constant basis functions for expansion are chosen. The current distributions are calculated by subdividing the linear antenna into 61 subsections.

Figure 5-10 shows the current distributions of a linear antenna with tapered resistive and capacitive loading as obtained from the moment-method approach.

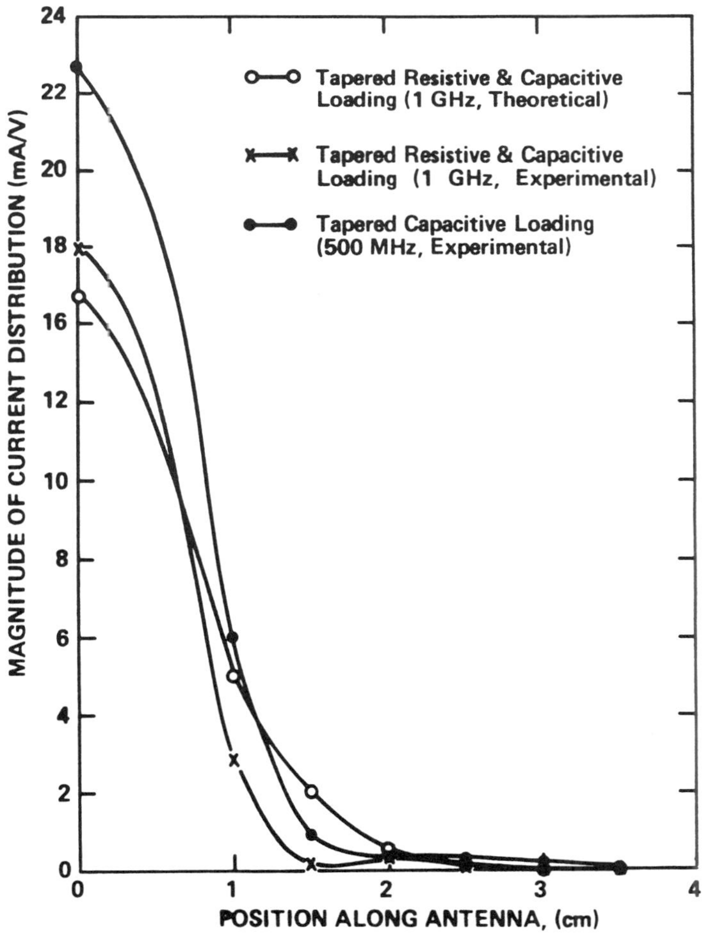

Fig. 5-10. Current distributions of the linear antenna with tapered resistive and capacitive loading.

For comparison, the current distribution of a linear antenna with tapered capacitive loading only is also shown. Since the current distributions are found to be basically traveling waves, the antenna is expected to have very broadband characteristics. Its impedance as a function of frequency is shown in Fig. 5-11. The typical resonances in the resistance and the reactance of the antenna have been eliminated. The experimental results for the receiving transfer function of the linear antenna with both resistive and capacitive loading are shown in Fig. 5-12; they indicate that the antenna does not exhibit any high Q resonances even at frequencies where $kh > \pi/2$ ($f > 2$ GHz).

Conical Antenna[4,7]

A conical antenna is one of the most popular types for use as a standard transmitting antenna. Expressions for the effective antenna length, $h_e(f)$, and the driving-point impedance, $Z_0(f)$, of a conical antenna are used to calculate its

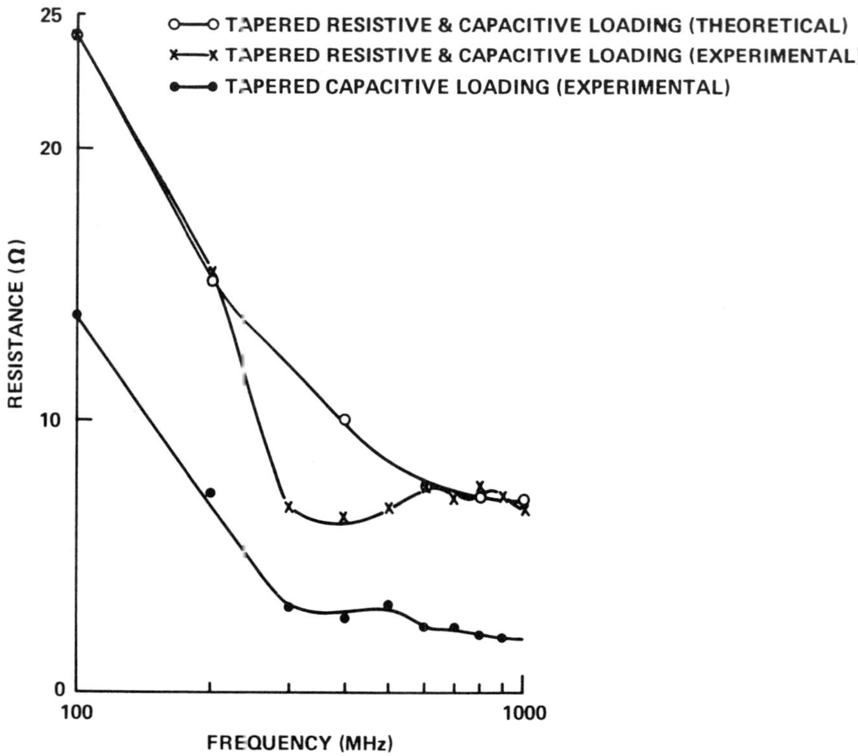

(a)

Fig. 5-11. Driving point impedance of the linear antenna with resistive and capacitive loading: (a) resistance, and (b) reactance.

receiving and transmitting characteristics in the frequency domain in order to obtain the corresponding transient properties.

The effective length, $h_e(f)$, of a conical antenna can be derived from its radiation field using the Rayleigh-Carson reciprocity theorem of Eq. 5-3. The driving-point impedance, $Z_0(f)$, of a conical antenna is given by the ratio of the frequency-domain representation of the voltage applied at the antenna terminal to the frequency-domain representation of the resultant current flowing into the antenna terminal. One thus arrives at the following formulas[7]:

$$h_e(f) = \frac{1}{k^2\ell\left(1 + \frac{\beta}{\alpha}\right)} \left(\frac{\beta}{\alpha} e^{jk\ell} - e^{-jk\ell}\right) \times \sum_{n=1}^{\infty} P_n(\cos\theta_0) P_n'(0) \frac{2n+1}{n(n+1)} \frac{j^{n+1}}{h_{n-1}^{(2)}(k\ell) - \frac{n}{k\ell} h_n^{(2)}(k\ell)} \tag{5-31}$$

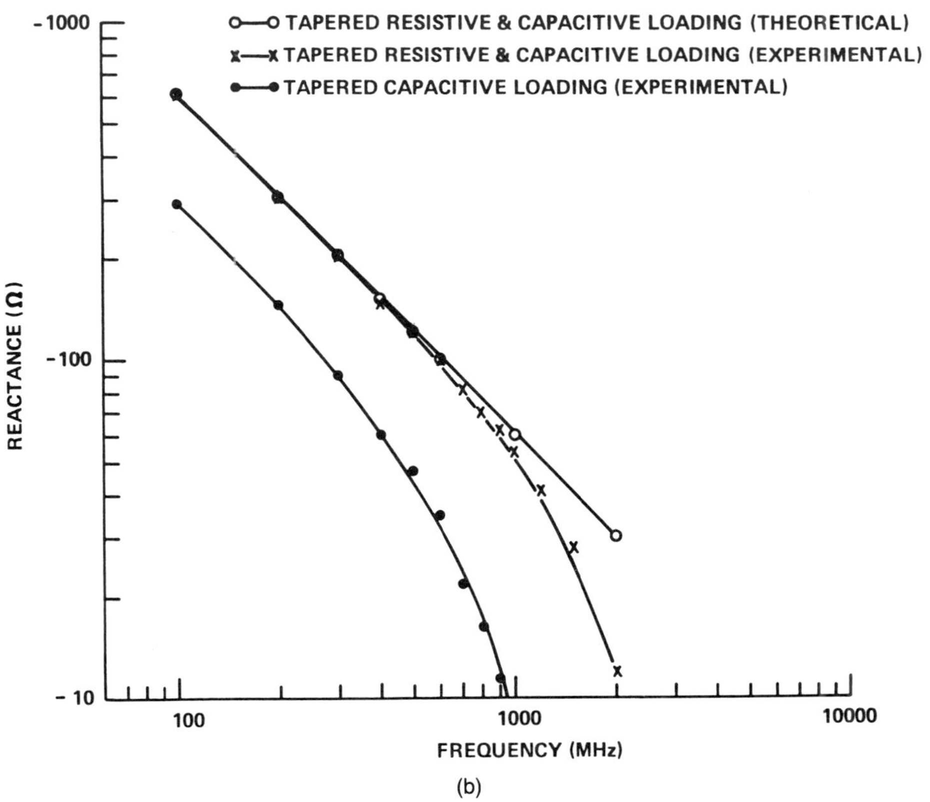

(b)

Fig. 5-11. Driving point impedance (*continued*)

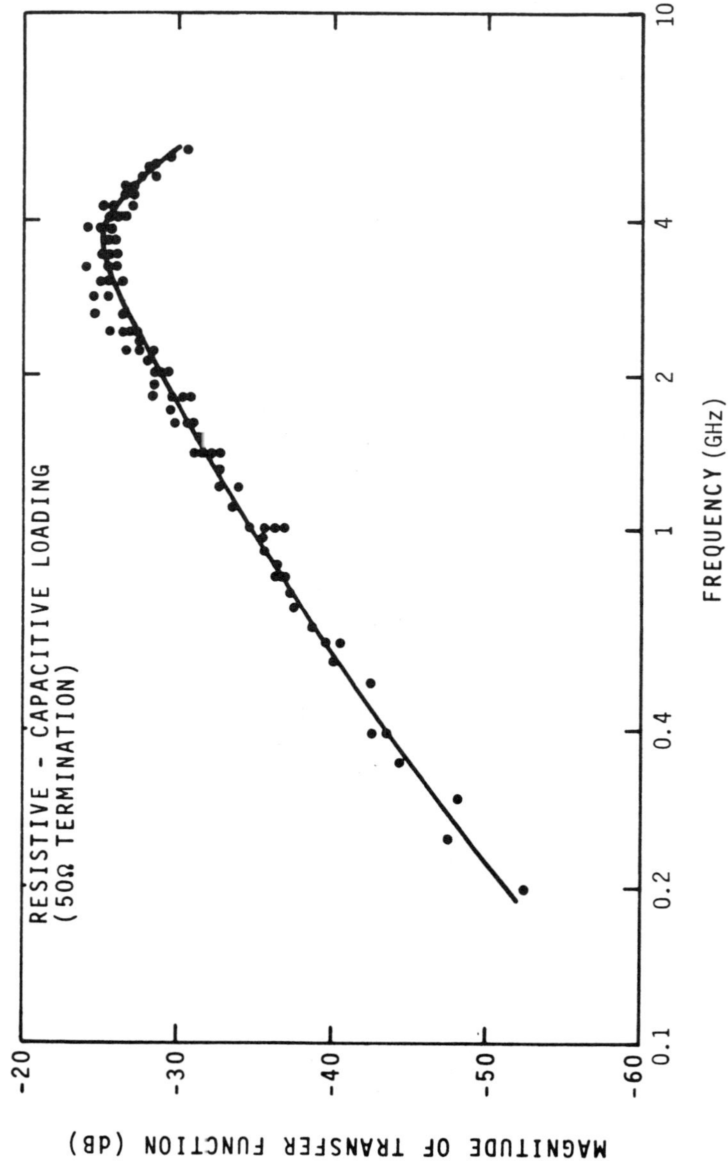

Fig. 5-12. Receiving transfer function of the linear antenna with resistive and capacitive loading.

$$Z_0(f) = Z_c\left(\frac{1 - \dfrac{\beta}{\alpha}}{1 + \dfrac{\beta}{\alpha}}\right) \tag{5-32}$$

$$Z_c = \frac{\zeta_0}{2\pi}\,\ell n \cot\frac{\theta_0}{2} \tag{5-33}$$

$$\frac{\beta}{\alpha} = e^{-j2k\ell}\,\frac{1 + j\dfrac{\zeta_0}{2\pi Z_c}\displaystyle\sum_{n=1}^{\infty}\frac{2n+1}{n(n+1)}[P_n(\cos\theta_0)]^2\,\zeta_n(k\ell)}{-1 + j\dfrac{\zeta_0}{2\pi Z_c}\displaystyle\sum_{n=1}^{\infty}\frac{2n+1}{n(n+1)}[P_n(\cos\theta_0)]^2\,\zeta_n(k\ell)} \tag{5-34}$$

$$\zeta_n(k\ell) = \frac{h_n^{(2)}(k\ell)}{h_{n-1}^{(2)}(k\ell) - \dfrac{n}{k\ell}h_n^{(2)}(k\ell)} \tag{5-35}$$

where

ℓ = the length of the antenna
ζ_0 = the free space characteristic impedance
θ_0 = half the apex angle
$P_n(\cos\theta)$ = the Legendre polynomial of order n
$h_n^{(2)}(k\ell)$ = the spherical Hankel function of the second kind

The results for the effective length and the driving-point impedance given by Eqs. 5-31 and 5-32, respectively, are used in Eqs. 5-1 and 5-2 along with the Rayleigh-Carson reciprocity theorem given in Eq. 5-3 to calculate the receiving and transmitting transient responses in the frequency domain. An FFT is then used to determine the transient fields for a known input waveform.

The transfer function of the transmitting characteristic for a conical monopole antenna 2.7 m long with an 8-deg apex angle is calculated theoretically and is given in Fig. 5-13. In this figure, there are two curves, one of which corresponds to a 57-term summation and the other to a 139-term summation in Eqs. 5-31 and 5-34. The curves agree with each other very well, indicating that the convergence of the 57-term solution is very good. Figure 5-13 shows that the transfer function of the transmitting characteristic for the conical antenna is relatively flat above 10 MHz.

The transmitting characteristic of the transient response for the conical antenna,

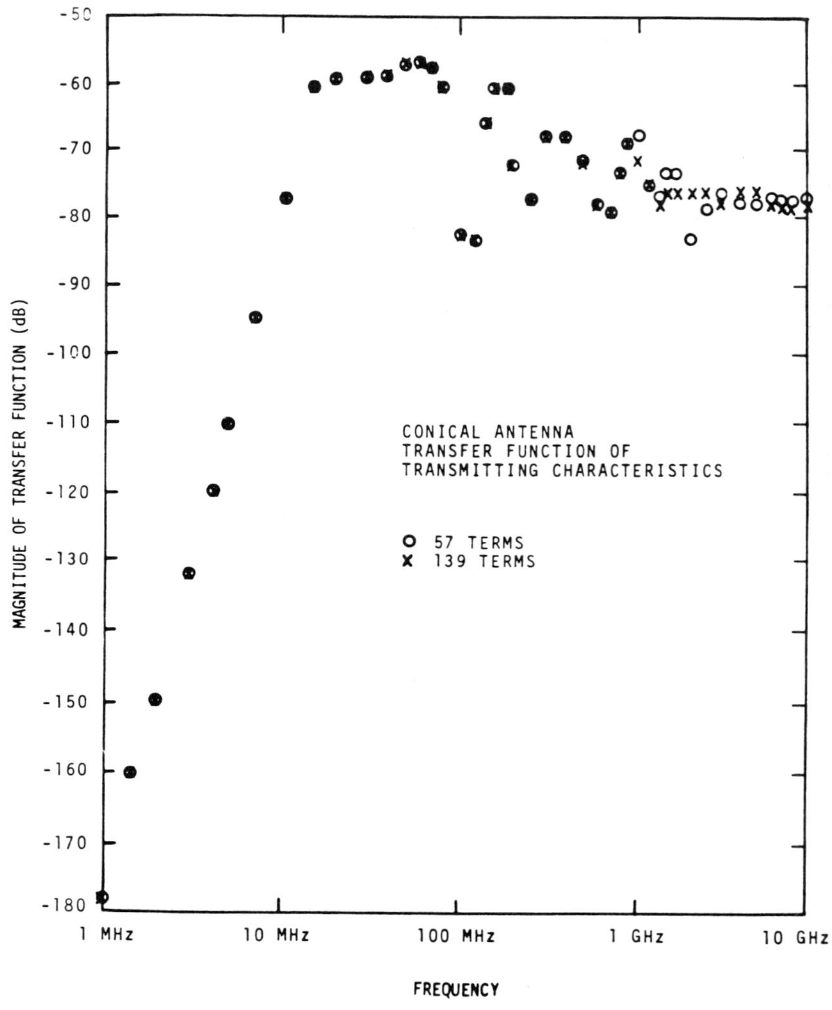

Fig. 5-13. Transmitting transfer function of the conical antenna.

also calculated theoretically, is given in Fig. 5-14 along with experimental results. Again, the theoretical calculation for the time-domain waveform using the 57-term summation agrees very well with that using the 139-term summation, and both compare well with the experimental data.

Asymptotic Conical Dipole[19,20]

The concept of the asymptotic conical-dipole sensor is based on the potential distribution resulting from some convenient equivalent charge distributions.[19]

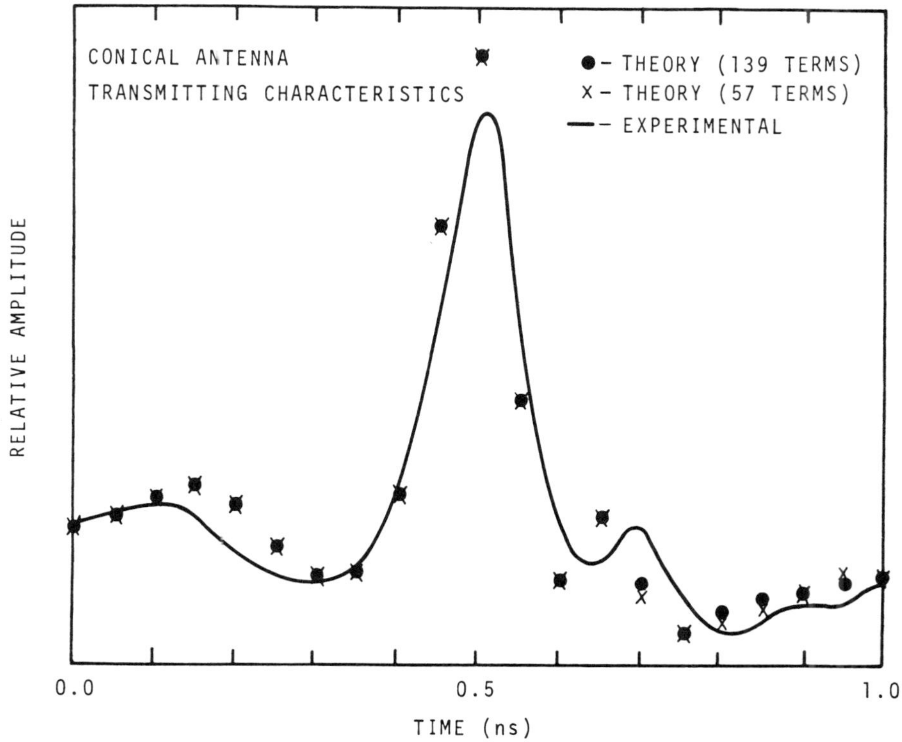

Fig. 5-14. Transmitting characteristic of the conical antenna.

Starting with a particular equivalent charge distribution, we calculate the resulting potential distribution. Several potential surfaces that result from this equivalent charge distribution are shown in Fig. 5-15. These are rotationally symmetric about the z axis. One of them is the zero-potential surface, which may be used as a conducting ground for the single-ended version of the sensor.

Also shown in Fig. 5-15 is the biconical surface, ϕ_s, which forms a 100-ohm transmission line when driven differentially at the origin. Half of this bicone is also a 50-ohm cone above the ground plane. The potential surfaces used for the sensors are that pair which are asymptotic to the 100-ohm bicone at the origin. These two surfaces are chosen so as to be geometrically asymptotic at some point in space (e.g., the origin of the coordinate system) to a transmission line geometry of known impedance. This arrangement eliminates reflections caused by an impedance discontinuity. Moreover, the potential of the transmission line geometry determines the potential of the particular surfaces chosen.

The charge distribution on the two antenna surfaces is calculated from the static potential distribution. The total antenna charge and dipole moment are then calculated by integration of the surface charge. The antenna capacitance

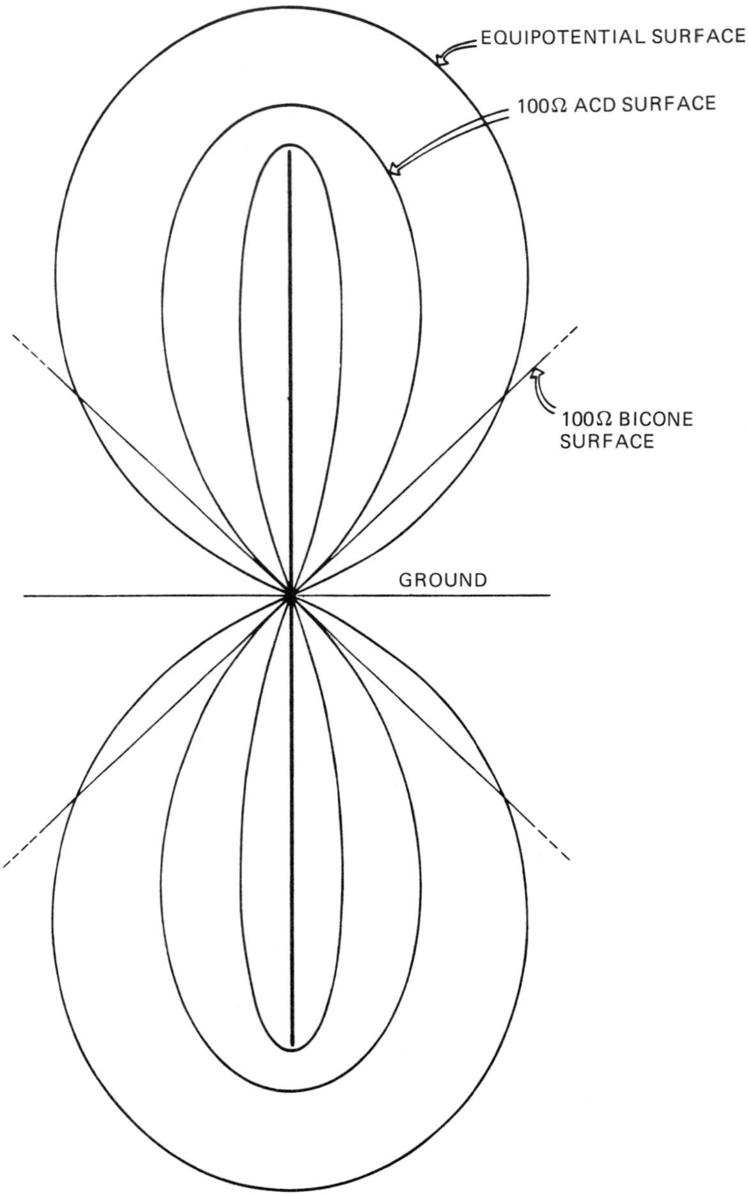

Fig. 5-15. Charge distribution of an asymptotic conical dipole.

and equivalent area can also be calculated. As an example, consider an asymptotic conical monopole with a length of 0.178 m. It has an antenna capacitance of 12 pF and a risetime of 0.94 ns between the 10- and 90-percent points of the final (settled) pulse height, as shown in Fig. 5-16. It also exhibits a 10-percent overshoot with its associated underdamped ringing.

TEM Horn[4]

In an effort to attain increased antenna directivity or increased antenna gain for a broadband and nondispersive antenna, many researchers have considered a TEM horn, which is a two-conductor end-fire traveling-wave structure. If the flare angle and plate widths of a TEM horn are properly chosen so that a constant, characteristic impedance can be maintained, the horn will guide a nearly pure TEM mode into free space. When the edge-diffraction effect and fringe fields are neglected, the linearly polarized spherical field at the horn aperture is given by[4]

$$E_y(x',y',0) = \frac{r_0^2}{\sqrt{r_0^2 + x'^2 + y'^2}} \frac{e^{-jk\sqrt{r_0^2 + x'^2 + y'^2} - r_0}}{\sqrt{r_0^2 + y'^2}} \qquad (5\text{-}36)$$

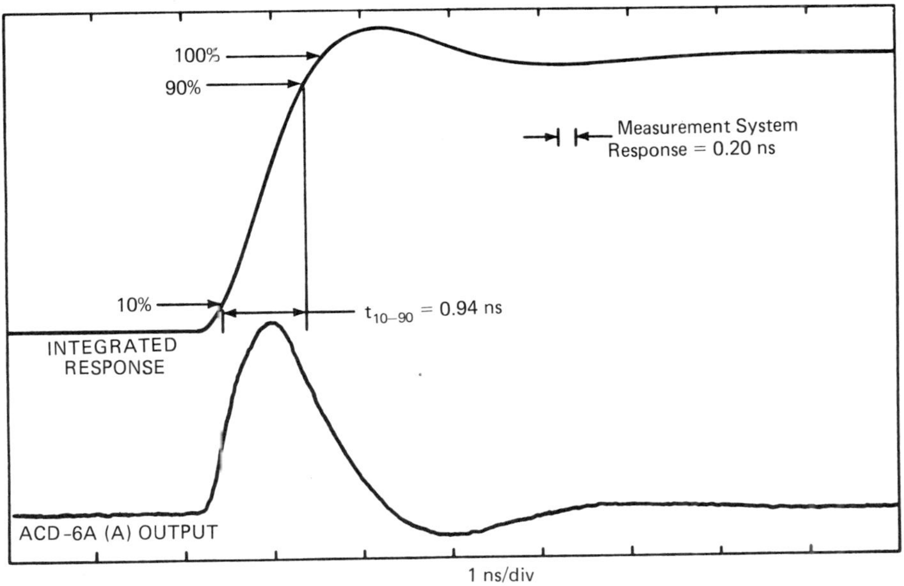

Fig. 5-16. Receiving characteristics of the asymptotic conical monopole.

Here, the source is assumed to be located at a distance r_0 from the center of the aperture. The coordinates in the antenna aperture plane are indicated by primes.

The radiated electromagnetic fields, $E^{\text{rad}}(x,y,z)$, outside the aperture (apt.) are then evaluated by use of the plane-wave spectrum analysis technique.[6] That is, since

$$\vec{S}(k_x,k_y) = \frac{1}{2\pi} \iint_{\text{apt.}} \vec{E}(x',y',0) e^{j(k_x x' + k_y y')}\, dx'\, dy' \qquad (5\text{-}37)$$

then

$$\vec{E}^{\text{rad}}(x,y,z) = \frac{1}{2\pi} \iint_{-\infty}^{\infty} \vec{S}(k_x,k_y) e^{-j(k_x x + k_y y + k_z z)}\, dk_x\, dk_y \qquad (5\text{-}38)$$

where $k_x^2 + k_y^2 + k_z^2 = k^2 = (2\pi/\lambda)^2$. These integrations are carried out using an FFT algorithm.

The other and perhaps more conventional technique for calculating the radiation field of a TEM horn is as follows. Given the aperture electric field, $\vec{E}$, in Eq. 5-36, one can determine an equivalent magnetic current distribution, $\vec{M}$, at the aperture by

$$\vec{M} = 2\vec{E} \times \vec{n} \qquad (5\text{-}39)$$

By use of an electric vector potential, $\vec{F}$, i.e.,

$$\vec{F} = \frac{1}{4\pi} \iint_{\text{apt.}} \vec{M} \frac{e^{-jkr}}{r}\, ds \qquad (5\text{-}40)$$

one can determine the electric and magnetic fields through the following relationships:

$$\vec{E} = -\nabla \times \vec{F} \qquad (5\text{-}41)$$

and

$$\vec{H} = j\omega\varepsilon_0 \vec{F} + \frac{1}{j\omega\mu_0} \nabla(\nabla \cdot \vec{F}) \qquad (5\text{-}42)$$

We find that for the electric fields

$$E_y^{\text{rad}}(x,y,z) = \frac{1}{4\pi} \int\int_{\text{apt.}} M_x(x',y')\left(jk + \frac{1}{r}\right) \frac{e^{-jkr}}{r} \frac{z}{r}\, dx'dy' \qquad (5\text{-}43)$$

and

$$E_z^{\text{rad}}(x,y,z) = \frac{1}{4\pi} \int\int_{\text{apt.}} M_x(x',y')\left(jk + \frac{1}{r}\right) \frac{e^{-jkr}}{r} \frac{(y - y')}{r}\, dx'dy' \qquad (5\text{-}44)$$

and for the dominant magnetic field,

$$H_x = \frac{1}{4\pi\mu_0} \int\int_{\text{apt.}} M_x(x'y') \frac{e^{-jkr}}{r} \left[-\frac{j\omega}{c^2} + \frac{1}{j\omega}\left\{\frac{(x - x')}{r}\right\}^2 \times \left\{\frac{1}{r^2} + \left(jk + \frac{1}{r}\right)^2 + \frac{\left(jk + \frac{1}{r}\right)}{r}\right\}\right] dx'dy' \qquad (5\text{-}45)$$

These two techniques are used to calculate the radiation field from a TEM horn in the frequency domain. The receiving characteristic of a TEM horn is determined from the transmitting characteristic by using the Rayleigh-Carson reciprocity theorem in Eq. 5-3; it is shown in Fig. 5-17.

Since the theoretical model given here is based on an assumed aperture field that neglects the edge diffraction and fringing field effects, a number of major limitations exist, although some useful engineering design concepts can be obtained. For example, the assumption that a TEM mode is basically guided from the throat to the aperture by the "nondispersive" TEM horn is somewhat questionable. Particularly at the high-frequency range of the spectrum, above 1 GHz for example, the aperture field may be quite different from a basic TEM mode because of the presence of higher-order modes. Even at the lower frequency ranges, below 1 GHz, the magnitude of the TEM mode at the aperture may not remain constant because of a high, frequency-dependent input reflection coefficient, e.g., 0.98 at 100 MHz.

In an effort to treat these aspects more rigorously, a formal solution to the TEM-horn problem using a vector integral-equation technique has been proposed. Using the equivalence principle,[21] the electric field $\vec{E}$ in free space generated by an unknown set of electric currents $\vec{J}$ on the TEM-horn plates is

$$\vec{E} = -\nabla \times \vec{F} - j\omega\mu\vec{A} + \frac{1}{j\omega\varepsilon} \nabla(\nabla \cdot \vec{A}) \qquad (5\text{-}46)$$

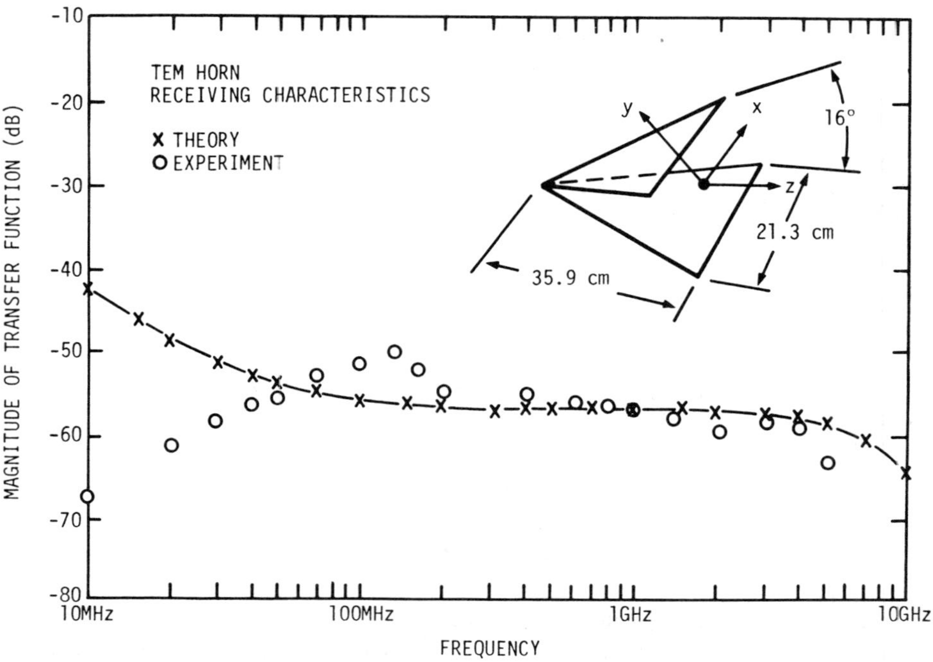

Fig. 5-17. Receiving transfer function of the conductive TEM horn (neglecting aperture diffraction).

Here the electric vector potential, $\vec{F}$, is given in terms of the known excitation equivalent magnetic current distribution, $\vec{M}$, by

$$\vec{F} = \frac{1}{4\pi} \iint_{\text{source}} \vec{M} \frac{e^{-jkR}}{R} ds \tag{5-47}$$

and the magnetic vector potential, $\vec{A}$, is given in terms of the unknown current distribution, $\vec{J}$, on the TEM horn plates by

$$\vec{A} = \frac{1}{4\pi} \iint_{\text{plates}} \vec{J} \frac{e^{-jkR}}{R} ds \tag{5-48}$$

where $R = |r - r'|$.

Boundary conditions are such that the tangential electric field at the TEM horn plates is zero, i.e.,

$$\vec{n} \times \vec{E}|_{\text{on the plates}} = 0 \tag{5-49}$$

Substituting Eqs. 5-46, 5-47, and 5-48 into Eq. 5-49 gives

$$\left[\vec{n} \times \left(-\nabla \times \frac{1}{4\pi} \iint \vec{M} \frac{e^{-jkR}}{R} ds\right)\right] + \left[\vec{n} \times \left\{-\frac{j\omega\mu}{4\pi} \iint \vec{J} \frac{e^{-jkR}}{R} ds \right.\right.$$
$$\left.\left.\left. + \frac{1}{j\omega\varepsilon} \nabla \cdot \left(\nabla \cdot \frac{1}{4\pi} \iint \vec{J} \frac{e^{-jkR}}{R} ds\right)\right\}\right]\right|_{\text{on the plates}} = 0 \tag{5-50}$$

Once this electric-field integral equation has been solved for the unknown $\vec{J}$, the other EM fields everywhere may be determined by using Eqs. 5-46, 5-47, and 5-48.

The derivation of the electric-field integral equation given in Eq. 5-50 is complete and straightforward. The solution may be obtained using the method-of-moments technique. The horn plates are divided into small patches. After the vector-integral equation in Eq. 5-50 has been decomposed into the appropriate coupled scalar-integral equations, two orthogonal components of current $\vec{J}$ with two unknown constants over each patch must be forced to satisfy the boundary condition. Although the matrix formulations of the problem are quite straightforward, their solution can be extremely formidable, particularly at higher frequencies where each plate has to be subdivided into patches that are small compared to a wavelength. Although some symmetry may be used to reduce the matrix order somewhat, treatment of a very large matrix is usually involved. In addition, because an FFT is used to obtain the transient characteristics of the TEM horn, the electric-field integral equation has to be solved for a good many frequencies, e.g., on the order of one thousand from 10 MHz to several gigahertz. This is a very expensive and time-consuming process.

To approximate the edge diffractions at the aperture and feed point of a TEM horn, one can examine a biconical antenna structure in which the apex angle and length of a bicone are adjusted to fit a TEM horn. The biconical model of a conducting TEM horn is shown in Fig. 5-18. For the frequency range over which the spectrum of the excitation has significant amplitude, a knowledge of both the driving-point impedance and the effective length of an antenna is required to obtain quantitatively the transient response of a transmitting and receiving TEM horn using the FFT technique. The effective length, $h_e(f)$, and the driving-point impedance, $Z_o(f)$, of a TEM horn based on this model are calculated using Eqs. 5-31 and 5-32, respectively.

The frequency-domain representations of the receiving transfer function $T_r(f)$ of a conductive TEM horn are shown in Fig. 5-19 for the dimensions given in Fig. 5-18. The receiving transfer function based on the model using the assumed aperture field is seen to be constant with frequency in Fig. 5-17, but it shows very unrealistic behavior in the lower frequency range below 100 MHz. The

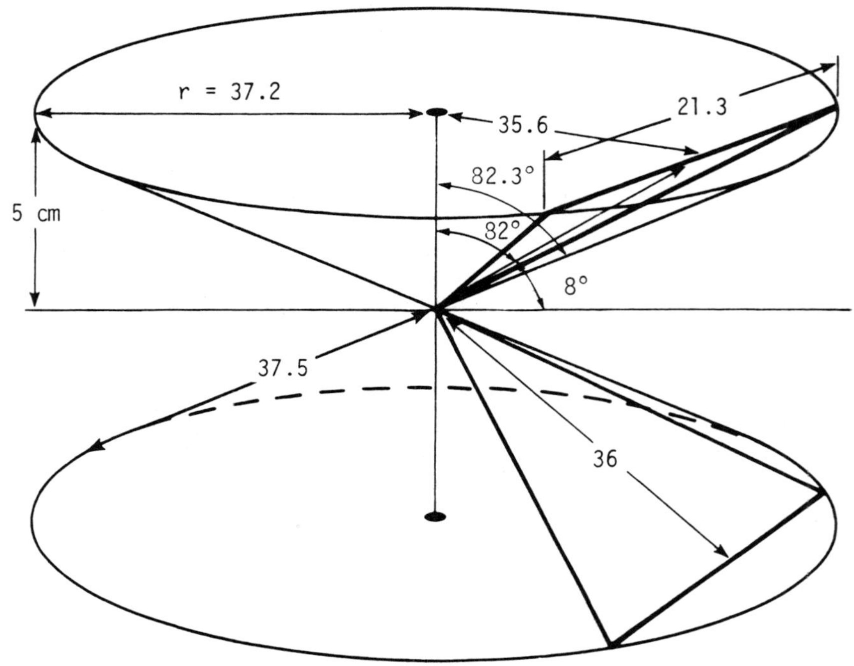

Fig. 5-18. Biconical model for a conductive TEM horn (dimensions in centimeters).

receiving transfer function based on the biconical antenna model as shown in Fig. 5-19, however, exhibits the expected multiple reflections between the feed point and the aperture of the conducting TEM horn.

The measured input reflection coefficient of the conducting TEM horn is shown in Fig. 5-20. It indicates that the reflection coefficient is very high, particularly in the low frequency range (0.98 at 100 MHz), and shows the oscillatory nature above 1 GHz because of interference between the reflection from the aperture and the reflection from the feed point.

The Cornell Aeronautical Laboratory (CAL) has designed a 1-m long TEM horn having a nonuniform line impedance matching from 50 Ω at the antenna input terminal to 377 Ω at the aperture. The nonuniform line impedance matching was achieved by forming the TEM horn plates empirically into a teardrop shape. A small resistance card termination was placed at the tip of the aperture to provide adequate current attenuation for a traveling wave. With these empirical modifications, it was possible to reduce the length of the TEM horn to 1 m while achieving a reasonable reflection coefficient (less than 0.48) for the frequency range of 100 MHz to 2 GHz, as shown in Fig. 5-20.

The receiving and transmitting characteristics of the transient responses for

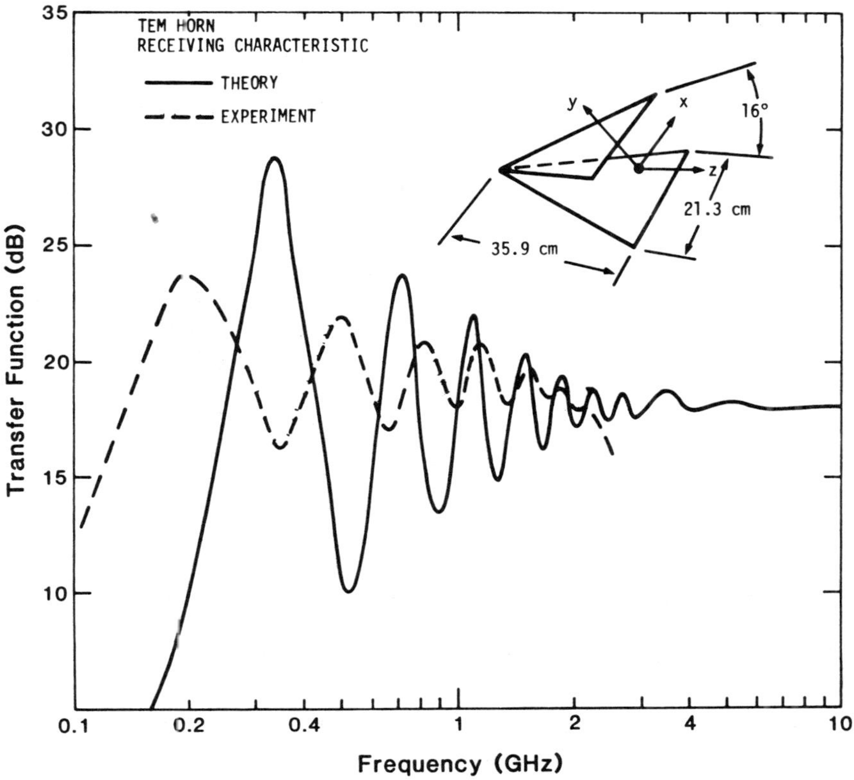

Fig. 5-19. Receiving transfer function of the conductive TEM horn (biconical model).

the conducting TEM horn are calculated using the biconical antenna model; they are shown in Fig. 5-21 along with the experimental results.

"TEM" Horn with Tapered Resistive Loading

As discussed in the previous section, in order to achieve a "smooth" transition—i.e., a transition with low reflection—the length of a conventional conductive TEM horn should be at least one-half wavelength at the lowest frequency of interest. This means that if the lowest frequency of interest is 100 MHz, a TEM horn may have to be as long as 1.5 m, which is often too long for practical applications. A "TEM" horn with continuously tapered resistive loading is considered here, therefore, for the purpose of developing a relatively short, broadband, and nondispersive antenna with high directivity.

Because of the presence of loss in TEM horn plates, the field patterns for EM

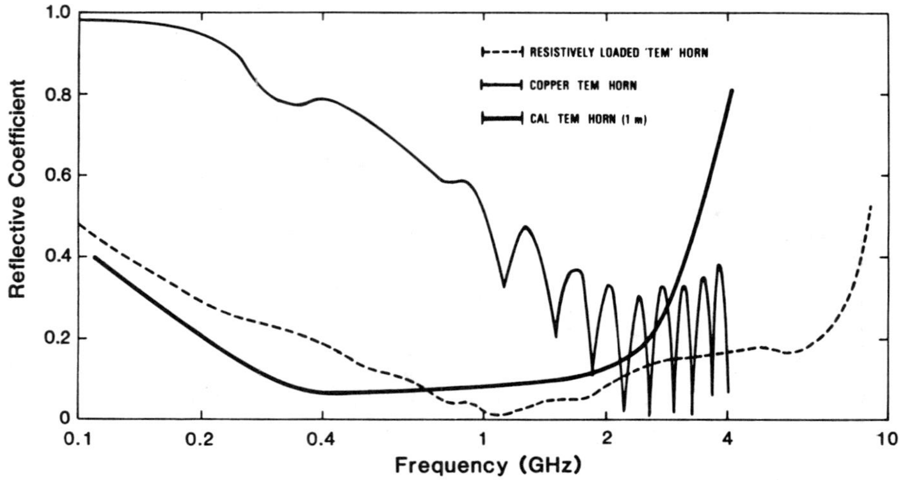

Fig. 5-20. Reflection coefficients of various TEM horns.

waves propagating between lossy plates deviate from a pure TEM mode. Since a resistively loaded "TEM" horn cannot support a pure TEM mode, the term "TEM" horn may not be truly appropriate but will nevertheless be used throughout this section.

If there are abrupt changes in the characteristics of the boundary (i.e., air to lossy TEM horn plate), the problem can be solved by finding general solutions for each of the regions and evaluating unknown constants by satisfying the boundary conditions across the plate interfaces. A transcendental matrix equation derived by matching the field across the plate interface is likely to be difficult to solve in general. Therefore, this section analyzes a one-dimensional model, such as the one that was used for the linear antenna with resistive loading, which ignores any transverse current components on the lossy plates. When the horn angle is constant and the plates taper smoothly, it is a reasonable assumption to neglect transverse components of current on the lossy plates. The shape chosen for horn plates in this study is therefore triangular. Using the one-dimensional model that neglects any transverse component of current on the lossy plate, we calculate the longitudinal component of the current numerically by using the method of moments.

Figure 5-3 shows a typical one-dimensional structure of a TEM horn modeled by a V antenna with a resistive section. Using a one-dimensional model, the current is assumed to be uniformly perpendicular to the figure and to flow in a single direction on each plate. The incident electric field, E^{inc}, associated with the current, I, and the charge density, ρ, on a one-dimensional plate are then given in terms of the usual vector and scalar potentials, $\vec{A}$ and ϕ, as given in

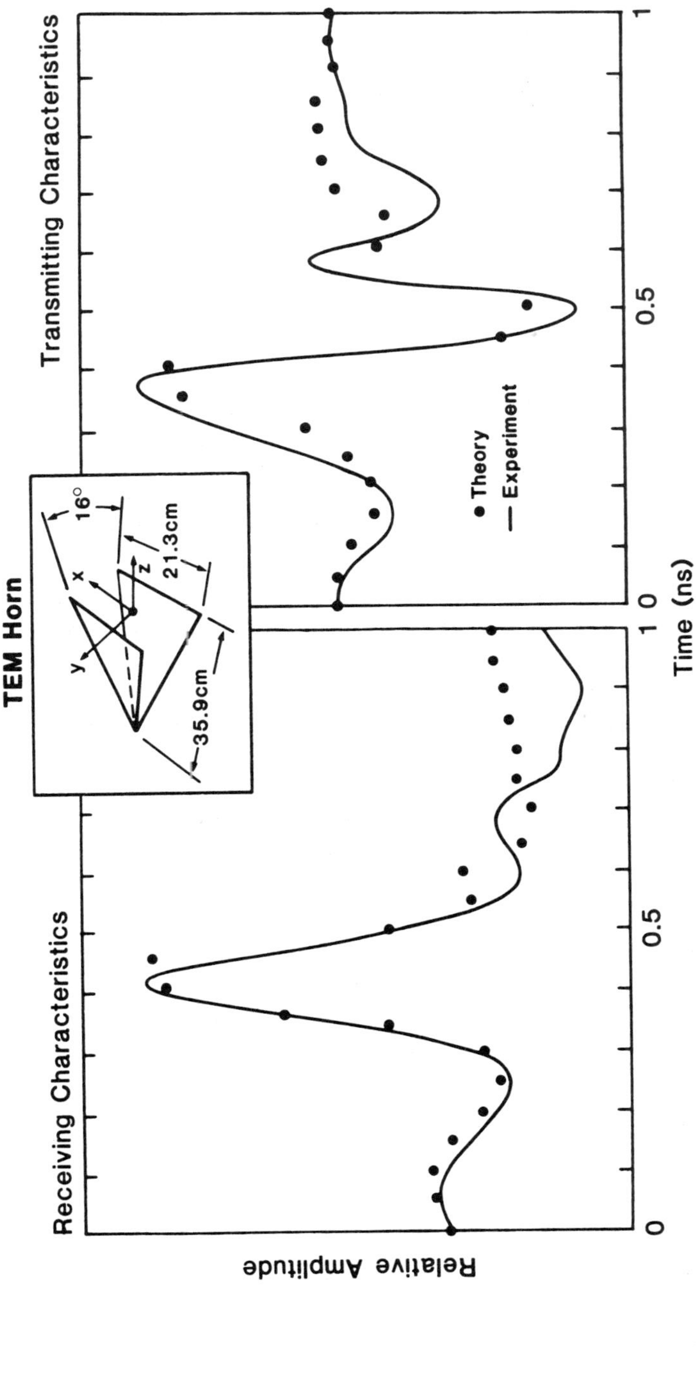

Fig. 5-21. Receiving and transmitting characteristics of the conductive TEM horn.

Eq. 5-17 and 5-18. Using pulse testing and expansion functions and applying a finite-difference approximation to compute the charge density, ρ, we can arrange Eq. 5-16 in terms of the usual impedance matrix, Z_{mn}, associated with the current and voltage matrix, I_n and V_m, in Eq. 5-22.

The resistive-loading profile $Z(s)$ in ohms per meter of a "TEM" horn is given by the expression,

$$Z(s) = \frac{Z_0}{1 - \dfrac{s}{\ell}} \ \Omega/m \tag{5-51}$$

where $0 \leqslant s \leqslant \ell$, Z_o is the resistance per unit length (Ω/m) at the driving point, and ℓ is the length of the "TEM" horn. Equation 5-51 shows that the resistive loading of the "TEM" horn should increase continuously from the value $Z_0(\Omega/m)$ at the driving point to infinity at the end of the "TEM" horn. This particular resistive-loading profile, which enables a "TEM" horn to handle extremely wide bandwidth signals without dispersion distortion, has been the profile of choice of many other researchers[9,10] as well as of the author.[11,12] In essence, there exists a critical value of Z_o for which a "TEM" horn can sustain a traveling-wave current from the driving point toward the end of the "TEM" horn. A traveling-wave structure supporting TEM-mode propagation is essential for transmitting and receiving a signal with large instantaneous bandwidth without serious pulse broadening or wide-angle side lobes.

A formal solution to the resistively loaded "TEM" horn problem can be formulated by modeling solid-plate surfaces as loaded wire grids or as loaded patches, as discussed before. Doing so extends the one-dimensional problem to a two- or three-dimensional problem by taking into account the transverse current components on the horn plates. The method of moments gives a reasonable solution for the frequency range in which each surface is smaller than a few square wavelengths. Above those frequencies, the moment method unfortunately needs many loaded wire or patch segments and therefore many unknown currents if a solid surface is to be modeled accurately. In this case, perhaps, the geometrical-theory-of-diffraction (GTD) technique would provide a more efficient solution to the problem.

In this section, a one-dimensional model for a triangularly shaped "TEM" horn with resistive loading has been developed and the method of moments used to solve for the current distribution. Physical dimensions of the "TEM" horn are shown in Fig. 5-22. If Z_o in Eq. 5-51 is chosen to be about 14 Ω/m, then a 36-cm long "TEM" horn can sustain a traveling-wave current along the antenna for the frequency range of 100 MHz to 8 GHz. When Z_o is less than 14 Ω/m, the current distribution along the antenna exhibits standing waves. On the other

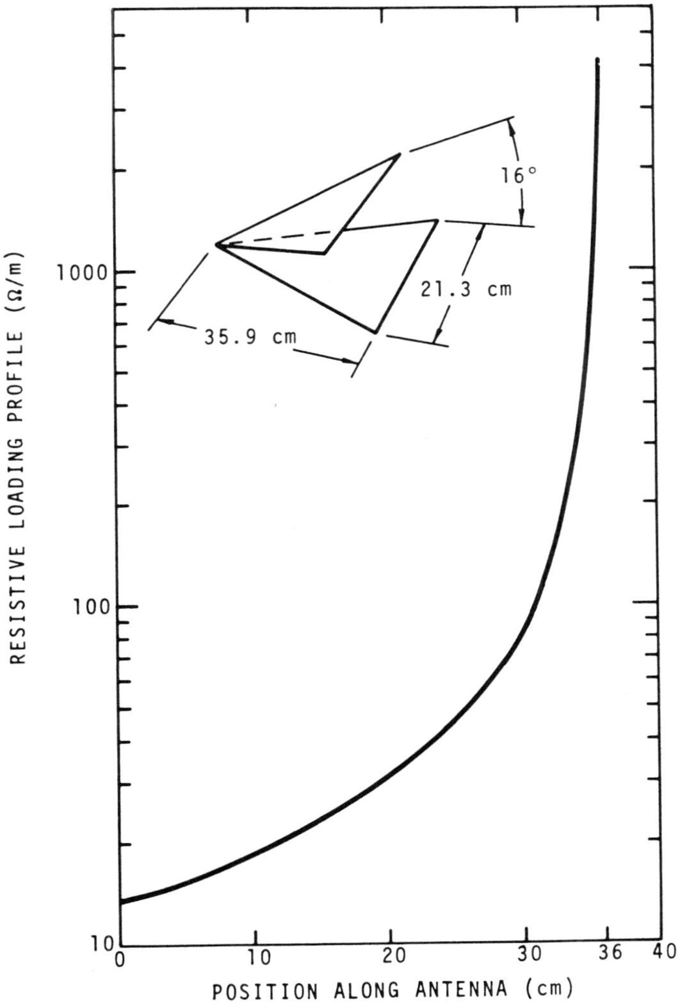

Fig. 5-22. Resistive loading profile of the TEM horn.

hand, when Z_0 is much higher than 14 Ω/m, the current distribution is over-damped, and the radiation efficiency of the antenna is expected to be very low because the antenna acts like a resistive termination. The resistive loading profile chosen for this study is also shown in Fig. 5-22.

The current distributions on the resistively loaded, 36-cm long "TEM" horn are calculated numerically using the method of moments for the frequency range of 10 MHz to 5 GHz. The current distribution at frequencies above 100 MHz is a travelling wave, whereas that below 100 MHz exhibits a standing-wave component because of a reflection from the end. Figure 5-23 shows the calculated current distribution at 1 GHz along with the experimental results. Figure 5-24 shows the driving-point resistance and reactance obtained from the method of moments along with the experimental results. The impedance of the resistively loaded "TEM" horn typically has a resistance of about 70 Ω and a capacitance of about 27 pF in a series configuration.

The measured input reflection coefficient of the 36-cm-long resistively loaded "TEM" horn typically stays below 0.5 for the frequency range from 100 MHz to 8 GHz, as shown in Fig. 5-20. For comparison, this figure also shows the reflection coefficients of a 36-cm-long conducting TEM horn and a 1-m-long CAL TEM horn and indicates that the reflection coefficient of the TEM horn has been significantly improved through resistive loading and that the reflection coefficient of the 36-cm-long resistively loaded "TEM" horn is comparable to that of a 1-m-long CAL TEM horn.

Once the current distribution of the resistively loaded "TEM" horn is determined, the effective length, $h_e(f)$, of the antenna can be calculated from Eq. 5-28. Using the effective length and the driving-point impedance of the antenna,

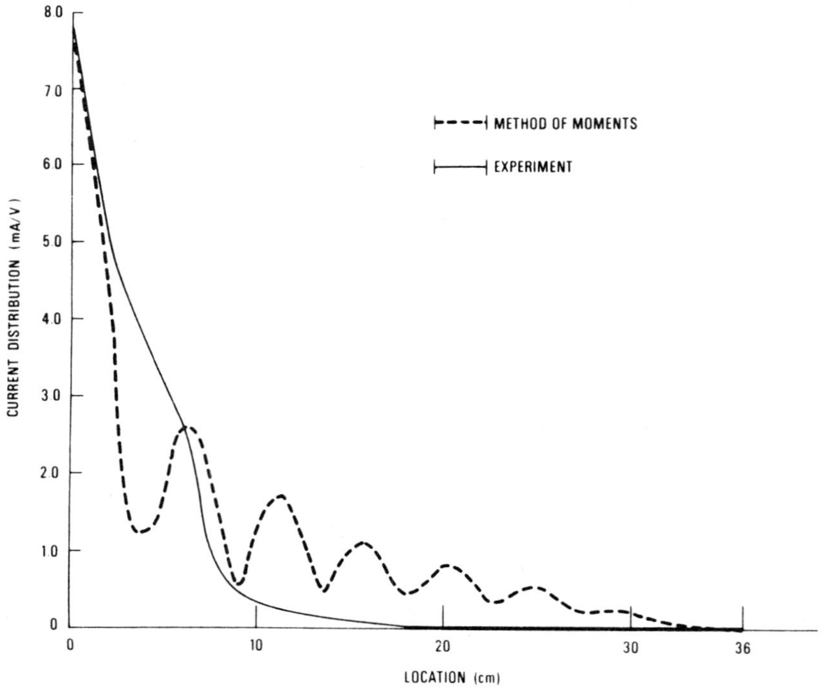

Fig. 5-23. Current distribution of the resistively loaded TEM horn.

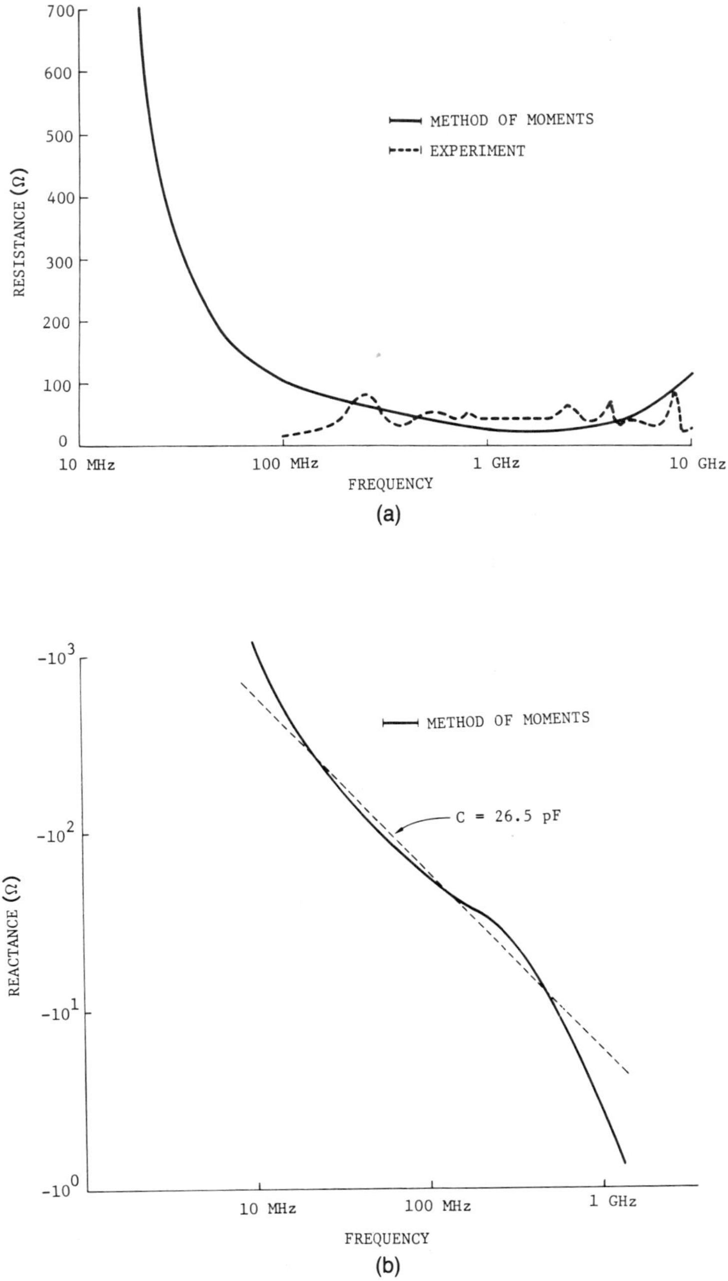

Fig. 5-24. Driving point impedance of the resistively loaded TEM horn: (a) resistance, and (b) reactance.

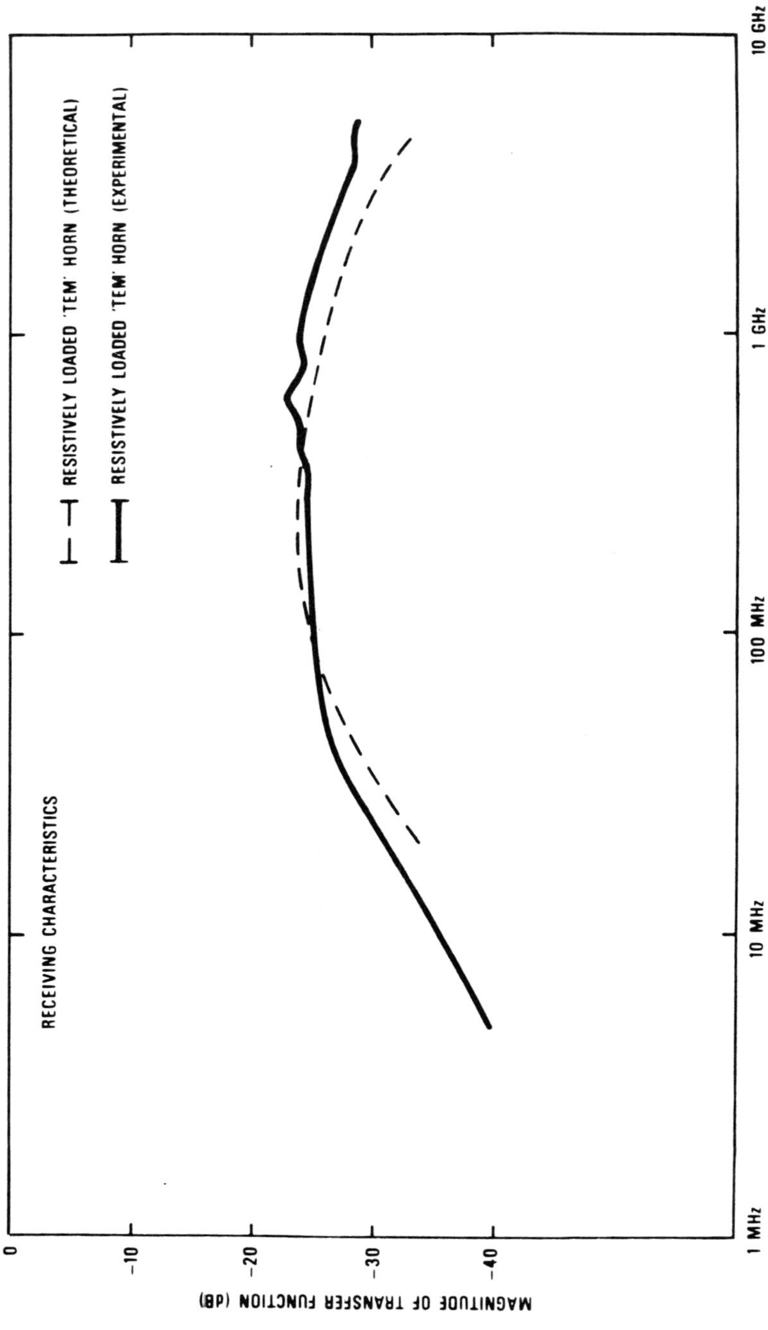

Fig. 5-25. Receiving transfer function of the resistively loaded TEM horn.

we calculate the frequency-domain representation of the receiving transfer function of the antenna using Eqs. 5-2 and 5-3. The receiving transfer function of the resistively loaded "TEM" horn is shown in Fig. 5-25, along with experimental results.

After obtaining the frequency-domain solution, transient fields for a known input waveform can be found using the FFT technique. Figure 5-26 demonstrates good agreement between the theoretical and experimental transient characteristics of the receiving and transmitting responses for the resistively loaded "TEM" horn.

5.3. MAGNETIC-FIELD SENSOR

In this section, we will discuss the receiving and transmitting characteristics of a loop antenna with uniform resistive loading.

Loop with Uniform Loading

Figure 5-27 shows the geometry and coordinate system of a loop antenna lying in the plane $z = 0$ and excited at $\phi = 0$ with a delta-function voltage source, V_o. Let us assume that a linearly polarized plane wave E^i impinges on the loop. The loop current $I(\phi)$ satisfies the following integral equation:[22]

$$V_0\delta(\phi) + bE^i_\phi(b,\phi) = \frac{j\zeta_0}{4\pi}\int_{-\pi}^{\pi} L(\phi - \phi')I(\phi')d\phi' \qquad (5\text{-}52)$$

where $E^i_\phi(b,\phi)$ is the ϕ component of the incident field, b is the loop radius, and ζ_0 is the free-space impedance. The kernel in Eq. 5-52 may be represented by a Fourier series as

$$L(\phi - \phi') = \sum_{-\infty}^{\infty} a_n e^{-jn(\phi-\phi')} \qquad (5\text{-}53)$$

with

$$a_n = a_{-n} = \frac{kb}{2}(N_{n+1} + N_{n-1}) - \frac{n^2}{kb}N_n \qquad (5\text{-}54)$$

$$N_n = \frac{b}{4\pi^2}\int_{-\pi}^{\pi} d\alpha \int_{-\pi}^{\pi} \frac{e^{jn(\phi-\phi')}e^{-jkr}}{r}\,d\phi' \qquad (5\text{-}55)$$

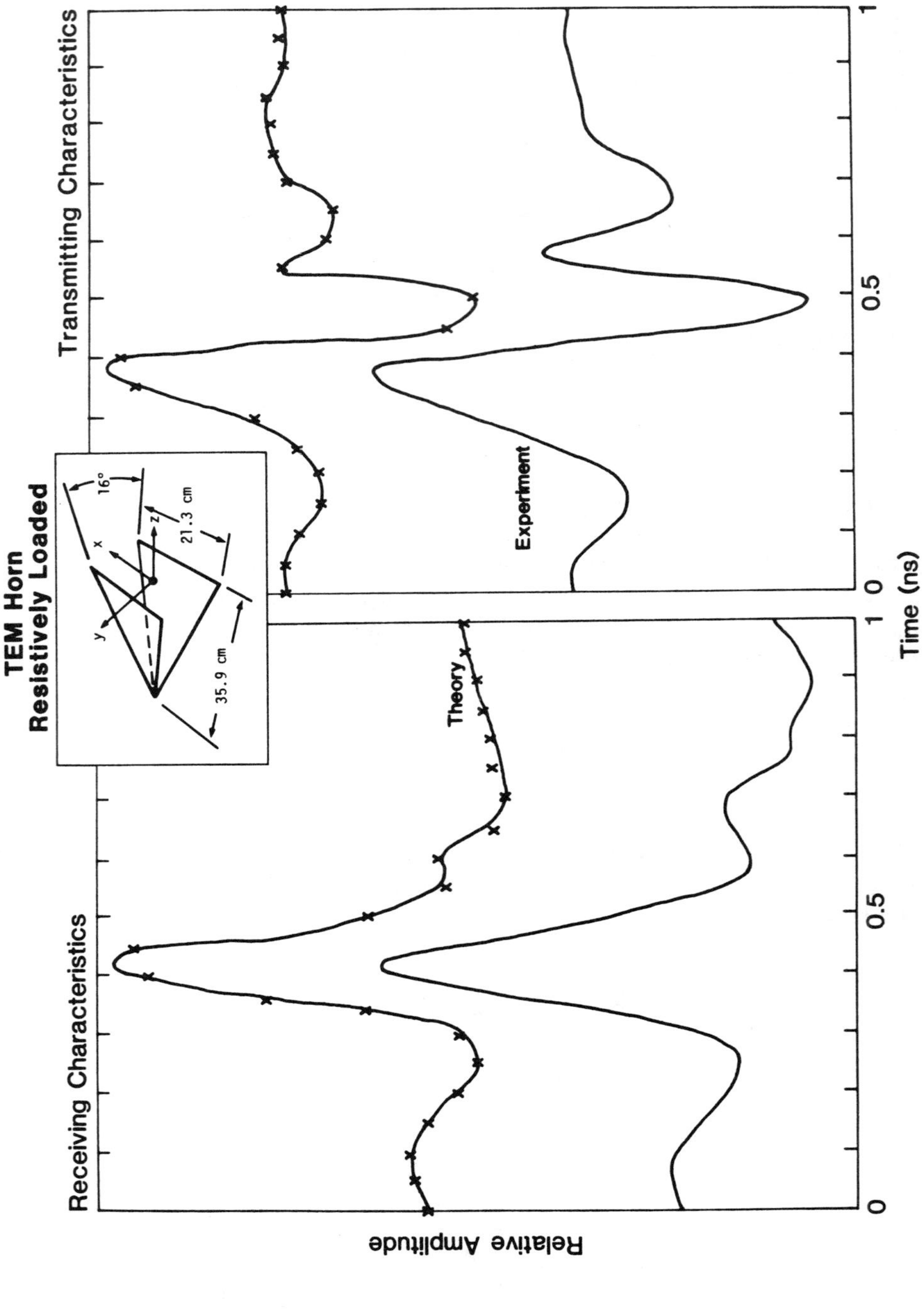

Fig. 5-26. Receiving and transmitting characteristics of the resistively loaded TEM horn.

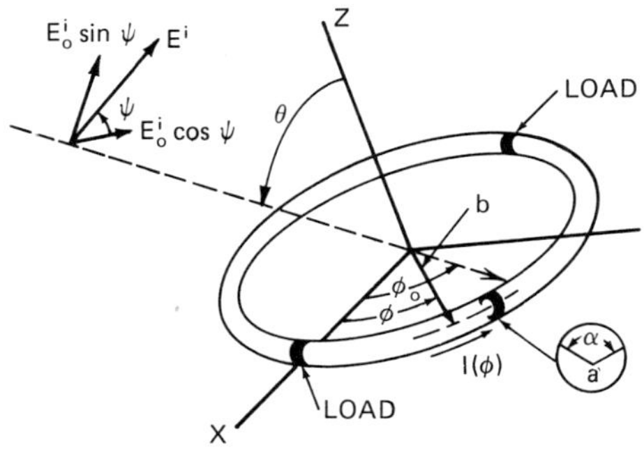

Fig. 5-27. Configuration of a loop antenna.

and

$$r = \sqrt{4b^2 sin^2[(\phi - \phi')/2] + 4a^2\ sin^2(\alpha/2)} \tag{5-56}$$

If $a < b$ and $ka < 1$, it has been shown[22] that N_n can be represented in terms of integrals of Bessel and Lommel-Weber functions,[23] as follows:

$$N_0 = \frac{1}{\pi}\ \ell n\ \frac{8b}{a} - \frac{1}{2}\left[\int_0^{2kb} \Omega_0(x)dx + j\int_0^{2kb} J_0(x)dx\right] \tag{5-57}$$

and, for $n \geqslant 1$,

$$N_n = N_{-n} = \frac{1}{\pi}\left[K_0\left(\frac{na}{b}\right) I_0\left(\frac{na}{b}\right) + C_n\right] - \frac{1}{2}\int_0^{2kb} [\Omega_{2n}(x) + jJ_{2n}(x)]\ dx \tag{5-58}$$

where

$$C_n = \ell n(4n) + \gamma - 2\sum_{m=0}^{n-1} \frac{1}{2m + 1} \tag{5-59}$$

$K_o(x)$ is the modified Bessel function of the second kind, $I_o(x)$ is the modified Bessel function of the first kind, $\Omega_n(x)$ is the Lommel-Weber function, $J_n(x)$ is the Bessel function, and $\gamma = 0.5772$ is Euler's constant.

The loop current $I(\phi)$ can be expanded in the Fourier series,

$$I(\phi) = \sum_{-\infty}^{\infty} I_n e^{-jn\phi} \tag{5-60}$$

and the electric field $E^i_\phi(b,\phi)$ in the Fourier series,

$$E^i_\phi(b,\phi) = E^i_0 \sum_{-\infty}^{\infty} f_n e^{-jn\phi} \tag{5-61}$$

According to Fourier's theorem,

$$f_n = \frac{1}{2\pi E^i_0} \int_{-\pi}^{\pi} E^i_\phi(b,\phi) e^{jn\phi} d\phi \tag{5-62}$$

It may be seen by examining Fig. 5-28 that the ϕ component of the incident field at ϕ on the loop is

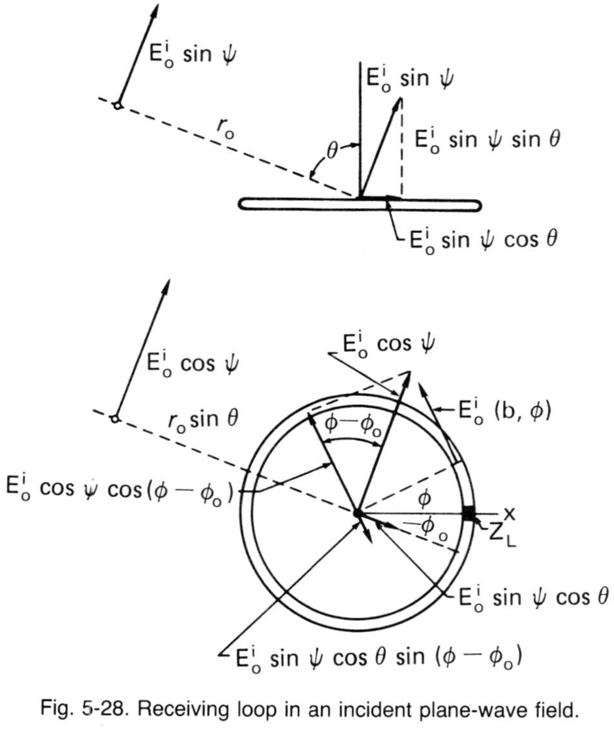

Fig. 5-28. Receiving loop in an incident plane-wave field.

$$E^i_\phi(b,\phi) = E^i_0 \,[\cos\psi \cos(\phi - \phi_0) - \sin\psi \sin(\phi - \phi_0) \cos\theta] e^{jkb\cos(\phi - \phi_0)\sin\theta} \tag{5-63}$$

Using the integral representation of the Bessel function,

$$j_n(z) = \frac{j^{-n}}{2\pi} \int_0^{2\pi} e^{jz\cos\phi}\, e^{jn\phi} d\phi \tag{5-64}$$

we obtain

$$f_n = j^{n-1} \cos\psi\, e^{-jn\phi_0} J'_n(kb \sin\theta) + j^n \sin\psi \cos\theta\, e^{-jn\phi_0} \frac{nJ_n(kb\sin\theta)}{kb\sin\theta} \tag{5-65}$$

Using the orthogonality property of the function, $e^{-jn\phi}$, we obtain from Eqs. 5-52, 5-53, 5-60, and 5-61, the Fourier coefficient of the current,

$$I_n = -\frac{j}{\pi\zeta_0}\,\frac{V_0 + 2\pi b E^i_0 f_n}{a_n} \tag{5-66}$$

By setting $E^i_0 = 0$ in Eq. 5-63, the admittance of the loop antenna is given by

$$Y = -\frac{j}{\pi\zeta_0}\sum_{-\infty}^{\infty}\frac{1}{a_n} = -\frac{j}{\pi\zeta_0}\left(\frac{1}{a_0} + 2\sum_{1}^{\infty}\frac{1}{a_n}\right) \tag{5-67}$$

Assuming that currents of higher order than $n = 1$ in Eq. 5-66 are negligible because of the small size of the loop, we can define the admittance for the magnetic-loop current ($n = 0$) as

$$Y_0 = -\frac{j}{\pi\zeta_0 a_0} \tag{5-68}$$

and the admittance for the electric-dipole current ($n = 1$) as

$$Y_1 = -\frac{j2}{\pi\zeta_0 a_1} \tag{5-69}$$

Let us consider a small loop with lumped resistive loads Z_m at ϕ_m in an incident plane-wave field, as shown in Fig. 5-27. Since the total current is the sum of the currents maintained by the electromagnetic forces at ϕ_m and by the incident field E_0^i,

$$I(\phi) = 2\pi b E_0^i u(\phi) - I(\phi_0)Z_0 v(\phi_0,\phi) - I(\phi_1)Z_1 v(\phi_1,\phi) \quad \ldots -I(\phi_{m-1})Z_{m-1}\, v(\phi_{m-1},\phi) \tag{5-70}$$

where

$$u(\phi) = -\frac{j}{\pi\zeta_0}\left(\frac{f_0}{a_0} + \frac{2f_1\cos\phi}{a_1}\right) \tag{5-71}$$

and

$$v(\phi_m,\phi) = -\frac{j}{\pi\zeta_0}\left(\frac{1}{a_0} + \frac{2\cos\phi_m\cos\phi}{a_1}\right) \tag{5-72}$$

Here it is assumed that currents of higher order than $n = 1$ in Eqs. 5-71 and 5-72 are negligible because of the loop's small size. By setting $\phi_m = 2\pi m/M$ ($m = 0, 1, \ldots, M - 1$), the currents $I(\phi_m)$ at ϕ_m can be determined by solving an $M \times M$ set of simultaneous equations. That is,

$$I(\phi_0)\,[1 + Z_0 v(\phi_0,\phi_0)] + I(\phi_1)Z_1 v(\phi_1,\phi_0) + \ldots = 2\pi b E_0^i u(\phi_0) \tag{5-73}$$

$$I(\phi_0)Z_0 v(\phi_0, \phi_1) + I(\phi_1)\,[1 + Z_1 v(\phi_1\phi_1)] + \ldots = 2\pi b E_0^i u(\phi_1) \tag{5-74}$$

$$\ldots$$

$$I(\phi_0)Z_0 v(\phi_0, \phi_{m-1}) + \ldots \quad I(\phi_{m-1})\,[1 + Z_{m-1} v(\phi_{m-1}, \phi_{m-1})] = 2\pi b E_0^i u(\phi_{m-1}) \tag{5-75}$$

We can rewrite these in matrix form as

$$[A]\,[I] = [B] \tag{5-76}$$

where

$$[A] = \begin{bmatrix} 1 + Z_0 v(\phi_0, \phi_0),\ Z_1 v(\phi_1, \phi_0),\ \ldots \\ Z_0 v(\phi_0, \phi_1),\ 1 + Z_1 v(\phi_1, \phi_1),\ \ldots \\ \ldots \\ Z_0 v(\phi_0, \phi_{m-1}),\ \ldots,\ 1 + Z_{m-1} v(\phi_{m-1}, \phi_{m-1}) \end{bmatrix} \tag{5-77}$$

$$[I] = \begin{bmatrix} I(\phi_0) \\ I(\phi_1) \\ \cdot \\ \cdot \\ \cdot \\ I(\phi_{m-1}) \end{bmatrix} \tag{5-78}$$

$$[B] = 2\pi b E_0^i \begin{bmatrix} u(\phi_0) \\ u(\phi_1) \\ \cdot \\ \cdot \\ \cdot \\ u(\phi_{m-1}) \end{bmatrix} \tag{5-79}$$

The current $I(\phi_m)$ at ϕ_m can be obtained by solving the matrix equation,

$$[I] = [A]^{-1}\,[B] \tag{5-80}$$

A wire loop antenna with a loop radius of 0.05 m and wire radius of 1.8 mm is examined in order to understand the effect of a uniform resistive loading in a loop antenna. Figure 5-29 shows the frequency-domain representation of the receiving transfer function of a resistively loaded loop antenna. The total impedances around the loop are chosen to be 800 Ω (i.e., 2.55 kΩ/m) and 4 kΩ (i.e., 12.7 kΩ/m). For comparison, the receiving transfer function of a conductive

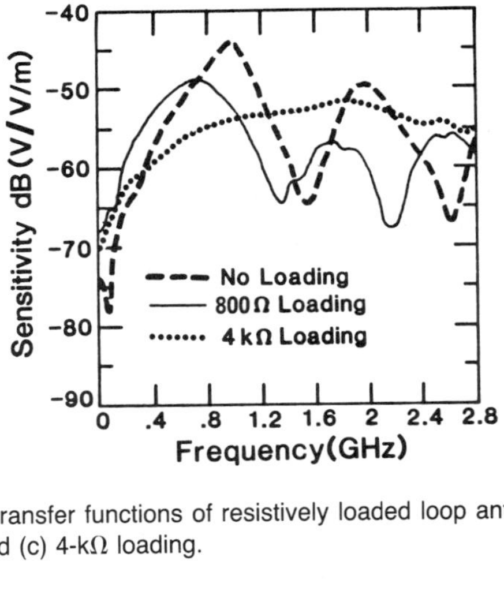

Fig. 5-29. Receiving transfer functions of resistively loaded loop antennas: (a) no loading, (b) 800-Ω loading, and (c) 4-kΩ loading.

loop antenna of the same size without resistive loading is also shown in this figure. Obviously, it is possible to achieve the broadband response of a loop antenna by uniform resistive loading. It has been noted, however, that the resistive loading reduces drastically the sensitivity of the magnetic response of a loop antenna, whereas that of the electric response is not much affected.[24,25] For example, it has been found that the response of a loop antenna with 4-kΩ resistive loading is due predominantly to the electric field and only slightly to the magnetic field, causing the loop antenna to behave as a resistively loaded dipole. Time-domain responses of loop antennas with 0, 800-Ω, and 4-kΩ resistive loading are shown in Fig. 5-30.

The development of the singularity expansion method (SEM)[26] permits one to describe the distributed parameter problems of electromagnetics by the same pole-zero representations used in the analysis and synthesis of lumped circuits and control systems. The SEM representation has been developed here for an investigation of analysis and synthesis problems for a uniformly loaded loop antenna.[27]

In the theoretical development given here, impedance loading is provided by lumped, discrete elements. Of course, as the number of such elements increases, the effect is to approximate uniform resistive loading. We now consider the effect of uniform loading with distributed impedances per unit length. Including the electromotive forces equivalent to these distributed impedances, and omitting the incident field contribution, the integral equation Eq. 5-52 becomes

$$\frac{j\zeta_0}{4\pi}\int_{-\pi}^{\pi} L(\phi - \phi')I(\phi')d\phi' + \frac{Z_L(s)}{2\pi}I(\phi) = V_0(s)\delta(\phi) \qquad (5\text{-}81)$$

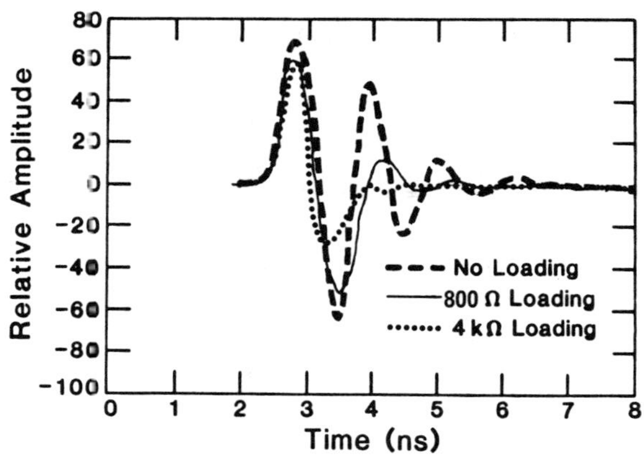

Fig. 5-30. Receiving characteristics of resistively loaded loop antennas.

where $Z_L(s)$ is the total impedance around the loop, i.e., $Z_L(s)/2\pi b$ is the impedance per unit length, where b is the loop radius.

Using the Fourier series expansions of the kernel given in Eq. 5-53 and the current given in Eq. 5-60, the loop admittance for the uniformly loaded loop is given by

$$Y_n = \frac{-j}{\pi\zeta_0}\left[\frac{1}{a_n(s) - jZ_L(s)/\pi\zeta_0}\right] \tag{5-82}$$

which has poles at frequencies, $s = s_{ni}$, such that

$$a_n(s_{ni}) = jZ_L(s_{ni})/\pi\zeta_0 \tag{5-83}$$

Using a partial fraction expansion for the admittance of a uniformly loaded loop,

$$\frac{1}{a_n(s) - jZ_L(s)/\pi\zeta_0} = \sum_i \frac{R_{ni}}{s - s_{ni}} \tag{5-84}$$

where the residues R_{ni} can be determined by

$$R_{ni} = \frac{1}{\left.\dfrac{d}{ds}(a_n(s) - jZ_L(s)/\pi\zeta_0)\right|_{s = s_{ni}}} \tag{5-85}$$

the SEM expansion of the current is then given by

$$I(s) = -\frac{j}{\pi\zeta_0}\sum_{n=-\infty}^{\infty}\sum_i \frac{R_{ni}e^{-jn\phi}}{s - s_{ni}} V_0(s) \tag{5-86}$$

Taking the inverse Laplace transform of Eq. 5-86 term by term, the time-domain expression for the uniformly loaded loop current is given by

$$i(t) = -\frac{j}{\pi\zeta_0}\sum_{n=-\infty}^{\infty}\left[\left(\sum_i R_{ni}e^{s_{ni}t}\right) * V_0(t)\right]e^{-jn\phi} \tag{5-87}$$

where the asterisk, $*$, denotes the convolution operation. Figure 5-31 shows the pole trajectories for resistive loading for the primary poles of a loop antenna with $\Omega = 2\ell n(2\pi b/a) = 10$. A procedure for synthesis of the loop response relies on using the resistive loading to shift the poles to new positions in order to obtain the desired response.

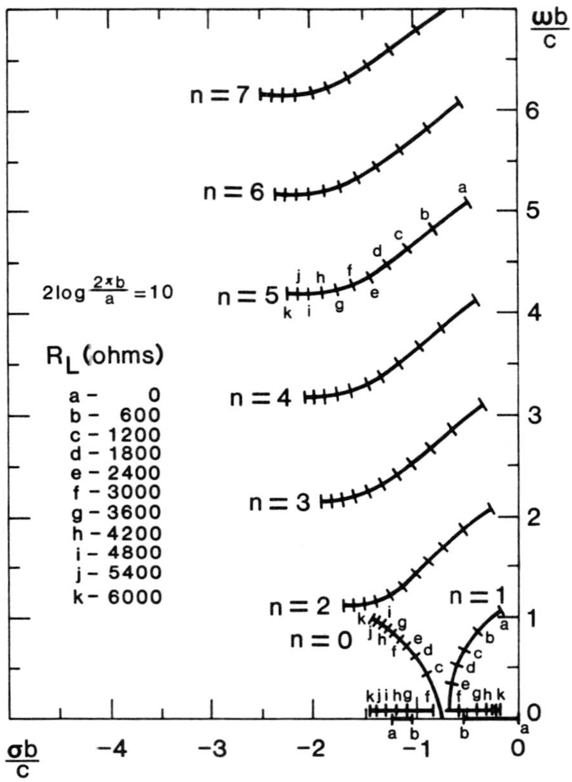

Fig. 5-31. Trajectories of the primary poles of the resistively loaded loop as a function of resistive loading.

5.4 CONCLUSIONS

This chapter has discussed various sensors and radiators commonly used for time-domain antenna measurements. The sensors and radiators discussed here are passive and analog devices that convert the electromagnetic quantity of interest to a voltage or current at their terminal ports. Moreover, they are primary standards in the sense that their transfer functions can not only be calculated from their geometries but also be made flat (constant) across a wide frequency range. One of the major requirements for these sensors and radiators is that the electromagnetic far field, transmitted or received, is a replica or high-fidelity derivative of the original pulse.* Note that the transmitting transfer function of

*Signal processing can relax this requirement—to some extent at least. (See Chapters 12 and 13.)

an antenna is proportional to the time derivative of the receiving transfer function of the same antenna, a relationship that follows from the reciprocity theorem.

For electric-field-strength measurements, linear antennas loaded nonuniformly and continuously with resistance or with both resistance and capacitance were also discussed, as were a conical antenna and an asymptotic conical antenna from the standpoint of improved characteristics. In order to achieve an improved directivity, various types of TEM horn were evaluated, e.g., a conducting TEM horn and a resistively loaded TEM horn. For magnetic-field-strength measurements, a loop antenna with uniform resistive loading was considered. These antennas are capable of measuring fast, time-varying, impulsive EM fields with minimal pulse-shape distortion caused by nonlinear amplitude or phase characteristics.

REFERENCES

1. Baum, C. E.; Breen, E. L.; Giles, J. C.; O'Neill, J.; and Sower, G. D. 1978. Sensors for electromagnetic pulse measurements both inside and away from nuclear source regions. *IEEE Trans. Antennas and Propagation* AP-26, No. 1: 22–35 and *IEEE Trans. Electromagnetic Compatibility* EMC-20, No. 1: 22–35.
2. Baum, C. E.; Breen, E. L.; Pitts, F. L.; Sower, G. D.; and Thomas, M. E. 1982. The measurement of lightning environmental parameters related to interaction with electronic systems. *IEEE Trans. Electromagnetic Compatibility* EMC-24, No. 2: 123–137.
3. Schmitt, H. J. 1960. Transients in cylindrical antennae. IEE Monograph No. 337E: 292–298.
4. Kanda, M. 1980. Transients in a resistively loaded linear antenna compared with those in a conical antenna and a TEM horn. *IEEE Trans. Antennas and Propagation* AP-28, No. 1: 132–136.
5. King, R. W. P. 1956. *Theory of linear antennas*. Cambridge, MA: Harvard University Press.
6. Kerns, D. M. 1981. Plane-wave scattering theory of antennas and antenna-antenna interactions. Nat. Bur. Stand. (U.S.) Monograph 162.
7. Harrison, Jr., C. W., and Williams, Jr., C. S. 1965. Transients in wide-angle conical antennas. *IEEE Trans. Antennas and Propagation* AP-13, No. 2: 236–246.
8. Lawton, R. A., and Ondrejka, A. R. 1978. Antennas and the associated time domain range for the measurement of impulsive fields. Nat. Bur. Stand. (U.S.) Tech. Note 1008.
9. Wu, T. T., and King, R. W. P. 1965. The cylindrical antenna with nonreflecting resistive loading. *IEEE Trans. Antennas and Propagation* AP-13, No. 3: 369–373. This paper was corrected by L. C. Shen and R. W. P. King: The cylindrical antenna with nonreflecting resistive loading. *IEEE Trans. Antennas and Propagation* AP-13, No. 6: 998.
10. Baum, C. E. 1969. Resistively loaded radiating dipole based on a transmission-line model for the antenna. Air Force Weapons Laboratory (AFWL), Electromagnetic Pulse (EMP) Sensor and Simulation Note 81.
11. Kanda, M. 1978. A relatively short cylindrical broadband antenna with tapered resistive loading for picosecond pulse measurements, *IEEE Trans. Antennas and Propagation* AP-26, No. 3: 439–447.
12. Kanda, M. 1980. The time-domain characteristics of a traveling-wave linear antenna with linear and nonlinear parallel loads. *IEEE Trans. Antennas and Propagation*. AP-28, No. 21: 267–276.
13. Altshuler, E. E. 1961. The traveling-wave linear antenna. *IEEE Trans. Antennas and Propagation* AP-9, No. 4: 324–329.

14. Harrington, R. F. 1968. *Field computation by moment methods*. New York: Macmillan.
15. Butler, C. M., and Wilton, D. M. 1975. Analysis of various numerical techniques applied to thin wire scatterers. *IEEE Trans. Antennas and Propagation* AP-23, No. 4: 534–540.
16. Wilton, D. R. 1978. *Digest of numerical and asymptotic techniques for electromagnetics and antennas*. University of New Mexico, Albuquerque.
17. Hallen, E. 1962. *Electromagnetic theory*. London: Chapman and Hall.
18. Kanda, M. 1980. The characteristics of a linear antenna with tapered resistive and capacitive loading. 1980 International IEEE/AP-S Symp. Digest, Quebec, Canada, pp. 696–699.
19. Baum, C. E. 1969. An equivalent-charge method for defining geometries of dipole antennas. Air Force Weapons Laboratory (AFWL), Electromagnetic Pulse (EMP) Sensor and Simulation Note 72.
20. Sower, G. D. 1980. Asymptotic conical dipole ACD-6A(A) sensor development. Albuquerque, NM: EG&G Report No. AG-1456.
21. Harrington, R. F. 1961. *Time-harmonic electromagnetic fields*. New York: McGraw-Hill Book Co.
22. Wu, T. T. 1962. Theory of the thin circular antenna. *J. Math. Phys.* 3:1301–1304.
23. Jahnke, E., and Emde, F. 1945. *Tables of functions*. 4th ed. New York: Dover Publications.
24. Kanda, M. 1984. An electromagnetic near-field sensor for simultaneous electric and magnetic-field measurements. *IEEE Trans. Electromagnetic Compatibility* EMC-26, No. 3: 102–110.
25. Kanda, M. 1984. Transients in a resisting loaded loop antenna. International Symposium on Electromagnetic Compatibility, Tokyo, Japan, pp. 286–290.
26. Baum, C. E. 1976. The singularity expansion method. In *Transient electromagnetic fields*, edited by L. B. Felsen. New York: Springer-Verlag.
27. Blackburn, R. F., and Wilton, D. R. 1978. Analysis and synthesis of an impedance-loaded loop antenna using the singularity expansion method. *IEEE Trans. Antennas and Propagation* AP-26, No. 1: 136–140.

6
Instrumentation for Time-Domain Measurements

J. R. Pressley and G. D. Sower

EG&G WASC, Inc.

6.0 INTRODUCTION

This chapter describes some of the instrumentation used to measure transient electrical signals in the time domain. These signals may come from any number of phenomena, either natural or man-made, which can be converted to an electrical signal by means of some sensor or transducer. A list of such phenomena can include blast and shock, electromagnetic effects (lightning, EMP), voltage and current, optical effects, ionization, radiation, or any of many other transient occurrences. The signals to be measured may be of an extremely fast nature, having transition times on the order of nano- or even picoseconds.

The instrumentation used will depend both upon the type of the signal to be measured and upon the intended use of the resultant data. The instrumentation system may therefore range from a single instrument to perhaps several rooms full of instruments making parallel measurements, with sophisticated signal processing, recording, and analysis of data.

The signal may be a single, isolated transient, such as from an underground nuclear test, or it may be in the form of a repetitive sequence of pulses. The latter case is, in general, simpler to measure and provides detection of much faster rates of signal change.

The intended use of the data may be quite simple, or it may be very complex. Perhaps all that is required is a quick look at single-pulse data such as can be seen on the CRT screen of a storage oscilloscope or on the video display of a transient digitizer. A permanent record of the data may be obtained by photographing the oscilloscope screen or by digitally plotting a digitized data record. Extensive processing of the data is sometimes required, dictating the use of transient digitizers and computers, this being particularly true in large facilities where the same type of data is obtained many times during a test program, and the same type of processing must be applied to many data sets. Inherent in this type of system is central storage of the data, both raw and processed, and perhaps access to it by a multitude of users. In such a system, many different but related phenomena may be simultaneously recorded by the data system. An example of such a system, where data system performance and reliability are critical, is an underground nuclear test.

An "ultimate" instrumentation system may then be described that performs all of the data acquisition and manipulations envisioned. Figure 6-1 shows a simplified block diagram of such a system that might also include control of the source that generates the phenomena of interest. The first row indicates one of many possible data channels in the system. The path of the analog data is indicated by single, solid arrows. The path of the digitized data is indicated by double-line arrows. Control signals are indicated by dashed arrows.

The requirements of a particular system may result in the use of a subset of the full system shown in Fig. 6-1. The simplest system that can be envisioned would consist only of a sensor, data link (e.g., cable), waveform recorder, and display. A very large system could consist of several tens of hundreds of data channels with extensive computer processing of the data.

Data systems may also be classified as either laboratory or field systems. A laboratory system is typically put together by an individual or small group and is used by those who assembled it. It will usually have several idiosyncrasies well known to the users but that might make it very difficult for an outsider to master. A field system or production system, on the other hand, must be capable of being operated by technicians or others who did not design and build it. It must, therefore, be simple and logical to operate. Furthermore, operations and maintenance manuals describing both hardware and software must be included.

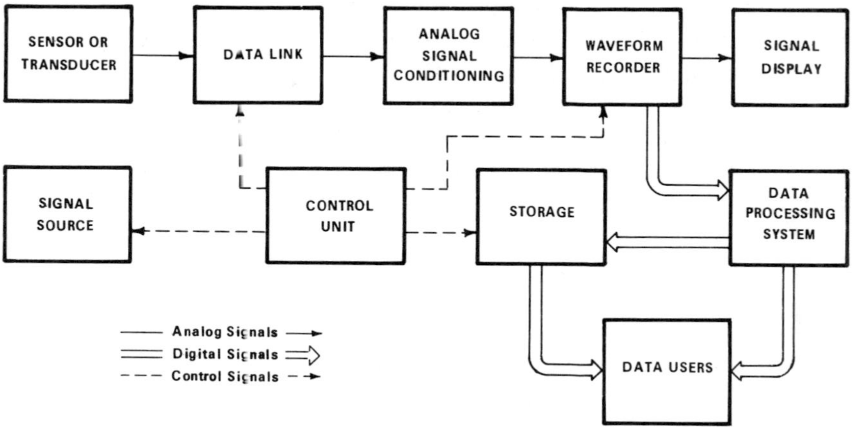

Fig. 6-1. Ultimate data instrumentation system.

The larger and more complex the system, the larger fraction of the total design effort that will be software.

The instrumentation discussed in this chapter will, in general, be the state-of-the-art of commercial hardware circa 1983. It may include some very recently developed instruments or ten-year-old equipment that has yet to be replaced by something better.

6.1 SIGNAL DETECTION

Signals from many different kinds of physical phenomena may be converted to electrical signals by means of transducers. Here, we will specifically discuss only those transducers (sensors) that measure some electromagnetic field component or electrical signal. There are four distributed quantities that are directly related by the Maxwell and constitutive equations, namely, magnetic field, voltage density, electric field, and current density. The electric signals of interest are voltage and current. An excellent overview of the theory and the particular sensors derived from that theory is contained in References 1 through 3.

This type of sensor is defined as a special type of transducer with the following properties:

1. It is an analog device that converts the electromagnetic quantity of interest to a voltage or current at its output terminals.
2. All electromagnetic quantities other than that of interest are rejected by the sensor.
3. It is a primary standard in the sense that its sensitivity for converting fields to electrical signals is determined only by its geometry; it is "calibrated by a ruler."
4. It is passive so that active circuit elements do not have to be calibrated.
5. Its transfer function is designed to be simple across a wide frequency band. This may mean volts per unit field or time derivative of field, for example, or some other simple mathematical relationship.
6. It is designed to have a specific, convenient sensitivity for its transfer function, such as a current probe with an impedance transfer of 5 ohms.
7. It is optimized for maximum bandwidth and sensitivity in a manner similar to gain-bandwidth product or in terms of energy extracted from the incident volume per sensor volume.

The nomenclature for these sensors, developed by the Air Force Weapons Laboratory, is included in their Sensor and Simulation Note publications. Comprehensive lists of references pertaining to these sensors are available.[1–3]

Electric-field and charge-density measurements are made with an electric dipole sensor. The shape of the dipole elements are calculated to give a known sensitivity, as for example, the Hollow-Spherical Dipole (HSD), Asymptotic-Conical Dipole (ACD), and Parallel-Plate Dipole (PPD). This sensitivity is characterized by an equivalent area, $\vec{A}_{eq}$, when the load impedance is small compared to the source impedance (Norton equivalent circuit), and the sensor measures the incident current density (including displacement current) by giving the output current:

$$I_o = \vec{A}_{eq} \cdot \frac{\partial}{\partial t} \vec{D} \tag{6-1}$$

When the load impedance is large compared to the source impedance (Thevenin equivalent circuit), the sensitivity is characterized by an equivalent length, $\vec{l}_{eq}$, and the sensor measures the incident electric field by giving the output voltage:

$$V_o = \vec{l}_{eq} \cdot \vec{E} \tag{6-2}$$

When the sensor is electrically small enough (dimensions much smaller than a wavelength), the equivalent length and equivalent area can be related by the dipole element capacitance, C, and the permittivity of free space, as follows:

$$\vec{A}_{eq} = \frac{C}{\varepsilon_0} \vec{l}_{eq} \tag{6-3}$$

Magnetic-field and current-density measurements are made with a magnetic dipole sensor, which usually takes the form of an extended-axis loop (a cylinder) in order to reduce the inductance and increase the bandwidth. Different methods of sampling the signals induced on the loop lead to different specific types of sensors, such as the Multi-Turn Loop (MTL), Multi-Gap Loop (MGL), and Cylindrical Moebius Loop (CML). When the inductance (source impedance) is small compared to the load impedance, the sensitivity is characterized by an equivalent area $\vec{A}_{eq}$, which measures the time-rate-of-charge of $\vec{B}$ as follows:

$$V_o = \vec{A}_{eq} \cdot \frac{\partial \vec{B}}{\partial t} \tag{6-4}$$

For large inductance, the sensitivity is characterized by an equivalent length $\vec{l}_{eq}$ which measures $\vec{H}$ from the equation,

$$I_o = \vec{l}_{eq} \cdot \vec{H} \tag{6-5}$$

The sensitivity quantities are related by the inductance and the free-space permeability, as follows:

$$\vec{A}_{eq} = \frac{L}{\mu_0} \vec{l}_{eq} \tag{6-6}$$

Currents can also be measured by a sensor that inductively couples to the magnetic field produced by the current. This inductive coupling is achieved by means of a toroidal configuration through which the incident current passes. For the case of an I-dot probe, where the inductive coupling is small, it is characterized by a mutual inductance, M, which related the sensor output voltage to the time derivative of the total current, I_t, through the center of the toroid, as follows:

$$V_o = M \frac{\partial}{\partial t} I_t$$

$$I_t = \vec{A}_{eq} \cdot \vec{J} \tag{6-7}$$

where $\vec{J}$ is the current density through a toroid of area $\vec{A}_{eq}$.

When the inductance of the toroidal cavity is large, which can be achieved by filling the cavity with a high-permeability material such as iron or a ferrite ceramic and by winding this material with many turns of wire, n, the sensitivity is characterized by the flux linkages of the primary and secondary windings, as follows:

$$I_o = \frac{1}{n} I_t \tag{6-8}$$

When this output current is driven into a standard load impedance (typically 50 ohms), the sensor is characterized by a transfer impedance, Z_T, calculated from the following:

$$V_o = Z_T I_t \tag{6-9}$$

The I-dot probes can be made to fit (1) around an existing structure, as in the Circular Parallel Mutual inductance sensor (CPM); (2) as part of the outer conductor of a coaxial cable, as in the I-dot One-Turn Insertion Unit (I1I); (3) as the outer tube of a particle-beam device or large coaxial pulsed-power source, as in the Outside Moebius Mutual inductance sensor (OMM); or (4) as an integral part of the current-carrying conductor, as in the Inside Moebius Mutual inductance sensor (IMM). For measuring currents normal to a conducting surface, such as photoelectric or Compton currents, there is the Flush Moebius Mutual inductance sensor (FMM).

Voltage measurements are sometimes made between conductors of a system for transient signals. This is usually done with probes that have input impedance high enough to be similar to electric field sensors. Care must be taken in making and interpreting voltage measurements because, at high frequencies, conducting surfaces may not be considered as equipotential surfaces; significant voltage differences may exist (and be measured) between different points on the same conductor.

6.2. SIGNAL TRANSMISSION

Once the selected sensor or probe has detected and converted the desired measurement parameter into an electrical signal, this signal must be transmitted to the recording instrumentation. Common transmission methods include shielded cables and fiber-optic, radio, and microwave telemetry links.

Where test-volume field strengths are low, shielded cables—such as shielded coaxial or twin-axial transmission lines—provide an adequate means of signal transmission, provided they do not cause any perturbation of the simulation environment or the test object responses. If the cable shield does not provide adequate shielding of the inner conductor(s) or if the experiment ground system geometry and/or power system permits any shield currents to flow, however, induced noise may mask the signal of interest.[4] To provide adequate isolation between the recording instrumentation and the test volume, telemetry links operating through free space (radio, microwave, or laser) or through dielectric waveguides (fiber optic or microwave) may be utilized.[5]

Line-of-sight radio, microwave, or laser links have the disadvantage of requiring transmission and reception antennas so placed as to provide reliable communications. The antenna and transmission lines themselves are subject to interference from the signal source. The center frequency of a radio-type link must be sufficiently high in frequency to be unaffected by high-frequency components of the signal source and to support the required wide bandwidth to pass all signals of interest.

The use of a microwave transmitter operating into a dielectric waveguide overcomes these particular limitations. Dielectric waveguides however, are relatively large and bulky, must be protected from the elements (moisture and dust), and must not interact with nearby surfaces. Microwave links operating in this manner sometimes require daily tuning because of the environmentally induced degradation in the waveguide and microwave beam-launch polarity sensitivity. It is thus imperative that a stable calibration source be included in the transmitter package.

The microwave, telemetry-link transmitter package uses a rechargeable battery pack. The internal microwave emitter requires a good deal of power, necessitating a large battery size. The transmitter must also be adequately shielded to prevent unwanted signal coupling between the transmitter electronics and the signal source and the test object.

The past decade has seen the introduction of wideband, fiber-optic telemetry links operating over small, lightweight, dielectric fiber-optic cables. Both light-emitting diodes (LED) and solid-state laser diodes are in use as emitters in fiber-optic (FO) transmitters. Today's LEDs and laser diodes require only a modest amount of bias current—typically 10–200 milliamperes, which reduces battery requirements.

Fiber-optics cables are easily handled with performance unaffected by nearby obstructions, moisture, etc. Although the use of FO telemetry links has eliminated most of the problems associated with microwave telemetry links, they have introduced a whole new set of problems, most of which are associated with the proper handling and maintenance of the FO cables and connectors.[6]

Step-index fibers are typically subject to high-attenuation and low-bandwidth characteristics and are best used for short and lower frequency links. Plastic-clad silica (PCS) fibers are also temperature sensitive, with the cladding becoming denser at low temperatures (approximately $-30°C$) to the point of no longer performing as a waveguide. At high temperatures (approximately $+55°C$), they turn soft and become fragile. The PCS fibers are also subject to pistoning of the core, which is a breaking of the cladding-to-core bond. They are, however, quite resilient and, when combined with a loose buffer jacket to prevent crushing of the fiber and strengthened with a tough flexible outer jacket, constitute a quite rugged cable. The PCS fibers have also been found to be the most radiation-resistant FO cables.[7] Glass-on-glass, step-index fibers have less temperature dependence and are less prone to pistoning. The glass-on-glass fiber is much more easily broken however, and most glass-fiber cables will snap when bent sharply.

Most step-index fibers have a large acceptance cone and a large cross-sectional area (100 to 200 microns) and are thus commonly used with LEDs, which have a large launch angle. Graded-index fibers have much lower attenuation characteristics and much wider bandwidth, but they also have a much smaller acceptance cone and smaller cross-sectional area (approximately 50 microns), decreasing the emitter-to-fiber optical energy transfer efficiency.

The primary problem with all types of FO cables used for test measurements is in making connections. PCS fibers are difficult to glue and require special preparation to prevent a weak bond between the cladding and the connector. Glass fibers usually must be etched to provide a proper surface for glueing. Plastic connectors tend to wear and are easily distorted, but when reproducibility is not a major concern, as in digital transmission, they are inexpensive and easily

(a)

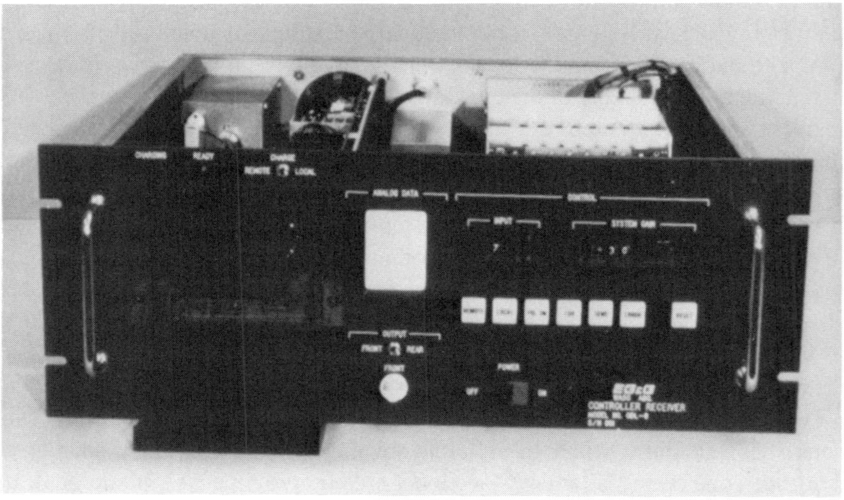

(b)

Fig. 6-2. Optical data link: (a) ODL-6 transmitter, and (b) ODL-6 controller.

terminated. For the smaller, graded-index fibers, various schemes are utilized to guarantee reliable and accurate alignment of fiber ends. Special care must be given to the termination ends on all FO cables, for example, covering the connector ends when not in use, and inspecting connector ends prior to mating.

Fiber-optic links add signal degradation between the sensor output and the recording-instrumentation input, as do the other types of data links. This degradation comes from noise added to the signal, bandwidth limitations, and band-pass ripple or variation of gain with frequency. Many of the data links available today include selectable attenuators (either manually or remotely controlled) to extend the dynamic range in the same manner as selectable vertical amplifier gains do in oscilloscopes. They extend the usable dynamic range from 30 to 40 dB to over 100 dB. Many data links also include built-in calibration sources as a means of accurately determining both pulse response and system gain. Some data links also include built-in baluns for interfacing to differential sensors and integrators for converting the time-derivative output of field sensors to the actual time variation of the signal being measured. Integrating before data transmission is advisable because of the limited dynamic range of the data link.

Remotely controlled data links allow the user to turn on the transmitter only when needed, thus extending the battery life. Utilizing a computer interface with the data-link control provides automatic setup and logging of data-link settings and also allows automatic calibrations to be performed. Software can then automatically correct later test data for attenuation and link performance.

Figure 6-2 shows a typical EMP-hardened, wideband, fiber-optic telemetry system. The transmitter unit in this particular system (a) provides eight separate 50-ohm inputs, any one of which may be selected via remote control. Also included are an internal bipolar-pulse-calibration generator, a TDR option for checking current-probe integrity, an external rechargeable battery pack, and local microprocessor control. The microprocessor-driven receiver/control unit (b) provides a single-ended, 50-ohm, unbalanced output; front-panel controls; verification displays; either an RS-232 or IEEE-488 computer interface for remotely controlling the transmitter; and a battery charger. The FO-link-signal bandwidth is 1 kHz to 130 MHz (half-power points), with a dynamic range of 30 dB (measured as the ratio of a 1-dB compressed, unipolar pulse to tangential noise) or 42 dB (measured as the ratio of a 1-dB compressed, peak-to-peak signal to rms noise). Tangential noise is defined as the amplitude of the envelope of the noise and is about 7 dB greater than the rms noise value.

6.3 SIGNAL CONDITIONERS

Special signal conditioning may be required prior to sending the detected signal to the recording instrumentation. Derivative sensor outputs are best integrated prior to digitization because of the limited dynamic range of both telemetry links and recording instrumentation. (Typical FO links and oscilloscopes have a dynamic range of approximately 32 dB and 8-bit digitizers have a dynamic range of 48 dB.) In addition, a time-derivative output may have extremely fast rise-times (<1 ns) whereas FO links may have risetimes of approximately 3 ns and recording instrumentation, of 1–3 ns.

Differential outputs of free-field sensors and differential voltage probes are converted to single-ended signals for input to telemetry links or to oscilloscopes and digitizers. Baluns are used to match impedances properly and to provide a single-ended unbalanced output. These may be mounted either internally or external to the telemetry links.

To extend the useful dynamic range of telemetry links, attenuators are often built into the link front ends. Preamplifiers are also often needed for the detection of extremely small signals. Such amplifiers are available both as stand-alone units and as integral components in FO links.

When long runs of cable are used for signal transmission, some sort of cable compensation is required because of the frequency-response degradation from

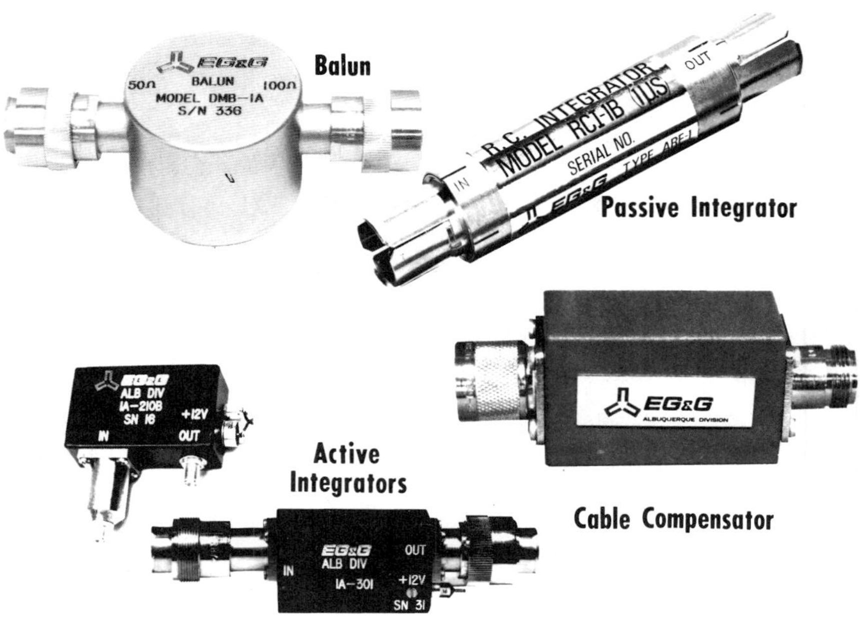

Fig. 6-3. Signal conditioners.

skin-effect losses. Standard three-section bridged-T compensators are available for this purpose.[8] They must be tailored for each particular cable type and length or customized for a given cable installation. Figure 6-3 depicts a number of typical signal conditioners.

Once the signal has arrived at the recording instrumentation, it may need to be sent to several instruments simultaneously, often to allow the recording of a single-shot event at several different vertical and/or time-base settings. Passive power-splitters may be used, but these introduce attenuation and do not have completely independent outputs. For these reasons, an active signal divider has been developed containing multiple distribution amplifiers.[9]

As shown in Fig. 6-4, the function of the distribution amplifier is to provide up to four outputs identical to the input signal regardless of the loads on the other three outputs. It isolates each output from the other three with impedances matched so that the impedance looking into any stage in either direction is 50 ohms. Symmetrical overload protection is provided in the input circuitry for each channel by Schottky diodes.

Although not a signal conditioner, another piece of equipment often utilized in a transient recording system is the Programmable Trigger Delay (PTD) unit. The PTD provides multiple, independently adjustable time-delay trigger outputs from a single trigger input. Various trigger delays are required to adjust the signal presentation on recording instrumentation, either for close examination of different positions of the signal (in time) or simply to compensate for long cable runs.

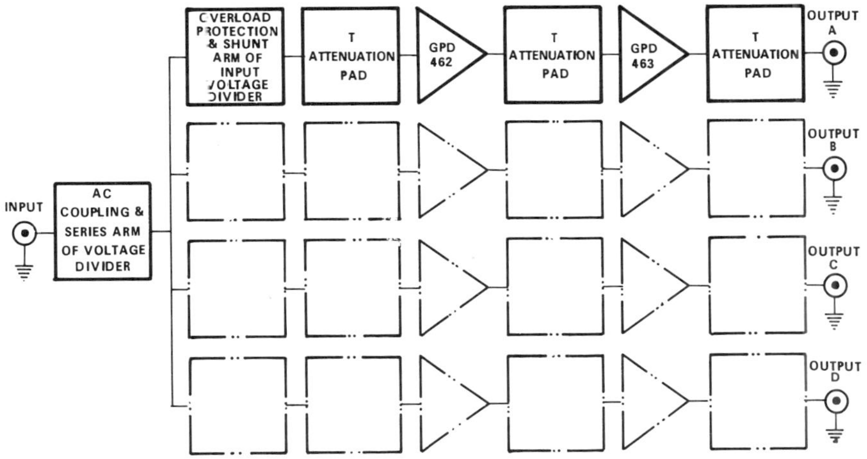

Fig. 6-4. Active signal-divider distribution amplifier.

6.4 WAVEFORM-CAPTURING INSTRUMENTATION

The next step in the measurement sequence is signal acquisition, which is accomplished with either an oscilloscope or some sort of analog-to-digital converter (ADC).[10] Digitizing of an acquired signal is required for storage and further processing. Although digitizing may not be required for many quick-look applications, some means of signal storage is usually desired and is mandatory for single-event transient signals.

A high-speed oscilloscope equipped with a camera and very-high-speed film will provide this capability but requires good operator technique to capture every desired signal successfully. A high-speed storage oscilloscope can eliminate much uncertainty once the proper intensity and persistence settings have been found for a particular combination of sweep speed and amplifier gain. A photo may then be taken of the stored trace with low-speed film.

For repetitive signals, digitizing oscilloscopes may be used. These may be either the sampling-oscilloscope variety, where the digitization is inherent in the signal capture but the display is usually analog, or the type where the signal is processed by conventional oscilloscope techniques, then digitized using analog-to-digital circuitry and stored and displayed as digital signals. The first technique is exemplified by the Tektronix 7S11/7T11 sampling plug-ins for the 7000 series of oscilloscope mainframes, and the second by the Tektronix 7854 mainframe which accepts most of the 7000 series plug-ins. The two techniques can even be combined by utilizing the 7S12 TDR plug-in as a sampling input in the 7854 mainframe (the 7T11 time-base is not compatible with the 7854 when the *stored* mode is used).

A wide variety of instruments are available for signal digitization. Table 6-1 contains a listing of ADC techniques available today. Most waveform-capturing instruments on the market utilize either successive approximation or flash converter ADCs. The others use scan-converter-tube technology, as in the Tektronix

Table 6-1. Analog-to-Digital Conversion Techniques.

ANALOG BANDWIDTH	SAMPLE RATES	DEVICE
dc	5–500 Hz	Integrating ADC
dc–1 kHz	1–500 kHz	Voltage-to-frequency converters
dc–100 kHz	0.1–2 MHz	Successive approximation ADC
dc–10 MHz	1–50 MHz	Flash converter ADC
dc–80 MHz	200 MHz	Electron-bombarded semiconductor (EBS) tube (Tektronix 7612D)
dc–500 MHz	100 GHz	Scan converter tube (Tektronix 7912AD)
dc–500 MHz	1 GHz (?)	Charge-coupled devices (emerging technology); an LLNL/LANL Q-dot instrument is being built in this range

Model No. 7912AD and the Invatsu Model No. TS8123 transient digitizers, or the Electron-Bombarded Semiconductor (EBS) tube used in the Tektronix Model No. 7612D.[11]

Waveforms acquired by the Tektronix 7912AD can be viewed directly on a video monitor in real time as they are acquired. Waveforms stored in memory can be displayed on an X-Y-Z monitor. The Tektronix 7912AD operates very much like a conventional oscilloscope. The input signal is connected to the input of a Tektronix 7000 series vertical plug-in unit to drive the vertical deflection amplifier, and a Tektronix 7000 series time base plug-in unit drives the horizontal deflection amplifier at calibrated sweep speeds. In place of the conventional oscilloscope CRT, the Tektronix 7912AD contains a scan converter tube in which the input waveform is written on and read from a diode matrix target as in a vidicon TV camera.[12] The scan-converter electron tube contains two electron guns at either end of the tube with the target between them. The beam from the reading electron gun continuously scans the target at a slow speed, charging each diode in the array. The beam from the writing electron gun is vertically deflected by the input signal and swept horizontally across the target by the time-base ramp. This electron bombardment discharges adjacent diodes in the target. When the reading beam again scans the target, the discharged diodes are recharged, thereby causing a signal current to be detected at those locations in the target. Wherever the target was not written, the diodes are still charged, and little or no current flows. The locations where the current does flow thus gives a digitized waveform.

Each of the two channels in the Tektronix 7612D contains an Electron-Bombarded Semiconductor (EBS) tube to perform the analog-to-digital conversion.[13] The EBS tube contains a target at one end and an electron gun at the other. The target is arranged as an eight-bit gray-code-pattern diode array. The beam is deflected vertically across this array by the corresponding analog input-signal level. Whenever the beam penetrates a diode junction, that diode conducts and generates an output current. The tube output is continuously sampled by memory at 5-nanosecond intervals. The time base then selects samples from this continuous stream and stores them in the waveform memory at the programmed sample rate. Since the tubes are continuously scanned and sampled, both pre- and post-triggering capabilities are provided.

The future of transient recorders appears to lie in the use of charge-coupled devices (CCD). Using a string of CCD charge wells and a fast clock to transfer charge from well to well permits fast analog storage of the input signal.[10] The charge levels can then be clocked out at a slower rate to a more conventional ADC for digitization. Problems associated with this technique include the amount of resolution available and charge losses in the CCD wells. Digitizing oscilloscopes based on this technology have been recently introduced, the Philips Model No. PM3310 and PM3311. As the CCDs improve, the problem of developing fast enough clocks arises. The National Bureau of Standards has recently developed and tested a 100-GHz Josephson junction binary counter.[11]

6.5 DATA STORAGE

The two primary techniques for storing acquired signals are a photograph of an oscilloscope trace or direct digitization of a signal. Photographs may be digitized through the use of a graphics tablet.[8] They are converted to a data file by locating points on the film to predetermined X–Y coordinates on the tablet. Once the operator enters X–Y coordinate scale factors, the digitized film data may then be treated as any other digitized file record.

After it has been digitized, acquired signal data may be stored on a number of different media—magnetic tape, diskettes, or disks. Typical systems first store the data on either a large removable disk or a removable diskette. For archival purposes, files are stored on magnetic tape or removable diskettes.

6.6 DATA PROCESSING (AND DISPLAY)

Various degrees of processing may be applied to digitized data.[8] The simplest form is the generation of raw data displays. The numeric data file is converted into a conventional X–Y plot resembling an oscilloscope display. Figure 6-5

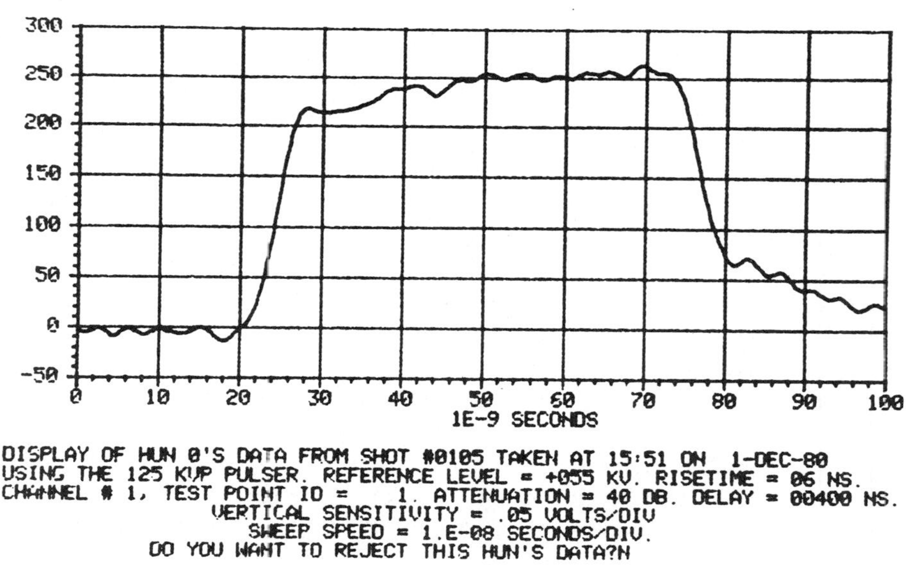

Fig. 6-5. Raw data display.

depicts an actual raw data plot of a transient signal recorded in an automatic data acquisition system.

The next level of processing includes proper scaling of the plot; e.g., converting the display into engineering units. The processing system may then remove (deconvolve) effects on data caused by the transfer function of each item in the data path, e.g., gain, attenuation, risetime, etc. Various user-specified processes may then be employed to manipulate the data further. In Fig. 6-6, the signal risetime has been calculated, the waveform has been compared to a known curve, and a best-fit algorithm has been utilized to define the signal more closely. Various standard math packages are available to provide other processing, such as Fourier transforms to give frequency-domain plots. (See Chaps. 12 and 13 for additional discussion of signal processing.)

Acquisition of Electromagnetic Pulse (EMP) data is a complex operation because of signal characteristics that can embody a fast risetime leading edge (<10 ns) and a slow decay (microseconds) in the form of a double-exponential or a damped-sinusoidal waveform. No one data-recording device possesses the ability to capture the thousand-to-one characteristic times of this signal. To overcome this recording limitation, it is necessary to employ multiple recorders set to different sweep speeds. For example, Fig. 6-7 identifies a method used to characterize a complete EMP simulator free-field environment using either transient digitizers or storage oscilloscopes. The signal is recorded on three

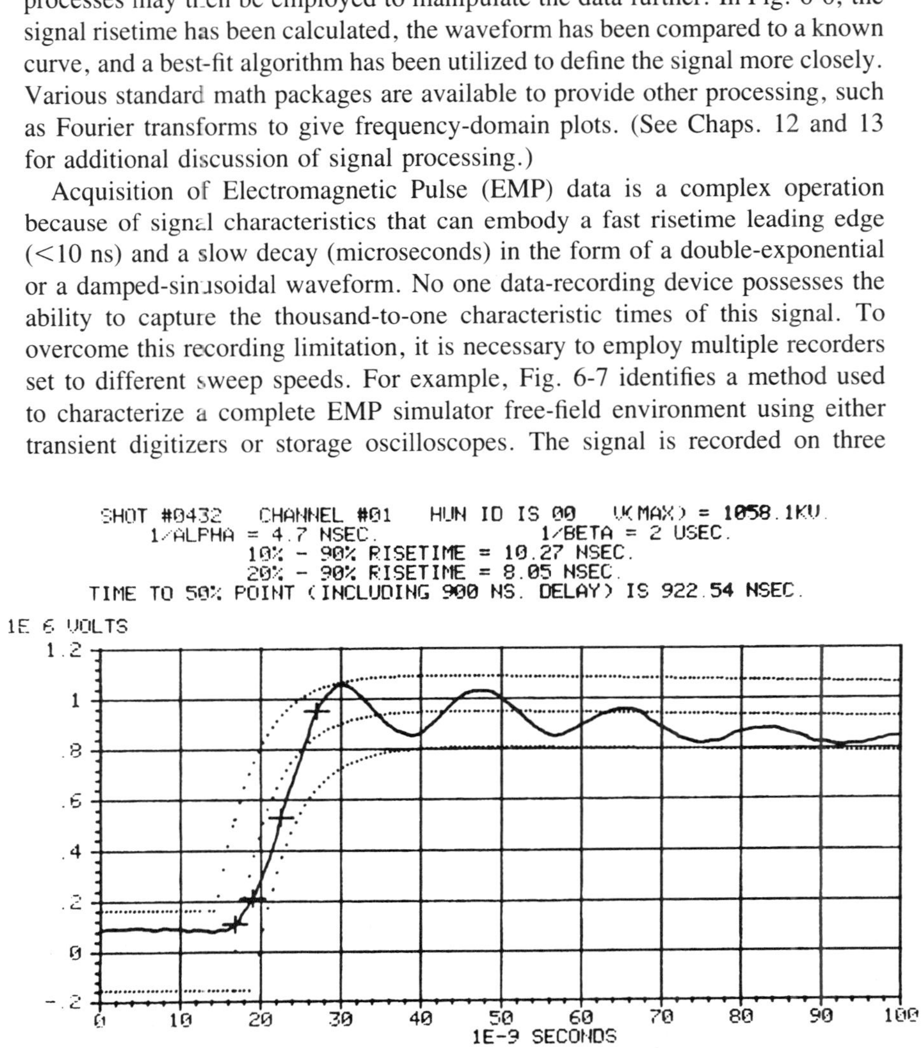

Fig. 6-6. Scaled data display.

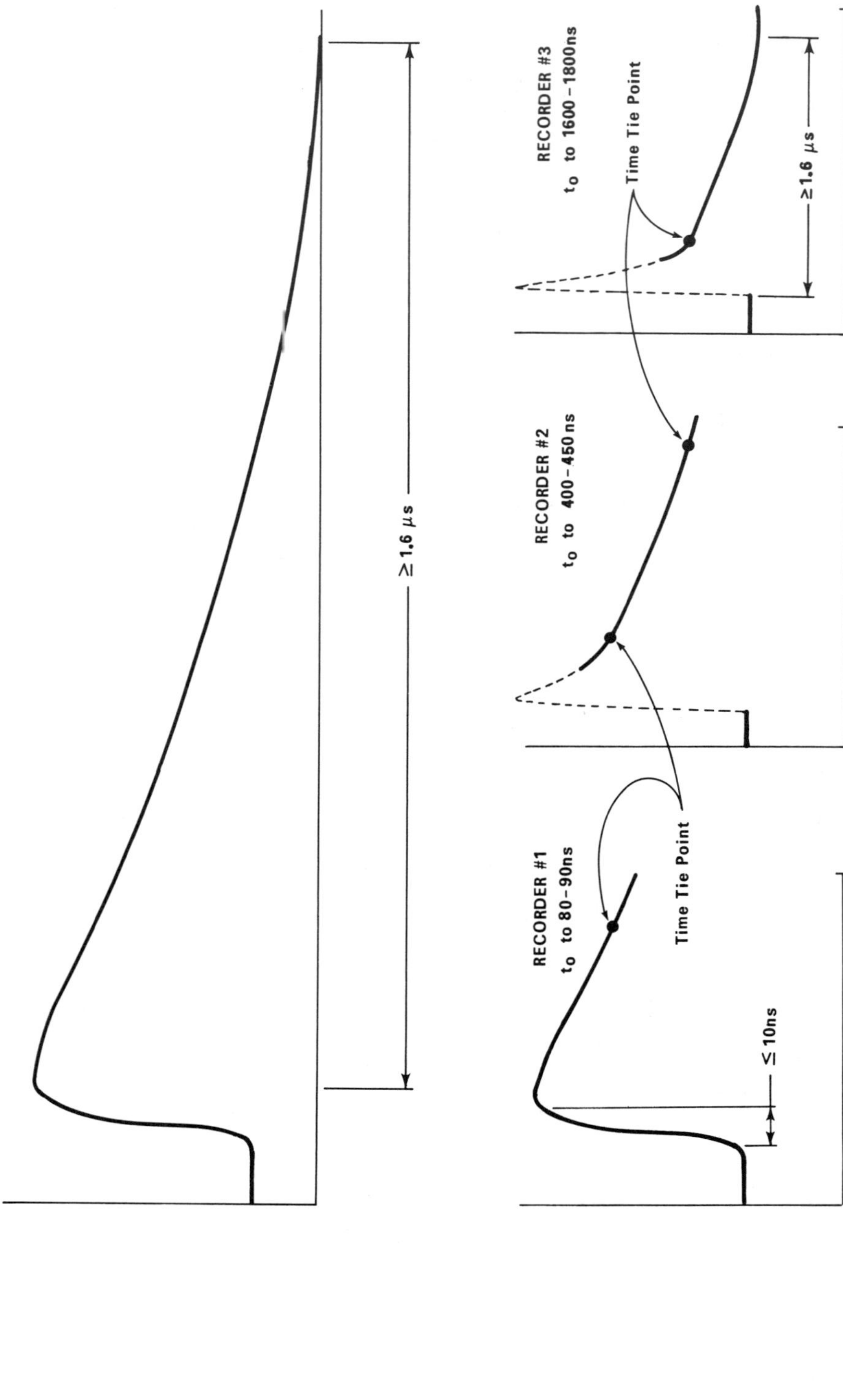

Fig. 6-7. Data time tie.

recorders, each with a different time-base setting. The recorded data is then processed to construct a single composite trace by matching segments of the three recordings ("time-tying"). It has been found that a difference from one time-base setting to the next by a factor of five is the limit for achieving consistently good reconstructions of recordings. Therefore, 10/50/250 ns/div or 20/100/500 ns/div settings are possible choices for recording signals of 1- to 5-microsecond durations. Note that two recorders are normally not adequate because of the long duration of the signal. The data processing system can then "time-tie" these traces to form an overall picture of the signal while retaining required time-sample resolution of the fast risetime.

6.7 CONTROL SYSTEMS

The discussion up to this point has been restricted to the signal acquisition portion of an experiment. If computers are incorporated into such a system, the next logical step is to provide for some means of automatic control of the instrumentation. To automate the data acquisition process, a control system can be implemented to provide command and control (C&C) signals to various system components.

Such a C&C system can provide automatic control of the instrumentation setup, e.g., digitizer vertical and time-base settings, telemetry link settings, etc. The instrumentation settings and calibration data may then be stored for further processing of the acquired data as previously discussed above. Most important, a true record of experiment conditions is logged, thus ensuring an accurate history of the experiment.

The C&C system can also control the pulse generator, providing automatic setup, warning, and firing of the source. In a complete system, the operator would input an experiment menu, including source settings, sensors to be utilized, telemetry link requirements, digitizer settings, and a time-sequenced order of events. The system would then automatically sequence warning indicators, set up the source and instrumentation, fire the source, trigger the instrumentation, and store the resultant data. Following data acquisition, the controller would in sequence turn off the pulser, telemetry links, and warning indicators. The data base may include calibration data, a sensor library containing appropriate sensor and signal conditioner functions and correction factors, and instrumentation settings that will allow for immediate and automatic correction of the acquired data. The corrected data would then be displayed in engineering units.

6.8 ISOLATION OF MEASUREMENT SYSTEM

In many types of tests, measurements must be made in severe electromagnetic environments. These include nuclear source regions (underground tests), electromagnetic pulse simulators, lightning storms, X-ray machines, etc. The fields generated in such environments can couple directly into the recording equipment and significantly distort the desired signal. Figure 6-8 shows one possible effect, in which not only is the amplitude of the waveform severely distorted, but the time base has actually reversed its direction. Such waveforms have actually been obtained in practice!

A proper data system, therefore, must prevent coupling from the test environment to the measurement system, affording screening from radiated electromagnetic waves, for example, and, more important, shielding of currents induced on test cables and power lines. The best method for screening the instrumentation is to place a barrier between it and the source that is impervious to electromagnetic waves propagated through space or along cables or other conductors.[14–17] The barrier must therefore be a closed topological surface, and all penetrations must be properly treated to prevent the passage of electromagnetic energy. Such barriers are usually called "screen boxes" or "screen rooms," depending upon their size. None of them are ever perfect in actuality but do supply a significant amount of shielding.

Penetration of electromagnetic energy into the enclosure can occur in three ways: diffusion of fields through imperfectly conducting walls, field penetration through holes or apertures in the walls, and current injection into the enclosure via penetrating conductors that are necessary to supply power to the instrumentation and supply input and output signals. If the walls are constructed properly, such as with steel or aluminum sheets or with double layers of metal screen wire, then the barrier effectiveness is determined not by the walls but rather by the apertures and penetrations.

The apertures include not only the doors, but also windows, air vents, seams and joints in the walls, and connector and pipe flanges. The doors must be electrically sealed when closed, usually by means of metal gaskets known as

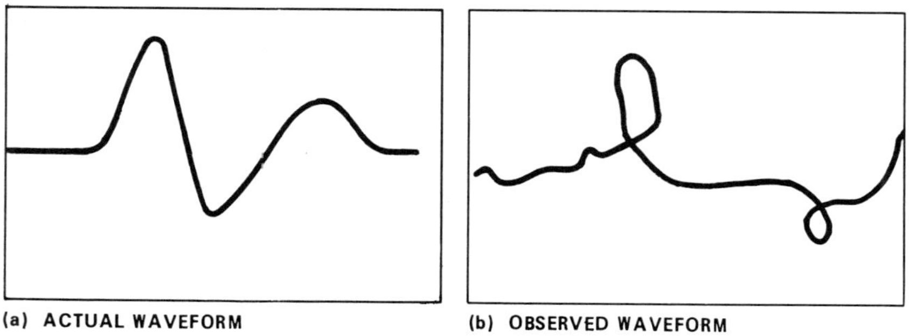

(a) ACTUAL WAVEFORM (b) OBSERVED WAVEFORM

Fig. 6-8. Possible effect of plain recording instrumentation in a hostile electromagnetic environment: (a) actual waveform, and (b) observed waveform.

"finger stock." Windows can be covered with conducting screen. Ventilation ports and penetrations for nonconducting lines, such as pneumatic lines, sewer lines, etc., can be treated by using waveguides beyond cutoff.

Cylindrical conducting tubes form transmission lines for microwave energy but possess a cutoff frequency below which energy will not propagate, that is, when the wavelength is about twice the cylinder diameter. Below cutoff, the attenuation of electromagnetic energy for each unit increase in the length-to-diameter ratio is about 30 dB (a factor of about 1,000 in power) for each length of the tube equal to its diameter. A length of five diameters will thus provide shielding of 150 dB, a figure usually considered adequate. Many such small tubes may be stacked together and electrically bonded to make a grill for ventilation, etc. It should be emphasized that any conductor that penetrates through one of these waveguides totally voids the shielding of the entire enclosure!

The penetrations of electrical conductors into the enclosure, called "deliberate penetrations," are normally the limiting factor in determining shielding effectiveness. It is sometimes possible to eliminate them completely by operating in a screen box and powering the instrumentation with batteries and an invertor. Most installations, however, require the use of 60-Hz power that comes from an outside source.

To prevent significant coupling of electromagnetic energy into the shielded enclosures on power lines, filters that attenuate high-frequency signals may be utilized. Enclosures that contain communications equipment such as radios have a more serious problem because filters cannot be used at radio frequencies. Such radios, or at least their RF sections, must be enclosed in a separate shielded enclosure. Several papers on the subject of shielding and shielding topology have recently appeared.[17–20]

Signal cables that enter the enclosure represent a special type of penetration. Most of these are high-frequency coaxial cables that penetrate the shield wall via bulkhead connectors. The outer conductor of these cables, that portion exterior to the enclosure, can be considered as an extension of the shield wall. If it is of good quality, it will not cause any degradation of the enclosure shielding, but care should nevertheless be taken in its selection and use.

A fact that may be surprising is that all cables have a certain amount of signal coupling from their outside to their inside.[21] In particular, even in good cables, a current on the exterior of the outer conductor will generate an open-circuit voltage signal between the inner conductor and the shield. This coupling can exist even in solid-jacket cable, where it is due to current diffusion through the imperfect conductor. If the cable shield is composed of a woven braid, an additional form of coupling exists as a result of the penetration of the small apertures of this braid by magnetic fields generated by the exterior current. The sum of these two forms of coupling is termed the *transfer impedance*, Z_T, which is the ratio of the internal open-circuit voltage to the external current. Yet another coupling mechanism is due to the penetration of the shield apertures by the electric field that results from the external charge density on the cable, which generates an internal short-circuit current. This is usually called the *transfer admittance,* Y_T, or the ratio of the internal short-circuit current to whatever external voltage generates the charge density.

These coupled currents and voltages do not usually create a shield-enclosure problem but rather an unwanted noise source. For this reason alone, they should be kept to a minimum. Good quality cables with adequate shielding should be used. Apertures or breaks in the cable shield should be avoided, especially at the connectors. External signals, particularly the current, should be reduced as much as possible by (1) routing the cable in low-field regions, (2) periodically shorting the cable current to ground, or (3) spoiling the high-frequency current flow by stringing ferrite beads periodically along the cable.

A separate, but no less important, subject is the treatment of multiconductor shielded cables. The treatment of the coupling problem with such cables historically arose in the power and audio worlds, where the wavelengths of interest are much longer than the cable lengths. As a result, the most common method used to reduce coupling is to float the cable shield at one end, i.e., "break the ground-return path." This is probably the worst single practice that can be committed when shielding for high frequencies. When a cable is several wavelengths long, the end with the shield opened is not at ground potential, and the voltage at that end drives a current onto the conductors; in essence, therefore, there is no shielding at all! Successful high-frequency shielding requires that the shield topology remain intact throughout the system. The cable shield must be circumferentially bonded to the connector shells, and all connectors must have conductive mating surfaces.

Another common practice in wiring cable—again, a holdover from low-frequency technology—is to "pigtail" the cable shield conductor through a connector on one of the pins.[22] Consider what happens to the current on the cable shield exterior. It now appears on one of the pins in the connector, where it inductively couples to the other pins and, hence, to the conductors. Thus, the shielding effectiveness of the cable shield can be reduced from several tens of decibels to about 6 dB!

The proper method of shielding for both high- and low-frequency signals is to treat the cable shield as a topological surface and not to allow any breaks in it. Exterior currents should be minimized by one or more of the three techniques discussed here. If properly done, the cable shield becomes, in essence, a part of the facility ground. Separations of the cable from the ground structure are minimized or eliminated so that the area of the ground loops thus formed are also minimized, as are the currents generated by magnetic fields penetrating a part of the facility ground. Separations of the cable from the ground structure are minimized or eliminated so that the area of the ground loops thus formed are also minimized, as are the currents generated by magnetic fields penetrating such loops.

There are certain hostile environments that merit special mention because of the coupling of noise to the data cables. In particular, ionizing radiation (protons, neutrons, X-rays, γ-rays) can penetrate into the cables and create electrical signals by interacting with the matter within. The usual approach in this case is to use special types of sensors that generate differential signals. The signals are brought out of such environments on two identical coaxial cables. The unwanted noise signals generated in the cables (and in the sensors) will then be nearly identical. The two cables are then connected to a balun (or some equivalent device or technique) that passes only the difference of the two signals (the differential signal) and rejects the common-mode signal (the noise).

All of the problems associated with signal cables (except possibly the ionizing radiation) can be eliminated by using nonconducting data transmission systems. Microwave data links utilizing a special dielectric waveguide were used in EMP testing for many years. These have now been largely supplanted with fiber-optic data links. For subnanosecond risetimes, however, coaxial cables must still be used.

6.9 PERTURBATIONS TO DATA

Care must be taken when interpreting data obtained on any data acquisition system. Each piece of instrumentation used in the signal path will introduce errors. These errors include less-than-ideal frequency response (usually seen as risetime degradation), wideband noise, nonlinearities, and amplitude uncertainty.[8] Digitizers introduce other errors such as aperture uncertainty and aliasing. Repeatability of instrumentation measurements is of utmost importance.

The idealized transfer function of each signal path component is easily unfolded from the data, but the real signal path transfer functions are not ideal. Each component has a definite bandwidth beyond which deviations become quite large. Even within its specified bandwidth the transfer function usually contains some amount of bandpass ripple.

The effects of risetime degradation can be removed by measuring the risetime of each signal path component and taking them in quadrature (square-root of the sum-of-the-squares) to find an overall signal path risetime. The final signal risetime recorded can then be taken in quadrature with the signal-path risetime to find the corrected risetime.[18]

Figure 6-9(a) shows the resultant video display of a transient digitizer driven by a Tektronix 109 pulse generator through a long (100-meter) coaxial cable, cable compensator, and active signal divider. Figure 6-9(b) shows the digitizer

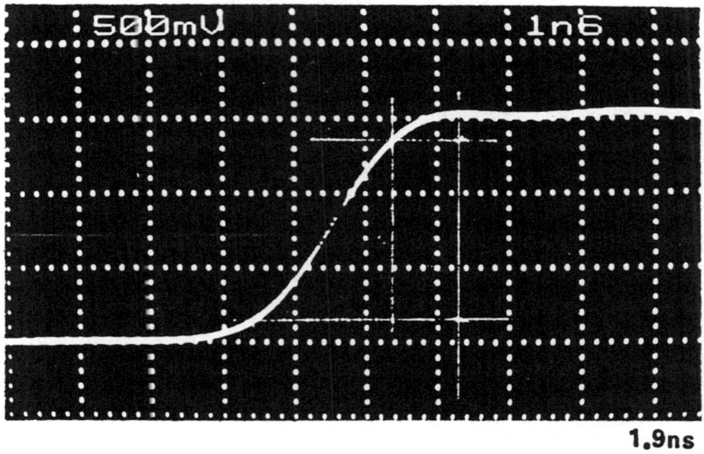

(a) Displayed System Pulse Risetime

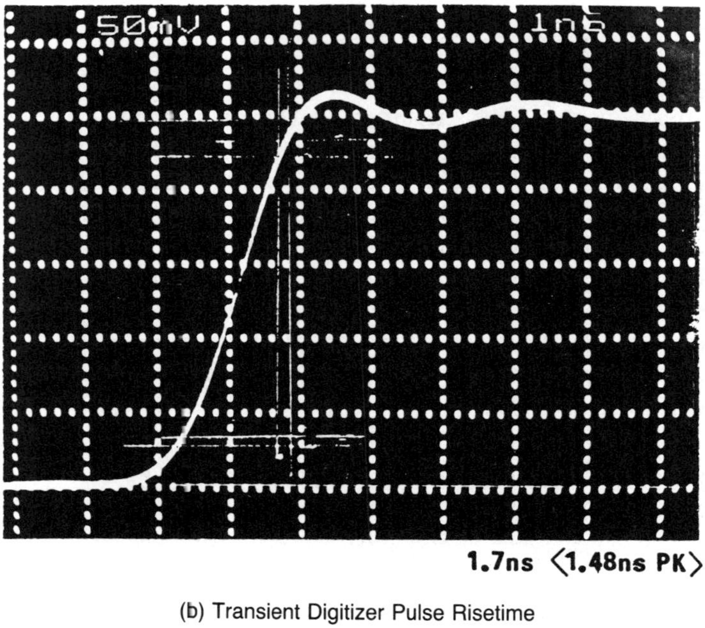

(b) Transient Digitizer Pulse Risetime

Fig. 6-9. Data system risetimes: (a) displayed system pulse, (b) transient digitizer pulse, and (c) system pulse response with in-line integrator.

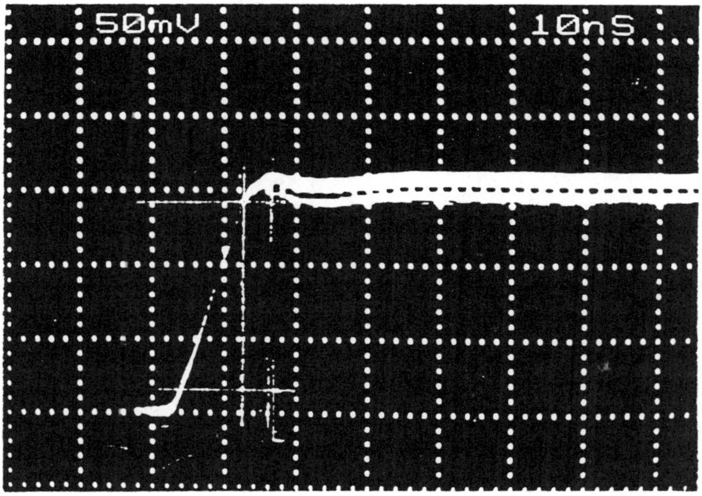

(c)

Fig. 6-9. (*continued*)

response when it is driven directly by a Tektronix 284 pulse generator.[19] From these photographs, the true risetimes can be calculated by the method of quadratures. Let

t_1 = risetime of transient digitizer
t_2 = combined risetime of cable, cable compensator, and active signal divider
t_3 = risetime of pulse generator
t_4 = total risetime of signal path

The method of quadratures then gives

$$t_4^2 = t_1^2 + t_2^2 + t_3^2 \tag{6-10}$$

To find t_1, let t_5 = measured risetime of pulse = 1.5 ns [see Fig. 6-9(b)]. For the Tektronix 284 pulser, t_3 = 70 ps maximum. Then,

$$t_1 = \sqrt{t_5^2 - t_3^2} = \sqrt{1.5^2 - .07^2} = 1.50 \text{ ns} \tag{6-11}$$

To find t_4, let t_6 = measured risetime of signal path = 1.9 ns [see Fig. 6-9(a)]. For the Tektronix 109 pulser, $t_3 = 0.25 \pm 0.05$ ns. Then,

$$t_4 = \sqrt{t_6^2 - t_3^2} = \sqrt{1.9^2 - (0.25 \pm .05)^2} = 1.88 \pm 0.06 \quad (6\text{-}12)$$

To find t_2,

$$t_2 = \sqrt{t_4^2 - t_1^2 - t_3^2} \quad (6\text{-}13)$$
$$= \sqrt{(1.88 \pm 0.06)^2 - 1.50^2 - (0.25 \pm 0.05)^2} = 1.1 \pm 0.1 \text{ ns}$$

Note that the same method must be utilized to find the true risetime of any measured signal. For example, if a pulse being displayed on the above system has a measured risetime of 10.0 ns, the actual pulse risetime, using the 1.88 ns data path risetime from above, is

$$t_3 = \sqrt{t_6^2 - t_4^2} = \sqrt{10^2 - 1.9^2} = 9.8 \text{ ns} \quad (6\text{-}14)$$

Figure 6-9(c) depicts the response of the signal path with a 1-microsecond active integrator. This is not a measurement of the integrator/data path risetime but rather of the integrated pulse width. The calibration/verification of the integrator must be made separately using either a very narrow, high-amplitude pulse (δ function) or a network analyzer in the frequency domain.

Still another problem influencing data interpretation is the introduction of noise into the signal path, especially in the telemetry links. Noise is often measured in terms of rms amplitude, a nearly meaningless way to make pulse measurements. In repetitive pulse measurements, a tangential noise measurement is more meaningful, but when considerating a single event pulse measurement, even this is not sufficient. A single noise burst or random excursion may mask data, as shown in Figs. 6-10 and 6-11.[5] In these actual measurements, noise has displaced the 10-percent and 90-percent amplitude points, yielding erroneous 10- to 90-percent risetime measurements. If all the signal-path and signal-source components have good repeatability, a number of measurements may be taken and averaged to reduce the baseline noise.[5]

6.10 TIME-DOMAIN REFLECTOMETRY

A special type of time-domain measurement that merits separate discussion is time-domain reflectometry (TDR), where measurements are made of signal reflections from impedance discontinuities (see Chaps. 1 and 10 for further discussion). The value of this type of measurement is that specific discontinuities have readily identifiable characteristic signatures. Also, the location of the discontinuity on a transmission line can be accurately measured by the two-way propagation time of the incident signal and reflection.

A simplified schematic of a TDR system is shown in Fig. 6-12. A pulse

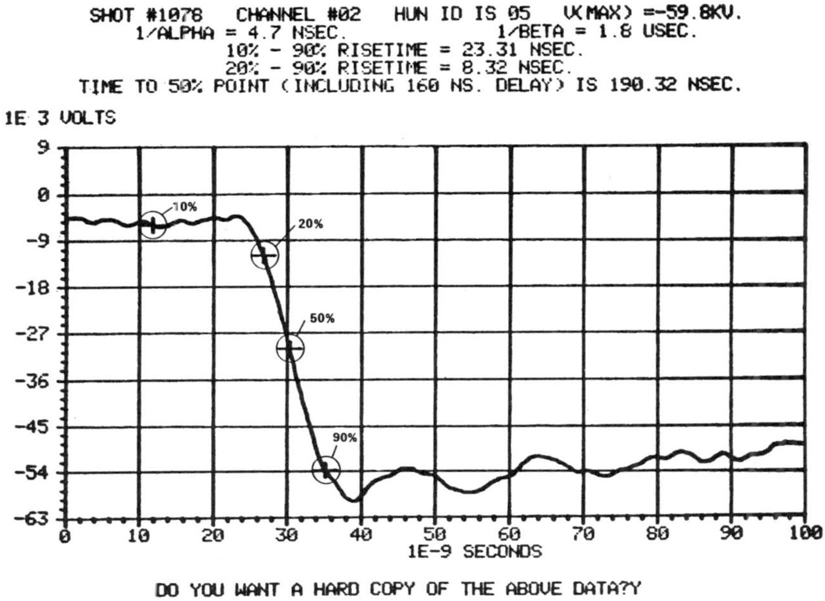

Fig. 6-10. Risetime degradation caused by noise on prepulse.

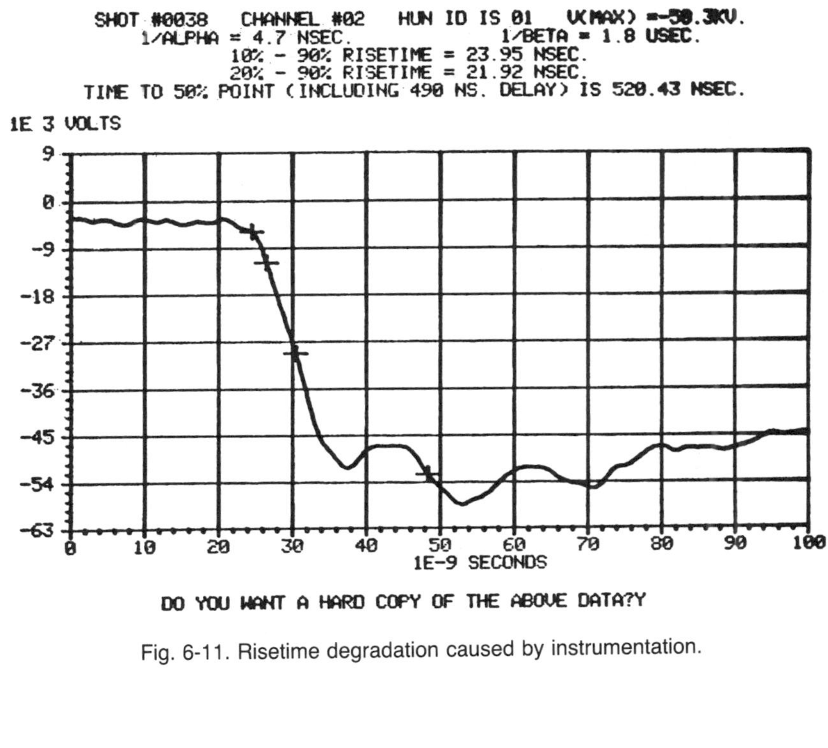

Fig. 6-11. Risetime degradation caused by instrumentation.

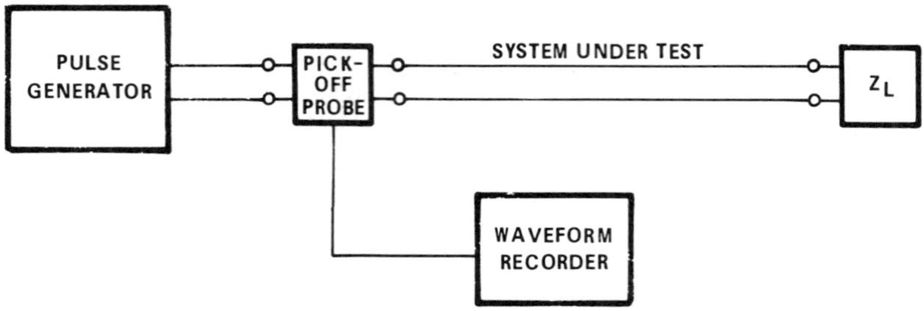

Fig. 6-12. Simplified TDR system.

generator produces the signal that is directed to the system under test. A step waveform is usually employed although certain applications use other waveforms, such as a short pulse that approximates an impulse or a short burst of RF (narrow-band TDR). A trigger signal may also be generated to trigger the waveform recorder. A pickoff probe is used to sample the incident and reflected waveforms, usually a high-impedance voltage probe of sufficient bandwidth for the timing accuracy required. Special applications may also employ current probes or I-dot probes. The waveform recorder must also be of sufficient bandwidth. For most applications, a low-level repetitive pulse generator is used so that equivalent time sampling systems can be used, allowing measurement of risetimes of less than 50 ps. A simple TDR can be made using only a square-wave oscillator (such as an oscilloscope calibration signal) and an oscilloscope with a high-impedance input by placing a coaxial "T" on the megohm input and connecting the source to the system under test through the "T."

The TDR measures reflections of traveling waves at impedance discontinuities. The pulse generator drives a signal of $U(t)$ onto a cable of characteristic impedance Z_0 (usually 50 ohms). The pulse generator should present this same impedance to prevent further reflections of the signal. The probe measures this incident signal, which is displayed on the recorder. The incident signal then travels down the system under test until it encounters a new impedance, Z_1, which gives a reflection coefficient of

$$\rho = \frac{Z_1 - Z_0}{Z_1 + Z_0} \qquad (6\text{-}15)$$

where Z_1, and hence ρ, may be complex. A thorough analysis of the problem requires that ρ be transformed into the frequency domain to determine the reflection coefficient of each frequency component of the incident signal.

Figure 6-13 exhibits several waveforms from specific load impedances. Figure 6-13(a) shows the incident waveform without reflections, such as would occur if the cable were to be terminated in its characteristic impedance. The unity step

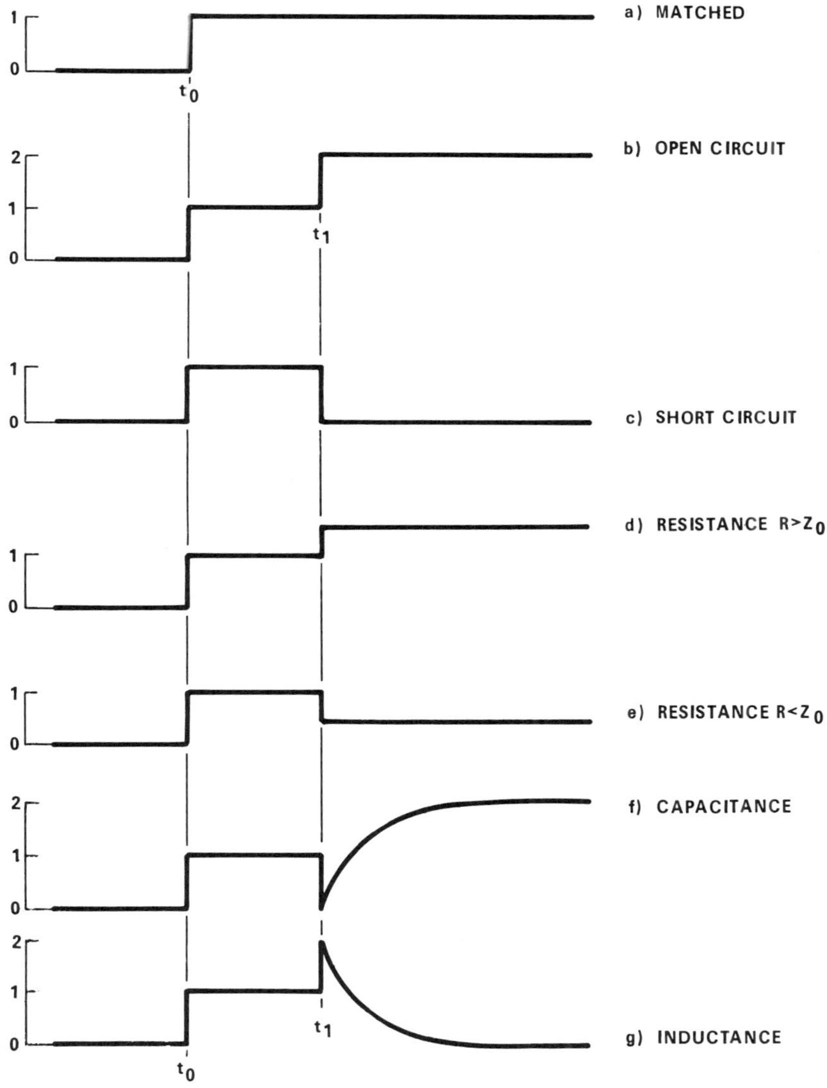

Fig. 6-13. Reflection signal from specific load impedance.

at t_0, $U(t - t_0)$, is the pulse generator output. Figure 6-13(b) shows a reflection from an open circuit at t_1. The reflection coefficient becomes unity as Z_1 becomes much larger than Z_0 so that the reflected voltage step at t_1 becomes equal in magnitude to the incident step at t_0. Figure 6-13(c) shows a reflection from a short circuit at t_1 for which $Z_1 = 0$ and $\rho = -1$. The reflection thus returns out of phase with the incident signal. Figures 6-13(d) and (e) show reflections from resistive loads larger and smaller, respectively, than Z_0. The magnitude and phase of the reflected signals are readily discerned.

Figure 6-13(f) shows the reflection from a purely capacitive load, $Z_L = 1/sC$, where $s(s = j\omega)$ is the Laplace operator. When the step first encounters it, the capacitor acts as a short circuit. At later times, it behaves as an open circuit, as can be seen in the behavior of the reflection coefficient,

$$\rho = \frac{\frac{1}{sC} - Z_0}{\frac{1}{sC} + Z_0} = \frac{1 - sCZ_0}{1 + sCZ_0} = \begin{cases} -1, & sCZ_0 \gg 1 \\ +1, & sCZ_0 \ll 1 \end{cases} \tag{6-16}$$

which goes to -1 at high frequencies and to $+1$ at low frequencies. The exponential risetime of the capacitive signature behaves as

$$V(t - t_1) = 2[1 - e^{-(t-t_1)/Z_0C}] \tag{6-17}$$

for times longer than t_1.

Figure 6-13(g) shows the reflection from a purely inductive load, $Z_1 = sL$. This load initially behaves as an open circuit to the step waveform, becoming a short circuit at later times:

$$\rho = \frac{\frac{1}{sC} - Z_0}{\frac{1}{sC} + Z_0} = \frac{1 - sCZ_0}{1 + sCZ_0} = \begin{cases} +1, & SL/Z_0 \gg 1 \\ -1, & sL/Z_0 \ll 1 \end{cases} \tag{6-18}$$

The reflection coefficient becomes $+1$ at high frequencies and -1 at low frequencies. Thus,

$$V(t-t_1) = 2e^{-(t - t_1)Z_0/L} \tag{6-19}$$

shows the exponential decay in the time domain. Reflections from series and/or parallel combinations of resistors, capacitors, and inductors can be analyzed in a similar fashion.

The nature of small impedance discontinuities can readily be determined from the TDR signature. For example, a short section of low-impedance line in the middle of a transmission line of characteristic impedance Z_0 (such as might occur at a connector or a crushed section of cable) will appear as a comparable decrease of the TDR signal, as shown in Fig. 6-14(a). A similar signature is obtained

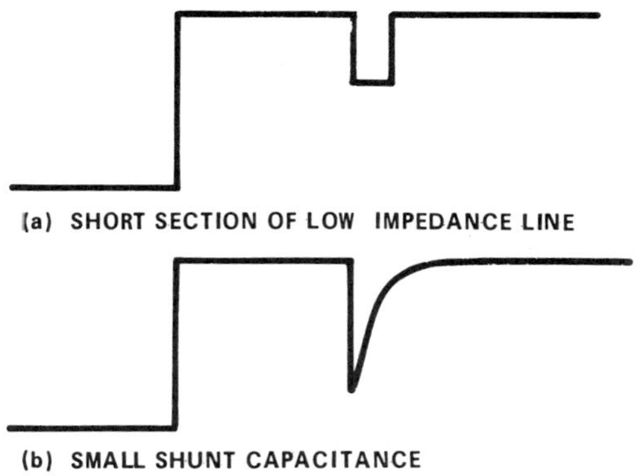

Fig. 6-14. Capacitance dips on TDR trace: (a) short section of low-impedance line, and (b) small shunt capacitance.

when a small shunt capacitance is placed on the line, as seen in Fig. 6-14(b). Thus, a small dip in impedance is usually called a "capacitance dip." Likewise, a small section of high impedance is termed an "inductive spike," whether due to high impedance or an actual inductance. This terminology is consistent with transmission line theory since the characteristic impedance of an ideal line is derived from the series inductance per unit length, L', and the shunt capacitance per unit length, C', as follows:

$$Z_0 = \sqrt{L'/C'} \tag{6-20}$$

An increase in capacitance results in a decrease in impedance, a decrease that can be termed "capacitive." Likewise, an increase in inductance causes an increase in the impedance, which is termed "inductive."

TDR analysis is also very useful for locating faults in a cable that manifest themselves as impedance discontinuities. Such faults may exist as breaks in a cable shield, as open circuits, or perhaps as water in a buried cable. Here the fact that the discontinuities are separated in time permits measuring the reflection time of each discontinuity. Knowledge of the propagation velocity, v, on the cable then gives the distance to each as

$$d = vt/2 \tag{6-21}$$

Four effects that may limit the resolution and usefulness of a TDR measurement are (1) the system risetime, (2) discontinuities that may be in front of the discontinuity of interest, (3) loss effects when making measurements over long and/or lossy cables, and (4) excessive noise on the cable.

The TDR-system risetime, t_r, precludes the resolution of discontinuities spaced closer than

$$d_r = vt_r/4 \tag{6-22}$$

The amplitude of small discontinuities will also be decreased by the system risetime, which acts as a pulse stretcher.

Multiple discontinuities on a transmission line create effects that complicate TDR analysis by virtue of the multiple reflections occurring between them. If the cable of interest is composed of several sections, each with its own distinct characteristic impedance, a part of the incident signal is reflected at each of these impedance discontinuities. Part of each reflected signal will then be again reflected at each subsequent discontinuity so that the amplitudes of the TDR signals beyond the second discontinuity are reduced.

For many practical measuring situations, the discontinuities are small. When they are, the magnitude of the multiple reflections rapidly approaches zero, and many small discontinuities can be analyzed without significant perturbation to each other.

The third effect that may distort TDR data is the loss characteristics of the transmission line under observation. For high-quality short lines, this may not be a problem because the effect is too small to distort the risetime. The problem may be significant for long, lossy lines, however, such as telephone wires (twisted pairs) or long power lines. A particularly bad example is the thermocouple wires that monitor the temperature inside nuclear reactor cores.

The loss on most transmission lines can be attributed to the skin effect caused by the finite conductivity of the electrical conductors. In the frequency domain, the attenuation factor becomes proportional to the square root of the frequency. As a result, the high frequencies become attenuated more than the low, resulting in a risetime degradation of the TDR signal.

The TDR waveform of a unit-step voltage traveling over a long line can be given by

$$V(t) = \operatorname{erfc} \frac{K}{\sqrt{t - t_0}} \tag{6-23}$$

where erfc (= 1-erf) is the complementary error function, K is a constant proportional to the length of line multiplied by the line loss per unit length, and t_0 is the signal propagation time. This distortion results in a risetime degradation with an initial apparent delay in the initial signal detection, a reasonably fast-rising section, and then a long tail with an apparent decrease in amplitude. Figure 6-15 shows these features, which can significantly affect the analysis of TDR data.

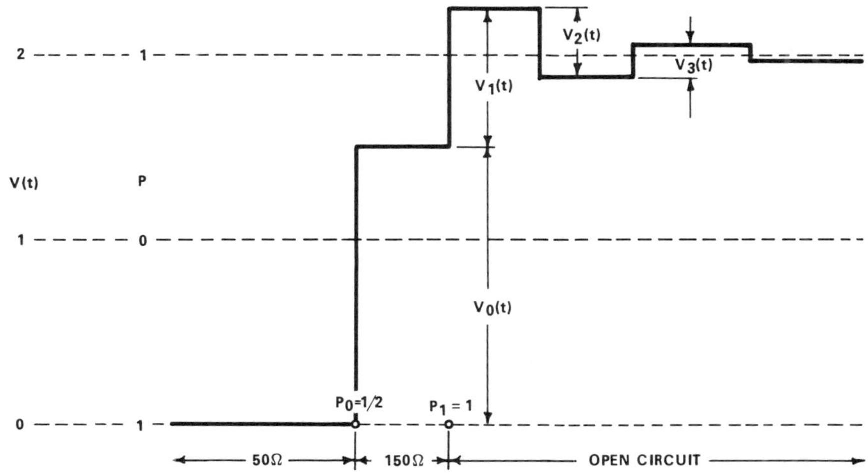

Fig. 6-15. Unit-step voltage waveshape after propagation over lossy line.

The final effect that makes TDR analysis difficult is noise on the transmission line on the system being analyzed. Even shielded cables may exhibit significant noise if damaged or if in proximity to power-carrying conductors. Open-wire transmission lines can pick up significant radio-frequency radiation from such sources as radio stations, mobile two-way radios, or electric arc welders. In many cases, this noise can be appreciable compared to the TDR signal level and, in some cases, may be so large as to damage the TDR unit. Since most commercial TDR units utilize a step waveform of about 200-mV amplitude and utilize tunnel-diode circuitry, they are quite sensitive and vulnerable. Static-charge buildup on a coaxial cable must be discharged before the cable is attached to this type of TDR, or the tunnel diode may be destroyed.

The solution to the problem of noisy lines is to increase the level of the TDR signal until it is significantly greater than the noise. Step generators of several tens of volts are now available with subnanosecond risetimes utilizing solid-state switches (see Chap. 4). Mercury-wetted reed relays may be used to deliver repetitive step waveforms of several hundred volts amplitude.

Figure 6-16 shows a TDR of a large transmission-line antenna that is a very efficient radio antenna. This TDR was done with a 400-volt signal and an appropriate attenuation of the signal after the voltage probe. The plot was made in the slow equivalent time output of the sampling system. Reference lines were obtained by driving various resistive loads instead of the antenna. The noise level recorded from the antenna is small enough to be negligible. The slow ramp at the front of the traces is due to the tunnel-diode sampling system. The step-voltage waveform was generated by charging and discharging a long cable into

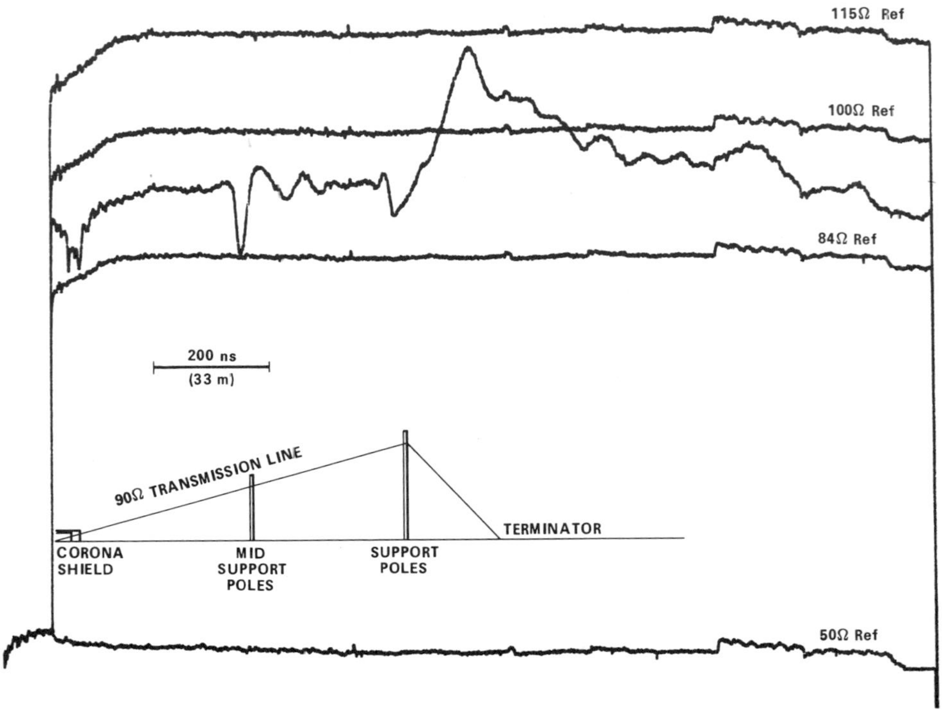

Fig. 6-16. TDR of a bounded-wave EMP simulator.

the test object with a mercury-wetted reed switch. For the very long pulse shown in Fig. 6-16 (1.5-microseconds total duration), several long coaxial cables were connected together to form the charge cable. The slight change in impedance between some of these can be seen in the 50-ohm reference line and is repeated in the other traces.

Noise reduction can also be achieved by averaging several TDR traces but requires some form of digitization of the TDR display. An alternative method that can be used with an equivalent-time sampling system is to use a very slow (external) sweep on the TDR time base and low-pass filtering on the TDR vertical output. A plotter often performs this function well since the slow pen response acts as the low-pass filter. This method of TDR testing requires that the charge cable be longer than the line under test. For very long lines, this requirement becomes impractical, and alternative pulse sources must be utilized even though they may not have subnanosecond risetimes.

For high-amplitude sensitivity, the TDR system requirements are (1) a large time derivative of $V_0(t)$, (2) a flat pulse top of $V_0(t)$, and (3) low noise detection. For high spatial resolution, the system must have a short system risetime.

6.11 SUMMARY

Instrumentation must be chosen to meet measurement needs. The requirements for survey-type experiments are usually quite minimal compared to those for large-scale, multiple-parameter, design or production-acceptance tests for which a large, general-purpose, automated system with data processing is often required.

To perform many laboratory-type experiments, a standard oscilloscope and pulse generator may well suffice. The addition of a camera with high-speed film will provide permanent data records. An oscilloscope equipped with CRT displays of scope settings will simplify recordkeeping since the scope settings will become part of the permanent photograph record of the signal trace. High-speed pulse work at low levels will require a fast-risetime pulser such as those described earlier or the Tektronix 109 or 284 Pulse Generators. Oscilloscope requirements will be driven primarily by experiment risetimes. For transient analysis, a high-speed storage oscilloscope may be desired. A good assortment of sensors, probes, cables, connectors, connector adapters, and attenuators will also be required.

If some level of data processing is required, either a graphics tablet for digitizing signal traces on photographs or a transient digitizer may be utilized. A large number of digitizing oscilloscopes are available for either slow-risetime or repetitive fast-risetime pulse work. A sampling oscilloscope may be required for the latter. Many parameters enter into the selection process: bandwidth, risetime, sensitivity, sampling rate, memory storage size, linearity, remote control features, etc. The means of accepting and storing the acquired data must be supplied along with storage media compatible with the data processing machine(s).

If a large number of measurements must be made, a more general-purpose data-acquisition system is required. Versatility and expandability requirements may need to be considered. Some level of automation is often desired, at least as a way to keep recordkeeping and repeatability requirements simple, but attempting to design a general-purpose system may add unnecessary complexity and expense.

The primary emphasis in the design of any data-acquisition system must be put on the overall system requirements. The functional requirements must be fully defined before selecting the equipment to implement the system. The temptation to select the equipment first and patch noncompatible gear together if the need should arise later must be avoided. The design of a data-acquisition system requires experienced system designers and project engineers who are not only familiar with the many hidden costs involved in the construction of what may appear on the surface to be a simple "rack 'em and stack 'em" job but who are also well versed in the ergonomics of the day-to-day operational requirements of a system.

The development of a large, sophisticated system requires an even more elaborate line of attack. A hierarchy of procedures has evolved in industry that is definitely applicable; it is as shown in Fig. 6-17.

The temptation to use instrumentation that advances the state-of-the-art must also be overcome. Taking a one-of-a-kind laboratory prototype and extending it into a production run of fieldable units is no trivial matter, as any manufacturer will attest.

In conclusion, the approach may be broken down as follows:

1. Define the system requirements.
2. Design the system as a whole based on these requirements.
3. Survey the available equipment required to implement the system.
4. Refine the system design based on available equipment.

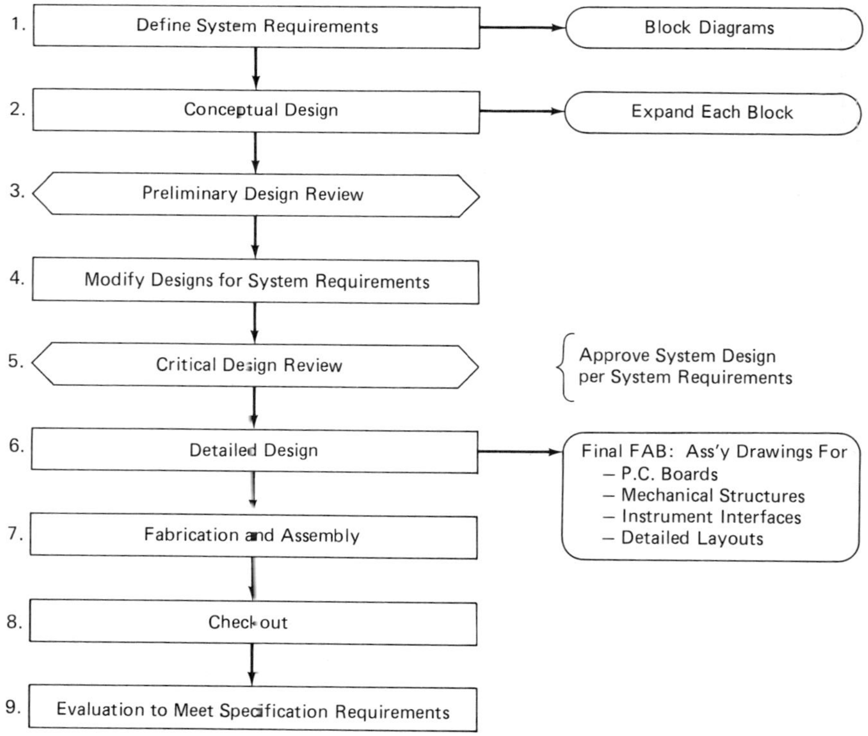

Fig. 6-17. System development hierarchy.

5. Select the equipment.
6. Design the final system integration.

REFERENCES

1. Baum, C. E.; Breen, E. L.; Giles, J. C.; O'Neill, J. P.; and Sower, G. D. 1978. Sensors for electromagnetic pulse measurements both inside and away from nuclear source regions. Sensor and Simulation Note 239, *IEEE Transactions on Antennas and Propagation,* pp. 22–35, and *IEEE Transactions on Electromagnetic Compatibility,* February 1978, pp. 22–35.
2. Baum, C. E. 1981, 1982. Sensors for measurement of intense electromagnetic pulses. Sensor and Simulation Note 271, June 1982, and Proceedings of 3rd IEEE International Pulsed Power Conference, Albuquerque, NM, June 1981.
3. Baum, C. E.; Breen, E. L; Pitts, F. L; Sower, G. D.; and Thomas, M. E. 1982. The measurement of lightning environmental parameters related to interaction with electronic systems. Sensor and Simulation Note 274, *IEEE Transactions on Antennas and Propagation,* and *IEEE Transactions on Electromagnetic Compatibility,* May 1982, pp. 123–137.
4. Gruchalla, M. E. 1982. *Electrical grounding in large instrumentation* systems.
5. Sower, G. D.; Bonham, A. J.; Gruchalla, M. E.; and Pressley, J. R. 1983. Instrumentation for EMP measurements. 5th Symposium and Technical Exhibition on Electromagnetic Compatibility, Zurich.
6. *Designer's Guide to Fiber Optics.* 1978. (*EDN* Magazine reprint collection.) Boston: Cahners Publishing Co.
7. Friebele, E. J. 1979. Optical fiber waveguides in radiation environments. *Optical Engineering* 18, No. 6: 552–561.
8. Pressley, J. R. 1982. Time domain instrumentation for EMP measurements. Presented at NEM 1982, Albuquerque, NM.
9. *ASD-1 active signal divider, O&M Manual.* 1980. AG-1432, EG&G Washington Analytical Services Center, Inc., Albuquerque, NM.
10. Sower, G. D., and Pressley, J. R. 1982. Instrumentation for time domain measurements. Presented at 1982 VRSI Meeting, Boulder, CO.
11. VonRoesgen, C. 1983. Digital oscilloscopes and waveform recorders. *Electronics Test,* March, pp. 48–63.
12. *7912AD programmable digitizer operators instruction manual.* 1978. Tektronix, Inc., Beaverton, OR.
13. *7612D Programmable Digitizer Operators Instruction Manual.* 1980. Tektronix, Inc., Beaverton, OR.
14. Vance, E. F. 1977. Shielding and ground topology for interference control. Interaction Note 306, Air Force Weapon Laboratory, Kirtland AFB, NM.
15. Vance, E. F. 1980. Electromagnetic interference control. *IEEE Transactions on Electromagnetic Compatibility* EMC-22: 319–328. Also Interaction Note 380.
16. Baum, C. E. 1981. Electromagnetic topology: A formal approach to the analysis and design of complex electronic systems. *Proc. EMC Symposium,* Zurich. Also Interaction Note 400.
17. Baum, C. E. 1982. Sublayer sets and relative shielding order in electromagnetic topology. Interaction Note 416.
18. Walker, R. M., and Wallnan, H. 1964. High-fidelity pulse amplifiers. *Vacuum Tube Amplifiers, Mat. Radiation Laboratory Series* 18:77, 78. Lexington, MA: Boston Technical Publishers, Inc.
19. *EMP simulator data acquisition system O&M manual.* 1981. AG-1462, EG&G Washington Analytical Services Center, Inc., Albuquerque, NM.

20. Digital storage scopes. 1983. *Test and Measurement World,* May, pp. 40–54.
21. Vance, E. F. 1978. *Coupling to shielded cables*. New York: John Wiley & Sons.
22. Paul, Clayton R. 1980. Effect of pigtails on crosstalk to braded-shield cables. *IEEE Transactions Electromagnetic Compatibility* EMC-22, No. 3: 161.
23. Ramo, S.; Whinnery, J. R.; and Van Duger, T. 1965. *Field and waves in communication electronics*. New York: John Wiley & Sons, pp. 286–303.

7
The LLNL Electromagnetic Transient Facility: Design, Test, and Applications of a Time-Domain, Ground-Plane Facility

R. M. Bevensee
Lawrence Livermore National Laboratory

E. K. Miller
University of Kansas

7.0 INTRODUCTION

This chapter describes the Lawrence Livermore Electromagnetic Transient Facility (EMTF), defines its basic capabilities and limitations, and evaluates its utility for data-base development in Electromagnetic Pulse (EMP) applications. A series of measurements and computations were made on a variety of test objects to calibrate and validate the EMTF. Other measurements were performed to assess the performance of various range components such as pulsers, connecting cables, sensors, and instrumentation (which are discussed in Chaps. 3, 4, and 5). The results presented here may be considered representative of what transient measurements are able to provide, but should not be viewed as testing the limits of what is possible.

7.1 DESCRIPTION OF EMTF OPERATION

Objectives

The LLNL Electromagnetic Transient Facility (EMTF), which is composed essentially of impulse pulser, transmitting antenna, sampling scope, and data processing equipment, is used to determine the time-domain transient responses of various antenna structures to an incident electromagnetic (EM) wave. For example, time-domain reflectometry (TDR) measurements enable us to compute the input impedance of a given structure in the frequency domain. Time-domain scattering measurements of the same structure excited as a receiver similarly enable us to compute its effective height in the frequency domain. Use is made of the equality of the input impedance as a transmitter and as a receiver. The Thevenin-equivalent circuit of the receiver structure thus obtained may now be scaled to any realistic life size, and a specified frequency-domain load can be attached. Then, the response to an incident temporal electromagnetic pulse can be computed, first in the frequency domain and then, by Fourier transformation, in the time domain. In this way, both the integrated responses (e.g., total energy absorbed, peak temporal load current) and differentiated responses (e.g., input impedance, effective height, load power spectrum in the frequency domain) may be determined from a combination of measured and computed values for a life-size structure.

Data Acquisition

A diagram explaining the data acquisition in both the TDR and scattering modes is shown in Fig. 7-1. In the TDR measurement, an impulse generator excites the test target through a precision power divider. The reflected signal then passes through the third arm of the divider to the scope and the data-processing equipment. The reflection coefficient computed in the frequency domain measures the complex input impedance, Z_I, of the test target. In a scattering measurement, the dashed lines in Fig. 7-1 are changed to the solid lines to indicate that the generator pulses the conical transmitter. Thus, the time-delayed scattered voltage is picked up from the test target in a 50-ohm coaxial line and relayed to the scope and processing equipment.

When transformed to the frequency domain, this scattered field (or voltage V_L across the Z_L load of 50 ohms in the target-equivalent circuit of Fig. 7-2) determines the effective height, $h_{\text{eff}}(f)$, since Z_I has been measured (assuming the incident field, $E^{\text{inc}}(f)$, is known). Because they can be easily scaled to represent the equivalent circuit of a larger structure, Z_I and h_{eff} are convenient parameters to define and measure. We can then use this scaled circuit to obtain the response to an arbitrary field, $E^x(f)$, for a specified load, $Z_L(f)$, when $E^{\text{inc}} = E^x(f)$ is applied to the input terminals.

All TDR and scattering measurements are taken during a 20-nsec "clear-time" interval before unwanted reflections occur (e.g., from the edge of the ground screen). Specific details of these measurements will now be given.

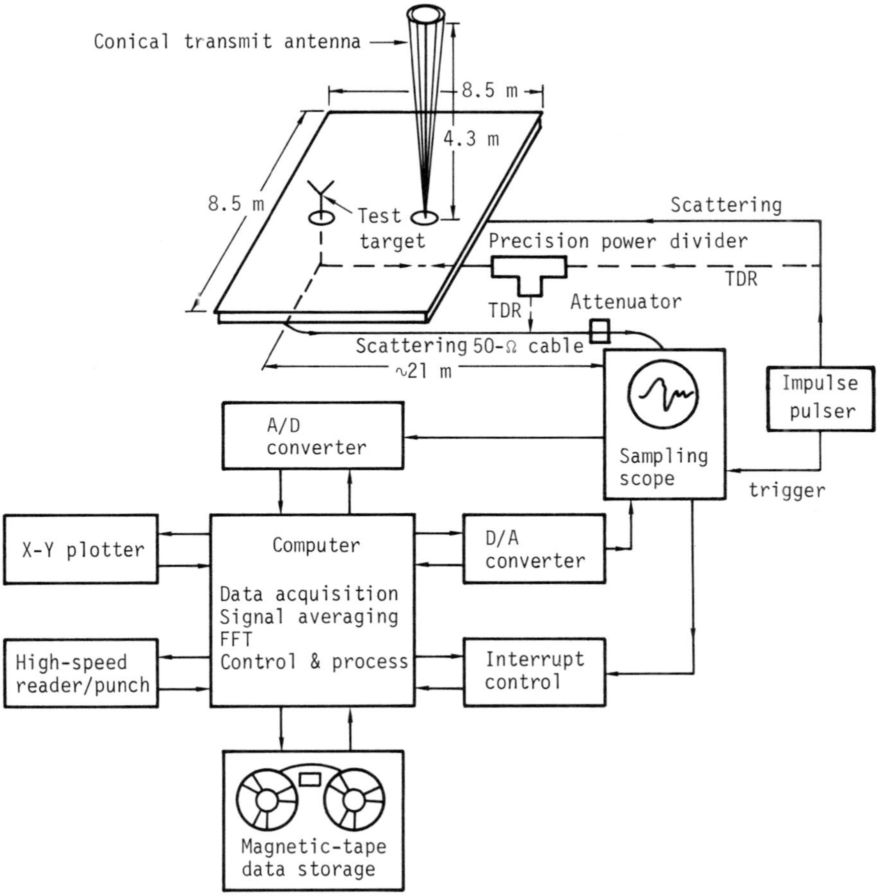

Fig. 7-1. Schematic diagram of the EMTF: The impulse pulser sends a narrow pulse either through a precision power divider to the test target, as in a TDR experiment, or directly to the conical transmitting antenna, as in a scattering experiment. In the TDR mode, the reflected signal returns to the sampling scope via the power divider; in the scattering mode, the signal received by the test target returns directly to the scope. From these, the signal passes to the data-processing equipment where it is averaged over repetitive pulses, Fourier-transformed to the frequency domain, plotted, and stored on magnetic tape.

7.2 MEASUREMENT OF CIRCUIT PARAMETERS

Input Impedance, Z_I

TDR reflected-voltage waveforms for both an open-circuit termination in Fig. 7-1 and the actual test target (as represented by its input impedance) are recorded

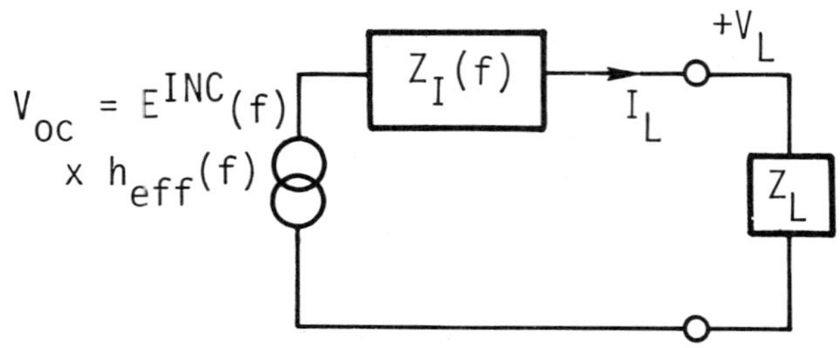

Fig. 7-2. Scattering equivalent circuit for the load Z_L across the test-target load terminals: The excitation, V_{oc}, is conveniently represented as the product of the incident electric field and the effective height, as determined from measurements; Z_I is the input impedance from a TDR experiment. This circuit is conveniently scaled to represent the reception of any scaled-up test target that has been excited by a specified temporal or spectral field.

and transformed to the frequency domain. The frequency-domain reflection coefficient, $\rho(f)$, is computed as

$$\rho(f) = -\frac{V_{\mathrm{LOAD}}(f)}{V_{\mathrm{OPEN}}(f)} \tag{7-1}$$

so that the test-target input impedance, $Z_I(f)$, is given by

$$Z_{\mathrm{in}}(f) = Z_I(f) = \frac{1 - \rho(f)}{1 + \rho(f)} Z_0 \tag{7-2}$$

where Z_0 is the coax characteristic impedance (50 ohms for the EMTF).

Effective Height, h_{eff}

The effective height, $h_{\mathrm{eff}}(f)$, in the equivalent circuit of Fig. 7-2 is determined by a temporal scattering experiment. The conical transmitter is pulsed (the solid lines in Fig. 7-1 indicate scattering), and the voltage pickup, $V_L(f)$, in the test target (loaded by the coax impedance, Z_0, of 50 ohms) is recorded. The target receiver is represented by the equivalent circuit in Fig. 7-2, in which the open-circuit voltage at the input terminals is conveniently represented by $E^{\mathrm{inc}}(f)h_{\mathrm{eff}}(f)$. Here, E^{inc} is the total incident field in the absence of the test target at a conveniently chosen distance, r_0, from the transmitter.

We define $h_{\mathrm{eff}}(f)$ because, like $Z_I(f)$, it can be scaled simply from the equivalent circuit for a scale-model test target to the circuit for a full-size target of proportional geometry. From this scaling, we can compute the EM response of a prescribed load in the full-size target.

The measured scattered-field response, $V_L(f)$, determines $h_{\text{eff}}(f)$ in the circuit of Fig. 7-2 to be simply

$$h_{\text{eff}}(f) = \frac{(Z_I + Z_L)V_L(f)}{Z_L\, E^{\text{inc}}(f)} \tag{7-3a}$$

It is convenient to write E^{inc} as the product of pulser voltage, V_p, and a calibration factor, $e^{\text{inc}}(f)$, so that

$$E^{\text{inc}}(f) = V_P(f)e^{inc}(f)$$

$$e^{\text{inc}}(f) \equiv E^{\text{inc}}/V_P \tag{7-3b}$$

presupposing that $e^{\text{inc}}(f)$ has been calibrated at reference distance r_0 from the transmitter. (This calibration is essentially a function of transmitter- and ground-plane geometries and is examined in more detail in the discussion of E^{inc} in the next section.)

7.3 VALIDATION, CALIBRATION, AND ERRORS

Validation

To validate the operation of the EMTF for both input impedance and effective-height measurements, field experiments were performed with an IKOR pulser and compared with the results obtained from computer models.

Input Impedance, Z_I. It is important to establish that the input impedance of the test target is being measured accurately, both for its use in the equivalent circuit of Fig. 7-2 and for calibrating the incident field, $E^{\text{inc}}(f)$. Thus, TDR measurements were made of the input impedance of a 30-cm monopole and 30-cm square (with image) loop (7.5-cm vertical arms, 15-cm horizontal arm) and compared with values computed with the TWTD time-domain computer code.[1] A single straight wire served as the transmitter in place of a conical antenna, and the computer code modeled the complete experimental setup of transmitter, ground plane, and test target without E^{inc} having to be computed separately. These results are shown as a function of frequency in Figs. 7-3 and 7-4.

We find that there is good agreement between calculated and measured values of Z_I up to $\simeq$2.4 GHz, where $L/\lambda = 2.4$ for the monopole (Fig. 7-3), except for a gradual shift of the calculated curves with respect to the measured ones. Such shifts appear to cause negligible errors in the computed integrated responses of whip antennas on practical structures because they tend to cancel on integra-

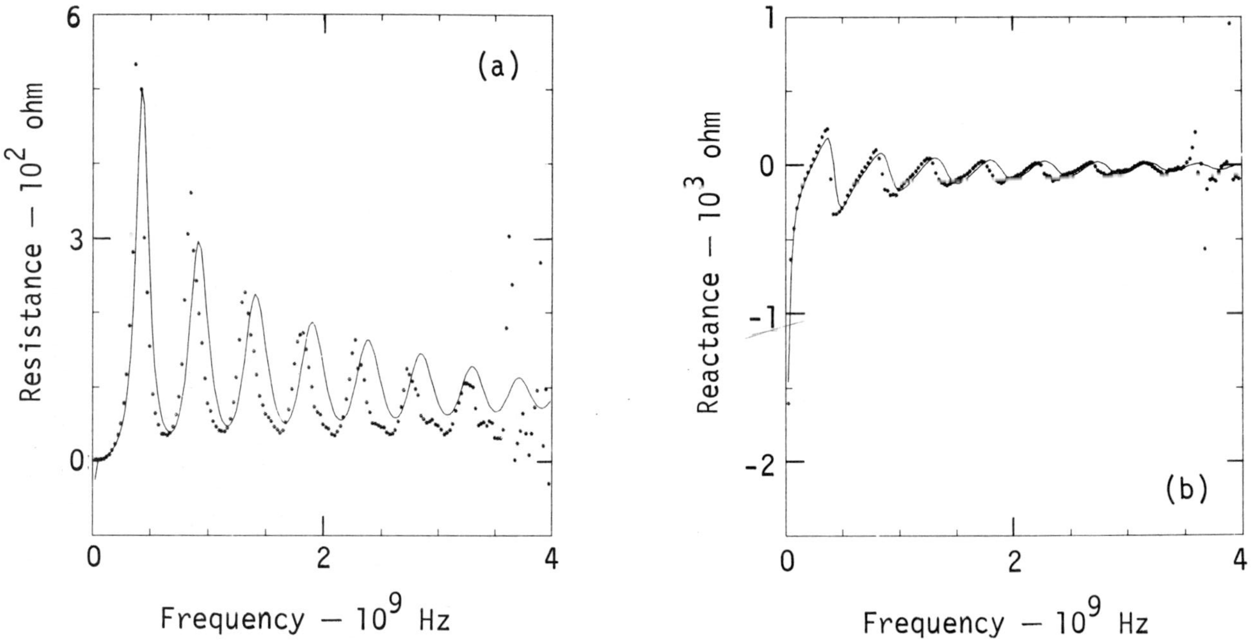

Fig. 7-3. TDR range measurments are validated by comparing the data-processed (**) real (a) and imaginary (b) parts of the input impedance, Z_i, of a 30-cm monopole on the ground plane with the impedance computed (—) from the time-domain computer code TWTD. The two curves gradually fall out of phase as frequency increases.

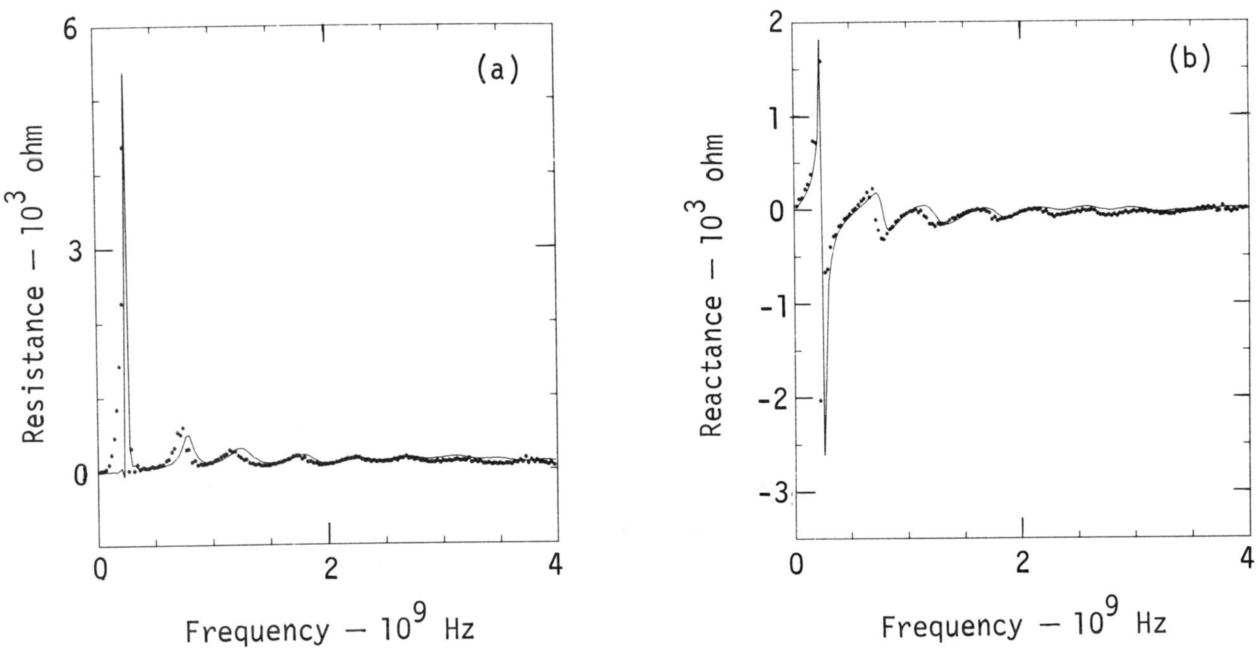

Fig. 7-4. TDR measurements are further validated by comparing data-processed (**) and computed (—) real (a) and imaginary (b) parts of the input impedance of a 30-cm loop on the ground plane. The incident field from the transmitter in Fig. 7-1 arrives "edgewise," with H^{inc} perpendicular to the loop, and the load terminals are at the "first arrival" point at the loop base.

tion. The good agreement between calculation and measurement extends to 4 GHz for the loop in Fig. 7-4, where L_p is the perimeter and $L_p/\lambda = 4$. At extremely low frequencies, the measured input resistance, R_I, has been fitted to an f^4 dependence [Fig. 7-4(a)]. This agrees with theory, and the curve rises to a maximum in a realistic way.

The narrower frequency response of the loop, in which the peak values of R_I and X_I are very large, implies a larger effect of the errors in measured $Z_I(f)$ on the derived EM response of a scaled-up structure. When we observe that approximately 99 percent of the EM energy absorbed by a loaded wire or loop usually occurs for $L/\lambda \lesssim 1$ or $L_p/\lambda \lesssim 1$, respectively (see the section on Range Limitations for elaboration), the errors in Z_I should have no appreciable impact on integrated response data.[2]

Effective Height, h_{eff}. Knowledge of the effective height, $h_{eff}(f)$, is required for the equivalent circuit in Fig. 7-2. This circuit is used to calibrate the incident field, $E^{inc}(f)$, as described subsequently, and is also scaled to compute the EM response of a life-size structure. Determining the effective height requires that the terminal current or voltage induced by a known incident field be obtained. The measurement or calculation by which this is accomplished can be performed in either the frequency domain or time domain. Therefore, by comparing measured and computed values of $V_L(t)$ or $V_L(f)$ for the same configuration, we can establish the relative validity of both.

In the present case, a 63.75-cm monopole was used for the transmitting antenna while a 30-cm monopole and a 30-cm square loop (7.5-cm vertical arms, 15-cm horizontal arm), respectively, were used as the targets. The sizes were chosen to permit practical computer modeling of the complete experimental setup using the time-domain code, TWTD, without separate calibration of $E^{inc}(f)$. The results obtained are shown in Figs. 7-5, 7-6, and 7-7 (two loop orientations were employed and both the transient waveform and frequency spectra plotted). Good agreement is seen in the time as well as the frequency domain out to $\gtrsim 1$ GHz ($L/\lambda \gtrsim 1$), where the EM temporal responses of the load energy absorbed and the peak current, for example, are essentially determined. This agreement validates the accuracy of EMTF measurements of load voltage induced on monopole and loop antennas of lengths greater than 10 cm.

We must note that these corroborations, while indicating that the EMTF does provide an accurate response of a given structure to a given incident field, do not ensure that an equally accurate measure of h_{eff} will be obtained. The reasons are that the EMTF source produces a field having wavefront curvature and $1/R$ attenuation, both of which can result in a response, and hence an h_{eff}, that is different from that which characterizes plane-wave excitation.

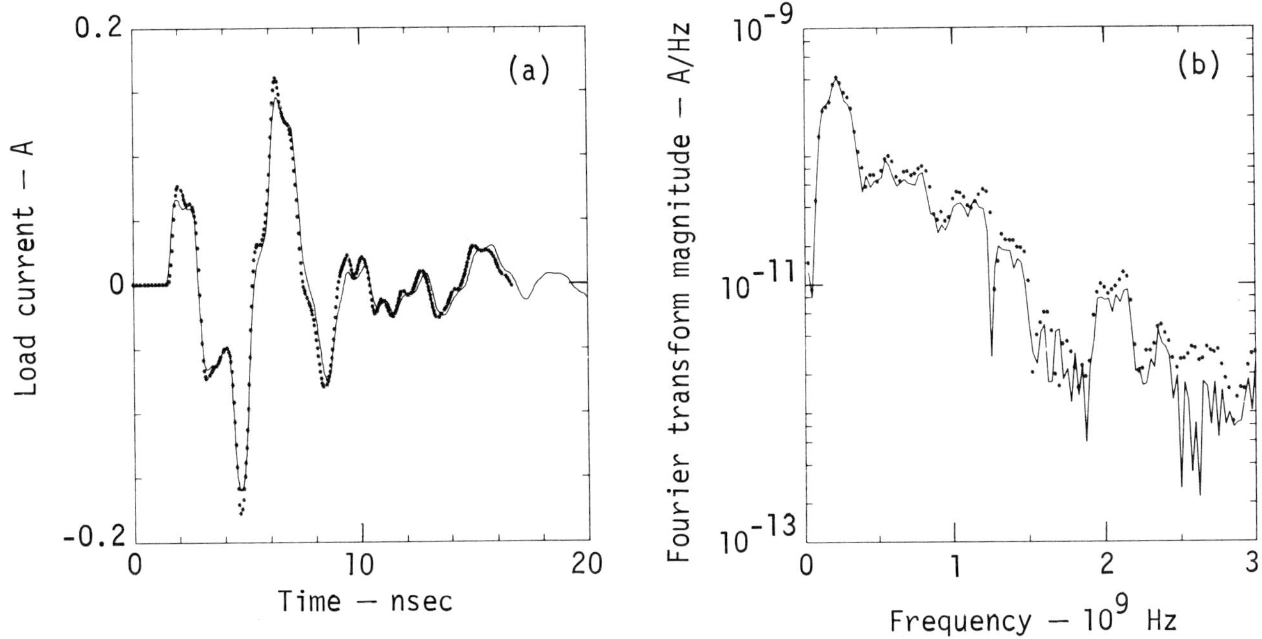

Fig. 7-5. The excellent validation of the scattered voltage measurements on the EMTF is shown by comparing, in (a), the measured short-circuit current, $I_L(t)$, in Fig. 7-2 with the current computed from the time-domain code for a 30-cm monopole on the ground plane. Their corresponding spectra also agree well in (b).

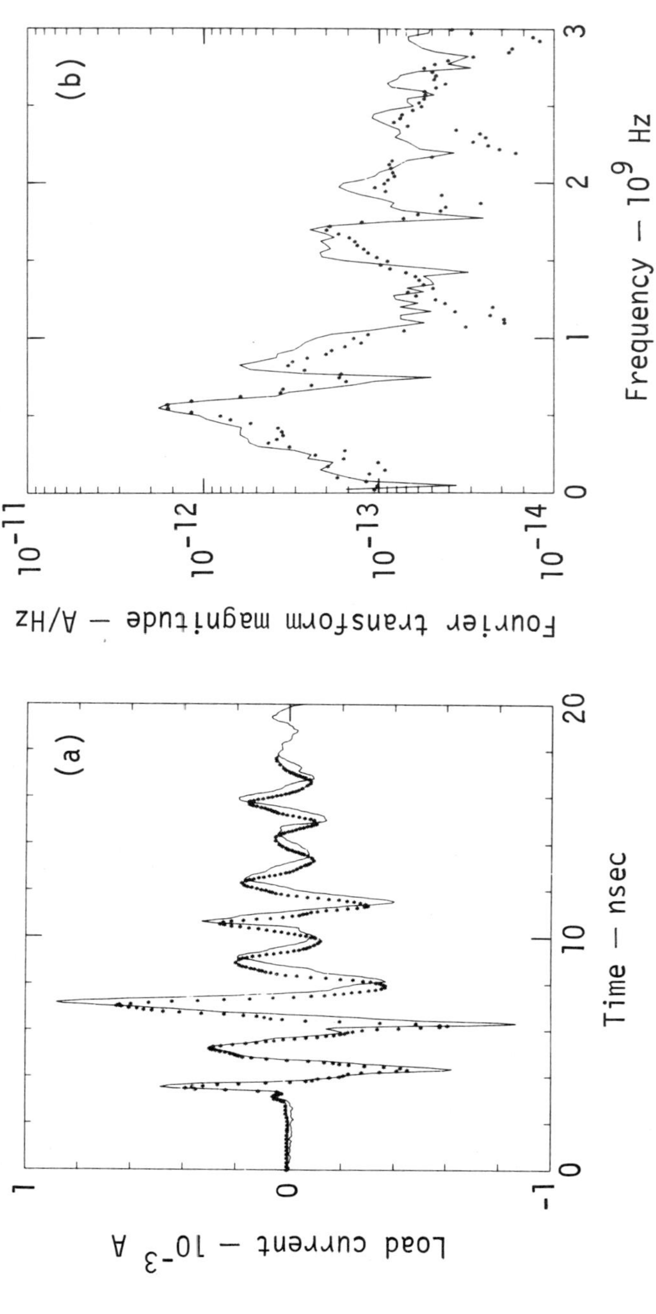

Fig. 7-6. Scattered voltage measurements on broadside-excited loops (H^{inc} from a straight-wire transmitter, parallel to the loop plane in Fig. 7-1) are validated by comparing, in (a), the measured and computed short-circuit current, $I_L(t)$, in Fig. 7-2 for a 30-cm loop on the ground plane. The spectra in (b) agree well out to 1 GHz where loop length/λ = 1.

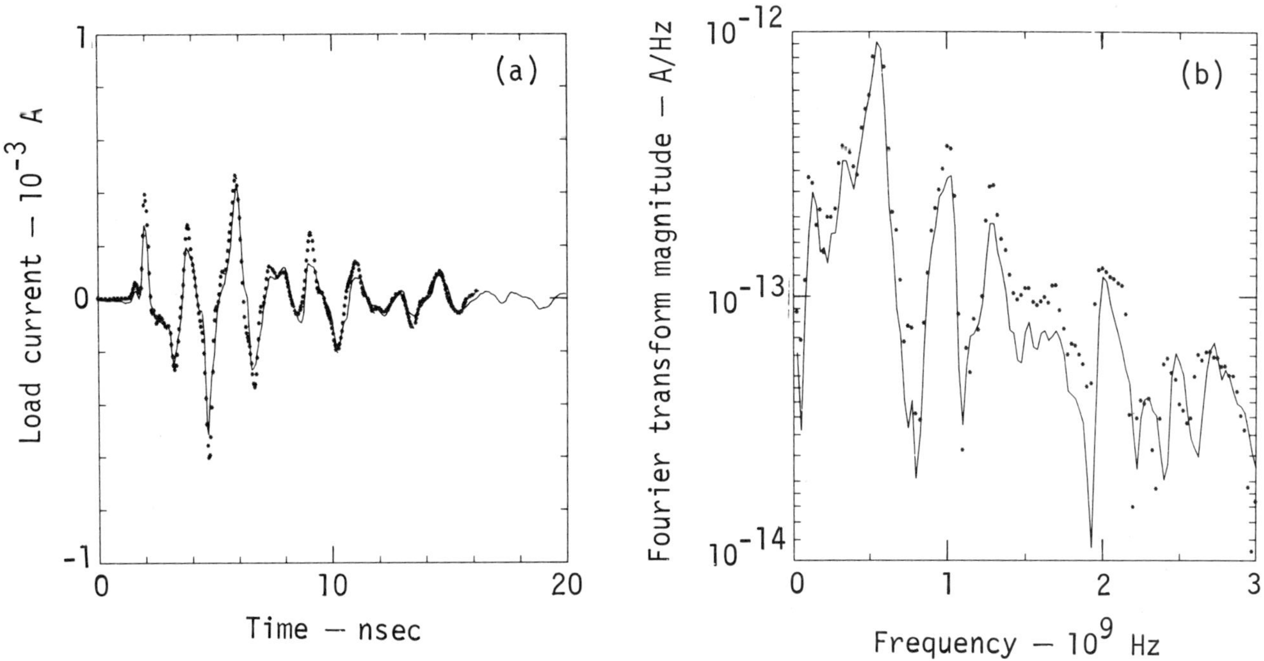

Fig. 7-7. Validation of the scattered voltage measurements on edge-excited loops (H^{inc} perpendicular to the loop plane in Fig. 7-1), similar to the validation of broadside-excited loops in Fig. 7-6. In (a), the measured and computed load currents in a 30-cm loop on ground plane are compared in the time domain; in (b), their spectra are compared.

Calibration of Transmitter Field, E^{inc}

Effective-Height Computation. The most direct way to calibrate $E^{inc}(f)$ at reference distance r_0 from the conical transmitter is to compute $h_{eff}(f)$ for a short monopole at r_0 to measure $V_L(f)$ and then to compute $E^{inc}(f)$ from Eq. 7-3, as follows:

$$E^{inc}(f)\bigg|_{cone} = \frac{(Z_I + Z_L)}{Z_L h_{eff}(f)} V_L(f) \tag{7-4}$$

where $Z_I(f)$ has also been computed (and is essentially the same as measured in the TDR experiment) and $Z_L = 50$ ohm.

When calibrating E^{inc} this way, we must allow for the fact that test antennas are placed about 1 m away from the transmitter; hence, for low frequencies, these antennas are in the near field of the transmitter. Therefore, E^{inc} will have the structure of a spherical wave field, and h_{eff} will represent this same field

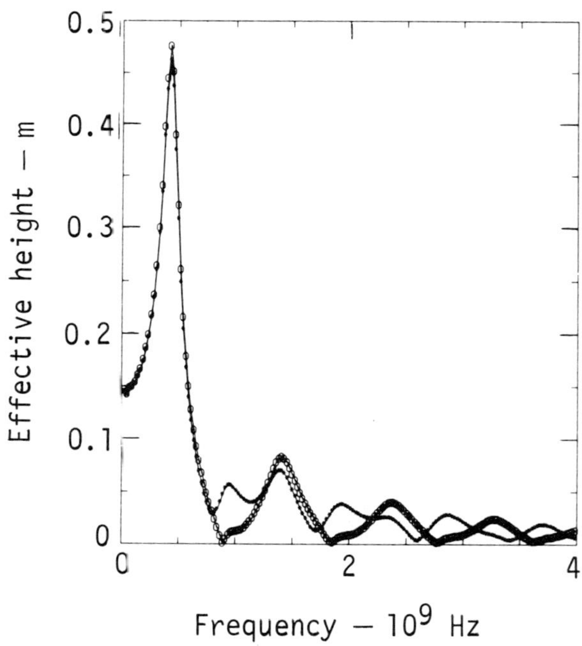

Fig. 7-8. There is a difference between the |effective height| of a 30-cm monopole on the ground plane as computed by the time-domain code for plane-wave and point excitation at the transmitter apex. Most EM integrated responses are determined in the range $\leq$ 1 GHz ($L/\lambda \leq 1$), where the error in processing effective height for point excitation is not serious.

structure. Thus, because nearly all the radiation comes from the conical transmitter apex, h_{eff} should be computed for a point-source field in the position of this apex. When measurements are made for only 20 nsec, radiation due to the reflection of current from the end of the transmitter is not seen at the test antenna. Because h_{eff} is computed for *point*-source radiation and the equivalent circuit is scaled up in size to estimate a *plane*-wave EM response by setting $E^{inc} = E^{EM}$ in the scaled circuit of Fig. 7-2, some error is introduced into the data-processing EM response. This error will now be discussed.

Effect of Field Curvature on h_{eff}. The differences between the effective height, $h_{eff}(f)$ computed by the TWTD code for point-source and for plane-wave illumination of the test target are shown in Figs. 7-8 and 7-9 for a 30-cm monopole on the ground plane and in Figs. 7-10 and 7-11 for a 30-cm rectangular loop (7.5-cm vertical arms, 15-cm horizontal arm) on the ground plane for edge incidence. The transmitter apex is located so that $r_0 \simeq 1.3$ m from the reference point for the field (the monopole position and the "first arrival" arm of the loop).

Figures 7-8 and 7-9 show that the differences between the magnitude and

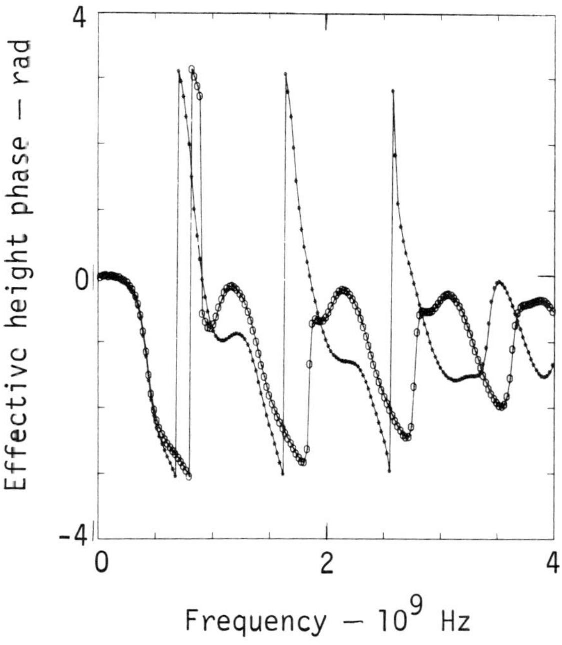

Fig. 7-9. The phase of the effective height of the 30-cm monopole of Fig. 7-8 also differs somewhat between plane-wave and point excitation, but the difference is not serious out to ≃1 GHz.

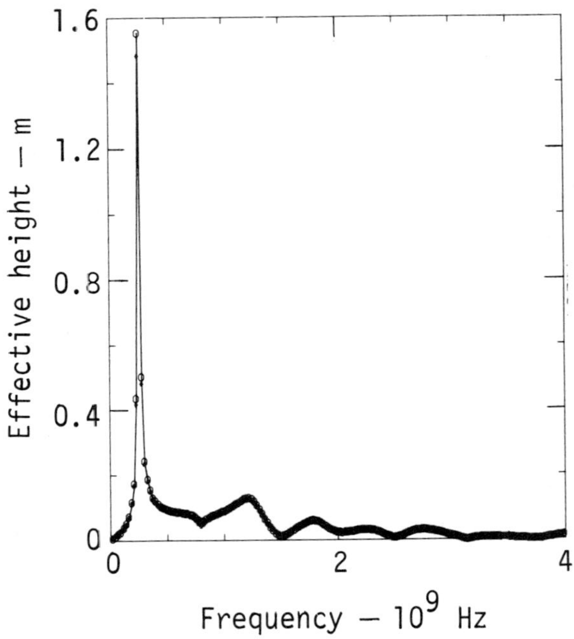

Fig. 7-10. There is essentially no difference between the |effective height| of the 30-cm loop of Fig. 7-7 computed for plane-wave and point edge-incidence, despite the fact that one vertical leg of the loop is 15 cm farther from the transmitter than the "first-arrival" leg.

phase of the monopole h_{eff} for the two illuminations are not appreciable in the interval $\lesssim$1 GHz ($L/\lambda \lesssim 1$) where the integrated responses are primarily determined.

Figures 7-10 and 7-11 show a much closer agreement between complex h_{eff} of the loop for the two illuminations, as far out as 4 GHz ($L_p/\lambda = 4$). This results, probably, because the height of the loop is only a quarter of the height of the monopole, making the field curvature a less important factor. These figures also imply that the near- and far-field variations as $1/R^3$, $1/R^2$, and $1/R$ negligibly affect h_{eff}, even though the loop extends 15 cm in the direction away from the transmitter (located at $r_0 \simeq 1.3$ m).

We find, indeed, that the integrated EM responses as computed from an equivalent circuit with $h_{eff}|_{point}$ differ negligibly from those computed with $h_{eff}|_{plane}$ for whip- and loop-antenna test targets on the EMTF. (We will develop this finding at greater length in the section on Range Limitations.)

Calibrating E^{inc}. Since $E^{inc}(f)$ of Eq. 7-4 is proportional to the pulser-voltage spectrum, $V_p(f)$, the calibration quantity

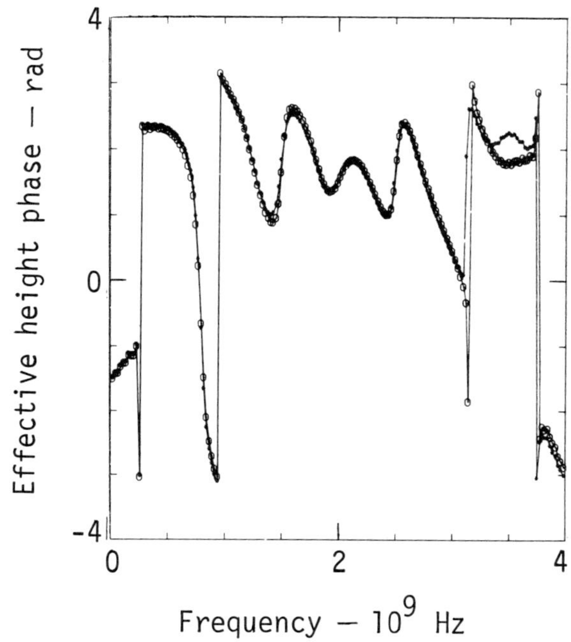

Fig. 7-11. The phase of the effective height of the 30-cm loop of Fig. 7-10 is also remarkably independent of whether plane-wave or point excitation is employed.

$$e^{\text{inc}}(f)|_{\text{cone}} = E^{\text{inc}}(f)|_{\text{cone}}/V_p(f) \tag{7-5}$$

obtained with the IKOR pulser but independent of it has been stored for data processing h_{eff} using Eq. 7-3. Its magnitude and phase are plotted in Fig. 7-12 and are observed to be well-behaved out to $\simeq$3.0 GHz. Also, e^{inc} is purely a function of transmitter geometry, scattering circuit geometry, and reference point r_0 (Fig. 7-1); it should be changed if either the transmitter or r_0 is changed. In addition, e^{inc} is implicitly dependent on the fact that scattering measurements are made only over a clear time of 20 nsec.

It is interesting to compare the calibrated $e^{\text{inc}}(f)$ of Eq. 7-5 with the value obtained from the theoretical biconical transmitter formula,

$$e_\theta^{\text{inc}}(r_0)|_b = E_\theta^{\text{inc}}(r_0)|_b/V_b = \frac{2}{\text{lncot}(\psi/2)} \times \frac{Z_{\text{bicone}}}{Z_{\text{bicone}} + 2Z_L} \frac{1}{r_0} \tag{7-6}$$

where $Z_{\text{bicone}} = 120 \text{ lncot } (\psi/2)$; ψ = cone half-angle ($\simeq 5°$ in this case); and Z_L = coax characteristic impedance. This formula includes the effect of the ground plane. Equation 7-6 is valid during the "clear" time of more than 20 ns

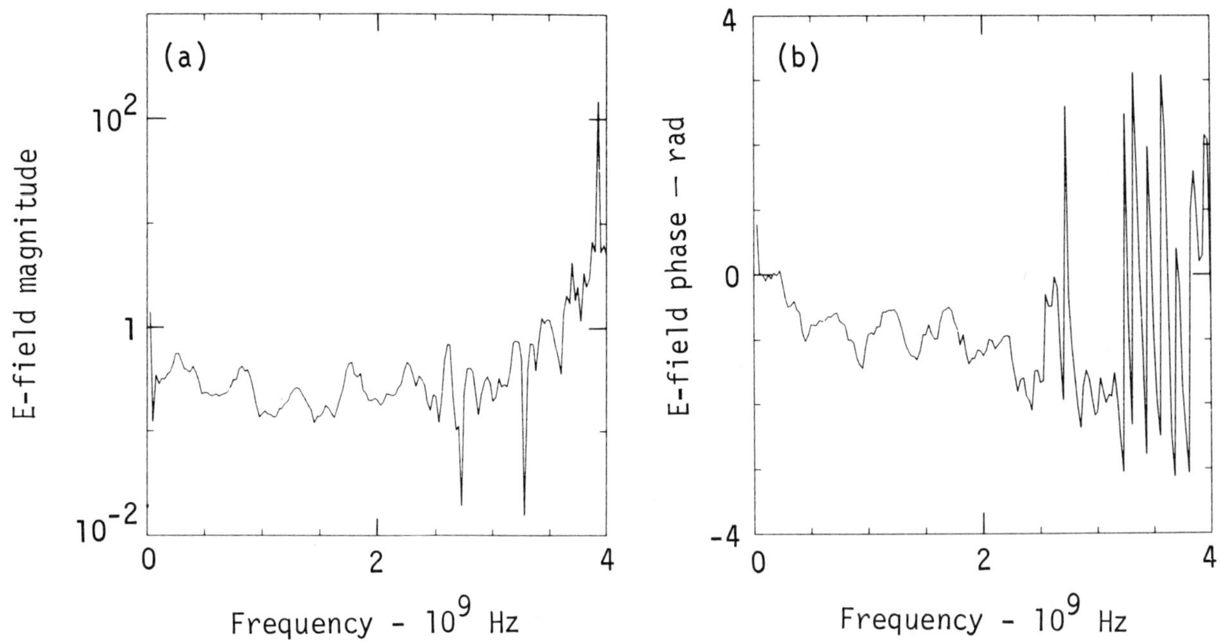

Fig. 7-12. The calibration factor, $e^{inc}(f)$, as obtained with the IKOR pulser, varies only moderately in magnitude (a) and phase (b) out to $\simeq$ 2.5 GHz; to obtain the EMTF incident electric field for effective height determination, e^{inc} may be used with any pulser of known spectrum.

before the field reflected from the end of the transmitting antenna reaches the test target. Although Eq. 7-6 predicts a frequency-independent $e^{\text{inc}}(r_0)|_b$, the measured value in Fig. 7-12 actually varies moderately out to $\simeq$2.5 GHz, beyond which point it becomes noisier.

Classification of Errors

The use of the EMTF to develop canonical geometry results that are intended for an EM data base raises several questions. Among them are the measurement accuracy attainable, the limitations due to EMTF constraints, and the relatability of scaled canonical data to real geometries. We will consider the first of these questions here; limitations will be discussed more fully in the sections that follow.

Error Sources and Types. Sources of measurement error can be divided into two categories, both of which may be further subdivided:

1. Systematic or deterministic errors, which are errors that result from cables (impedance, attenuation), components (couplers, attenuators, power divider), and sensors.
2. Statistical or nondeterministic errors, which are errors that result from pulser amplitude fluctuations, pulser time-base jitter, system noise, A/D (analog-to-digital) conversion, and interference.

Deterministic errors are those whose influences are measurable and removable from the data. Cable attenuation and phase shift are typical error sources of this kind. Examples of nondeterministic errors include interpulse variations in the pulser output and random "glitches" in the A/D conversion.

Both input impedance and transfer admittance (or effective height) are affected in similar ways by these errors. We can discuss their influence out of context of the measurement with the few exceptions noted below and consider systematic and statistical errors in turn.

Systematic Errors. Before discussing specific systematic error types, we should first re-examine how impedance and effective-height measurements are made to

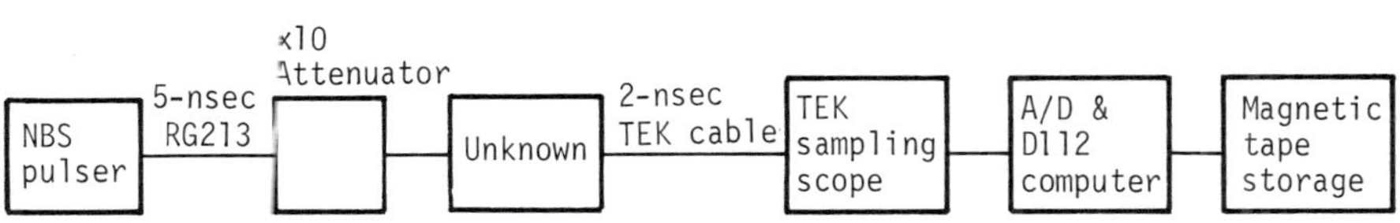

Fig. 7-13. The scheme for measuring frequency-domain insertion loss of a number of attenuators, coaxial cables, and the power divider. The signal is processed with and without the element in place and measures the attenuation in decibels.

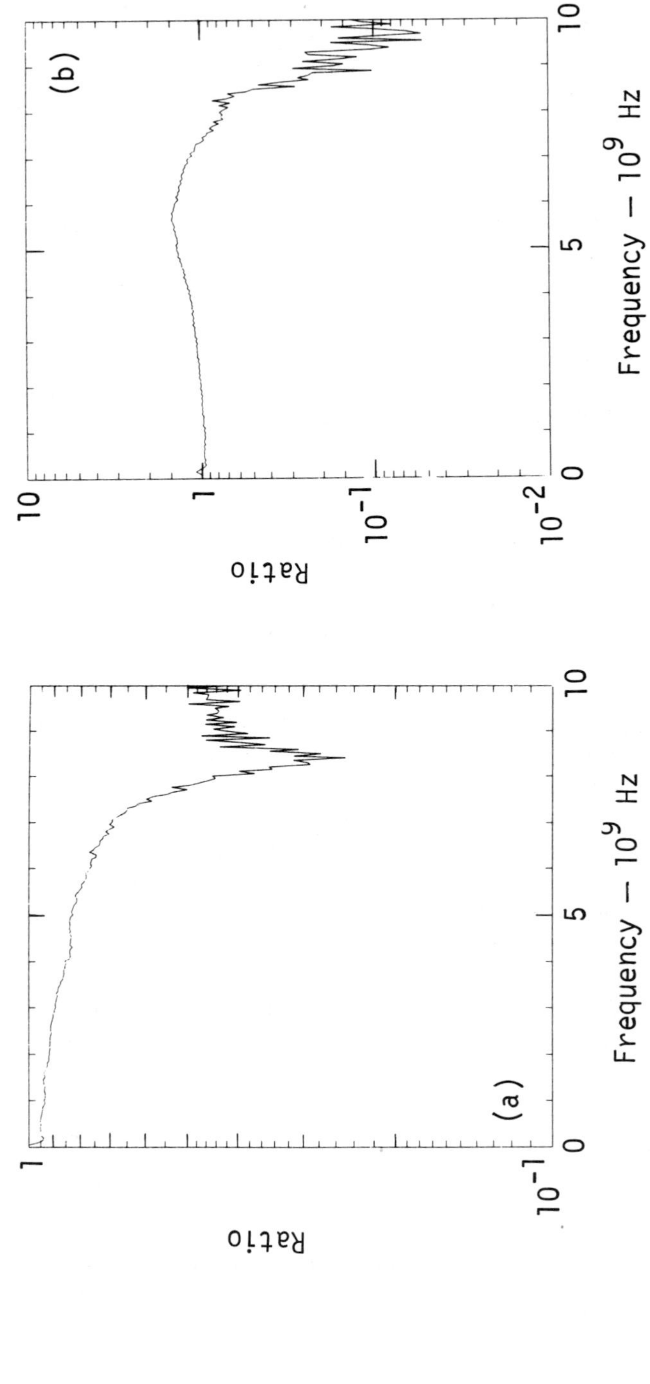
(a)
(b)
Ratio
Ratio
Frequency — 10^9 Hz
Frequency — 10^9 Hz
1
10^-1
10
1
10^-1
10^-2
0
5
10
0
5
10

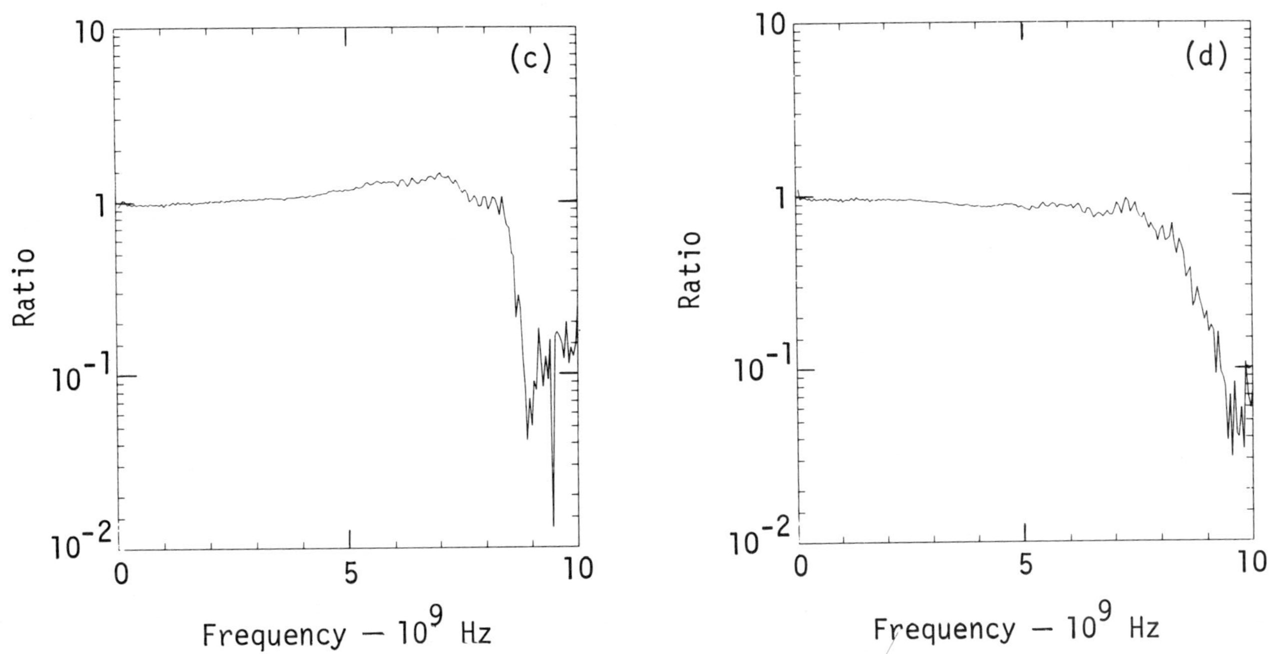

Fig. 7-14. Attenuation by the field-transmission factor relative to 1 (no attenuation) for (a) a 5 NS R6 213 cable, (b) $\times 10$ attenuator, (c) $\times 5$ attenuator, and (d) $\times 2$ attenuator.

determine how they may be affected by the various components involved. The typical arrangements used for such measurements were illustrated in Fig. 7-1. The 70-ft delay line used in both applications is required to permit pretriggering the sampling scope approximately 70 nsec before the arrival of the actual measured pulse. The input-impedance determination involves measuring the signal reflected from an open circuit and from the antenna being tested, using the same setup for both. Doing so automatically compensates for both cable and component phase shift and attenuation (see Eq. 7-1). One could also make the effective-height system identical for the two measurements required to obtain the pulser output and antenna response, but, for experimental convenience, we did not do so. The effects of the extra attenuator and cable must therefore be determined.

The characteristics of the various components were measured in the time domain with a simple insertion-loss arrangement (see Fig. 7-13). Results from these measurements are presented in Figs. 7-14 and 7-15. We observe that (1) loss due to the extra section of cable used for the scattering measurements is $\lesssim 1$ dB up to $\simeq 4$ GHz; (2) the attenuators are flat to within 1.6 dB of their specified values up to 4 GHz; and (3) the signal splitter is flat to within 1.0 dB up to $\simeq$ 4 GHz.

In most scattered-field measurements of effective height, the circuitry between the test target and scope was the same as for the calibration; thus, no systematic error was introduced into the measurements from these components. The error in V_L does affect the calibration of $E^{\mathrm{inc}}(f)$ in Eq. 7-4. For an unknown antenna configuration, however, the error affects V_L in a determination of $h_{\mathrm{eff}}(f)$ in exactly the same way:

$$h_{\mathrm{eff}}(f) = \frac{(Z_I(f) + Z_L)}{Z_L} \times \frac{V_L(f)}{V_p(f)\; E^{\mathrm{inc}}(f)/V_p(f)} \tag{7-7}$$

Thus, there is no error in the measured h_{eff} due to systematic errors in this situation. Since these errors are absent in a TDR measurement of $Z_I(f)$, we can then ignore deterministic errors on the computed EM responses of real-life configurations.

In a few measurements with an NBS pulser, however, one or a combination of the following alterations were made: (1) the signal splitter was removed, (2) the attenuator was changed, or (3) one section of cable was removed or added. According to the above error values, these changes created a systematic error of $\lesssim 3$ dB in the frequency domain.

Statistical Errors. Statistical errors are ones caused primarily by pulser amplitude fluctuations and time-base jitter, sampling scope noise, and converter or computer processing noise. Statistical errors are not only more difficult to assess than systematic errors, but usually cannot be removed from the data as well.

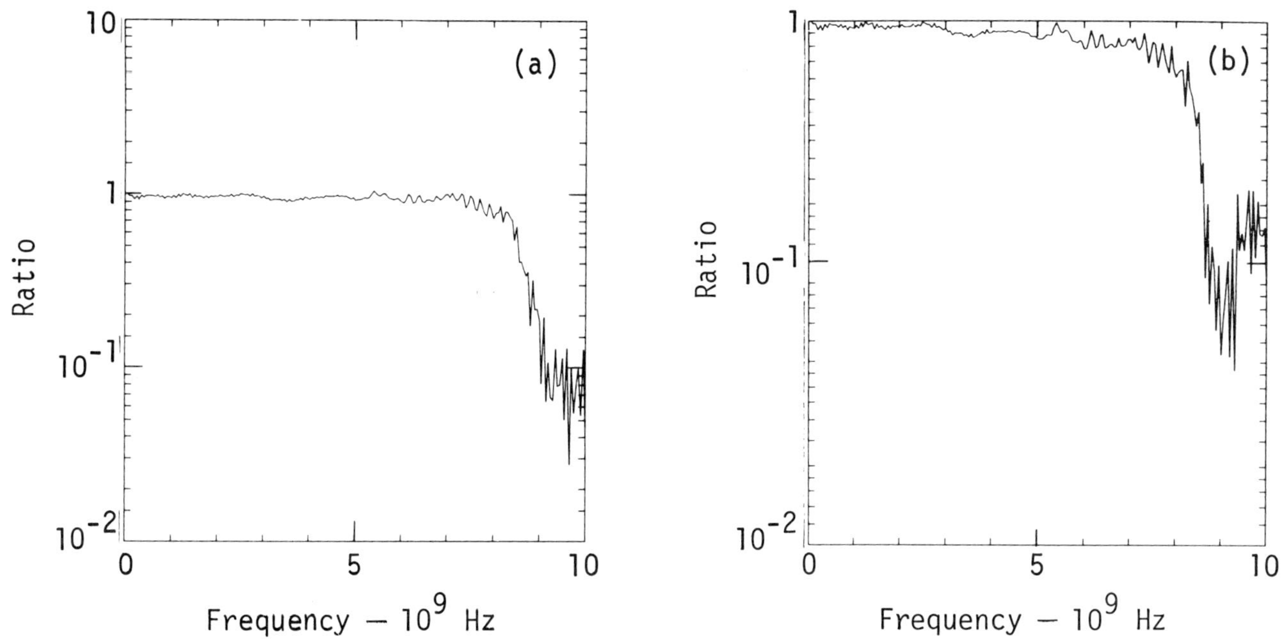

Fig. 7-15. Attenuation by the field transmission factor relative to 1 (no attenuation) of the power divider in Fig. 7-1: (a) the forward path in a TDR measurement, and (b) the backward, right-angle path.

Possible exceptions are "glitches" resulting from external interference or the dropping of bits or words by the computer, an example of which can be seen in the scattered voltage measurement of Fig. 7-16 (at $\simeq$14 nsec). In such cases, a cursory examination of the raw data, which is routinely plotted after each measurement, will reveal the problem, and the measurement can then be repeated. Glitches that are less obvious, however, might not be so identified and thus remain uncorrected in the data. Glitches such as those shown in Fig. 7-16, however, occur so infrequently and have such a high-frequency spectrum that they invariably have negligible effect on the EM response derived from EMTF data in the important spectral range $L/\lambda \lesssim 1$.

Pulser fluctuations affect the EM response in three independent ways: In the measurements of scattered voltage $V_L(f)$ (see Eq. 7-7), in the measurement of pulser voltage itself, and in the calibration of $E^{\text{inc}}(f)$ (see Eq. 7-4).

Evidently, a potentially severe source of statistical error is the pulser itself. Because the sampling scope develops waveforms by taking a succession of N evenly spaced samples, a minimum of N pulses are required to complete the measurement. The variations that occur in successive pulses can thus appear as noise in the measured waveform. These variations include the amplitude and shape of the pulse as well as the possible jitter in the time reference. The time

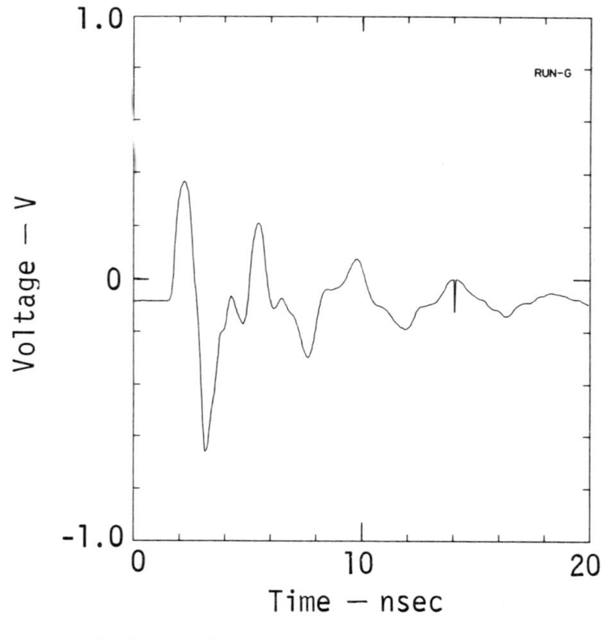

Fig. 7-16. This scattered voltage picked up by a 30-cm monopole shows a downward "glitch" at about seven units of time, or 14 nsec.

reference is established by using a portion of the pulse to trigger the sampling scope. Consequently, jitter can arise from variations in the leading edge of the pulse relative to the trigger level. Signal processing can reduce the influence of both such effects, as demonstrated subsequently.

We performed a series of measurements to evaluate the statistics of the two pulsers (an IKOR EM model and an NBS-built pulser) most often used for EMTF operation. Assuming constant amplitude pulses (or an average over the independent fluctuations in amplitude), we found that for both pulsers, the measured time-base jitter was negligible compared to the pulse width of ≃300 psec for the IKOR pulser and ≃150 psec for the NBS pulser. The standard deviations were about ±2.6 and ±2.2 psec, respectively. Consequently, we discuss only amplitude statistics (see Figs. 7-17 through 7-20).

First, comparing the amplitude statistics of Figs. 7-17 and 7-18, we find that the pulse amplitude variance of the NBS pulser is vastly less than that of the IKOR pulser. Not only is the NBS pulser more stable on a pulse-to-pulse basis, but also it has a much greater repetition rate, adjustable up to 50×10^3 pulse/sec, compared with ≃200 pulse/sec for the IKOR. Thus, a measurement can be made in a substantially shorter time. The single disadvantage of the NBS pulser is its

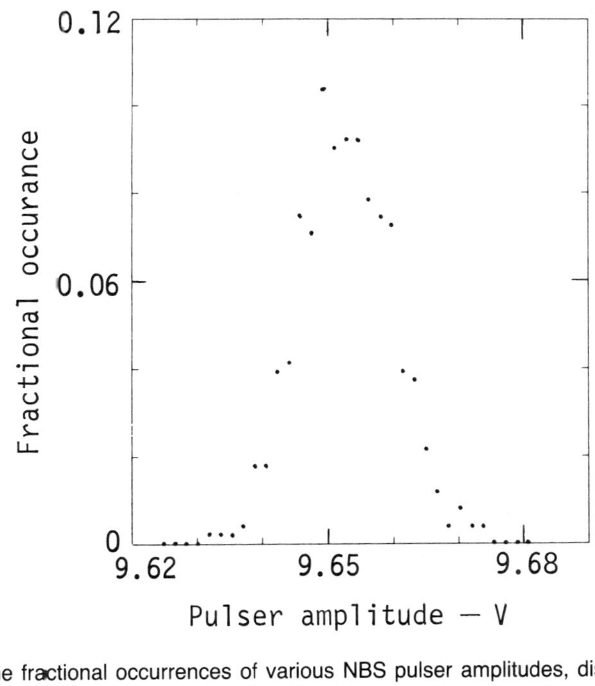

Fig. 7-17. The fractional occurrences of various NBS pulser amplitudes, distributed about an average of approximately 9.65 V. About half the time, the voltage lies within ±0.01 of the average.

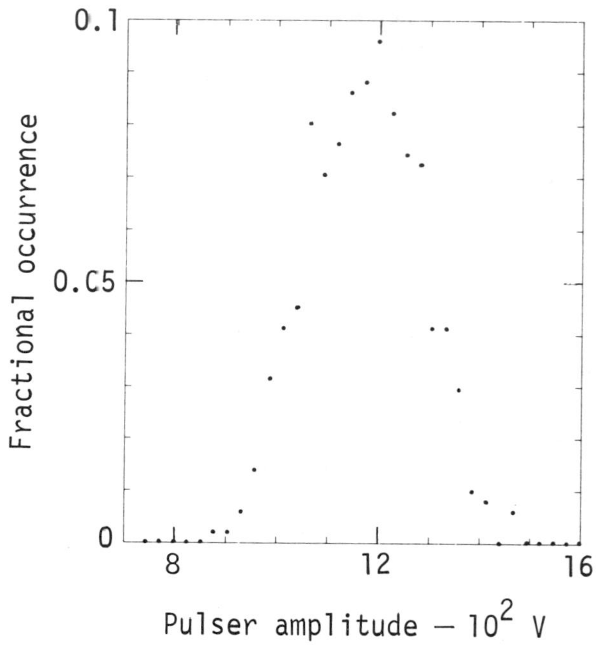

Fig. 7-18. The fractional occurrences of various IKOR pulse amplitudes, with an average of about 1.2 kV. Approximately half the pulse amplitudes lie in an interval of ±0.1 kV of the average.

lower pulse amplitude, which, at a ≃10-V peak, is a factor ≃100 times less than that of the IKOR pulser. Considering, however, that for most applications, attenuators are needed to reduce the input to the sampling scope to accommodate an acceptable peak signal, this lower pulse amplitude is generally not a serious limitation.

Second, upon examining the corresponding pulses and spectra of the two pulsers (Figs. 7-19 and 7-20), we observe that the NBS pulser (Fig. 7-20) provides more than twice the frequency coverage, extending past 8 GHz, compared with the 4-GHz spectrum available from the IKOR pulser (Fig. 7-19). Also demonstrated in these figures is the effect of signal averaging upon the measurement. When making wideband transient measurements where pulse-to-pulse variations can be substantial, preprocessing of the waveform can be very advantageous.

In Figs. 7-19(a), 7-19(b), 7-20(a), and 7-20(b) we have developed a composite waveform by averaging nine individual measurements M times for each point in the temporal sampled signal. Results are given for $M = 4$ and 16, although we do observe that M need not be prespecified but can be established by continuing to sample a given point until the variance falls below a desired value. (Note that since the sampling scope provides one sample per pulse, an N-point sampled waveform would require MN total pulses.) Each of the nine individual pulser-time waveforms was measured and Fourier-transformed to the frequency domain, where the sample mean, expressed as

$$\overline{V}_p(f) = \frac{1}{9} \sum_{i=1}^{9} |V_{pi}(f)|$$

at each frequency, and sample variance expressed as

$$\begin{aligned} \sigma &= (V_p(f))_{\text{rms}} \\ &= \left(\frac{1}{8} \sum_{i=1}^{9} |V_{pi}(f) - \overline{V}_p(f)|^2 \right)^{1/2} \end{aligned}$$

at each frequency, were computed. The pulse waveforms and spectra were subsequently plotted as in Figs. 7-19 and 7-20 to show the mean values and the $\pm 0.67\ \sigma$ limits. These illustrate the bounds within which 50 percent of the measurement points would fall. We observe that the consequent uncertainties in the waveforms and spectra are considerably less for the NBS pulser, as earlier implied by the amplitude statistics. We also see that increasing the number of samples per data point from $M = 4$ to $M = 16$ decreases the standard deviation of the IKOR pulser by more than a factor of 2 above 2 GHz and decreases the σ of the NBS pulser dramatically in the range of 3 to 6 GHz, with significant improvement out to 9 GHz.

For most measurements on test targets, it is important that the spectrum of pulser voltage have a negligible error out to $\lesssim$3 GHz. This is necessary because most configurations measured on the EMTF have antennas $L \geqslant 10$ cm long and the EM responses of interest (e.g., load energy absorbed) are essentially determined in the range $L/\lambda \leqslant 1$ (see Sec. 7.3). The EM-response data are currently processed from $M = 16$-point averaged temporal waveforms; Figs. 7-19(b) and 7-19(d) indicate that the IKOR pulser voltage has the requisite error; and Figs. 7-20(b) and 7-20(d) show that the NBS pulser also satisfies the desired error criterion. The curves in these figures illustrate the advantage of simply averaging the raw temporal data from the EMTF.

One additional observation should be made concerning the NBS pulser spectrum. When M is increased from 4 to 16, the variance of the measured spectrum decreases, and the high-frequency content of the spectrum also is appreciably

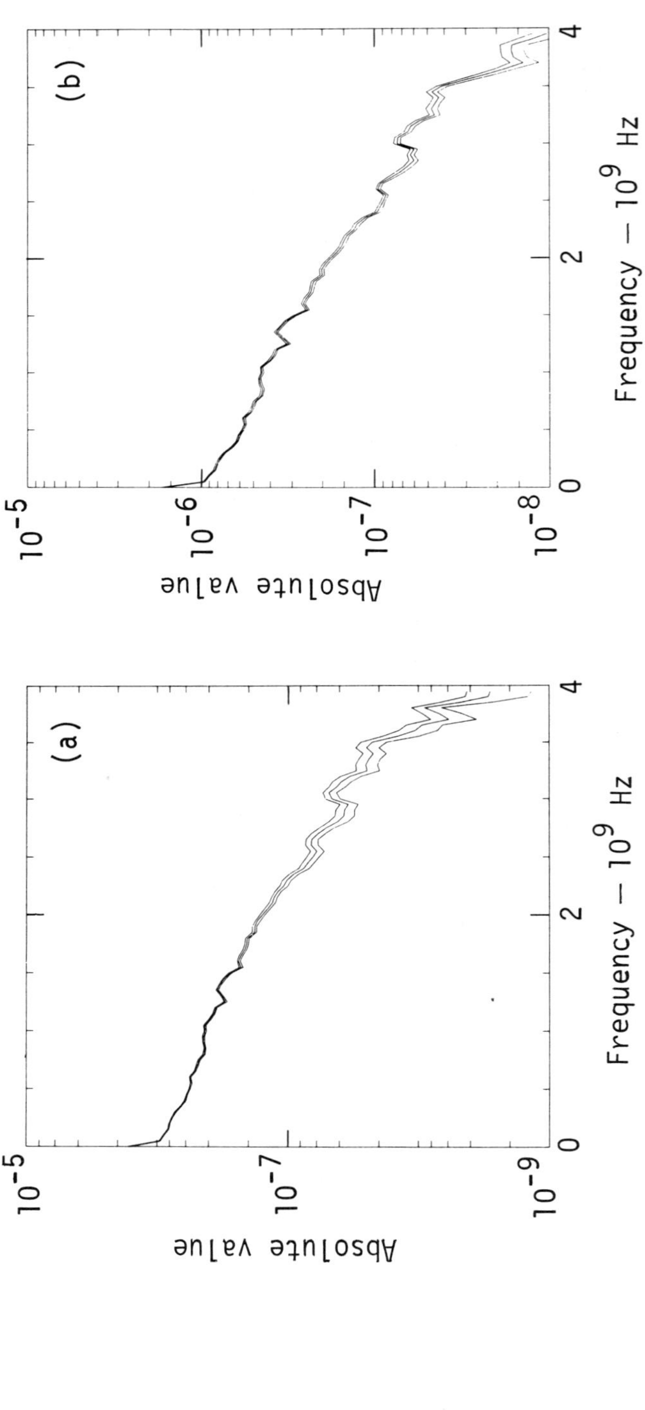
(a)
(b)
Absolute value
Frequency — 10^9 Hz
10^-5
10^-6
10^-7
10^-8
10^-9
0
2
4

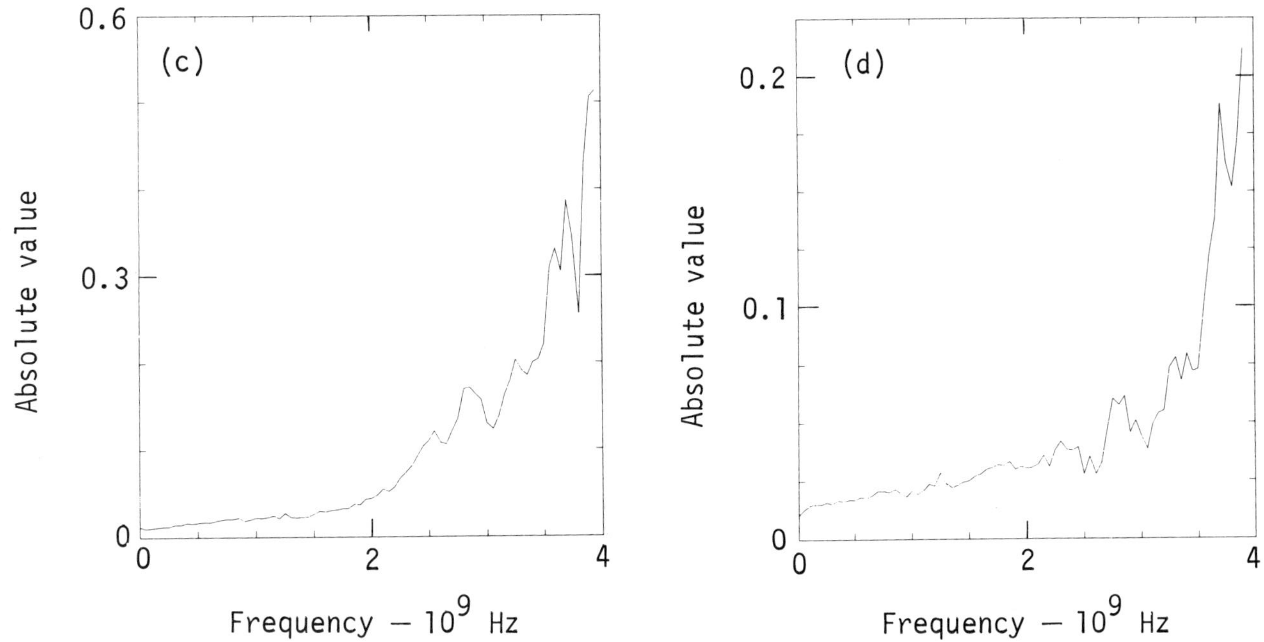

Fig. 7-19. Statistical variations of the IKOR pulser waveform appear as variations of its spectrum. In (a), we show the results of nine independent measurements of the spectrum magnitude, where each temporal point has been averaged four times to reduce statistical variations. The middle curve is the average spectrum; the upper and lower curves show the $\pm 0.67\sigma$ points at each frequency, where σ is the sample standard deviation. The spectrum in any experiment will lie between the upper and lower curves about 50 percent of the time. In (b), we show the same information as in (a), but each pulser waveform has been averaged 16 times at each time point. The additional averaging has reduced the spectral uncertainty. In (c), we show the statistical uncertainty in (a) in another way; the one-way spread is divided by the average spectrum vs frequency. Plot (d) shows the one-way spread divided by the average spectrum for (b) and indicates the reduction in uncertainty resulting from the increased temporal averaging of points on each pulser waveform.

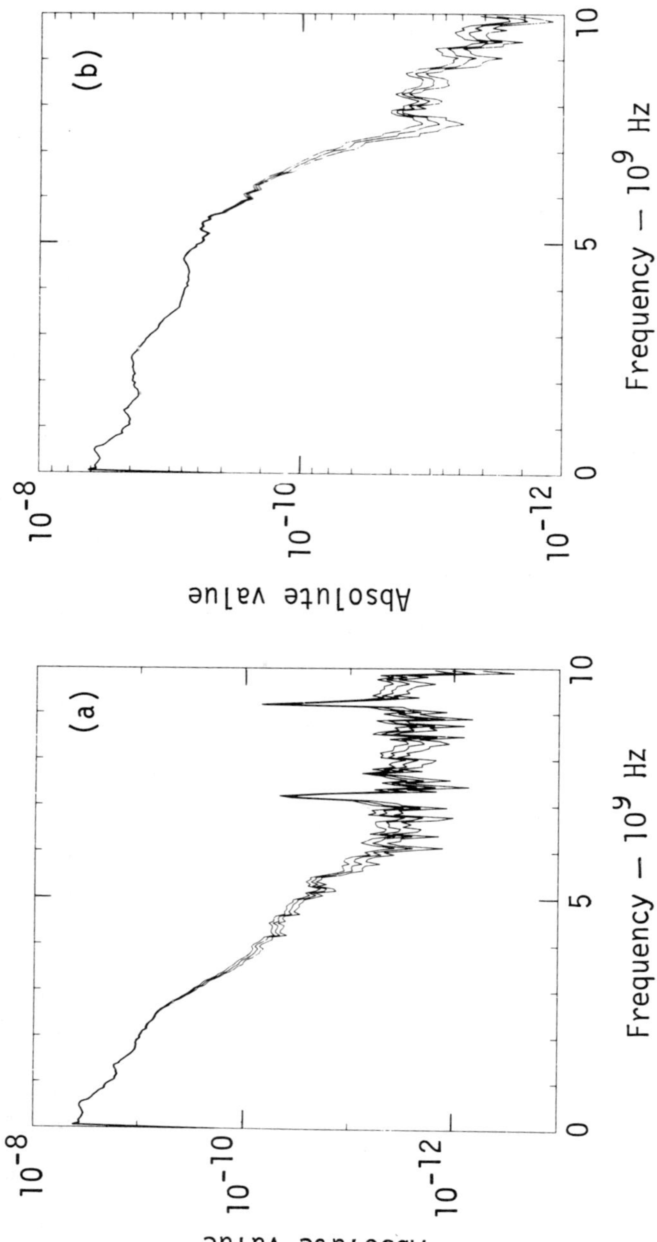
(a)
(b)
Absolute value
Frequency — 10^9 Hz
10^-8
10^-10
10^-12
0
5
10

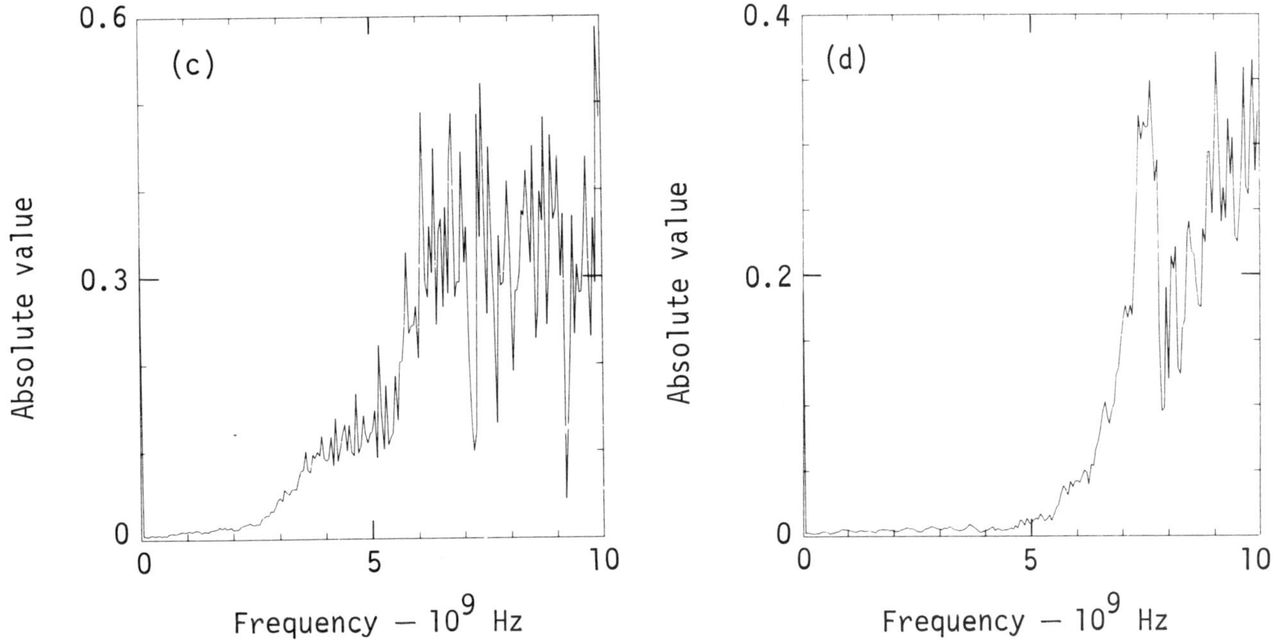

Fig. 7-20. The same statistical variations shown for the IKOR pulser in Fig. 7-19 are here seen in the NBS spectrum. In (a), the average spectrum magnitude and its $\pm 0.67\sigma$ points are plotted, as determined by the sample standard deviation over nine measurements; each temporal measurement has been averaged four times. Plot (b) shows the reduction in statistical uncertainty when we average each time point 16 times instead of four. Plot (c) shows the one-way spread in (a) divided by the average spectrum. The striking reduction in this "relative" one-way spread when we increase the temporal averaging from 4 to 16 is shown in (d).

enhanced [compare Figs. 7-20(a) and 7-20(b)]. The reason for this is not clear but appears to be related to the operation of the sampling scope.

System noise is exemplified by sampling scope noise. The signal-to-noise or S/N ratio of the scope is estimated by first measuring the noise spectrum N of the scope [Fig. 7-21(a)] and then by measuring S on the scope with the signal scattered from a 5-cm monopole excited by the NBS pulser [Fig. 7-21(b)]. The ratio of the two spectra (S/N) is shown in Fig. 7-21(c), and, over a range of about 0.5 to 4.5 GHz, the mean S/N ratio exceeds 10. The ratio drops below 1 around 0.25 GHz, however, and can appreciably distort the EM responses of a life-size structure derived from a scale model, with strong reponse in the neighborhood of 0.25 GHz. Figure 7-21(c) represents a "worst-case" S/N in that most monopoles and loops measured in various configurations on the EMTF are longer than 5 cm and pick up sufficient signal to maintain an S/N ratio greater than 10 in the frequency range of strong antenna response.

For longer whips and loops in various configurations, this excessive low-frequency noise introduces noticeable variations in the day-to-day, integrated response data.

Errors that may arise from the other components listed (e.g., the computer and A/D converter) are somewhat more obscure as to their origin and influence. One of the most obvious problems is the random glitches already mentioned. Another potentially more serious problem is caused by the limited word size of the computer and converter. To demonstrate this effect, Fig. 7-22 shows the reduction in the dynamic range of the spectrum of a Gaussian waveform due to a smaller word size with white noise added. Figure 7-22(a) shows the spectrum of the discretized Gaussian of piecewise-linear behavior,

$$g_n = 4000e^{-(n-50)^2/25} \tag{7-8}$$

for $0 \leq n \leq 100$, as obtained from a fast Fourier transform with a CDC 7600 floating-point word size (48 bit mantissa).

Figure 7-22(b) shows the spectrum as obtained with only a 12-bit word size to simulate the operation of the digital components on the EMTF. The dynamic range has shrunk from $\simeq$12 decades with the longer word length to $\simeq$4 decades out to $\simeq$5 GHz with the 12-bit word length. The addition of a small amount of white noise to the Gaussian waveform (noise amplitude about 0.12 percent of signal amplitude in the time domain, a value characteristic of the EMTF system) reduces the dynamic range with 12-bit words to $\simeq$2.5 decades out to $\simeq$4 GHz [see Fig. 7-22(c)].

These figures show the extreme importance of operating the sampling scope so that the peak-to-peak excursion of the measured waveform provides a full-scale deflection on the scope face. Otherwise, a rather drastic reduction of the available dynamic range will occur.

Range Limitations

Four limitations of the EMTF for assessing the EM plane wave response of full-sized structures have now been discussed. Three pertain to the nonplane-wave nature of the incident field: wavefront curvature, geometric attenuation ($1/R$ effect), and near-field components ($\sim 1/R^2$, $\sim 1/R^3$). The fourth is the bandwidth restriction to 3 GHz to avoid high-frequency errors in the data recording and processing.

The bandwidth, f_{max}, of the EMTF is determined by the pulser spectrum, the transfer function of the transmitting antenna, and noise in the A/D conversion of a processed waveform. Figure 7-19 shows that the IKOR pulse amplitude becomes negligible beyond 3 GHz and the noise in its spectrum becomes excessive. Figure 7-20 shows the same to be true for the NBS pulser if each measurement is averaged four times at each time point, but we observe that the spectrum is increased to $\gtrsim$6 GHz if each time point is averaged 16 times. The behavior of the transfer function in Fig. 7-12, $e^{inc}(f)$, independent of pulser, is sensibly constant out to $\simeq$3 GHz, which means that the transmitting antenna departs from its ideal biconical model at higher frequencies. We see also in Fig. 7-22 that the effect of 12-bit words in the A/D conversion and noise on the processed waveforms limits the bandwidth to $\simeq$3 GHz. For these various reasons, f_{max} is considered to be essentially 3 GHz.

The following limitations are discussed relative to the temporal EM excitation,

$$E^{EM} = E_0[e^{-\alpha t} - e^{-\beta t}] \tag{7-9}$$

where $E_0 = 52.5\text{kV}$, $\alpha = 4 \times 10^{+6}$, and $\beta = 4.76 \times 10^8$.

Nonplane-wave Incident Field. To assess the overall effect of these three factors on the EM data of structures with whip and loop antennas, the following experiments were conducted. The input impedence, $Z_I(f)$, of a 30-cm monopole on the ground plane was determined from measurement of the equivalent circuit of Fig. 7-2. Also, $h_{eff}(f)$ was computed for theoretical point excitation at the apex of the transmitter using the time-domain computer code TWTD. (Any convenient temporal excitation with sufficiently high spectral content will suffice for computing h_{eff}.) The circuit was scaled up to the full-size monopole length in the range $1 \leq L \leq 100$ m. The EM response was obtained for resistive loads in the range, $1 \leq R_L \leq 10^6$, and contour plots of W_L and the other integrated responses were plotted in the R_L–L plane.

Next, the whole procedure was repeated, starting with the measured Z_I but using $h_{eff}(f)$ as computed for the theoretical plane-wave excitation at the apex of the transmitter. The differences in the final response contours serve to estimate

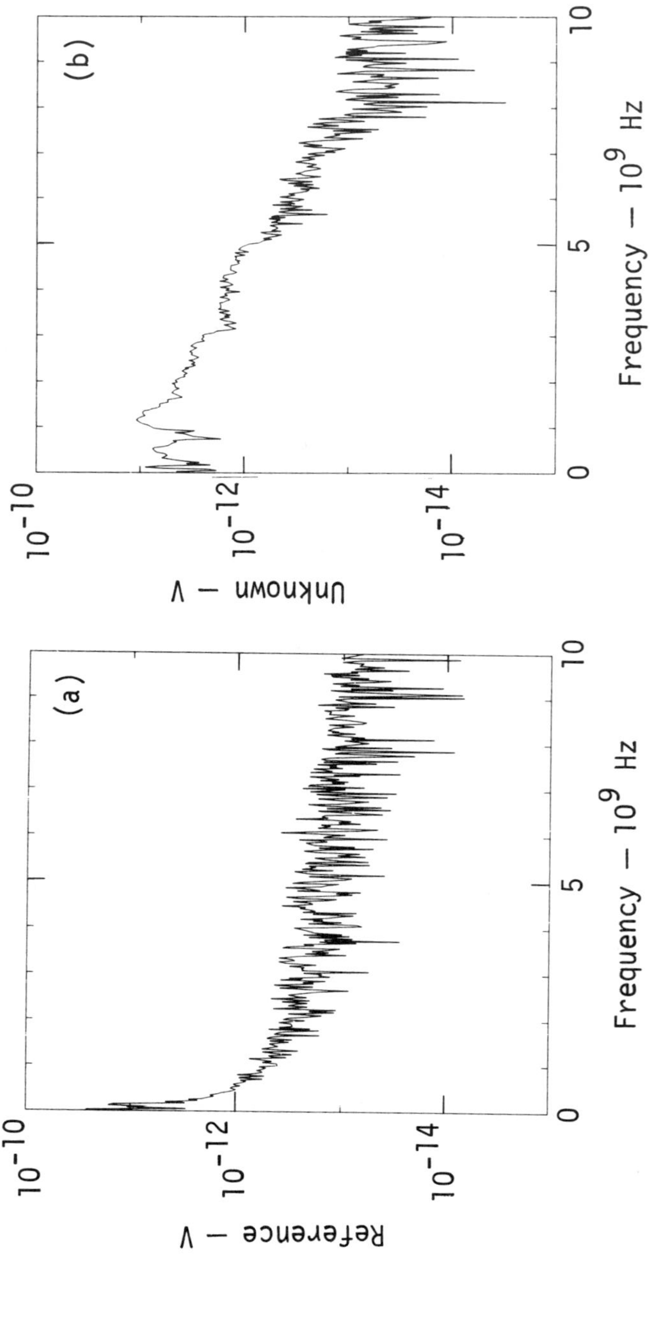
(a)
Reference — V
Frequency — 10^9 Hz
(b)
Unknown — V
Frequency — 10^9 Hz

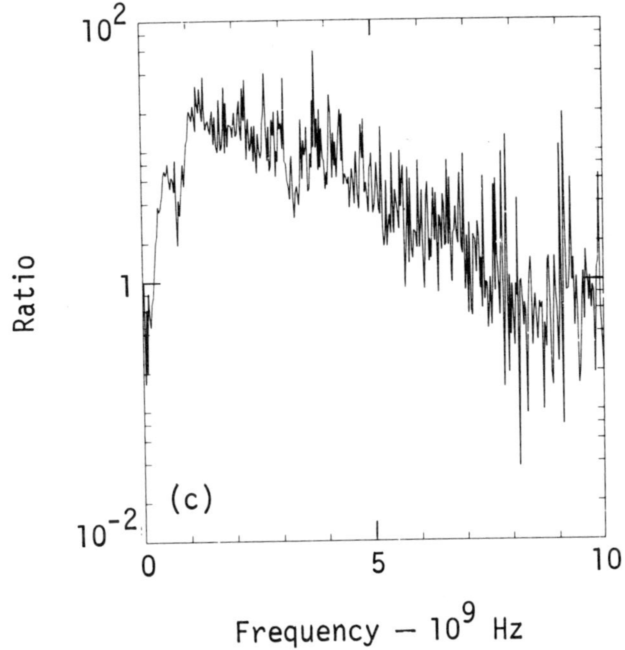

Fig. 7-21. When using the NBS pulser, the signal-to-noise, or *S/N*, ratio is low in a narrow frequency range of ≃0.25 GHz. The scope-noise spectrum, *N*, is shown in (a) and the scattered-voltage signal, *S*, of a 5-cm monopole appears in (b). Their quotient, *S/N* in (c), dips below 1 at low frequencies but is generally greater than 10, out to 4.5 GHz.

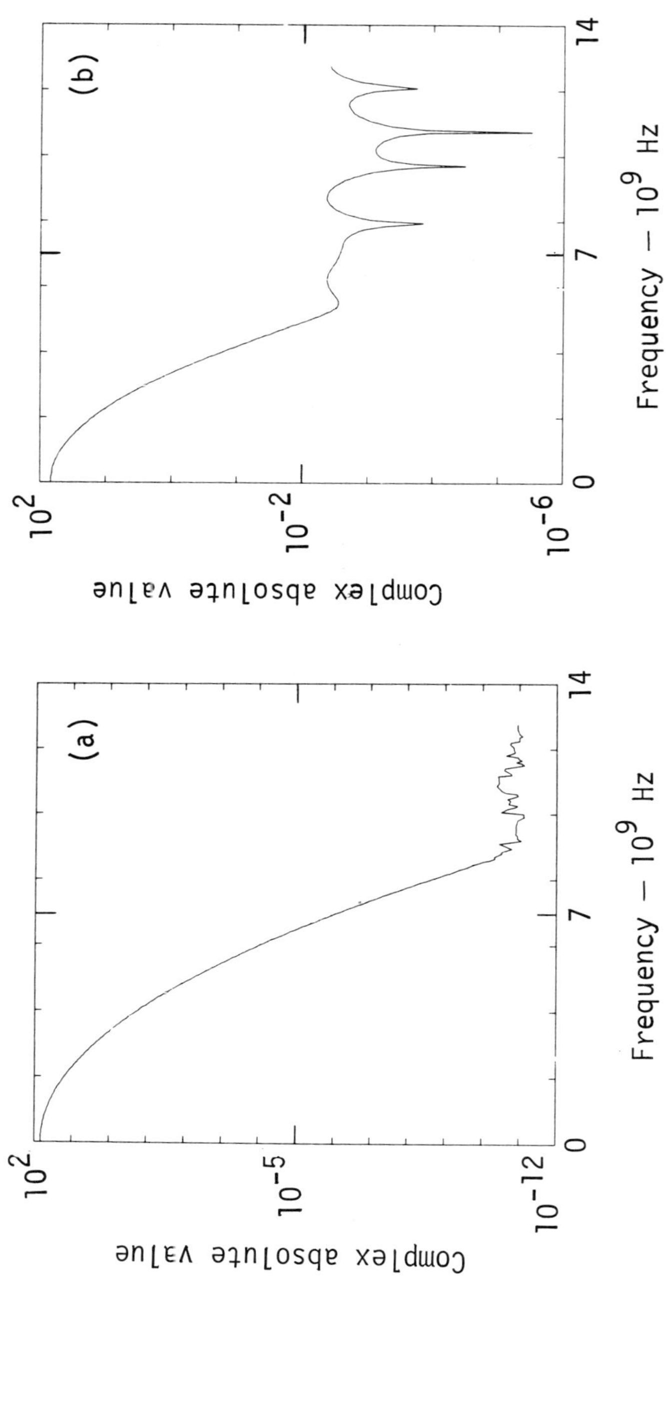
(a)
Complex absolute value
10^2
10^{-5}
10^{-12}
0
7
14
Frequency — 10^9 Hz
(b)
Complex absolute value
10^2
10^{-2}
10^{-6}
0
7
14
Frequency — 10^9 Hz

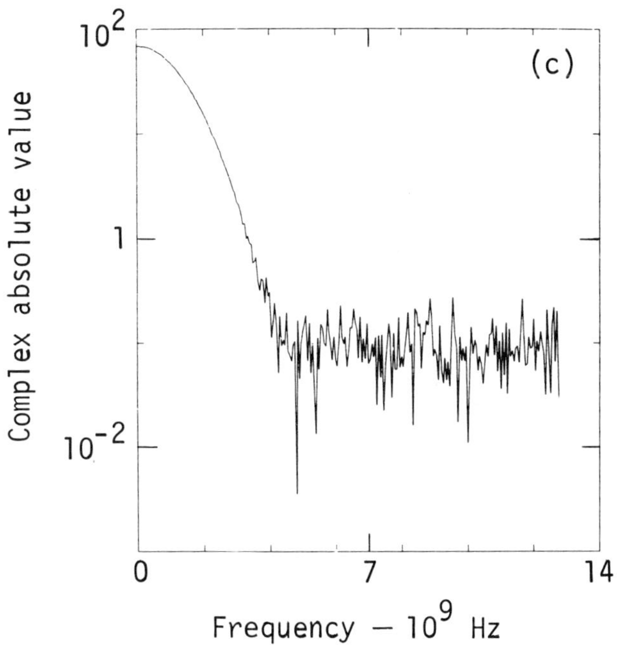

Fig. 7-22. Reduction in the dynamic spectral range of a processed waveform, caused by both limited computer word size and noise: If we used a 48-bit computer word size, we would have the spectrum of a discrete Gaussian waveform shown in (a). Use of 12-bit words corrupts the spectrum above ≃5 GHz in (b). Addition of a representative amount of white noise further reduces the useful dynamic range to ≃4 GHz in (c).

the effect of using an h_{eff} measured on the EMTF and scaled up to represent h_{eff} for plane-wave EM excitation of a full-size antenna configuration.

The differences in the response contours of this monopole are entirely negligible for all the responses of interest, indicating that phase differences for point and plane-wave excitations are not important, despite the relatively tall monopole (30 cm) compared to the distance from the transmitter apex ($r_0 \simeq 1.3$ m).

The above procedure was repeated for a 30-cm rectangular loop (15-cm horizontal arm, 7.5-cm vertical arms), connected for "edge-on" incidence (H^{inc} perpendicular to the plane of the loop) and with the load at the "first arrival" point in the test-target position (Fig. 7-1). The EM response contours were plotted in the R_L–L_p plane (L_p = full-scale loop length) and compared for h_{eff} computed by point and by plane-wave excitation.

Again, the discrepancies are negligible, as exemplified by the energy contours in Fig. 7-23. Evidently, the geometric attenuation and near-field components cause negligible error in these EM responses, despite the fact that the loop extension away from the transmitter (15 cm) was about 10 percent of the distance from the nearest arm to the apex ($r_0 \simeq 1.3$ m).

Effect of Truncating the Frequency Response. To determine the L/λ range in which the EM integrated responses (energy dissipated, W_L, peak current, voltage, power, average power, and 0.1- to 0.9-W_L pulse width) are accurately determined, we reprocessed the data to obtain the EM responses of the whip-on-box and loop-on-box. Instead of including the frequency response out to 4 Ghz, we truncated it at 1 GHz, where $L/\lambda = 1$. Comparison of the full-frequency response contours in the R_L–L plane for various independent EMTF measurements on the scale models indicates very small differences. The largest differences are seen for peak current, and Fig. 7-24 shows that, for the 30-cm whip-on-box, even these are small. (The differences for the 30-cm loop-on-box were comparable.)

These comparisons indicate the very important fact that the integrated EM responses of full-size structures with antenna length L in the range $1 \leqslant L \leqslant 100$ m, are essentially determined by the responses for $L/\lambda \leqslant 1$. In appreciation of this fact, Fig. 7-25 plots the EM current responses in the frequency domain for a short monopole ($L = 1$ m) and loop ($L_p = 1$ m perimeter) on the ground plane. Even for these short antennas, the effect of the high-frequency roll-off of the EM spectrum seen in Fig. 7-25(a) is small. We have also plotted $2|E^{inc}(f)|$ vs f, twice the nominal EM spectrum of Eq. 7-10 because of ground-plane reflection. The magnitude of the transfer function,

$$T(f) = \frac{I_L(f)}{E^{inc}(f)} = \frac{h_{eff}(f)}{Z_l(f) + 300} \tag{7-10}$$

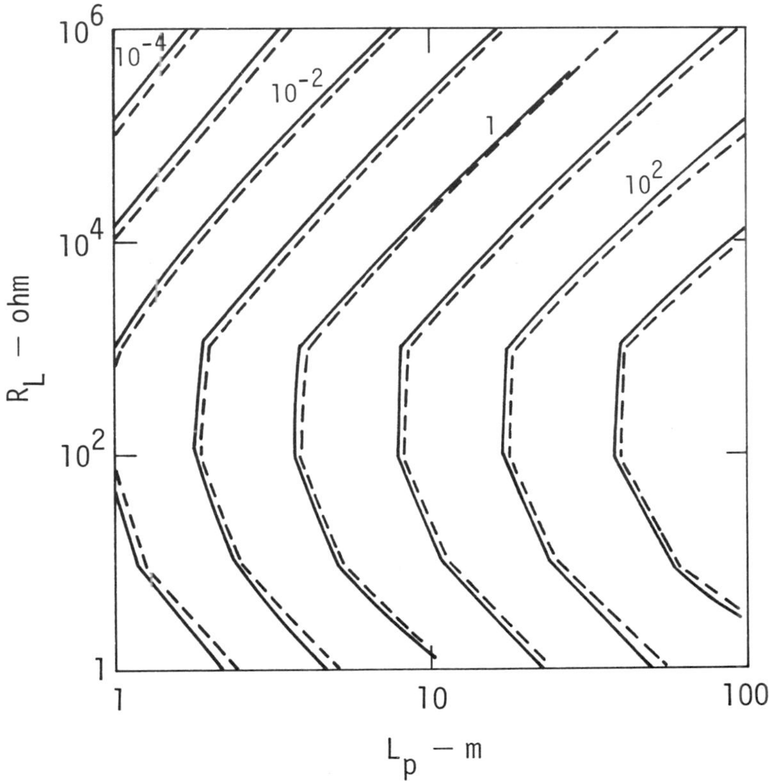

Fig. 7-23. Energy contours (joules) illustrating the effect of wave-front curvature, field $1/R$-variation, and near-field components on h_{eff} and EM-integrated response data: Solid-line energy contours are derived from scattered voltage measurements of a 30-cm loop on a ground plane using h_{eff} computed for the plane-wave incidence in Fig. 7-10. The differences in the contours show that the errors in h_{eff} caused by the nonplane-wave nature of the incident field have a negligib e effect on the derived EM response of full-size antenna configurations.

is plotted in Fig. 7-25(b) for the monopole resistively loaded by 300 ohm. This is about the "worst" case load for maximum load energy and peak load power absorption in most whip and loop antenna configurations studied.[2] We see that $|T|$ diminishes steadily above $L/\lambda = 1$ ($f = 0.3$ GHz). The product of the curves in Figs. 7-25(a) and (b) is current, $|I_L(f)|$, as shown in Fig. 7-25(c). Above $L/\lambda = 1$, the spectrum is more than two orders of magnitude below the low-frequency peak and diminishes rapidly as f increases.

We now consider an extremely long monopole of $L = 1000$ m, again loaded with 300 ohm. According to the scaling laws $T_{L=1000}(f) = 1000T_{L=1}(1000f)$.

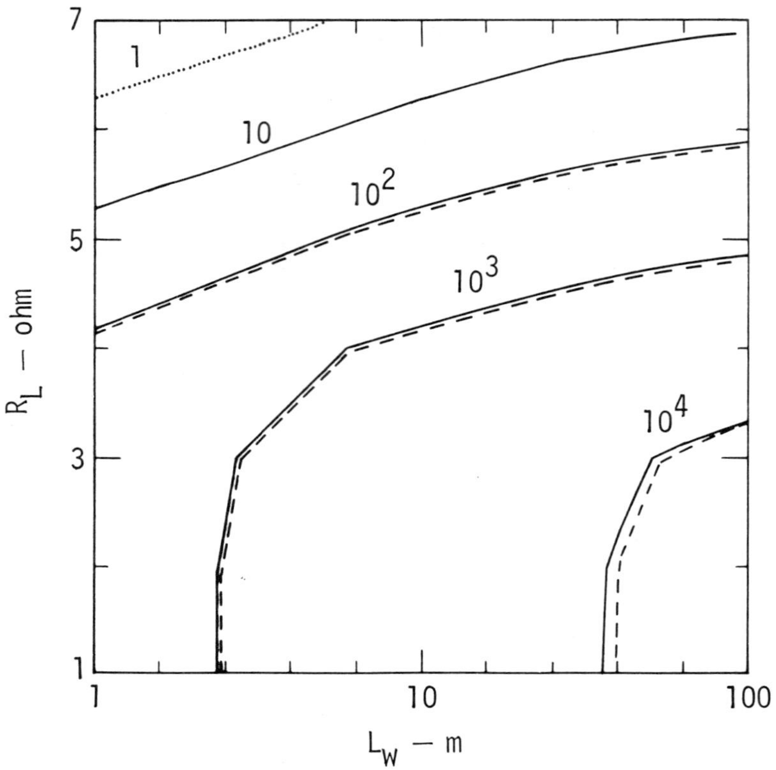

Fig. 7-24. Peak current contours show the dependence of integrated EM responses of full-size antenna configurations on the frequency response where $L/\lambda < 1$. EMTF measurements (whip-on box, Fig. 7-26) were processed to obtain the EM contours in the $R_L - L_w$ plane, according to full-frequency behavior and to the $L_w/\lambda \leqslant 1$ response of the scaled-up equivalent circuit. The largest changes are seen for the peak-current response for solid to dashed curves. Effects of spectrum truncation on loop-on-box EM responses (Fig. 7-26) are negligible.

This means that T for the 1000-m monopole has the same relative shape as T of Fig. 7-25b, but now the scale on the frequency axis is divided by 1000. The $L/\lambda = 1$ point occurs at the same relative position on the scale. At $L/\lambda = 1$, E^{inc} is almost at its low-frequency peak; at the end of the scale, $f = 1.2$ MHz and E^{inc} is about half its peak value. Therefore, the current spectrum looks similar to that in Fig. 7-25(b) with a slightly faster roll-off above $L/\lambda = 1$. Since the integrated EM responses are essentially determined by the $L/\lambda \leqslant 1$ spectrum, evidently there is much cancellation of high-frequency contributions to these responses, even for $L = 1000$ m.

Similar remarks apply to loops of $L_p = 1$ m and 1000 m. Figure 7-25(d)

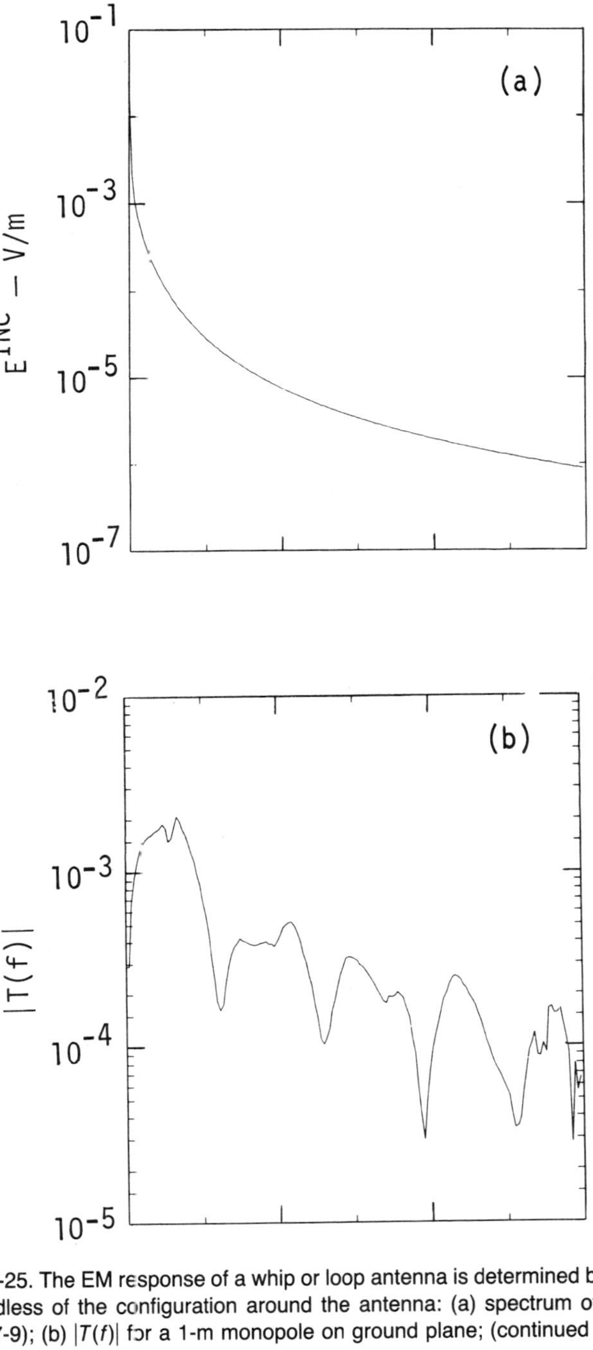

Fig. 7-25. The EM response of a whip or loop antenna is determined by its $L/\lambda \lesssim 1$ behavior, regardless of the configuration around the antenna: (a) spectrum of incident nominal EM (Eq. 7-9); (b) $|T(f)|$ for a 1-m monopole on ground plane; (continued on next page).

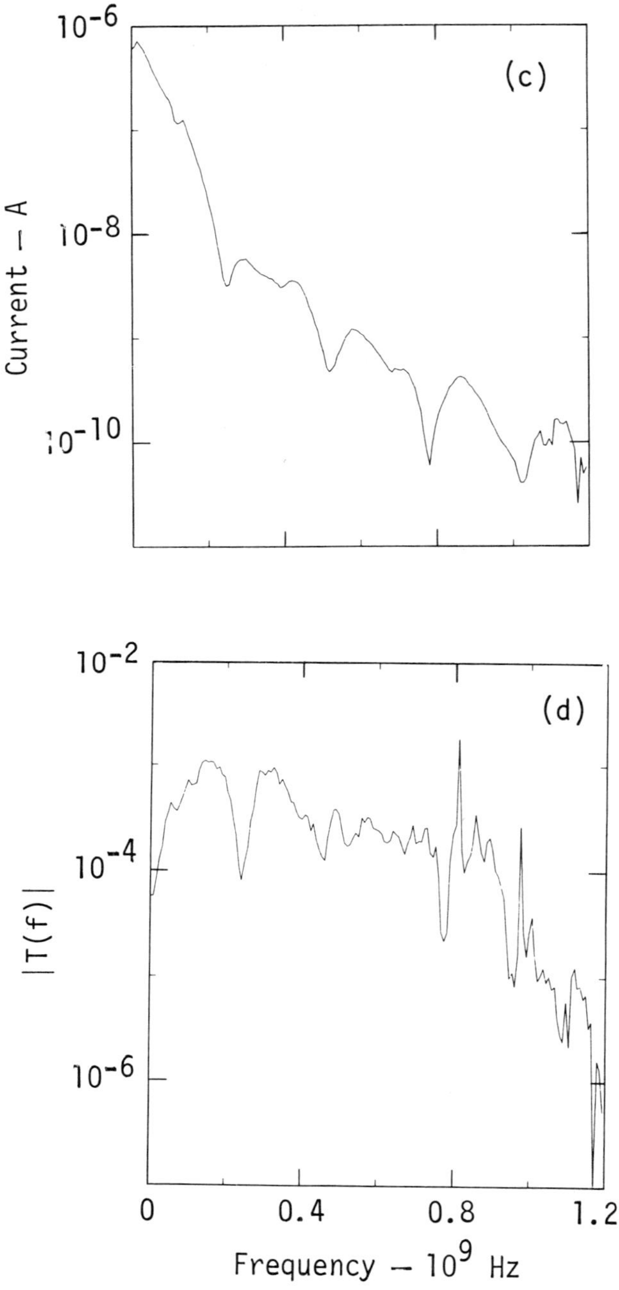

10-6
(c)
Current — A
10-8
10-10
10-2
(d)
10-4
|T(f)|
10-6
0
0.4
0.8
1.2
Frequency — 10^9 Hz

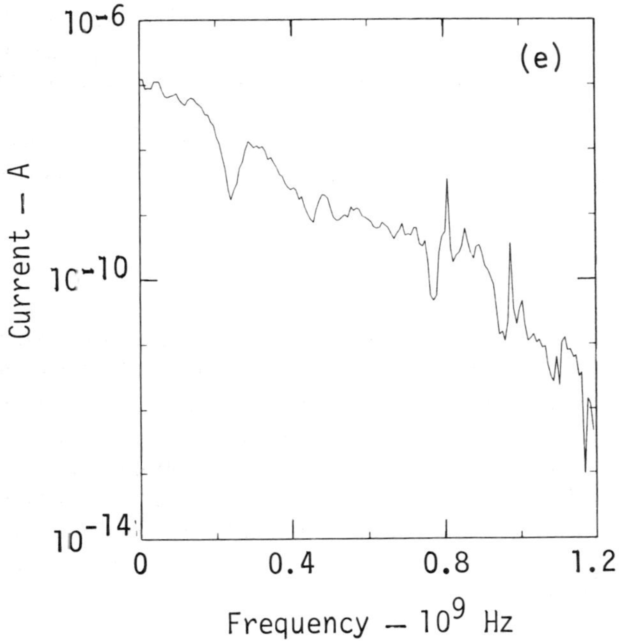

Fig. 7.25 (cont'd). The EM response of a whip or loop antenna is determined by its $L/\lambda \lesssim 1$ behavior, regardless of the configuration around the antenna: (c) load-current spectrum as the product of (a) and (b); (d) $|T(f)|$ for a 1-m loop on ground plane; and (e) load-current spectrum as the product of (a) and (d).

shows that the spectrum of $T(f)$ for the 1-m loop falls off more slowly above $L_p/\lambda = 1$ but that the current response in Fig. 7-25(e) is an order of magnitude below peak at $L_p/\lambda = 1$ and diminishes rapidly as f increases. For the 1000-m loop, the T-variation in Fig. 7-25(d) would occur over the range, $f \leqslant 1.2$ MHz, and the relative current response looks essentially the same out to 0.9 MHz ($L_p/\lambda = 3$) because $E^{\text{inc}}(f)$ rolls off very little in this range. Even though the current spectrum rolls off so slowly above $L_p/\lambda = 1$, the fact that the EM-integrated responses are essentially determined by the $L_p/\lambda \leqslant 1$ behavior indicates much cancellation of the high-frequency components.

7.4 APPLICATIONS

After calibrating the EMTF equipment and incident field of the transmitting antenna and validating the computations of equivalent circuit parameters from TDR and scattering data, one can proceed with confidence to measure scale models. These have included generic objects, such as whip and loop antennas on boxes, and models of real-life tanks, ships, and airplanes. One can also process transient data to obtain the complex Singularity Expansion Method (SEM) poles that characterize a body.

We will now give examples of applications after assessing the effect of statistical errors on the integrated responses of a simple structure (whip-on-box).

Statistical Errors

To estimate the effect of all statistical variations on the response to the EM wave of Eq. 7-10 of antennas mounted on configurations of interest, we selected a scale-model whip-on-box and a loop-on-box. The box was set on the ground plane (Fig. 7-26) in the test target position on Fig. 7-1. We made six independent EMTF measurements (one per day on successive days) on each scale model and processed each set of TDR and scattering data independently to obtain six sets of EM responses for each full-sized configuration. (The scale factor was such that the full-sized antenna lies in the range of 1 to 100 m on the abscissa in Figs. 7-26 through 7-30.) One of the runs on each model gave response curves

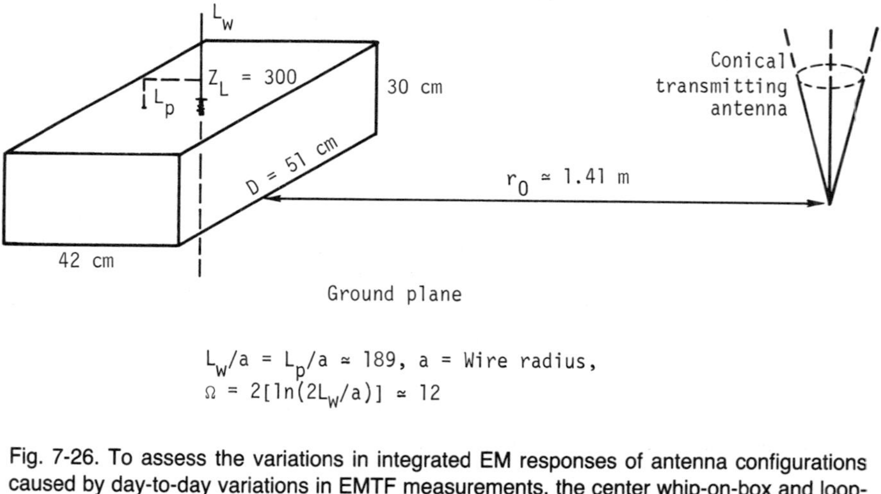

$L_W/a = L_p/a \simeq 189$, a = Wire radius,
$\Omega = 2[\ln(2L_W/a)] \simeq 12$

Fig. 7-26. To assess the variations in integrated EM responses of antenna configurations caused by day-to-day variations in EMTF measurements, the center whip-on-box and loop-on-box configurations were measured repeatedly. Each set of TDR and scattered-voltage measurements was processed independently to obtain EM responses of full-size structures. The load, $Z_L = 300$, corresponds to about the "worst-worst" value (response maximized relative to both D/L_w, or D/L_p, and to R_L). The whip length, L_w, and loop length, L_p are each 30 cm; the horizontal arm of the loop is 15 cm. The EM response variations are shown in Figs. 7-27 through 7-30.

that were grossly in error relative to all the others; reasons for the error are unknown and these curves were discarded.

The EM-integrated responses have been plotted as contours in the load resistance-antenna length plane with load reactance equal to zero. These responses are load-energy dissipated, W_L; peak current, $(I_L)_p$; peak voltage, $(V_L)_p$; and peak power, $(P_L)_p$. Each figure shows the variation in three of the contours for one of the responses.

In Figs. 7-27 through 7-30 for the 30-cm monopole on the box, the horizontal variation in each contour for a fixed EM-load resistance, R_L, is primarily a result of scope noise in the range, $R_L \lesssim 100$; in the higher range, Z_I variations predominate. Variations in the EM response from the true value caused by calibration error, δE^{inc}, however, do not appear in the figures; the same calibration factor was used for all the data processing. The horizontal variations are not serious for this configuration because the scattered-voltage/scope-noise voltage ratio is greater than 10 in the important frequency domain below $L/\lambda = 1$. The antenna response is also sufficiently broad to minimize the frequency-domain errors in Z_I.

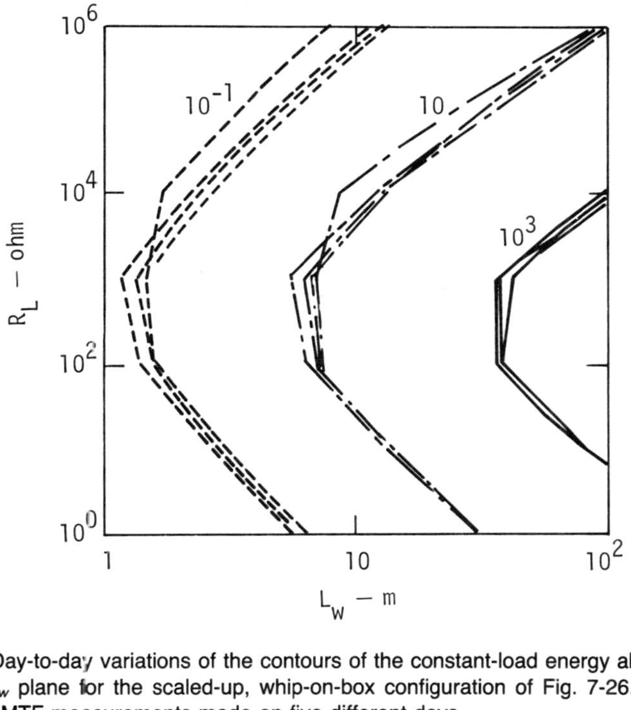

Fig. 7-27. Day-to-day variations of the contours of the constant-load energy absorbed, W_L, in the R_L-L_w plane for the scaled-up, whip-on-box configuration of Fig. 7-26. Curves are based on EMTF measurements made on five different days.

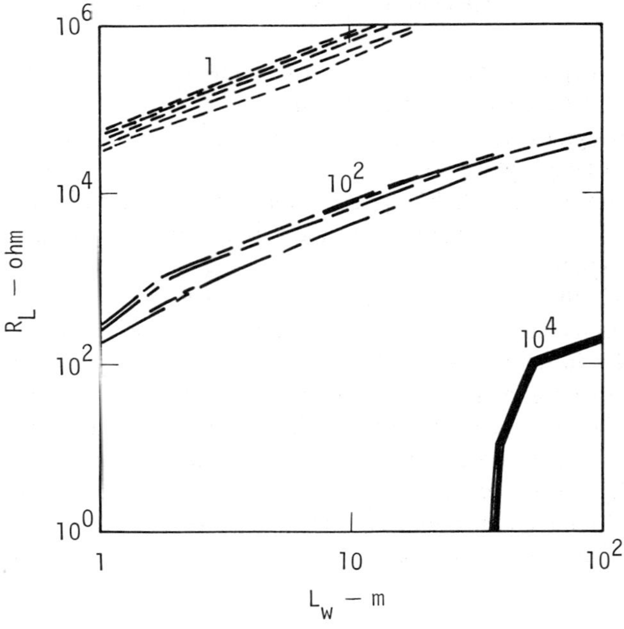

Fig. 7-28. Day-to-day variations of the contours of constant-peak current, $(I_L)_p$, in the R_L-L_w plane for the scaled-up, whip-on-box configuration of Fig. 7-26.

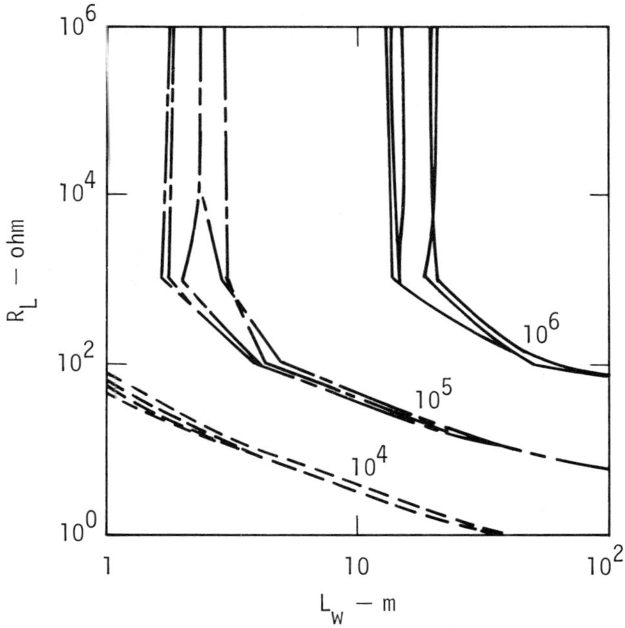

Fig. 7-29. Day-to-day variations in the contours of constant-peak voltage, $(V_L)_p$, in the R_L-L_w plane for the scaled-up, whip-on-box configuration of Fig. 7-26.

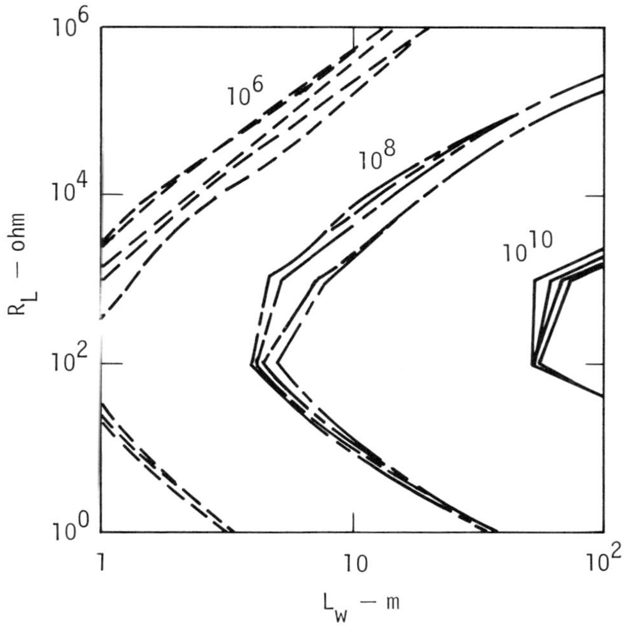

Fig. 7-30. Day-to-day variations in the contours of constant-peak-load power, $(P_L)_p$, in the R_L-L_w plane for the scaled-up, whip-on-box configuration of Fig. 7-26.

Complex Pole Evaluation

There are other ways of obtaining frequency-domain information from transient data. A particularly intriguing approach is to use an exponential curve-fitting process based on Prony's algorithm.[3] This technique is based on the premise that time-dependent data can be fit by a sum of complex exponentials, i.e.,

$$f(n\Delta t) = \sum_{i=1}^{K} A_i e^{\alpha_i n \Delta t} \qquad (7\text{-}11)$$

where $n \leq N$; α_i for $i = 1, 2, \ldots K$ ($K \leq N/2$) are the complex poles of a polynomial equation used in the Prony solution; and A_i is the complex amplitude or residue found in fitting the data with a least-squares-curve fit.

Equation 7-11 has several unique properties that make it useful. First, it allows extrapolation of the time-sampled function to times later than those measured. Thus, it is possible to extrapolate the time response to times greater than the normal 20-nsec experiment sampling window or the free time of the ground plane. It is even possible to extrapolate backwards in time. Examples of both forward and backward extrapolation are shown in Fig. 7-31 for results derived from a calculated waveform. Note, however, that the extrapolation only works over the time interval where the exciting pulse is no longer present.

Another feature of this exponential form (Eq. 7-11) is the simple frequency-domain transform that results from using the complex exponents and coefficients given in the following equation:

$$F(s) = \sum_{i=1}^{K} \frac{A_i}{(s - \alpha_i)} \tag{7-12}$$

where $s = j\omega$.

Figure 7-32 shows the frequency-domain transform given by this method. The comparable FFT for the same transient response is shown in Fig. 7-33. One advantage of Eq. 7-12 over the conventional FFT method is that this transform permits continuous frequency variation rather than the discrete sampling imposed by FFT. A penalty is paid with the use of this method, however, in terms of computation time.

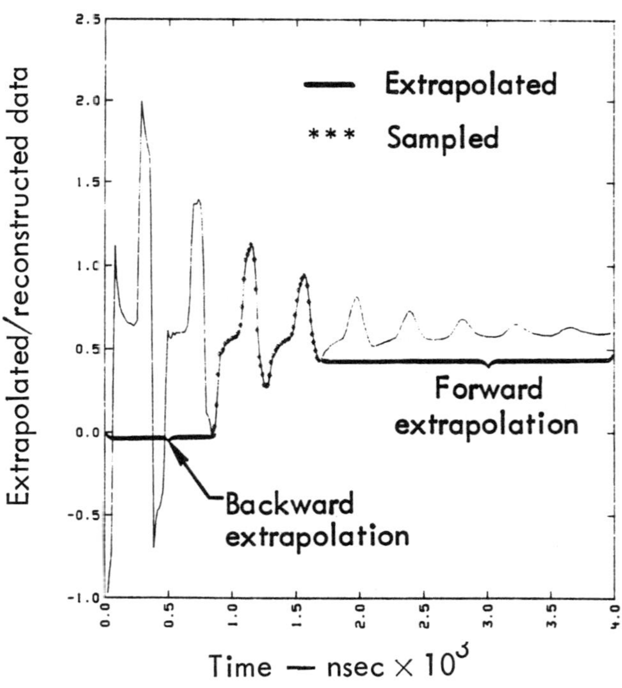

Fig. 7-31. Example of forward and backward extrapolation of a finite data sample.

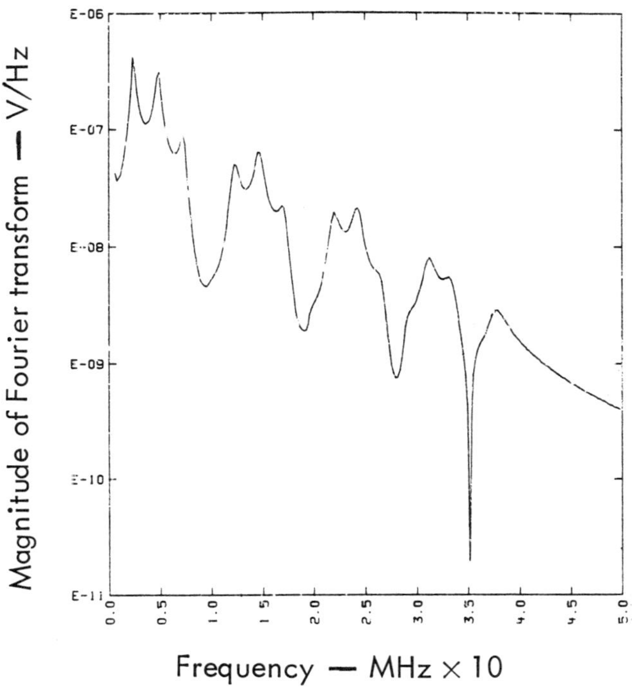

Fig. 7-32. Frequency-domain transform as determined by the Prony method.

A comparison of complex pole sets, α_i, obtained for a straight wire from a computed waveform response is shown in Fig. 7-34. The locus of these poles is comparable to those found via the Singularity Expansion Method (SEM).[4]

Structure-antenna Response (Whip Antenna on Tank)

Geometry. The scaled-up model tank analyzed, with its whip antenna, is shown in Fig. 7-35.* The model was thoroughly sprayed with a highly conducting silver-impregnated paint. The integrated EM responses for the transient excitation of Eq. 7-10 are shown in Figs. 7-38 through 7-43 as solid lines for the scaled-up tank. For comparison, the EM responses of two similar structures, the scaled-up whip-on-large-box of Fig. 7-36 and scaled-up whip-on-short-box of Fig. 7-38, are shown in Figs. 7-38 through 7-43.

*The model tank and whip-on-boxes in this section were excited with the NBS pulser.

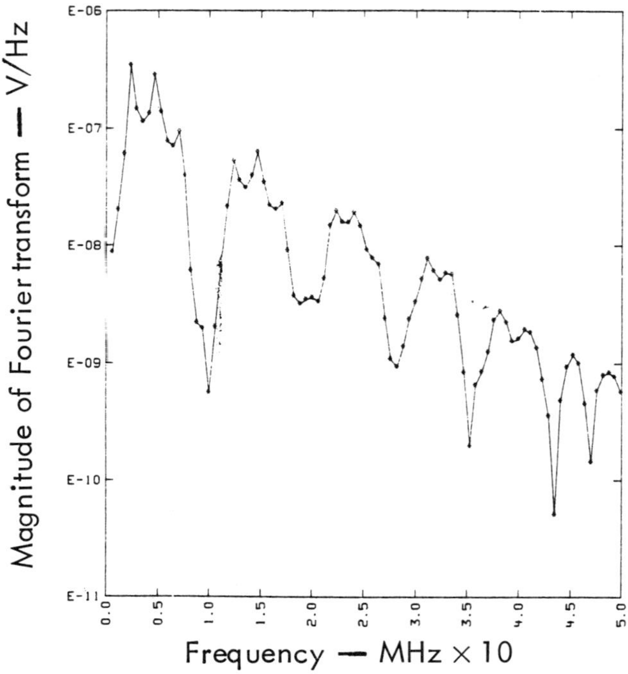

Fig. 7-33. Frequency-domain transform as determined by the FFT method.

Method of Data Generation. EMTF data were measured on a scale model tank 25 times smaller than that shown in Fig. 7-35 and were processed to obtain the equivalent circuit parameters $Z_I(f)$ and $h_{\text{eff}}(f)$ in the frequency domain, as explained earlier. The equivalent circuit was then scaled by a factor of 25 to represent the actual whip on tank. Integrated responses to the EM spectrum of Eq. 7-10 were computed from this circuit for various resistive loads in the range, $0 \leqslant R_L \leqslant 10^6$ ohm, and were then transformed to the time domain. The antenna fatness parameter, Ω, defined as $2 \ln(2L_w/a)$—a being the wire radius—was $\simeq 12$. Data for the whip-on-large-box of Fig. 7-36 and whip-on-small-box of Fig. 7-37 were processed similarly for comparison of responses.

The large box of Fig. 7-36 is taller than the tank body in Fig. 7-35 and deeper in the direction away from the incident wave, but only slightly longer. The small box of Fig. 7-37 is about half as long and seven-tenths as deep as the tank body but has the same height.

Integrated Data, Tank Model. The processed EM responses, W_L, $(P_L)_p$, $(I_L)_p$, $(V_L)_p$, $(P_L)_{av}$, and T_w, all vs R_L (with $X_L = 0$), are shown for the full-size tank model in Figs. 7-38 through 7-43 for $\phi = 0$ (broadside incidence),

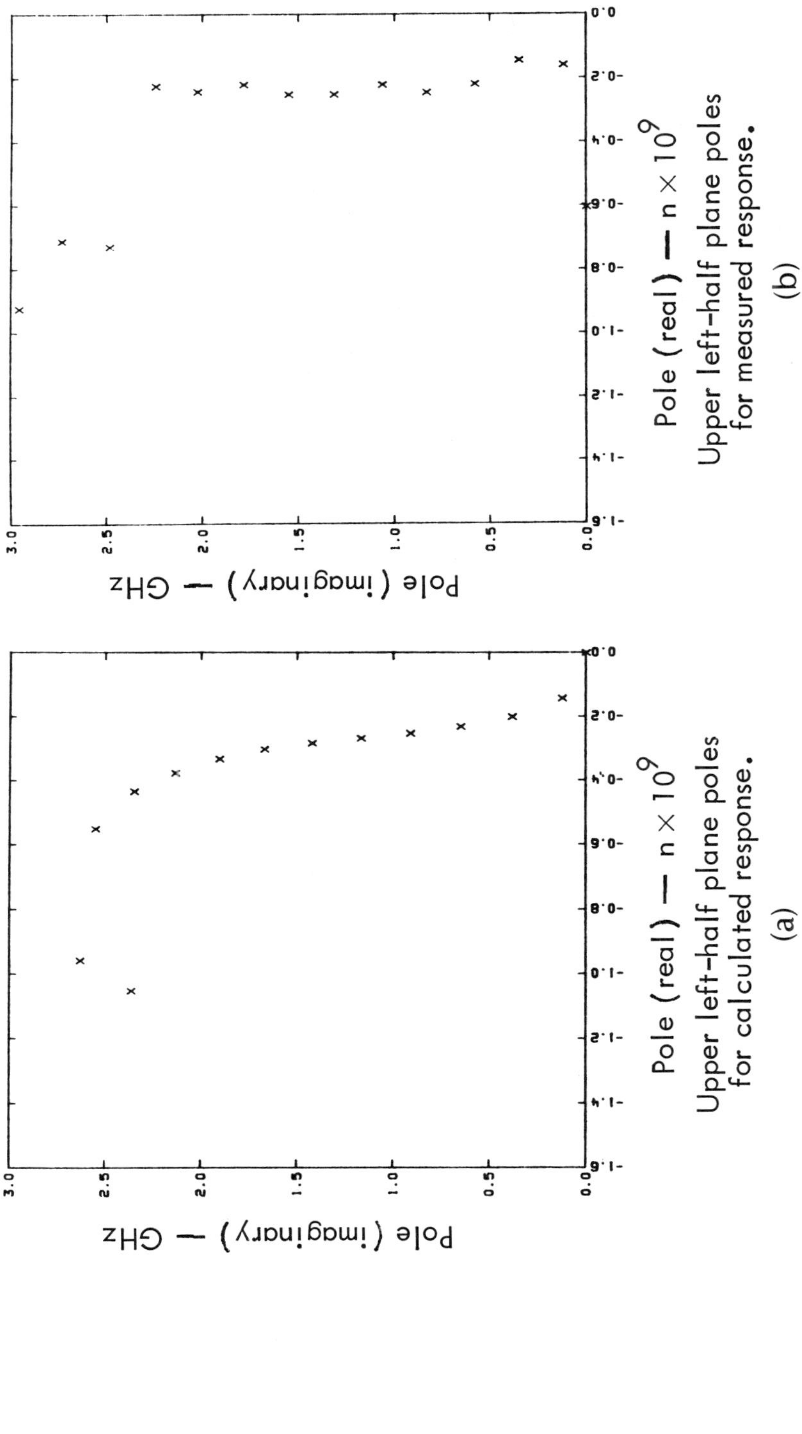

Fig. 7-34. Comparison of measured versus calculated pole sets for a straight-wire target.

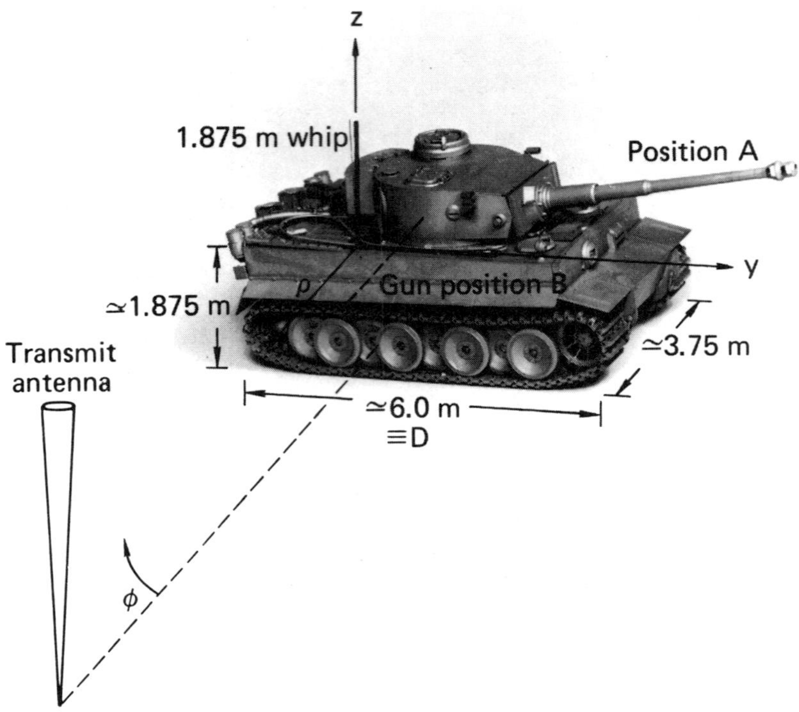

Fig. 7-35. A representative life-size tank with whip antenna, measured in miniature on the EMTF; gun is in position A or B.

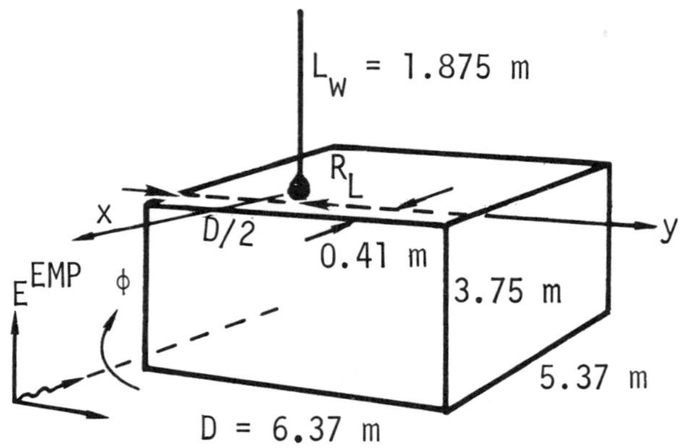

Fig. 7-36. A scaled-up whip-on-large-box to simulate the life-size tank model of Fig. 7-35.

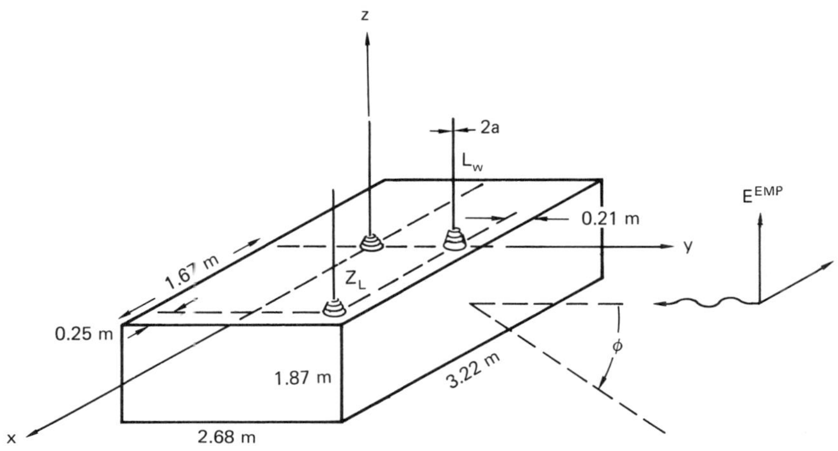

Fig. 7-37. A scaled-up whip-on-small-box to simulate the life-size tank model of Fig. 7-35.

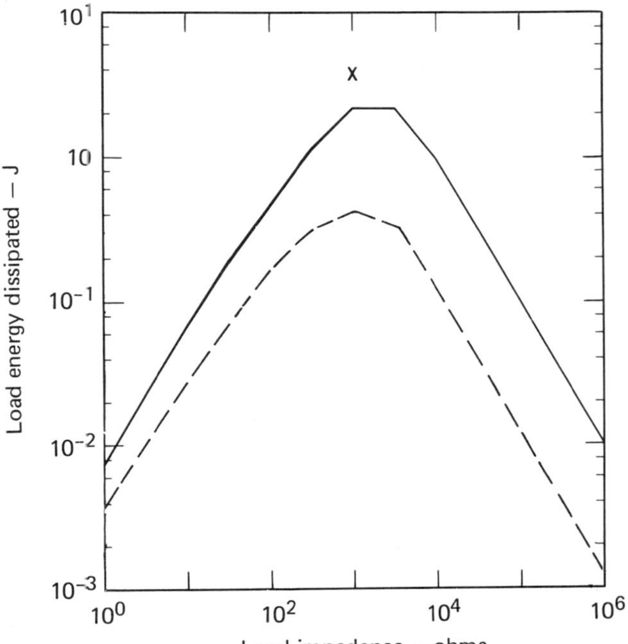

Fig. 7-38. W_L vs R_L ($X_L = 0$) of the tank of Fig. 7-35 for both gun positions, A and B. Dashed line shows the response of the whip-on-large-box of Fig. 7-36. The X denotes the response of the comparable small whip-on-box of Fig. 7-37.

both with the gun in the front position perpendicular to the direction of broadside incidence (position A) and in the side position (position B).

We notice that all the EM responses except $(I_L)_p$ maximize for $R_L \simeq 10^3$ ohm. All the responses are remarkably insensitive to the gun position, front or side. All responses except $(P_L)_{av}$ are somewhat higher for the front position (A).

Approximations by Whip-on-box Models. The responses for the comparable whip-on-large-box of Fig. 7-36 are indicated by dashed lines in Figs. 7-38 through 7-43. Despite the fact that the box is about twice as tall as the tank body, that it is considerably deeper in the $\phi = 0$ direction, and that it has no turret, the whip-on-box responses are remarkably similar to those of the whip-on-tank. For all the responses except W_L, the whip-on-large-box peak response is within a factor of 2 of the corresponding tank response; however, the peak box load energy dissipated is only 0.15 as much as the tank load energy.

The nearly peak responses for the whip-on-small-box of Fig. 7-37 are shown

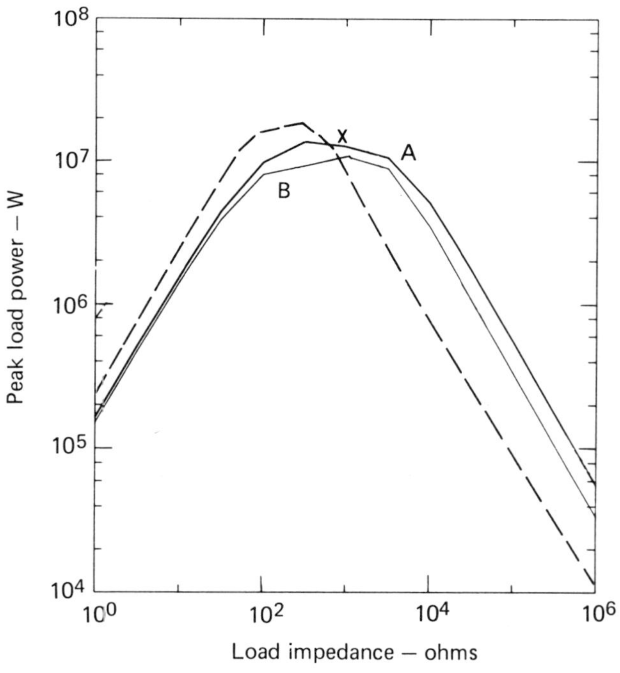

Fig. 7-39. $(P_L)_p$ vs R_L ($X_L = 0$) of the tank of Fig. 7-35 for gun positions A and B. Dashed line shows the response of the whip-on-large-box of Fig. 7-36. The X denotes response of the comparable whip-on-small-box of Fig. 7-37.

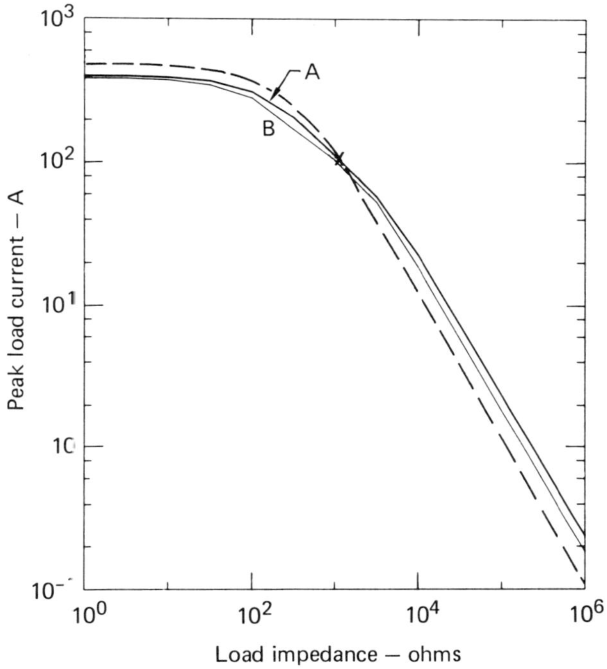

Fig. 7-40. $(I_L)_p$ vs R_L (X_L = 0) of the tank of Fig. 7-35 for gun positions A and B. Dashed line shows the response of the whip-on-large-box of Fig. 7-36. The X denotes the response of the comparable whip-on-small-box of Fig. 7-37.

as "x's" in Figs. 7-38 through 7-43. Even though the box is about half as long and half as deep as the tank body, but of the same height, its responses are all remarkably close to those of the whip-on-tank. The small-box peak load energy absorbed is about 3 dB higher than that of the tank, whereas the large-box peak load energy underbounds that of the tank.

Observations on These Box Approximations to the Tank Model. All the EM responses of the full-size detailed tank model are very insensitive to the gun orientation, whether forward or broadside, a plausible eventuality considering the barrel is nearly perpendicular to E^{inc}. The R_L load that essentially maximizes any integrated response is 300 ohms.

The comparable whip-on-small-box, scaled to have the edge whip length and height of the full-size tank model, has nearly the same "worst case" EM responses ($R_L \simeq$ 300 ohm) as the full-size tank model, except for W_L. The whip-on-small-box has about 3 dB more peak load energy.

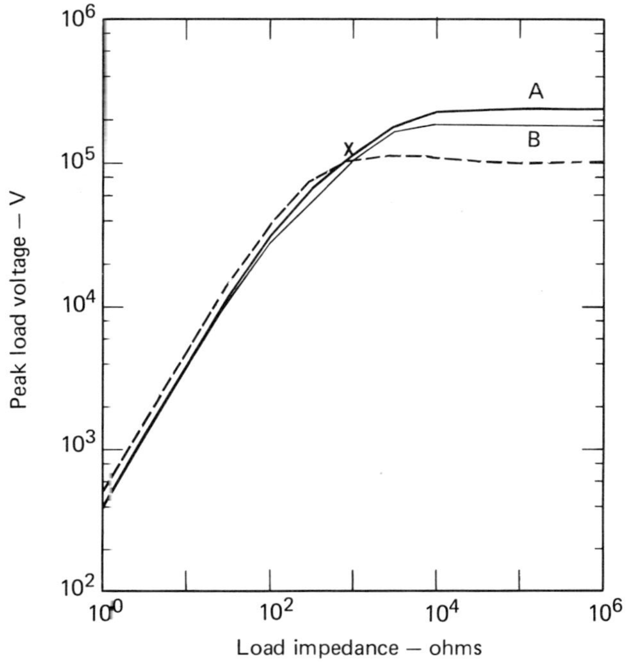

Fig. 7-41. $(V_L)_p$ vs R_L ($X_L = 0$) of the tank of Fig. 7-35 for gun positions A and B. Dashed line shows the response of the whip-on-large-box of Fig. 7-36. The *X* denotes the response of the comparable whip-on-small-box of Fig. 7-37.

The whip-on-large-box is not as good an estimator of EM response of the full-size tank model, particularly regarding load energy W_L. The reasons are not obvious but may be due to the fact that the large-box whip was twice as far away from the incident side of the box as was the small-box whip.

There is no real whip-on-tank simulation data presented here to compare with the model data above.

7.5 CONCLUSIONS

We have described how a practical EMTF can be used effectively to evaluate the equivalent circuit parameters of a test antenna at its load port without resorting to sophisticated system modeling algorithms. This frequency-domain circuit yields the integrated or differentiated response of a load on a full-size antenna exposed to an incident field with a spectrum which, when scaled to the test antenna, falls within the operating range of the EMTF and of the test load response.

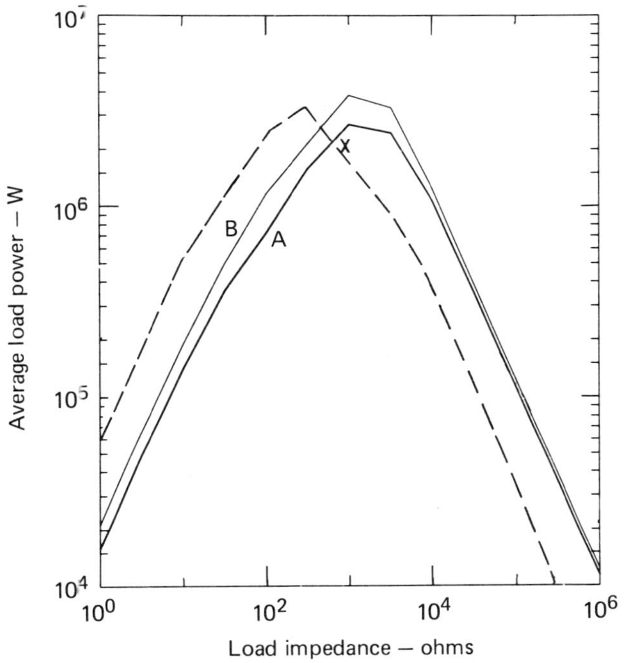

Fig. 7-42. $(P_L)_{av}$ vs R_L (X_L = 0) of the tank of Fig. 7-35 for gun positions A and B. Dashed line shows the response of the whip-on-large-box of Fig. 7-36. The X denotes the response of the comparable whip-on-small-box of Fig. 7-37.

Certain measurement constraints must be observed, in particular, the frequency sampling interval, $\Delta f = 1/T_{\max}$ [$T_{\max}$ being the clear time of observation of the waveforms, sometimes extendable by "zero filling"] and $\Delta t = 1/2f_{\max}$. Bandwidth $(-f_{\max}, f_{\max})$ is established by factors summarized below.

TDR measurements can be validated by comparing $Z_I(f)$ derived from measurements for simple monopole and loop antennas on the ground plane with direct computer code computations. Scattered voltage measurements can be validated similarly, without involving $h_{\text{eff}}(f)$ explicitly. After validation, one may calibrate the incident field by a short monopole of known $h_{\text{eff}}(f)$ and known (or measured) $Z_I(f)$ and the measured scattered field.

The effects of both systematic (deterministic) and statistical errors on antenna response should be considered. On the EMTF discussed, the systematic errors can essentially be calibrated out of the data with careful measurements. Statistical errors are more serious. Noise from pulser voltage fluctuations can be minimized by repeated averaging of the TDR and scattered voltage temporal waveforms.

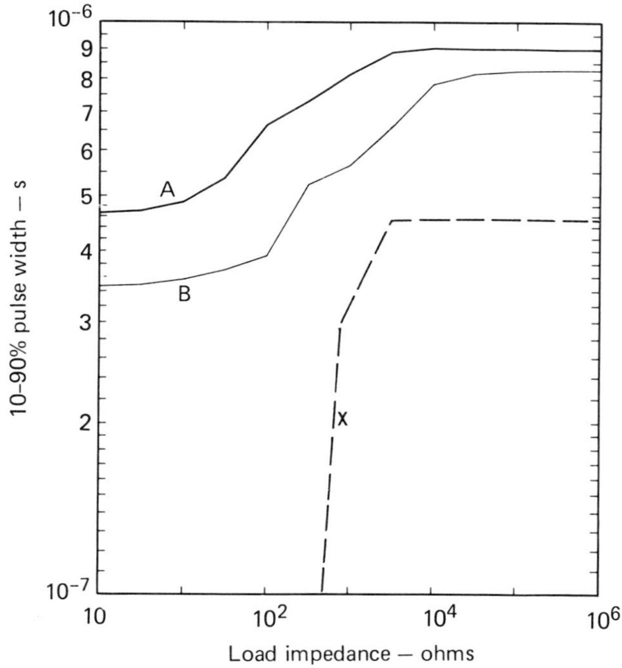

Fig. 7-43. T_w vs R_L ($X_L = 0$) of the tank of Fig. 7-35 for gun positions A and B. Dashed line shows the response of the whip-on-large-box of Fig. 7-36. The X denotes the response of the comparable whip-on-small-box of Fig. 7-37.

The discretization noise from rather short (12-bit) computer words and system noise can both restrict the dynamic range of an EMTF to about 2.5 decades, out to 4 GHz or so. For these and other reasons mentioned, the effective frequency range of the EMTF under discussion is 3 GHz.

Excessive sampling scope noise in the low-frequency range ($\lesssim$0.25 GHz here) causes variations in test-object Z_l and h_{eff}. To minimize these, one should operate the scope with full-scale deflection of each measured waveform on the scope face.

The nonplane-wave nature of the incident field (waveform curvature, $1/R$-attenuation, and near-field components) negligibly affects the EM integrated responses of typical whip- and loop-antenna configurations. As a rule, these responses are essentially determined in the frequency interval $L/\lambda \lesssim 1$ (where L is the physical antenna length), provided the incident field spectrum rolls off above this interval.

To show the net effect of all errors, we presented the integrated response data processed on five independent days for a scaled-up whip-on-box test object.

Although the variations in each response were certainly not negligible, they allow for a meaningful assessment of the responses expected for the type of transient excitation involved.

To illustrate applications, we have reliably evaluated the complex poles of a test object and the integrated responses of a whip antenna on a full-size tank exposed to transient excitation. Comparison of these responses with those of two comparable whip-on-box structures indicates that crude models of an actual test object will somtimes yield useful estimates of integrated responses.

Acknowledgements. The authors gratefully acknowledge the contributions made by many of their co-workers at Lawrence Livermore National Laboratory to the development, validation, and calibration of the LLNL Electromagnetic Transient Facility. They especially want to express their appreciation to F. J. Deadrick, H. G. Hudson, J. A. Landt, and J. T. Okada.

REFERENCES

1. Landt, J. A.; Miller, E. K.; and Van Blaricum, M. 1974. WT-MBA/LLL1B: A computer program for the time-domain electromagnetic response of thin-wire structures. (Name applies to improved version of TWTD.) UCRL-51585, Lawrence Livermore National Laboratory.
2. Bevensee, R. M.; Cabayan, H. S.; Deadrick, F. J.; Egbert, R. W.; Martin, L. C.; Miller, E. K.; Okada, J. T.; Poggio, A. J.; and Willows, J. L. 1978. External coupling of EMP to generic system structures. M-090, Lawrence Livermore National Laboratory.
3. Prony, R. 1795. Essai expérimental et analytique sur les lois de la dilatabilité de fluides Élastiques et sur celles de la force expansive de la vapeur de l'alkoal, a différentes températures. *J. l'École Polytech.* 1(2):24–76.
4. Tesche, F. M. 1973. On the analysis of scattering and antenna problems using the singularity expansion technique. *IEEE Trans. Ant. Prop.* Ap-21:53–62.

8
Measurements of System Responses to Transient Excitations

J. P. Castillo and L. Marin

8.0 INTRODUCTION

The interaction of transient (or broadband) electromagnetic fields with different electronic systems is a very complicated one. The continuing dependence on increasingly more sophisticated, interconnected electronics makes systems more and more vulnerable to electromagnetic noise environments. For systems to function in such environments requires that special measures be taken to protect the equipment. The soundness of these protective measures needs to be verified upon the construction of the system prototype. To maintain confidence in life-cycle protection will require additional testing, but clearly economics demand that the need for such testing be kept to a minimum.

The measurement techniques discussed in this chapter are those commonly used during the experimental assessment of the level of protection to be afforded to various systems exposed to broadband or transient electromagnetic field environments. In this chapter, it is assumed that such fields are created outside the system, for example, the nuclear electromagnetic pulse and transient fields associated with lightning discharges.

Two quantities must be assessed to determine the degree of immunity of a system to various electromagnetic noise environments. On the one hand, there is the stress generated at various interfaces in a system by the extraneous environment. On the other hand, there is the strength (or susceptibility) of the system at the same interfaces. A safety margin, defined as the strength-to-stress ratio, can then be determined for each interface at various positions in the system, for example, between electronics black boxes and the cables connected to them or between the walls surrounding a communications facility and all penetrating lines such as power cables or antenna connections.

In planning any measurement program, it is important to have a model in mind. The interaction of transient EM fields with large systems is complicated not only by the physical and electrical complexity of the system as a whole but also by the electrical properties of many of its components that are not well understood throughout the entire spectrum of interest. In order to guide the test planning and analysis it is therefore often conceptually convenient to subdivide the problem into a set of subproblems, each one of which is simpler to obtain appropriate test data for. The response of the entire system can then be interpreted in terms of the behavior of the parts.

One way of thinking of the EM stress problem is outlined in Fig. 8-1. The concept of shielding topology can be useful both as a means of actually protecting a system and as a way to understand its response to an incident broadband electromagnetic field. For each surface, one may define a number of fundamental problems that will determine the amount of energy penetration into the shielded

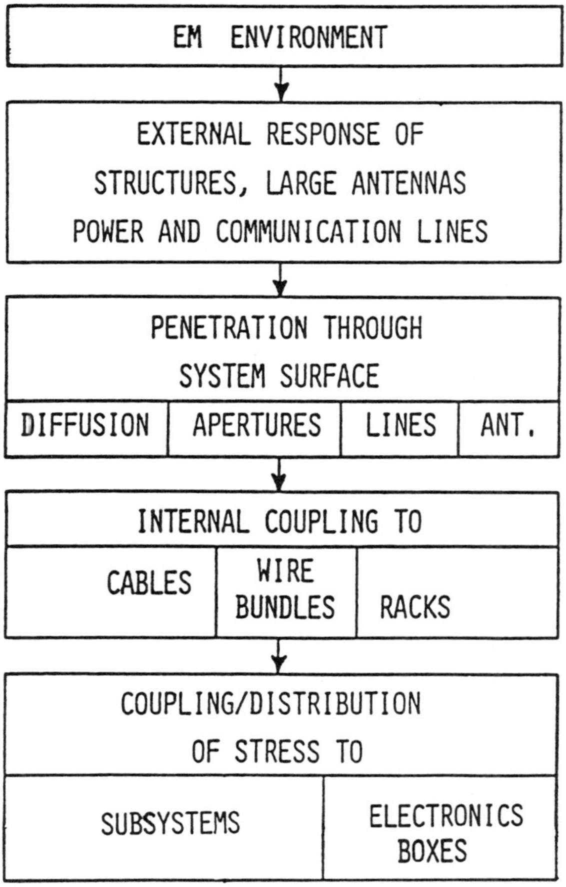

Fig. 8-1. Electromagnetic stress interaction with electronic systems.

region. These problems include field penetration through apertures, direct energy penetration along conductors piercing the shield, and field diffusion through imperfectly conducting shield surfaces.

The discussion in this chapter is limited to methods commonly used to measure transient responses (stress) of a system to a pulsed electromagnetic field. These can consist of measurements of electric or magnetic fields; of current or charge densities induced on the external surface of a system, such as the skin of an aircraft; or of current or voltage measurements on various cables, wires, or other lines inside a system. This chapter does *not* discuss the measurement of the strength (susceptibility) of a system.

8.1 MEASUREMENTS AND MEASUREMENT TECHNIQUES

This discussion will be limited to the types of sensors used in measuring such quantities as electric and magnetic fields and currents and voltages on conductors. Equipment used to record the measured signals are discussed elsewhere in this book (see Chaps. 2 and 6).

The measurements of free fields and surface fields are separated into the following categories:

1. Magnetic fields and surface current densities
2. Electric fields and surface charge densities
3. Monitoring techniques for free-field and surface-field sensors

The measurements of currents and voltages discuss the following:

1. Probes
2. Monitoring procedures

Magnetic-Field Sensors

The basic principle involved when designing magnetic-field sensors is the voltage induced in a loop by a linking magnetic flux. Thus, all electrically small magnetic-field sensors can be characterized by an inductance L and an equivalent area A_e (relating the time rate-of-change of the magnetic field to the induced open-circuit voltage V_{oc} in the loop). The Thevenin equivalent circuit for the sensor is shown in Fig. 8-2(a).

Sometimes it is of interest to relate the H-field to the short-circuit current I_{sc} induced in the magnetic-field sensor. The parameter that relates these two quantities is referred to as the "equivalent length," l_e [see Fig. 8-2(b)]. The equivalent area and length are related as follows:

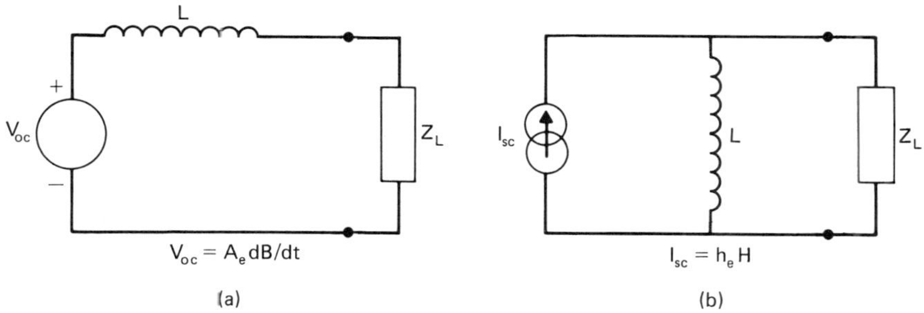

Fig. 8-2. Equivalent circuits for a magnetic-field sensor: (a) Thevenin equivalent circuit, and (b) Norton equivalent circuit.

$$\mu_0 A_e = L l_e \tag{8-1}$$

One requirement for the sensor to be useful is that L and A_e be well-defined and frequency independent. This, of course, implies that the sensor's linear dimensions are electrically small enough for the field it measures to be regarded as uniform across the sensor.

Another quantity of importance is the sensor's internal impedance relative to the line connected to the sensor. The output voltage, V_L, from the sensor is given by

$$V_L = \frac{Z_L}{Z_L + j\omega L} V_{oc} \tag{8-2}$$

For the recorded response, V_L, to be equal to V_{oc}, it is required that $|Z_L| \ll \omega L$. This condition defines an upper limit on the frequency content of the measured signal. It should be pointed out, however, that Eq. 8-2 can be used to deduce V_{oc} (and hence the magnetic field) from V_L.

One common type of magnetic-field sensor used for transient EM test purposes is the multigap loop (MGL). The sensor has the form of a right circular cylinder. On the cylindrical surface there are four distinct metal plates separated by a gap (the configuration, when unfolded from the cylindrical surface, is shown in Fig. 8-3). The signals from gaps 1 and 3 are combined and the signals from gaps 2 and 4 are combined to yield a differential signal output. The purpose of doing so is to minimize undesired E-field responses, even at the expense of a lower sensitivity. The transmission-line impedances and the gap angles are adjusted to obtain an optimum high-frequency performance from the sensor.

Two different versions of this sensor exist, as is true of many other magnetic-field sensors. The version used to measure magnetic fields in free space, shown

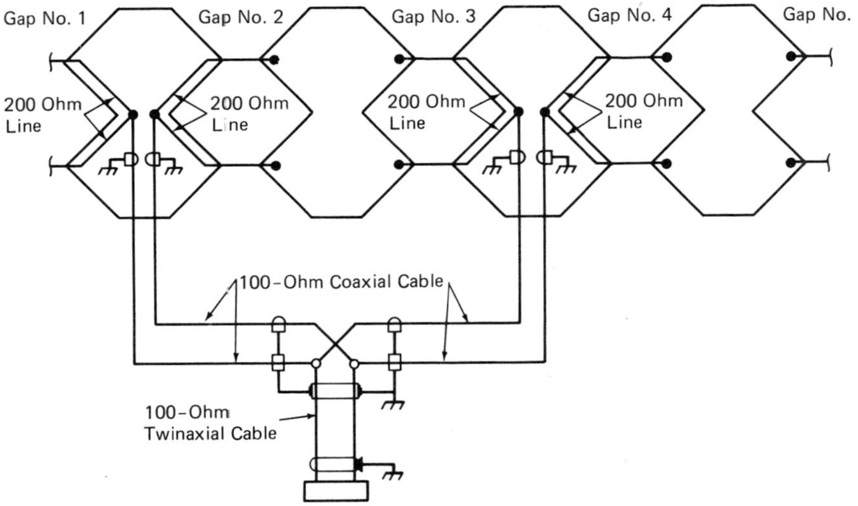

Fig. 8-3. MGL B-dot sensor.

in Fig. 8-4(a), consists of a full cylinder and employs the principle of operation described above. The other version, shown in Fig. 8-4(b), is used to measure tangential magnetic fields (or, equivalently, surface current densities) on a conducting surface. It consists of a half cylinder mounted on a ground plane and has two gaps with a topological arrangement, as shown in Fig. 8-5. This sensor is placed on (and electrically bonded to) the surface whose current density is

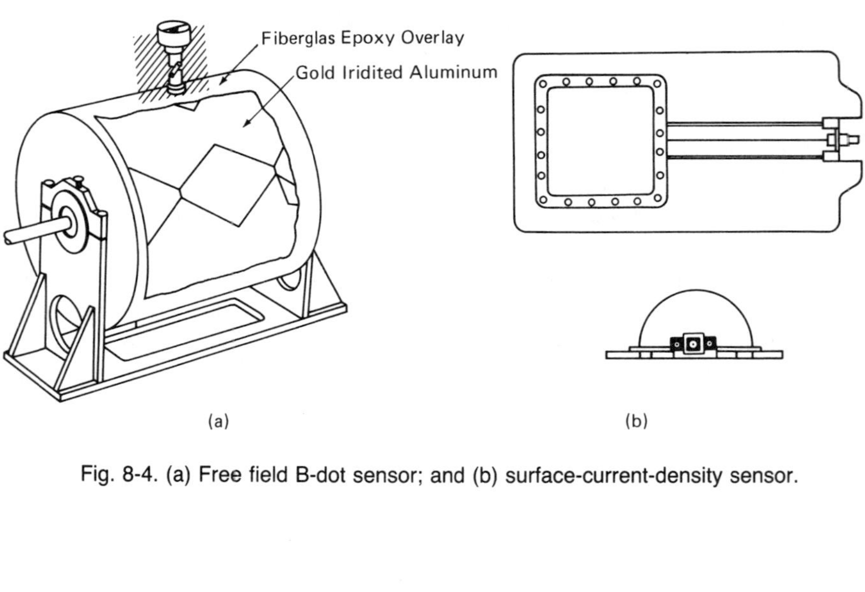

Fig. 8-4. (a) Free field B-dot sensor; and (b) surface-current-density sensor.

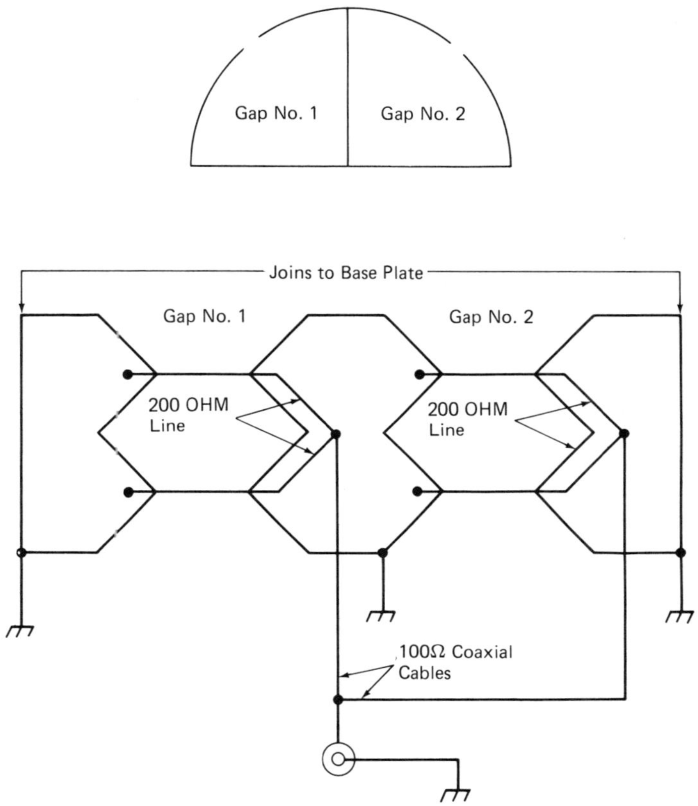

Fig. 8-5. Surface-current-density MGL sensor.

being monitored. The output signal from this type of sensor is usually an unbalanced signal, i.e., it can be directly connected to a coaxial cable.

Electric-Field Sensors

The design principle upon which electric-field sensors are based is that of inducing a voltage across a gap separating two conductors. Thus, electrically small electric-field sensors can be characterized by a capacitance C and an effective height h_{eff} (relating the electric field to the induced open-circuit voltage across the gap). The Thevenin equivalent circuit for such sensors is shown in Fig. 8-6(a).

The open-circuit voltage is characterized by the sensor's effective height. The short-circuit current, I_{sc}, on the other hand, can be characterized by an effective

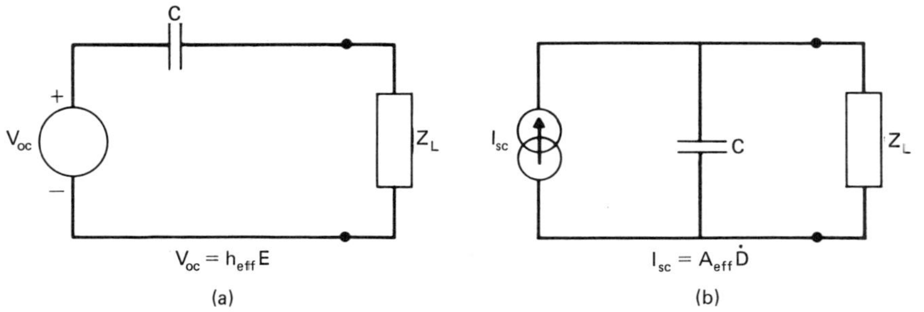

Fig. 8-6. Equivalent circuits for electric-field sensor: (a) Thevenin equivalent circuit, and (b) Norton equivalent circuit.

area, A_{eff}, that relates I_{sc} to the incident D field ($D_{inc} = \varepsilon_0 E_{inc}$), as shown by Fig. 8-6(b). The effective heights and areas are related as follows:

$$\varepsilon_0 A_{eff} = C \times h_{eff} \tag{8-3}$$

As is the case with magnetic-field sensors, electric-field sensors must be electrically small so that the field they measure can be considered quasistatic. This condition makes their size depend upon the highest frequency component that is to be recorded.

The internal impedance of the sensor influences the signal that is measured by a recording system whose matched impedance is Z_c (usually 50 Ω). In this case,

$$V_L = \frac{j\omega\varepsilon_0 E_{inc} A_{eff} Z_c}{1 + j\omega C Z_c} \tag{8-4}$$

From this formula, it is clear that an accurate knowledge of the sensor's effective area is of importance. An accurate knowledge of the internal capacitance may be less critical, especially if the associated impedance is large compared to Z_c.

The electric-field sensor most commonly used in transient EM field tests is the hollow spherical dipole (HSD) sensor, as shown in Fig. 8-7(a). This sensor consists of a sphere with a narrow slot around its equator. A symmetry ground plane is inserted between the two hemispherical surfaces. The signal picked up by each hemisphere flows to the ground plate through four equally spaced 200-Ω strip lines. The four lines from each hemisphere are then joined to provide a

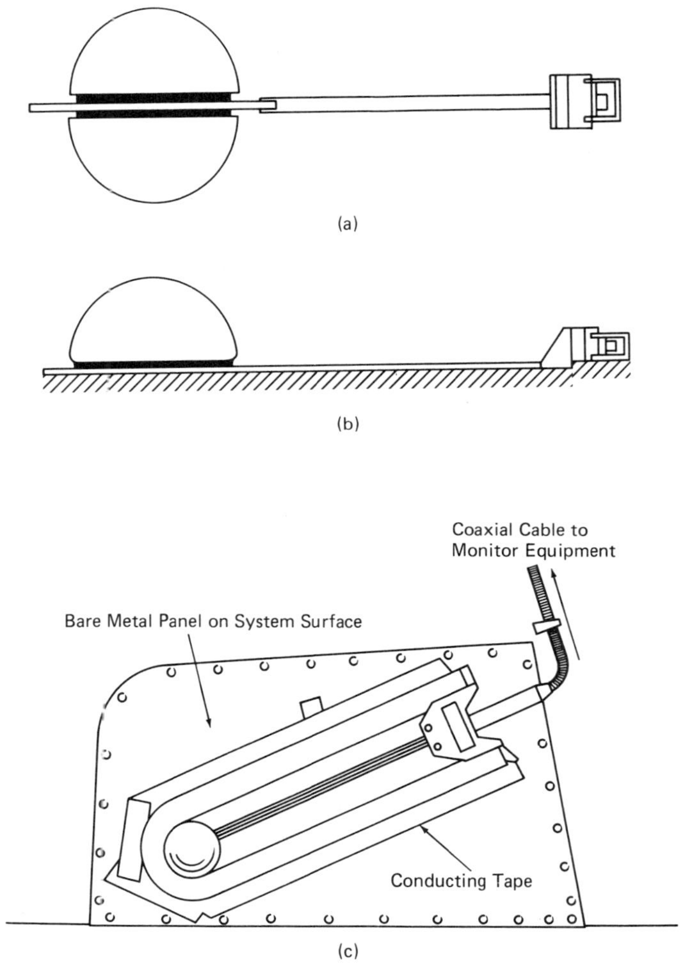

Fig. 8-7. Electric-field and surface-charge-density sensors: (a) free-field sensor, (b) surface-charge sensor, and (c) typical measurement arrangement.

50-Ω output source. A balanced (differential) signal source is then obtained by combining the outputs from each hemisphere.

Measurements of charge densities can be made using a single hemisphere mounted on top of a ground plate, as illustrated in Fig. 8-7(b). This plate, in turn, is placed on top of the surface where measurements are to be made using the arrangement of Fig. 8-7(c). The sensor then picks up the local normal electric field (or charge density) provided the field is properly bonded to the surface, as discussed in the next section.

Techniques For Mounting Sensors

When acquiring high-quality electromagnetic data, every effort must be made to control the instrumentation topology. Probes, cables, and recording devices must be laid out in a way that will minimize perturbation of the electromagnetic fields. Various methods to minimize measurement errors introduced by the presence of instrumentation will be discussed, including the following:

1. Sensor placement and grounding methods
2. Placement of shield boxes housing recording equipment
3. Cable routing and grounding methods.

Sensor Placement and Grounding Methods. One of the most important considerations when making electromagnetic-field measurements is the electrical bonding or grounding of the field sensor and associated cabling to the surface where measurements are to be made. Figure 8-8 shows a field sensor mounted on a surface where the field quantity is to be measured. In this particular case, the surface is metallic (highly conducting), and the sensor ground plane makes continuous electrical contact with the surface. The cable is routed directly through the surface with a minimum loop and peripherally bonded (via a connector) where it penetrates the surface. The part of the cable that penetrates beyond the surface need not be either shielded or bonded and can be routed to other equipment in the most convenient manner. Every effort must be made to keep the external cable loop size (circumference) small compared to wavelengths of interest. Continuous bonding of the cable along the sensor ground and along the surface is preferred. If the cable cannot be routed through the surface locally, as in Fig. 8-9, the cable must be either continuously or periodically bonded to the conducting surface. Again, the loops formed when the cable is periodically bonded must be smaller than the wavelengths of interest.

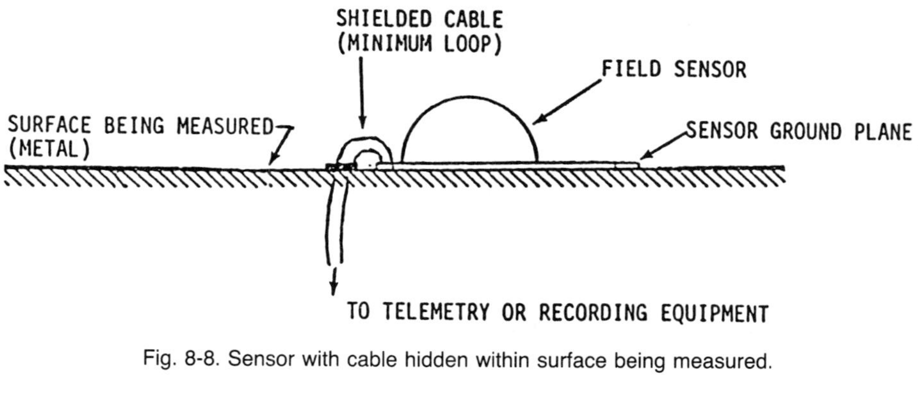

Fig. 8-8. Sensor with cable hidden within surface being measured.

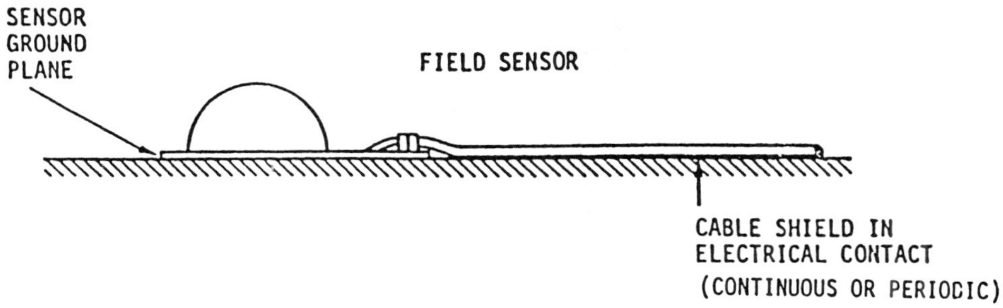

Fig. 8-9. Local sensor grounding topology.

In Fig. 8-10, current and voltage probes are shown in the preferred electrical topology. The cables leading from the probes to the recording instruments are continuously or periodically bonded in accordance with the same rules as those discussed for the sensors. Every effort must be made to reduce the amount of perturbation to the system topology.

Experiments have been conducted using the 200-Ω cone ground-plane setup shown in Fig. 8-11 to determine practical means for accomplishing this require-

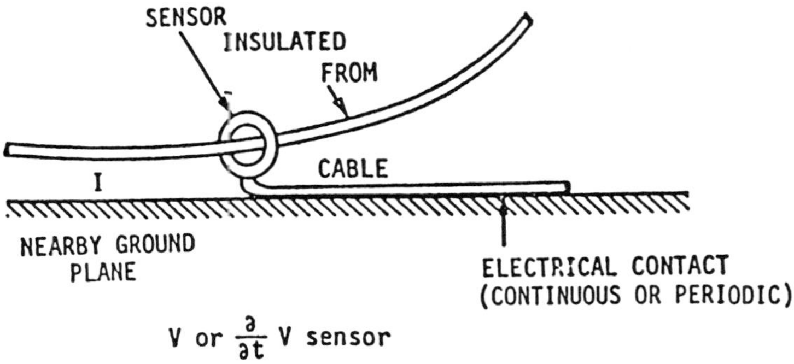

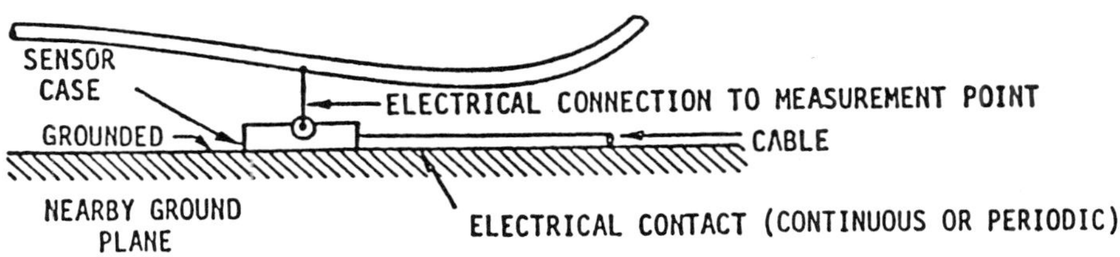

Fig. 8-10. Probe grounding.

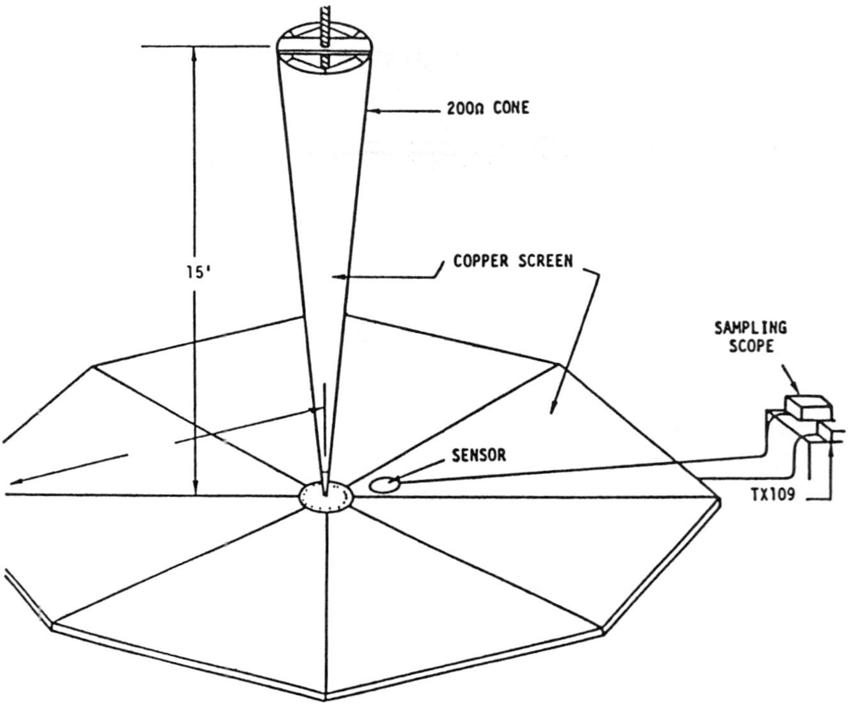

Fig. 8-11. 200-Ω cone and ground plane.

ment.* The pulse source is located at the cone apex. An HSD sensor is used to measure the cone pulse output, first, by properly attaching (grounding) the sensor to the ground plane, and, second, by isolating the sensor with a thin mylar sheet from the ground plane. The results are shown in Fig. 8-12. As can be seen from this figure, the introduction of the insulation changes the measured pulse form substantially.

Measurement-Cable Grounding. Figure 8-13 depicts a sensor and cable mounted on a conducting surface. The sensor is electrically connected to the surface with either conducting tape or metal screws. The distance between cable grounds close to the sensor is approximately 30 cm (loop resonance frequency about 500 MHz) and approximately 50 cm (loop resonance frequency about 300

*E. L. Breen, The application of B-dot and D-dot sensors to aircraft skin surfaces, AFWL Instrumentation Development Memo 3, July 1974.

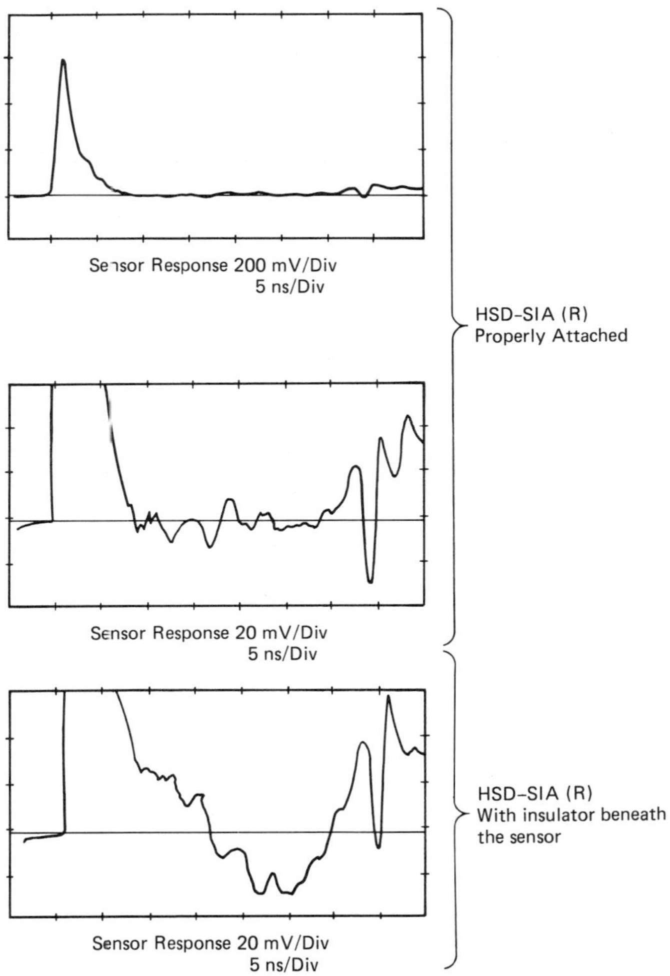

Fig. 8-12. Results of sensor bonding experiment: properly attached HSD and response at (a) 200 mV/div., 5 ns/div.; and (b) 20 mV/div., 5 ns/div., HSD isolated by insulator and response at (c) 20 mV/div., 5 ns/div.

MHz) at greater distances from the sensor. These distances are selected so that the resonant frequencies introduced by the loops formed by the bonds are higher than the maximum frequencies of interest. For the example shown in Fig. 8-13, the maximum frequency of interest is approximately 100 MHz. Higher frequencies would require closer spacing between the cable grounds.

If a dielectric layer (such as from a coat of paint) separates the conducting surface from the sensor and cable, it should be removed to allow for good metal-

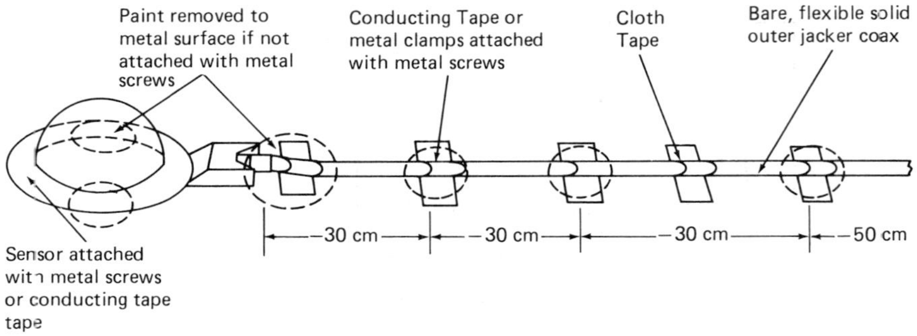

Fig. 8-13. Acceptable sensor and cable grounding.

to-metal contact. Of course, if metal screws are used, this may not be necessary. If the dielectric layer is not removed, one can expect the following approximate capacitance between the sensor and conducting surface:

$$C = \frac{S\varepsilon_r\varepsilon_0}{\Delta} \tag{8-5}$$

where S is the sensor plate area, ε_r is the relative dielectric constant for the dielectric, ε_0 is the permittivity of free space, and Δ is the thickness of the dielectric layer.

Sensor Location. When making surface current or charge density measurements, great care must be taken in placing the sensors. Considerations include the size of the sensor and its distance from edges or other discontinuities. Ideally, sensors should be located at least 10 wavelengths away from an edge or discontinuity, but experience has shown that acceptable measurements usually result if the sensor is located two to three sensor diameters away. For making certain measurements, this may mean selecting smaller sensors with associated lower sensitivities.

The requirement for placing sensors some distance away from edges or discontinuities is due to two facts: (1) The current and charge densities near these regions change very rapidly as a function of distance, and (2) the field distribution is more likely to be modified as the sensor is positioned closer to them. However, it is often not necessary to obtain detailed measurements of the current and charge densities close to edges or discontinuities; rather, only knowledge of the global values of a large structure, such as a wing or fuselage, is desired. In this case, every effort must be made to locate the sensors as far from edges or discontinuities as possible.

Cable Routing. Figure 8-14 shows two examples of cable routing that should be avoided. The original topology in both examples is seriously modified by the introduction of a conducting instrumentation cable. If this type of cable layout is unavoidable, a fiber optic or other dielectric transmission system should be used. Figure 8-15 shows an acceptable method for routing conducting cables for the same geometries shown in Fig. 8-14.

Placement of Shield Boxes Housing Recording Equipment. Local field distortion will also occur when large metallic items—e.g., a shield box—are introduced. Although it is sometimes desirable to locate the recording system remotely, it is often necessary to locate it nearby. Not only must the system be isolated from local fields, it must also be placed to minimize its perturbation of the fields being measured. The "best" configuration, one in which the recording system is "hidden" within the system (object) where the measurements are being made, is shown in Fig. 8-16.

In the many cases when it is not possible to access holes with feed-through connectors, the recording system must be located outside the object in the mea-

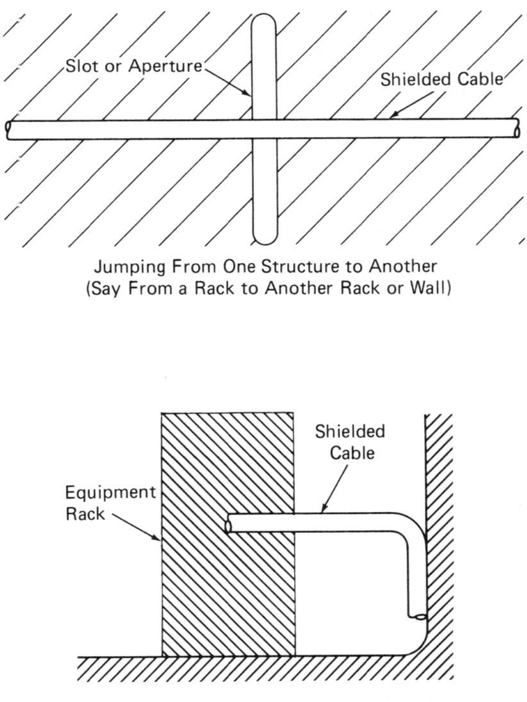

Fig. 8-14. Unacceptable cable routing over slot or other aperture.

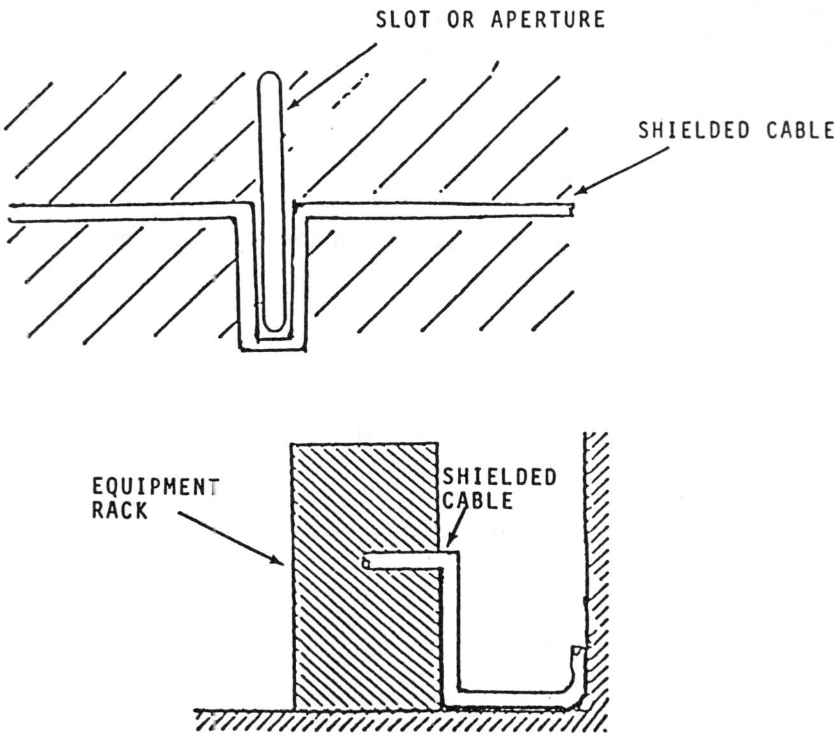

Fig.8-15. Acceptable cable routine.

surement volume itself, a configuration shown in Fig. 8-17. If the system under measurement contains continuous metal surfaces, then the cabling connecting the sensor to the recording system should be extended so that the latter is located away from any incident (excitation) field. For transient measurements, the recording system should be located far enough from the sensor so that reflections from it to the sensor occur at times that are long compared to the measurement time of interest. In the steady state, a recording system so located will always cause problems by changing the resonant frequencies of the object system. In this case, either the recording system should be made as small as physically possible or ferrite beads should be inserted on the cable to damp out the resonances.

If the surface between the sensor and the recording system is not continuous, a fiber optic or dielectric microwave link should be considered as a means of completing this connection. Care must also be taken at the final connection between the cable from the sensor and the shield box containing the recording

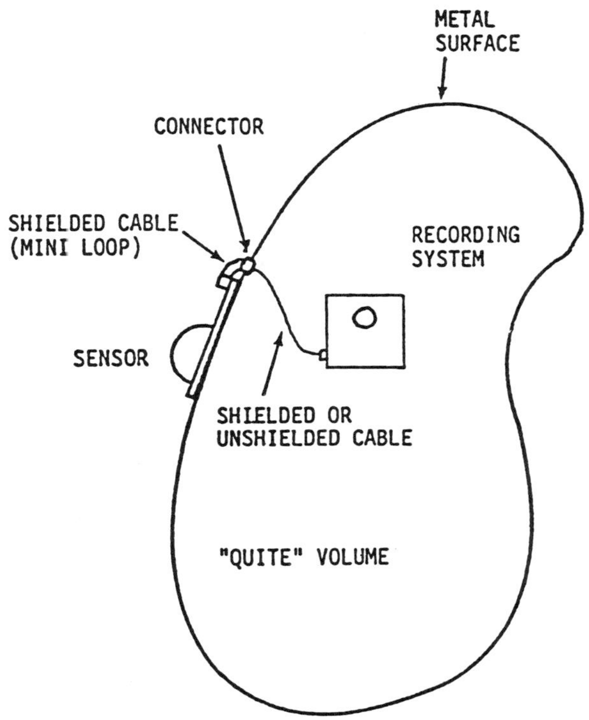

Fig. 8-16. "Hidden" recording system.

equipment. Figure 8-18 shows a typical transition from the shielded cable (or sensor) to the shielded box. An extra coaxial shield is introduced at the adaptor (a) and is fanned out to mate with the feedthrough connector at the shield box (b). The points of contact between the shielded cable and the adaptor and between the feedthrough and the shield box are electrically bonded peripherally. The shielded cable is then continued through the shield box to the oscilloscope or other recording equipment (c). The oscilloscope is isolated from the shield box further to reduce any electromagnetic interference with the recording system.

Current Measurements

Field measurements are important when it comes to characterizing the environment to which a system is being subjected. Similarly, current and charge density measurements are most important when it comes to characterizing the external response of an object to an incident electromagnetic field. In determining the

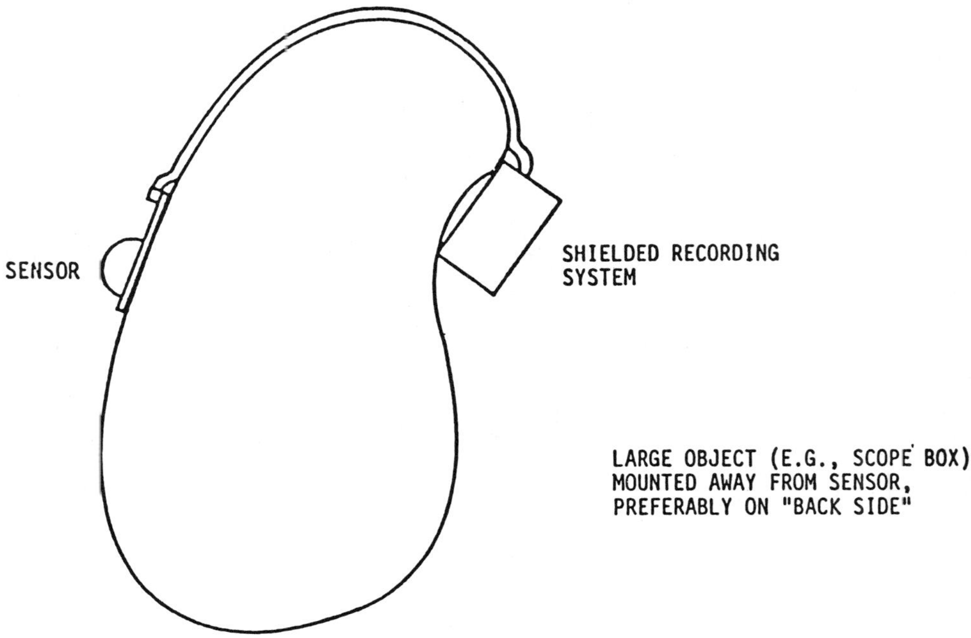

Fig. 8-17. "External" current and charge measurements.

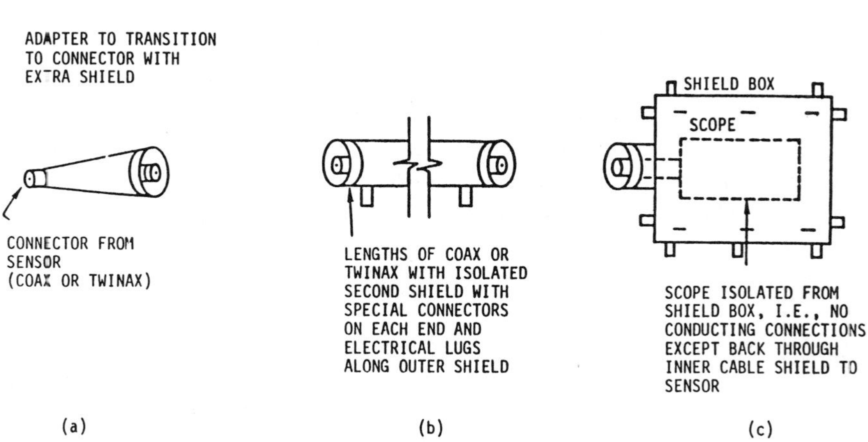

Fig. 8-18. Shielded-transmission/shield-box system.

amount of electromagnetic energy that penetrates through various surfaces to reach electronic equipment located inside a system, it is necessary to measure currents and voltages on various lines.

Probes commonly used for measuring line currents usually consist of a toroidally shaped core of high-permeability material. A coil is then wound around the toroidal core. Thus, the current probe can be viewed as a transformer in which the primary side (the line being measured and around which line the probe is mounted) has one turn and the recorded probe signal is the output from the secondary side. Current probes are usually characterized by their transfer imped-

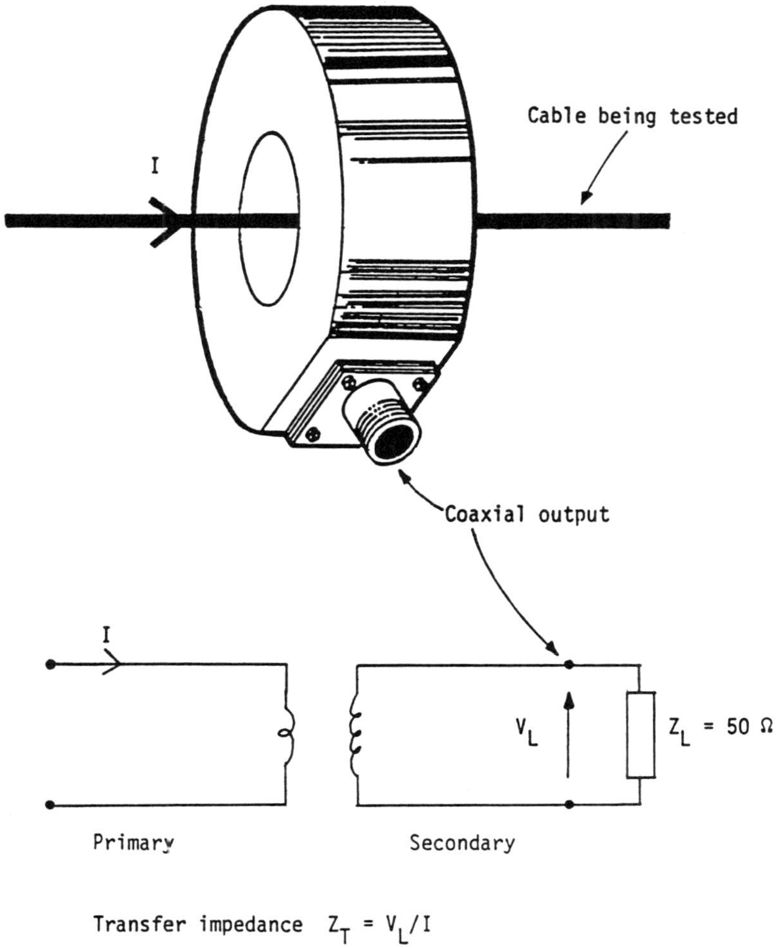

Fig. 8-19. Schematic of current probe.

ance, which relates the current being measured to the output voltage of the probe when it is connected to a 50-Ω recording system. Most probes now in use have an almost frequency-independent (flat) transfer impedance between 0.1 and 100 MHz. A schematic diagram of a current probe and the frequency variation of a typical probe transfer impedance are shown in Figs. 8-19 and 8-20.

The output of a probe can be connected either to a coaxial cable or a fiber optics transducer. The coaxial cable guides the probe-induced RF signal to the data-acquisition equipment. Installation procedures for interconnecting cables have already been discussed.

Break-Out Boxes

It is often of interest to measure the current on wires located inside an overall cable shield. These measurements must be performed in such a way that the integrity of the shield is not compromised while they are being performed. The technique commonly used is to install a break-out box and break-out cable. This assembly can be installed naturally wherever the cable terminates in a connector and connects to an electronics box or rack. A diagram of the break-out cable/box is shown in Fig. 8-21.

The continuity of the shield on the cable being measured is maintained by

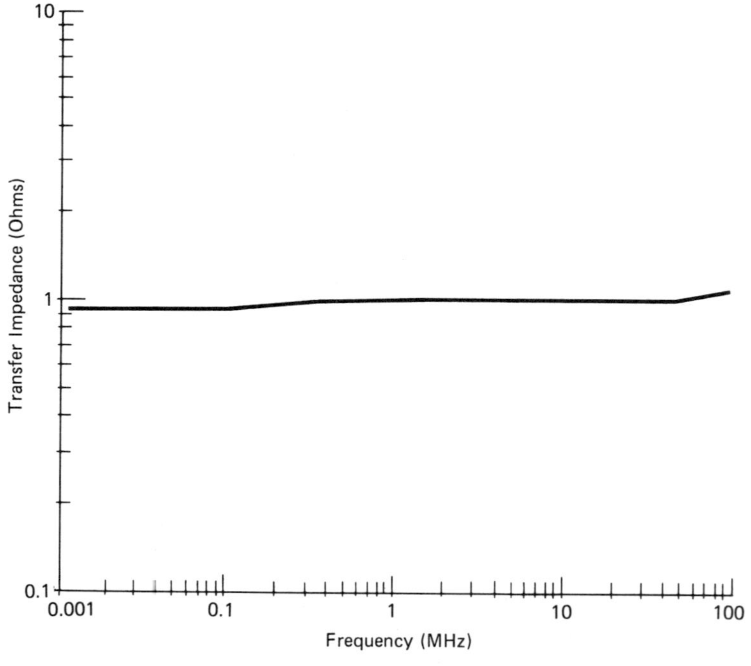

Fig. 8-20. Typical current-probe transfer impedance.

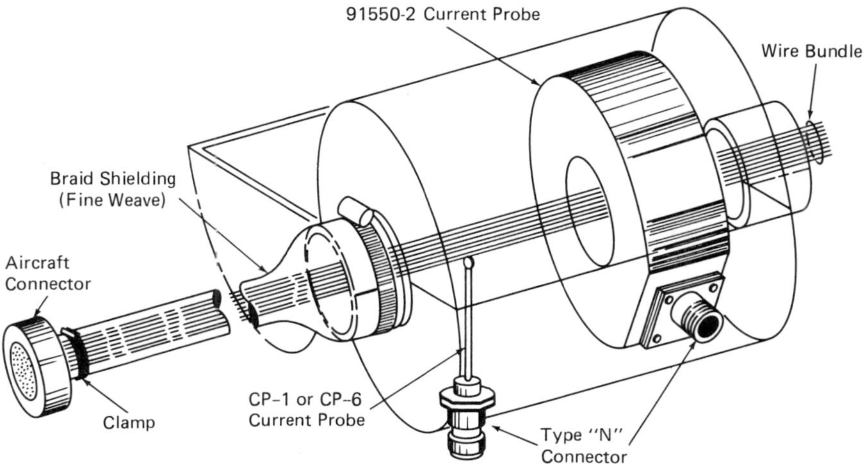

Fig. 8-21. Typical break-out cable/box assembly.

way of the break-out cable shield and the break-out box walls. The current probes are placed inside the break-out box with special feed-through connectors mounted on the break-out box. Access to the inside of the box is accomplished by making the box from two half cylinders that are hinged together on one side of the cylindrical surface. For the box to maintain its shielding properties, the joint between the two halves needs to be sealed electromagnetically (such as covering it with conducting copper tape). As shown in Fig. 8-21, the probes that are used to measure currents can either be small ones (allowing for the measurement of the current in an individual wire) or large ones (allowing for the measurement of the total current in a wire bundle).

Voltage Measurements

Voltage measurements of large signals such as those usually present on HF-antenna-system leads are straightforward to perform. Open-circuit voltages on coaxial cables connected to HF antennas can be measured by connecting a high-voltage probe to the end of the cable, as shown in Fig. 8-22. The probe, in turn, is connected to the recording equipment (such as the oscilloscope shown in Fig. 8-23).

Break-out boxes, as described in current measurements, are also used in voltage measurements to gain access to an individual wire inside a shielded cable without disturbing the connection between the cable and the equipment to which it is connected. Since the induced pin voltages are relatively small, sensitive

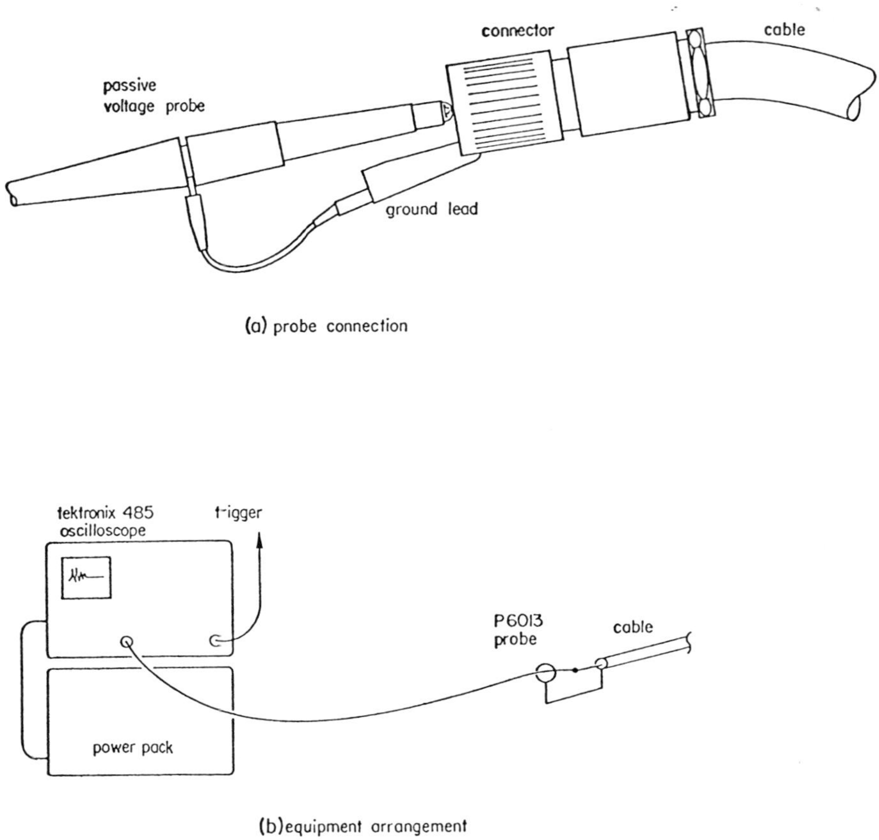

Fig. 8-22. Open-circuit voltage measurements on coaxial cables: (a) probe connection, and (b) equipment arrangement.

voltage probes are commonly used for these measurements. These probes are equipped with an E-Z hook at the probe tip to catch the wire to be measured. Since the probes are fabricated without any surrounding shield, the noise picked up by the probe is usually unacceptably large compared to the level of the measured signal, and it is therefore necessary to modify the probe by surrounding it and its cable by a braided cylindrical shield. On one end of the cylinder, the braid is tightly bonded to the break-out box, and, on the other end, it is bonded to the backshell of the oscilloscope connector. The tight bonding can be achieved by using conducting copper tape. The shielding integrity of the voltage probe can be verified by recording the noise level with the probe short-circuited and placed near the wire to be measured.

As illustrated in Fig. 8-23, the load voltage for a wire can be measured by

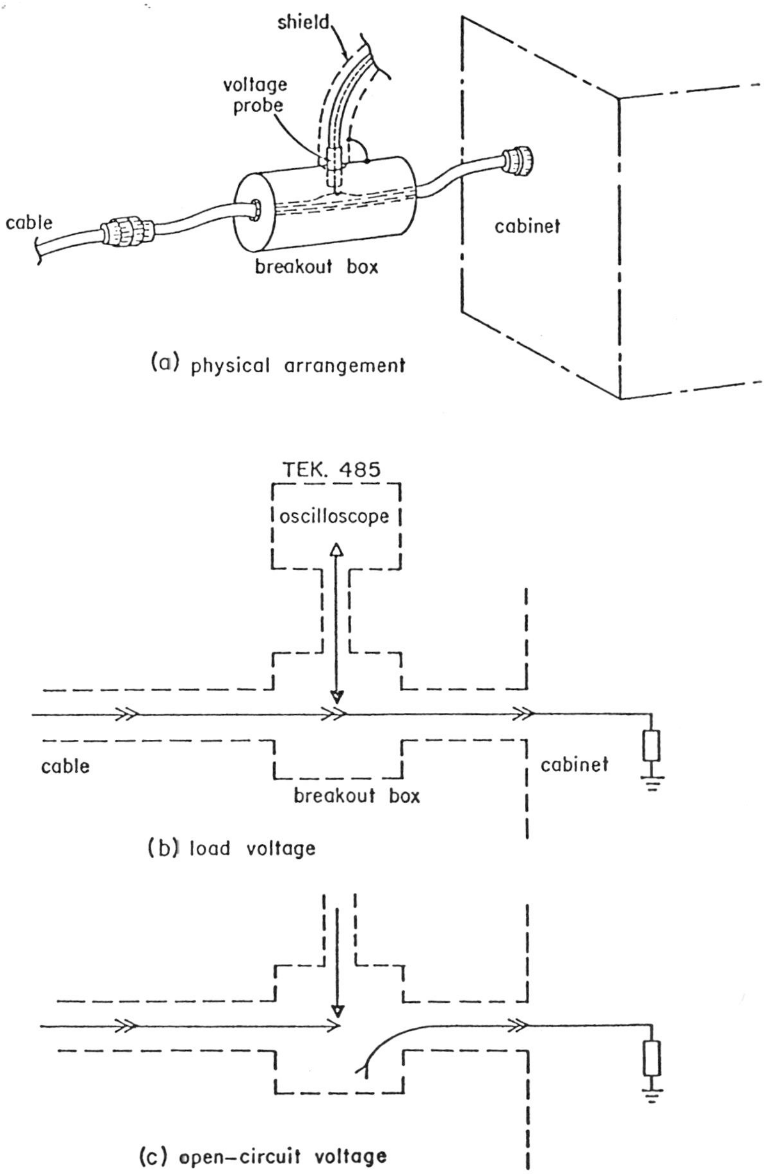

Fig. 8-23. Open-circuit and load-voltage measurements with break-out boxes: (a) physical arrangement, (b) load voltage, and (c) open-circuit voltage.

attaching the probe to the wire without breaking it. The wire inside the breakout box is usually prefabricated by breaking it at the middle, stripping out small portions of insulation near the breaking point, and welding a socket connector to each side of the broken wire. The wire can then be reconnected simply by snapping the socket connector together.

The open-circuit voltage induced on the wire can be measured, as illustrated in Fig. 8-23, by disconnecting the wire (all other wires in that bundle remain connected) and attaching the probe to its cable side.

8.2 THE USE AND ANALYSIS OF DATA

As mentioned in the introduction, the measurements of transient signals are made for different purposes. The underlying reason stems from the fact that small-sized electronic circuits, whether they use discrete components or integrated circuits, are susceptible to malfunction or damage caused by electromagnetic transient interference. The problems are particularly common in data-processing circuits because these circuits often cannot distinguish between a spurious transient and a legitimate signal because they are designed for small switching levels in order to conserve power and reduce heat dissipation problems. Logic levels are often a few volts or a few tens of milliamperes in these circuits.

On the other hand, transients associated with different electromagnetic interference environments such as lightning, EMP, and power switching commonly have peak currents and voltages that are significantly larger than those tolerated by logic circuits. Thus, if small-signal electronic circuits are to be operated in the adverse conditions created by these EM noise environments, it is necessary for the transient currents and voltages induced by them on various lines to be reduced to a level that is unharmful to the electronic equipment. One purpose of the time-domain measurements is to establish the transient coupling levels at different locations inside the system.

The signals recorded during EMP tests have a frequency content that spans roughly four decades or more (from around 10 kHz to around 100 MHz). These signals can be recorded using available equipment (i.e., oscilloscopes for analog display and transient digitizers for digitally formatted data). Thus, the techniques for capturing the data are quite straightforward, and the problems associated with these tests are not the measurement of a single current/voltage/current density/charge density. The problem one usually encounters lies with the total number of measurements needed to characterize the electromagnetic response at a given set of system interfaces. These responses are needed to evaluate the system hardness to extraneous electromagnetic field environments. Furthermore, since the spatial characterization of the overall system response requires making a large number of measurements at different locations, it is necessary to capture the temporal characteristics of the signal with a few scalar quantities. Examples will be given in this section of quantities that can be used to characterize the system's temporal and spatial response characteristics.

Measured Waveforms

The direct output of the equipment used during the tests is a time-domain waveform, as shown in Fig. 8-24, from which all information about the measured response is to be derived. Properties that can be determined directly from it without much processing are peak values of the measured response and estimates of major resonance frequencies. Peak values, which can also be measured directly using special peak detectors, can serve as a first rough estimate of the overall magnitude of the measured signal but do not yield any information about the overall energy content, duration, rise time, or oscillating behavior of the measured signal. In certain applications, the peak value may be all the information that is needed, such as when verifying whether a response is below a given current level. Even though it is possible to determine the frequencies of the major resonances directly from the time-domain waveforms, it is usually difficult to determine their respective magnitudes to any reasonable degree of accuracy.

Spectral Density of Measured Waveforms

The resonances contained in a measured signal are uncovered most efficiently from its Fourier transform. This fact is one reason why much of the study and characterization of system responses takes place in the frequency domain even though it is transient phenomena that are of main interest. Examples of spectral densities for measured signals are shown in Fig. 8-24. The few examples shown in this figure indicate that measurements made at different locations have vastly different spectral characteristics. These results suggest a view of the system as a large filter, with the filter characteristics being a function of the location selected inside the system.

The frequency-domain representation of a measured response also allows one to estimate the system's response to an environment other than that used during testing. This procedure is simple if the two environments have the same spatial characteristics but differ only in their temporal (or spectral) behavior. It is often very difficult to reproduce in a test facility a field with all the spatial properties of a given specified field. Estimates of the system's response to the specified field based on extrapolated test responses become increasingly more uncertain as the test field deviates from the specified field.

Frequency-domain representation can also be useful in cases where there is a need for an approximate reproduction of a certain current/voltage waveform at the connector of an electronic box. This need could originate from a requirement of being able to test the performance of a certain box in the laboratory to

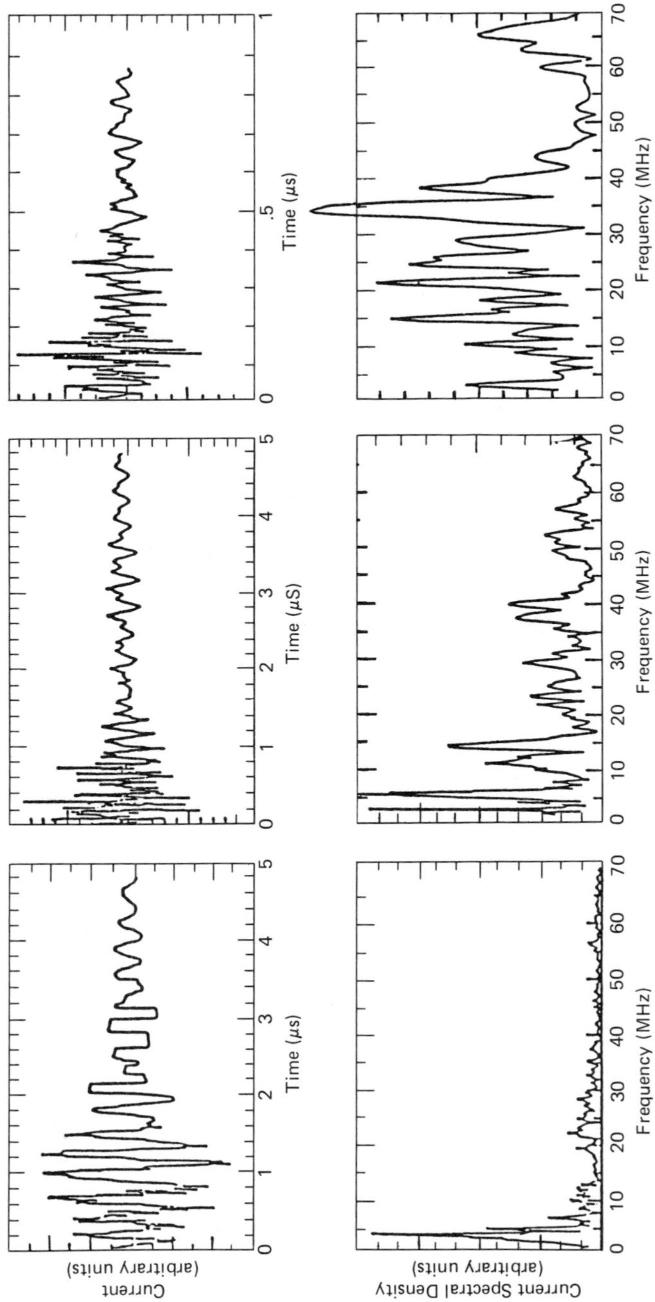

Fig. 8-24. Measured time-domain waveforms and their corresponding spectral densities of system-wave responses.

extraneous pulses that are similar in nature to those found during system tests. One way of reproducing the waveform is to construct it from a superposition of a few damped sinusoidal oscillations. The resonance frequency, damping constant, and amplitude of these oscillations are usually obtained from the spectral density of the transient pulse at hand.

Scalar Quantification

Scalar quantities are often introduced so that one can quantify, by a single number, the sensitivity (or immunity) of various equipment to the field environment considered. Thus, at a given interface (say a box connector), the scalar quantity is used to quantify both the strength of the equipment and the stress caused by the incident field environment. The strength-to-stress ratio, referred to as the "margin," is usually expressed in terms of decibels.

In defining the scalar quantities, it is important to incorporate those characteristics of the pulse that can cause equipment malfunction. To determine these characteristics requires that the equipment failure modes be understood. Since the failure modes are equipment specific, it is difficult to define general rules for defining the scalar quantities. From a practical point of view, on the other hand, it is necessary to have a few quantities that can be universally applied to an entire system with all its various types of equipment. Quantities that have been or are now in use are described below.

Peak Values. The time-domain peak values are the crudest way of characterizing the measured signal. Such rough estimates are sufficient in many applications when order-of-magnitude estimates only are needed, for example, measurements made at certain interfaces before and after the installation of a particular protection device. A well designed and installed device often results in the reduction of the induced response by at least a few orders of magnitude. Peak-value readings tend to capture the important features sought for under these circumstances.

Historically, peak values were of chief importance before extensive use came to be made of digital-data recording equipment. When only analog data (in the form of oscilloscope traces) were available, the automatic processing of large amounts of data was not feasible. The peak value was one characteristic that could be readily recorded and used for further analysis. With digital data now directly available, one can perform different operations on it to define various scalar quantities. This ability has led to the use of some other scalar quantities.

Total Energy and Peak Power in Pulse. A common type of failure that can occur in a circuit is component failure caused by overheating. If all the heat phenomena involved in the component are adiabatic (so that there is no transfer of heat away from the critical region of the component), then the important quantity describing the failure mechanism is the total energy in the pulse. It should be pointed out that the pulse discussed here is the current/voltage pulse arriving at the critical component. Since it is not possible to make measurements at the component level in most applications, one has to make them at the box connector/pin level. Even though box-susceptibility levels vary from one circuit design to another, the pulse power at the pin still points towards one useful quantity.

In cases where component failure is caused by overheating but heat transfer in the bulk material is of importance, the instantaneous power delivered to the component can be used to quantify its susceptibility level. The largest power present during any period of the pulse is called the "peak power." Peak-power quantification of the susceptibility level assumes that the heat transfer is efficient so that the temperature rise is proportional to the instantaneous power.

When quasi-adiabatic heat processes are involved, the fact that the temperature increase is a function of the entire pulse shape makes it more difficult to define a single scalar quantity of the failure level. The two limiting cases of total power and peak power tend to limit the susceptibility levels caused by quasi-adiabatic heat transfer.

Time Rate of Change and Time Integral of Pulse. The failure mechanisms related to pulse power and energy are primarily thermally caused failure modes. Other malfunctions can be caused by voltages occurring at a critical point. Mechanisms that contribute to this voltage are voltage drops along leads and voltage build-ups across narrow gaps or junctions. In the first case, the voltage is an inductive one caused by the time rate of change of current in the lead. In the second case, the voltage is a capacitive one resulting from a build-up of charges and is proportional to the time integral of the current.

The overvoltages thus created in a circuit may not cause any physical damage to the components involved but may trigger logic circuit elements and thereby cause transient upsets. These upsets may or may not result in overall equipment malfunction.

Arrival Time of Pulse at Different Locations. The scalar quantities discussed so far primarily refer to those used in determining the immunity of electronic equipment to various transient interference signals. In this use of the data, it is the local properties that are of importance. The relationship between current pulses at different locations within the system is not of concern in this application. In other applications, establishing the relative arrival time of current pulses at different locations is of value. By measuring this quantity together with the overall magnitude of the response at these locations, it is possible to establish coupling mechanisms. Doing so is of special value when it is necessary to locate points where electromagnetic energy leaks into a system, such as through the skin of an aircraft.

9
Applications of Subsurface Transient Radar

Leon Peters, Jr. and Jonathan D. Young

The Ohio State University

9.0 INTRODUCTION

The use of transient electromagnetic remote sensing for subsurface probing was begun in the early 1960s with Army-sponsored work at M.I.T. devoted toward nonmetallic land mine and tunnel detection.[1] Since that time, the technique has spread to several more applications requiring the detection or mapping of nonmetallic subsurface features, including buried utility lines, archeological sites, subsidence in soil, and plastic pipe and nonmetallic drain tile location.

For all of these applications, the system configuration follows the general outline shown in Fig. 9-1. (Specific systems based on this block diagram are being used by at least four organizations: Geophysical Survey Systems, Inc., ENSCO, Inc.; Microwave Associates, Inc.; and Southwest Research Institute.) An impulse generator emits a synchronized impulse that is coupled to the ground by a very broadband antenna. The impulsive electromagnetic energy propagates through the lossy, nonhomogeneous, dispersive soil and is partially reflected by any localized contrast in the electrical parameters of the medium. This reflected energy is received by another antenna on the surface, and the signal is detected and processed in the time domain by the receiver block.

Several features inherent in the purpose and environment of these systems favor the transient approach, including: (1) target identification, (2) soil propagation, (3) target range and resolution, (4) undesired signal characteristics, and (5) environment variability.

Target Identification

It is assumed for all these applications that many varied regions of local electrical parameter variation exist in the soil for depths of a few inches to tens of feet.

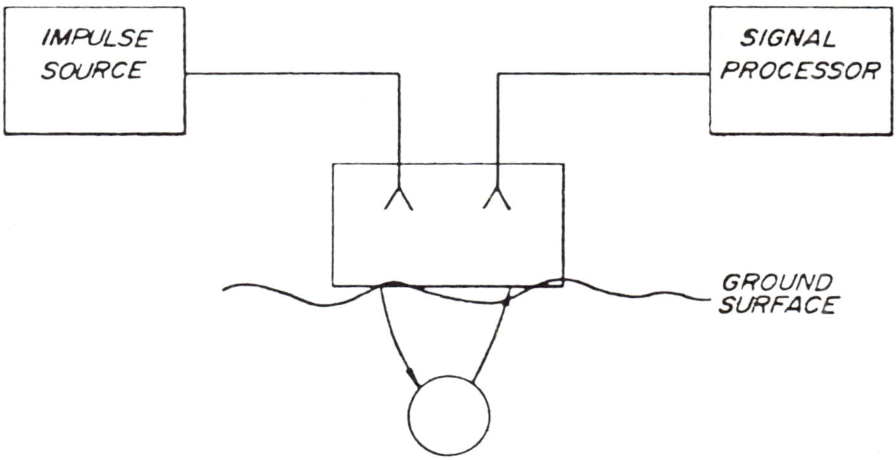

Fig. 9-1. Block diagram of subsurface-probing radar.

Some are likely to be naturally occurring, such as the water table, layers of differing soil, and soil nonhomogeneities. Others may consist of a wide variety of man-made targets, both metallic and nonmetallic. In virtually all real-world applications of this technology, target identification or discrimination is required in addition to simple echo detection. Information related to target identity is available by recording either echo vs. system location, or echo vs. frequency, or both. For surface remote-sensing systems where it is impossible to "get on the other side of the target" (as in a tomographic approach[2]), frequency spectrum information takes on increased importance. Those frequencies known to be most useful for identification have wavelengths comparable to the size of the object or smaller, with the smaller and smaller wavelengths (higher frequencies) giving more and more resolution. Considering the targets of interest and typical soil properties, frequencies for these systems thus range from about 1 MHz to about 3 GHz. Either a swept frequency or a transient system could be used to cover such a frequency spread.

Soil Propagation

For the frequencies of interest, soil can be characterized as a lossy, dispersive medium.[3–5] Water and salt content have the most important effect on its propagation properties.[6] Energy will penetrate much deeper and travel much faster in dry, sandy soil than in most clay. In almost all applications, however, there is a sufficient range of target depths that the effect of soil properties on the echo signature of buried targets must be considered. Both echo size and echo shape are affected. Furthermore, the effects vary as different (unknown) soil conditions are encountered. Experimental procedures have been developed for soil-parameter measurements in situ, and the variability of soil propagation vs. location, depth, and weather has been demonstrated.[7–9] If a transient system is used, the receiver sensitivity can be controlled to compensate for the loss vs. penetration depth (time delay) characteristics of the soil. This system is a significant advantage over a swept frequency approach.

Target Range and Resolution

The expected depth of the targets and the resolution desired in order to identify them properly are important factors when comparing the transient system approach to a more conventional radar technique, the pulsed-carrier signal system. A normal pulsed-carrier radar system transmits a burst of from about ten to hundreds of cycles of its carrier frequency with a repetition rate that is determined by the desired target range, normally several miles. The range of the subsurface targets of interest begins at a depth of just a few inches, and resolution of a foot or less is desirable. Also, the "carrier" frequency of subsurface systems is limited to values less than 1 GHz by soil properties, as has already been discussed. This combination of constraints leads to a pulsed-carrier system with just one half cycle of carrier—in other words, to a transient radar system—as the best choice for subsurface probing applications.

Undesired Signal Characteristics

In addition to the target echos, several sources of undesired signal in a real-world subsurface probing system impact on its design. One of these is random noise, as found in any radar system. Also, there is direct-coupled energy from the transmit to the receive antenna as well as echos from nearby objects on the surface. Finally, for many applications, the echos from soil nonhomogeneities could be classed as undesired signals. For sensing applications involving targets less than 30 feet deep and with sensing antennas near the soil boundary, for example, it has been found that undesired returns that are random with respect to location predominate over the noise components. Therefore, the time gating of a transient system makes it more suitable than a frequency system for eliminating such signals. The frequency system, on the other hand, would perform better against noise by its use of narrow-band filtering.

Environmental Variability

The impulse is a simple waveform in the time domain and has the property of degrading gracefully, as it propagates through the lossy soil, to an approximately

Gaussian shape of longer than original duration. Even when the soil loss characteristics vary, its shape stays the same, a property that makes the results much easier to interpret in widely different soil media.

9.1 SOIL PROPAGATION CHARACTERISTICS

Since each of the preceding topics has alluded to the soil medium in which the subsurface sensing radar must operate, it is obvious that an understanding of electromagnetic propagation in soil and rock is fundamental in a discussion of sensing system design concepts and performance.

For most buried-target radar-sensing applications, the soil can be characterized as a nonmagnetic ($\mu = \mu_0$) lossy dielectric medium having a frequency-dependent relative dielectric constant ε_r and conductivity σ. The concept of a low frequency window (LFW) and high frequency window (HFW) has been introduced to simplify the description of propagation behavior in typical geological media. It is best defined in terms of the curves of Fig. 9-2, in which $|E_{\theta 0}|$ is the transverse dc electric-field intensity at the observation point shown and $|E_\theta|$ is

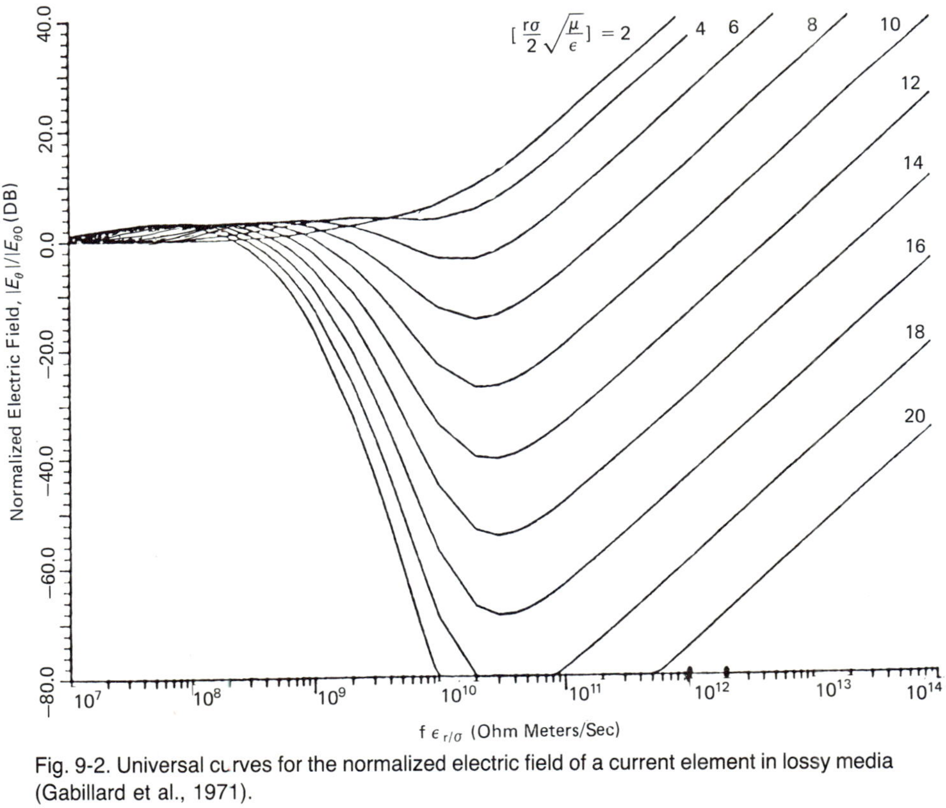

Fig. 9-2. Universal curves for the normalized electric field of a current element in lossy media (Gabillard et al., 1971).

the electric-field intensity as a function of frequency for an infinitesimal dipole.[10] The frequency at which the normalized quantity $|E_\theta|/|E_{\theta 0}|$, is at -3 dB is the cutoff frequency of the LFW. The frequency at which the normalized field again reaches 0 dB (as frequency is increased) is the cutoff frequency for the HFW. The major difference that occurs when a definite length of antenna is introduced consists of a resonance behavior in which the level in the HFW attains a peak as the frequency is increased. If the operating point in Fig. 9-3 is to the left of a particular slanted line, then detection in HFW is practical at the range indicated. For example, at $\sigma = 10^{-3}$ Si/m and $\varepsilon_r = 4$, the (X) would indicate that detection is possible using the HFW at a depth of 30 m.

Pulse Propagation in the HFW

For propagation between parallel dipoles in the HFW, the transmission transfer function is more strongly dependent on frequency than in the LFW. Consequently, there is significant distortion of the input pulse. In particular, because insulated antennas will inevitably be used for HFW radars, there is effectively no transmission at low frequencies, and the received signal will have no average (dc) value. To obtain efficient transmission in the HFW when using dipole type antennas, it is necessary to choose the antenna length so that antenna resonance occurs in the frequency band of operation of the radar. A design goal is to minimize this resonance so that the antenna has a wide bandwidth, but the same constraints on the length of the antenna will still be expected to exist. Improve-

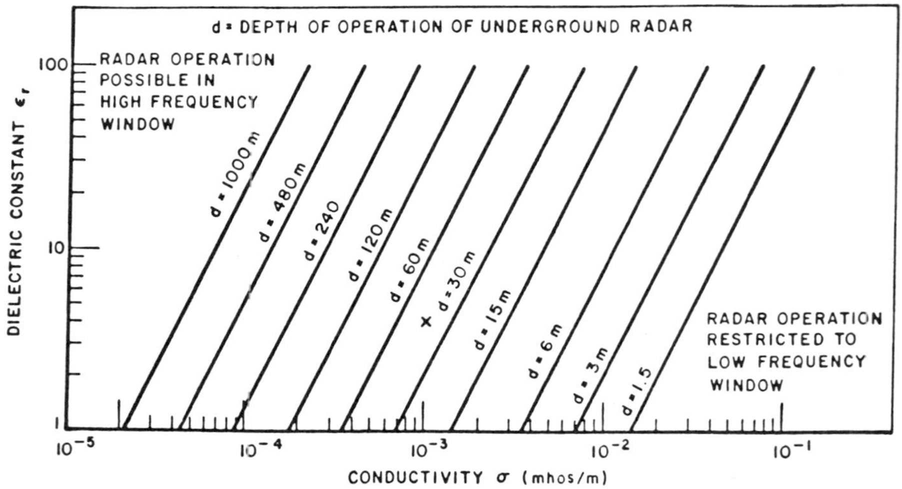

Fig. 9-3. Design graph to determine whether radar operation should be HFW or LFW.

ments in the antenna design will be discussed in a later section; only propagation between simple dipoles is considered here.

There are several advantages in operating a radar in the HFW, among them the following:

1. The antenna is portable (because it does not have to be in conducting contact with the ground) and considerably smaller in general than an LFW radar antenna.
2. The band of frequencies that can be transmitted is often many decades of frequency higher than that passed by the LFW, with the result that short input pulses can be used and a much wider separation in time obtained between the transmitted and received signals (called a "clear range window"). From a signal separation viewpoint, this clear range window means that it is possible to use the same antenna for transmitting and receiving.
3. High-frequency scattering information means that identification of targets whose characteristic dimensions are of the order of one or two meters is feasible.

Operation in the HFW does not imply that the received waveform is a reduced replica of the transmitted waveform even though the attenuation factor is not explicitly a function of frequency and the propagation factor is a linear function of frequency. Of a number of reasons for this waveform shape change, most prominent and important are the changes introduced by the target, which form the basis for target identification capability. The antenna also introduces changes in waveform shape. There is also the possibility that the electrical parameters are frequency-dependent, which would introduce implicit frequency dependence in the attenuation and propagation factors.

The concept of an attenuation factor has been retained in this discussion in spite of the received signal not being, in general, a near replica of the input signal. Its usefulness is that it relates the peak amplitudes of the transmitted and received signals. The transmitted signal will usually be a simple signal, such as a pulse or doublet, and is readily described by its peak amplitude. The peak value of the received signal is useful as a measure of the "visibility" of the signal to the receiver.

It was shown that for propagation to be practical in the HFW, the dip in the frequency response curve for the elemental dipole (see Fig. 9-2) should be less than about 10 dB. If the radar antennas are insulated, then the frequency band of operation of the radar should be on that part of the curve where the electric field of Fig. 9-2 slopes upward at 6 dB per octave of frequency. If it isn't, then significant signal loss will occur because the conduction currents are not strongly excited when the dipole is insulated (Fig. 9-2).

Since the upward sloping part of the frequency response curve occurs when α is constant, we conclude that an HFW radar should operate in a frequency band where α is constant. The lower limit of this band can be determined from Fig. 9-4. It should be noted that a radar can be operated through the transition region providing the antennas are not insulated. Operation at higher frequencies with insulated antennas seems more desirable, however, and consequently the transition region will not be further considered here.

For observation of tunnels, the desired frequency band ranges from 20 to 60 MHz, depending on tunnel parameters. Figure 9-4 shows that the systems are operating in this frequency band except for the highest conductivities.

Design Curves for HFW Radars[11]

When α is constant over the frequency band of operation of the radar, the attenuation factor A (which relates peak amplitudes of signals) can be written as

$$A = A_0 \frac{e^{-2\alpha d}}{2d} \tag{9-1}$$

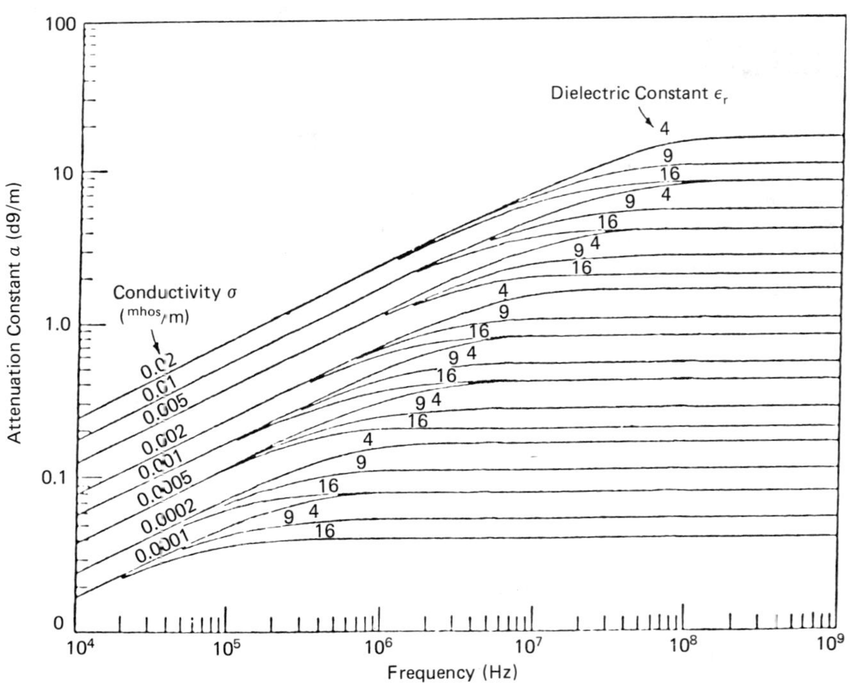

Fig. 9-4. Attenuation constant, α, plotted as a function of frequency for lossy media.

A_0 is a constant that is determined by numerical experiment for a particular antenna geometry, medium, generator impedance, and input signal. The factor, $e^{-2\alpha d}$, accounts for the medium attenuation over the distance $2d$, and the $2d$ in the denominator accounts for the signal loss caused by the spatial attenuation (or the spherical expansion) of the fields of the dipole. The only effect of α is to introduce signal attenuation.

Figure 9-5 shows the system responses for two 1-m long dipoles separated by 60 m in a homogeneous medium where σ = 0.001 mhos/m and ε_r = 9. The dipole radii are 0.001 m, and each is coated with a 0.0015-m radius layer of polyethylene (ε_r = 2.3). The upper plots show the input reflection coefficient

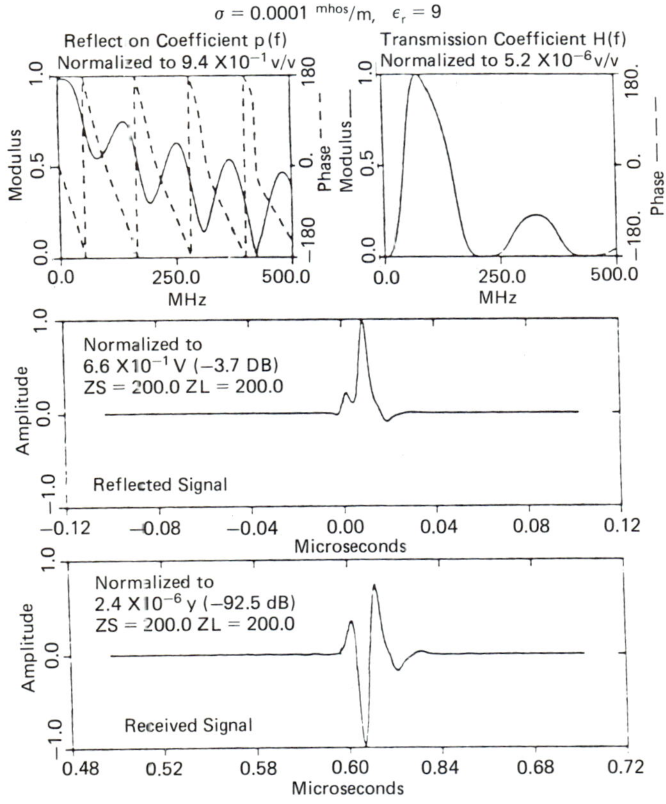

Fig. 9-5. System responses for two 1-m-long, 0.001-m-radius dipoles separated by $2d$ = 60 m in a homogeneous medium of σ = .001 mhos/m and ε_r = 9. The dipoles are coated with a 0.0015-m radius layer of polythene insulation (ε_r = 2.3). The upper plots are the system frequency responses: the lower plots, the signals obtained when the frequency responses are weighted by the spectrum of the input signal and transformed into the time domain. The input signal was a 5-ns Gaussian pulse.

ρ(f) and dipole to dipole transmission coefficient H(f), and the lower plots show the time-domain signal reflected from the terminals of the transmitting antenna and the signal observed on the receiving antenna when a 5-ns gaussian pulse is transmitted. The source and load impedances were set at 200 ohms. Note that the transmission coefficient plot $H(f)$ shows that significant energy is transmitted when the antenna is about a half wavelength long in the medium (≈ 100 MHz). Significant energy is also transmitted when the antenna is about 1.5 wavelength long (300 MHz), but the transmission efficiency is less than that for the first resonance because the effective length of the receiving antenna when it is 1.5 wavelengths long is about a third the effective length when it is 0.5 wavelength long. Little energy is transmitted when the antenna is 1 wavelength long because of symmetry.

The peak received signal is 92.5 dB below the transmitted signal, i.e., A is equal to 92.5 dB. Observe also that the scale is shifted for the received voltage waveform and delayed by a time corresponding to approximately 100 transmitted pulse widths so that there is a very clear range window. It should be practical

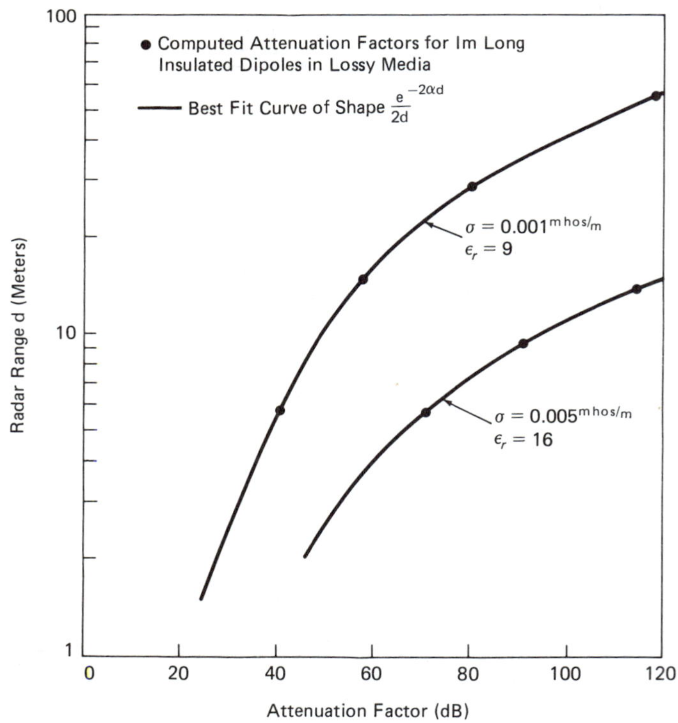

Fig. 9-6. Attenuation factors for 1-m-long, 0.001-m-radius insulated dipoles computed for two different media with best-fit curves.

to observe the reflected signal on the transmit antenna if there are no reflections from parallel layers present.

Figure 9-6 shows the same antenna system but where $2d = 50$ m. Note that the shape of the received signal is identical to that shown in Fig. 9-5, but the amplitude has changed by 22.1 dB. Figure 9-7 shows computed values for A for this medium and also for a medium where $\sigma = 0.005$ mhos/m and $\varepsilon_r = 16$. The excellent agreement wih the best fit curves of the shape, $\exp(-2\alpha d)/2d$, confirms Eq. 9-1).

Equation 9-1 can be expressed in decibels as follows:

$$A = A_1 + A_F \tag{9-2}$$

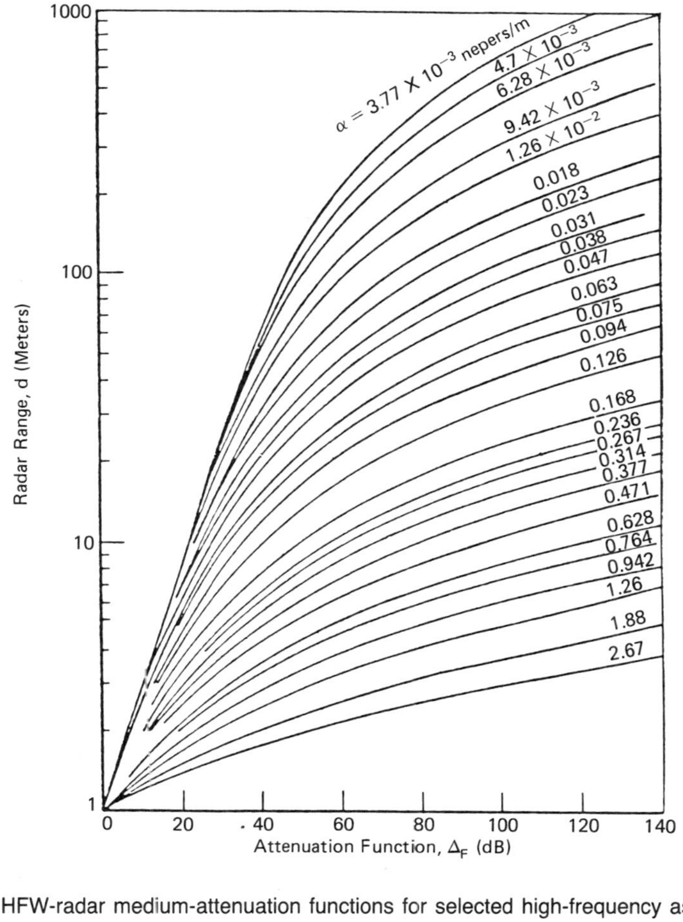

Fig. 9-7. HFW-radar medium-attenuation functions for selected high-frequency asymptotic values of α.

where A_F is given, in dB, by

$$A_F = 20 \log_{10} \frac{e^{-2\alpha(d - 1)}}{d} \tag{9-3}$$

and the constant A_1 is given, in dB, by

$$A_1 = 20 \log_{10}(A_0) \tag{9-4}$$

where A_0 is the value taken when $d = 1$ meter. A_F, the medium attenuation function, is normalized to have unit value when $d = 1$ meter. Figure 9-8 is a plot of A_F for media having different values of α. Table 9-1 shows the high frequency forms for α for a wide range of media constitutive parameters. Figure 9-4 is useful as a gauge of the frequency at which α reaches its high-frequency asymptotic form.

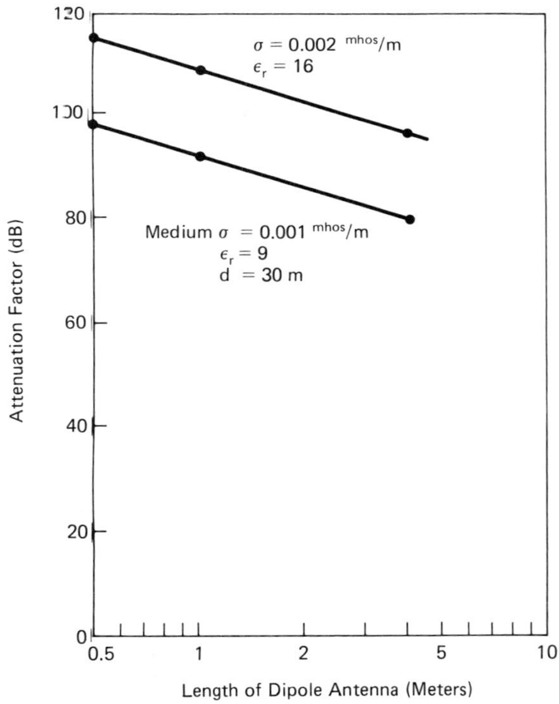

Fig. 9-8. Illustrating the variation of attenuation factor with antenna length for HFW radars. The input signal duration is scaled together with the antenna length and radius.

TABLE 9-1. HIGH-FREQUENCY ASYMPTOTES FOR α (NEPERS/M).

	σ mhos/m							
ε_r	0.0001	0.0002	0.0005	0.001	0.002	0.005	0.01	0.02
2	1.333×10^{-2}	2.666×10^{-2}	0.067	0.133	0.267	0.0666	1.33	2.67
4	9.42×10^{-3}	1.88×10^{-2}	0.047	0.094	0.188	0.471	0.942	1.88
9	6.28×10^{-3}	1.26×10^{-2}	0.031	0.063	0.126	0.314	0.628	1.26
16	4.71×10^{-3}	9.42×10^{-3}	0.023	0.047	0.094	0.236	0.471	0.942
25	3.77×10^{-3}	7.54×10^{-3}	0.018	0.038	0.075	0.188	0.377	0.754

The constant A_1 is a function of antenna length, antenna type, medium constitutive parameters, generator source impedance (and receiver load impedance), and input signal. In this analysis, we have restricted ourselves to the simple dipole, and we have further reduced the amount of data required by choosing an antenna 1 meter long (it will be shown later that antenna length can be simply accounted for). This data is given in Table 9-2.

The choice of a generator source impedance and an input signal required considerable deliberation. The input impedance of the straight, insulated dipole is not constant over a wide frequency range. One essential choice of generator source impedance is a constant resistance, because this best simulates the transmission lines that will inevitably be used to feed HFW radar antennas. The value selected, 200 ohms, satisfies two criteria. The first is that this value can be physically realized as a balanced transmission line by connecting two pieces of 100-ohm coaxial cable in series. The second criterion is that 200 ohms approximately match the surge impedance of a dipole antenna so that the reflection of the input pulse from the terminal region of the dipole is relatively small (observe

Table 9-2. Values for Constant A_1 for 1-m Long, Insulated Dipoles in Lossy Media.

		σ (mhos/m)				
ε_r	INPUT SIGNAL	.0002	.00054	.001	.002	.005
2	5 – ns pulse	34.3 dB	34.7 dB	37.8 dB	—	—
4	5 – ns pulse	30.6 dB	31.1 dB	32.9 dB	34.5 dB	—
9	5 – ns pulse	29.6 dB	30.3 dB	31.2 dB	32.8 dB	37.0 dB
16	5 – ns pulse	37.7 dB	29.6 dB	31.2 dB	31.5 dB	33.5 dB
25	5 – ns pulse	30.5 dB	30.9 dB	30.7 dB	31.8 dB	—

The Dipole radius is 0.001 m and the outer radius of the polythene insulation (ε_r = 2.3) is 0.0015 m. A conjugate match was used for the optimum input signal; a 200-ohm match, for the other input signals.

that in Figs. 9-5 and 9-6 the first peak in the reflected signal, which is the pulse reflected from the terminals, is about −18 dB of the input signal amplitude). It has been observed that the level of the received signal is not significantly affected by relatively large changes in the source resistance (little change was observed as Z_s varied from 50 ohms to 200 ohms).

The choice of a constant resistance for the generator source impedance does not give a reflection free match nor maximum power transfer. Because we are constructing our signals from data tabulated at many frequencies, we can choose different matching impedances at different frequencies. These could correspond to some physically realizable network, or they could just as easily represent a nonphysical network. If a reflection free match, $Z_S = Z_A$, is modeled, a network can be constructed that will have a frequency response similar to that of the dipole. There are advantages to achieving such a reflection free match in terms of reducing the multiple reflections that would cause ringing into the range window of the radar. In practice, however, such a match over a broad frequency band can be achieved only by matching out reflections at the point at which they occur. In the studies on antenna structures,[12,13] this has been the approach attempted.

There have been several attempts at pulse shaping and optimization procedures to enhance the received signal. An initial approach was modifying the Gaussian pulse to that of a Gaussian doublet.[11] This choice was made since this type of signal has no dc component. As can be seen in the transmission coefficients, $H(f)$, of Figs. 9-5 and 9-6, the received signal for HFW radars contains no low-frequency components. Thus, the doublet is apparently better tailored to the system. Since it introduced only a 4-dB transmission gain, however, it was decided that such a small gain was not justified in view of the more complex source that would be needed. Ignoring such complications as physical realizability, complex waveform shape, etc., and optimizing all possible parameters, the received signal peak of Fig. 9-5 could be increased by an order of magnitude (23 dB), but the required pulse shape could not possibly be achieved in practice. More recently, the K-pulse concept, which essentially controls the pulse shape so that ringing is eliminated, has been developed.[14] The aim of this concept is to reduce clutter other than increase the signal level. The ultimate goal, of course, is to increase the signal to clutter level.

Target recognition is an important feature of underground radar. Information on target resonances is required for recognition schemes based on complex resonances.[15–19] The underground radar designer knows the type and approximate dimensions of the target, although its precise electrical composition may be considerably uncertain. Consequently, the resonant frequency of a target can be predicted with reasonable accuracy, and the frequency band of operation of the radar selected accordingly. The latter determines the length of the dipoles.

Figure 9-8 shows how the attenuation factor depends upon the dipole length, providing two examples with different ground parameters. In the calculations used to prepare Fig. 9-8, the frequency range of the data, the pulse duration, and the antenna radius were scaled in proportion to the antenna length. For example, for the 4-m long antenna, data was collected to 125 MHz, and a 20-ns Gaussian input pulse was used. In each case, a 200-ohm resistive match was chosen to correspond with the surge impedance of the antenna, which is frequency-independent.

9.2 COMPONENTS OF A TRANSIENT UNDERGROUND RADAR SYSTEM

This section will discuss the components of a typical system (shown in Fig. 9-1 in more detail) and relate component design and performance parameters to system application goals.

Impulse Generator

The impulse generator puts out a train of periodic video pulses of assumed Gaussian shape with a voltage amplitude, V_{max}; a pulse duration between points 3 dB down from a peak voltage, T_d; and a repetition rate, F_{rep}. All of these parameters assume that the output port is terminated in a matched load, R_ℓ.

Since a variety of commercial impulse generators fitting the above description is available, there is some choice of parameter values to best fit the underground sensing application. Note that this class does not necessarily include all pulse generators that allow the independent control of pulse width as well as amplitude and repetition rate. For purposes of maximum resolution and broad frequency spectral coverage, it will be assumed that all sources, including the adjustable pulse width type, are set to approximate a true periodic impulse output waveform as closely as possible. A survey of such instruments has been presented and also summarized elsewhere in this book (see Chaps. 2 and 4).

The choice of pulse duration, T_d, depends primarily on the depth and position resolution desired (or permitted by soil propagation at the system-design sensing depth). For a soil medium having a characteristic relative dielectric constant of ε_r, the maximum resolution, D_{res}, is given as

$$D_{res} = T_d c/\sqrt{\varepsilon_r} \tag{9-5}$$

where T_d is in seconds and c is the speed of light in meters per second. For a typical example of sensing buried utilities with a desired resolution of 10 cm in soil having an ε_r of 16, it can be seen that a pulse duration of approximately 1 1/3 nsec is desired.

The above resolution degrades with increasing depth in lossy, dispersive soil.

In such cases, a short pulse duration may still be used so that shallow clutter echoes will have minimum duration and may be removed from the echo time history by time gating, even though the resolution is limited for the deeper targets.

The pulse repetition rate is governed by the maximum expected target depth as well as by system factors such as processing speed, efficiency, and return processing approach. The time delay for the return from a target at maximum unaliased depth, D_{max}, is

$$T_{del} = 2^*D_{max}\sqrt{\varepsilon_r}/c \tag{9-6}$$

If a repetition rate higher than $1/T_{del}$ is used, it is possible for the return from a target at D_{max} to be aliased and appear to be at a shallower (maybe zero) depth. For a system to sense to a depth up to 4 ms in soil having ε_r values as high as 25, the maximum repetition rate would be approximately 10 MHz. Since tube-type pulsers usually cannot obtain such higher repetition rates, this frequency will have to be reduced if higher pulse voltage is desired. On the other hand, sampled returns from several successive pulses might be averaged together by the receiving processor if a low-voltage, solid-state pulser is used at this high repetition rate, thus giving greater effective system sensitivity with some sacrifice in system processing speed.

The pulse-amplitude choice depends on desired system sensitivity, soil propagation parameters, target echo strength, and the performance parameters of the other system components. It will be examined at greater length along with the overall system transfer function.

All of the preceding discussion has assumed that an approximate impulse is the signal for use in underground radar probing. It would be possible to use step generators or ramp generators as periodic transient sources having a different distribution of spectral energy, but no example of this having been tried yet is known. Recent research has identified a special transient waveform called the "kill pulse"[14]; this may be well suited for use in underground sensing with a known antenna performance and expected target signature spectrum.

Antennas

The antennas of an underground radar system are normally designed to operate above, but in close proximity to, the ground surface so that they can be conveniently moved about to probe the region directly beneath their physical aperture. The advantage of close proximity to the ground is that with a properly designed antenna, the energy can be more efficiently coupled to the below-ground region. The disadvantage is that the ground surface is rough and soil parameters vary widely subject to geography and weather, so that an optimum antenna design is not possible. For the initial discussion, the general case of separate transmit and receive antennas will be assumed.

As for the transmit antenna, its task is to transfer as much of the impulse generator signal toward the underground target region with as little distortion as possible, with as little reflection of energy back to the generator as possible, and with as little illumination as possible of regions that do not contain the target (such as in the air above the antenna). These goals are to be met for a range of soil and surface roughness parameter values, and with a structure of reasonable size, weight, and complexity.

The receive antenna tasks are similar. The echo signal from the target is to be received with as much sensitivity and as little distortion as possible, whereas received energy from other sources is to be minimized and the transfer of energy from the antenna to the receiver maximized. Again, antenna size, weight, complexity, and adaptability to varying environments are secondary but significant factors.

An additional set of goals for the antenna pair includes the minimization of energy directly coupled from the transmit to the receive antenna, and an arrangement of the antennas physically in such a way that they have a common target-sensing region and the polarization of the transmitted field to the target couples strongly to the target while the resulting echo field polarization couples strongly to the receiving antenna.

There is no recognized set of rigorous and succinct performance parameters that characterizes antenna performance under these constraints and in these conditions. The analysis and design problems are still the subject of research effort. Traditional frequency-domain parameters may be applied if a careful accounting of the soil interface is made;[20] these are cumbersome, however, if a transient radar system is involved. Time-domain parameters that would be more appropriate will be mentioned wherever they pertain.

Using the concept of reciprocity, it can be shown that the receiving antenna parameters and design goals are the same as for the transmitting antenna. One important factor, however, is not normally encountered in narrow-band antenna designs. It has been shown that the receiving-antenna transfer function for a given configuration is the integral of the transmitting transfer function,[21] or

$$F_{\mathrm{rax}}(t) = \int F_{\mathrm{tax}}(t) \tag{9-7}$$

(see also Chapter 5).

No optimized antennas for transient underground radars have yet been designed, but there are two configurations that are commonly used. A single resistivity-loaded dipole is applied to the problem of general subsurface sensing. Both transmitter and receiver are connected to the same antenna feed port, with

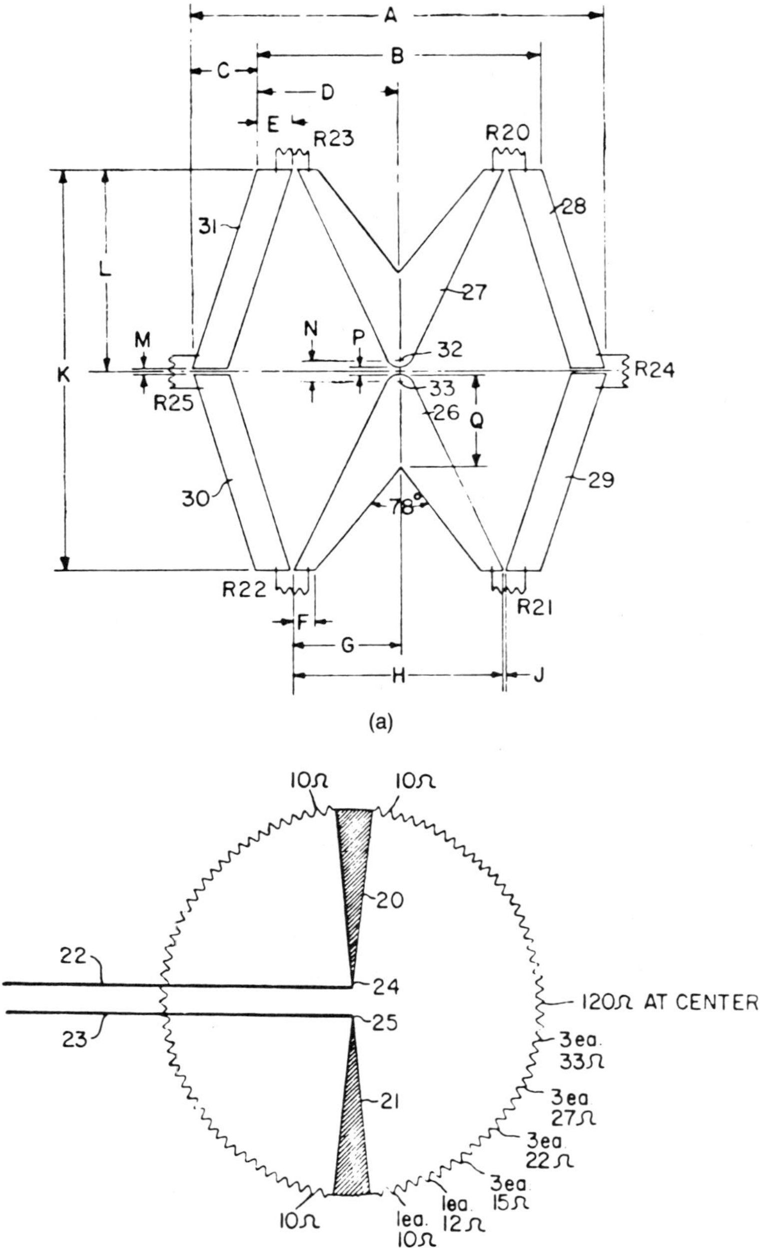

Fig. 9-9. Antennas for subsurface-sensing transient radar.

a fast-response limiter protecting the receiver from the transmitted impulse signal. The loading minimizes ringdown, so that echos from buried targets can be detected. Sometimes an additional compensating circuit is added between the generator and receiver to cancel out the unvarying portions of the antenna ringdown signal. Two single-antenna configurations described in one of the most important patents on subsurface radar[22] are shown in Fig. 9-9.

In some subsurface sensing applications, horizontal layers are not to be detected; in this situation, an orthogonal loaded dipole pair has been successfully applied.[23,24] The performance of such antennas has been analyzed[25] and techniques for objective time- domain measurements of target-detection performance devised.[26] A typical loaded orthogonal-dipole-pair antenna used in a pipe-location system is shown in Fig. 9-10.

An experiment for evaluating the sensing performance of radar antennas in a pipe-locating application is shown in Fig. 9-11. A plastic pipe has been inserted into a bored hole beneath a grass lawn at a very shallow angle with respect to the horizontal and its path very carefully surveyed. The test antenna is located on the ground surface, and the source antenna is inserted inside the slant pipe. The source antenna is fed through a coaxial cable from the pulse generator. The pulse signal for these tests was a pulse of 0.15-nsec width and 1000-V peak voltage, with a repetition rate of about 600 Hz.

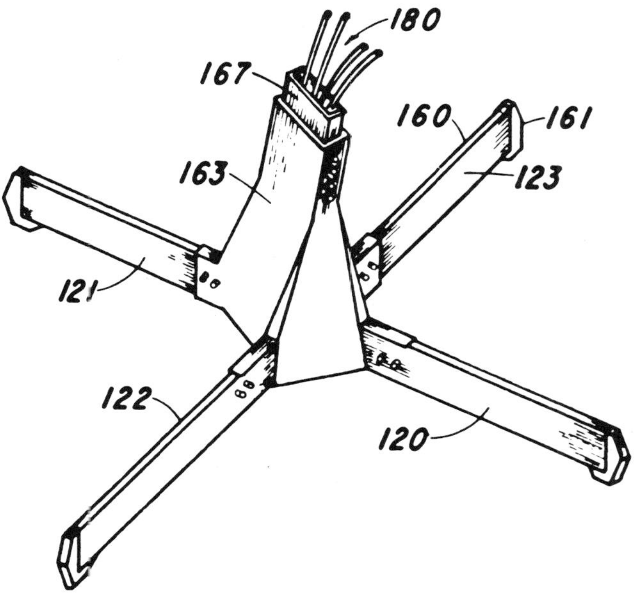

Fig. 9-10. A cross-polarized antenna for subsurface-sensing transient radar.

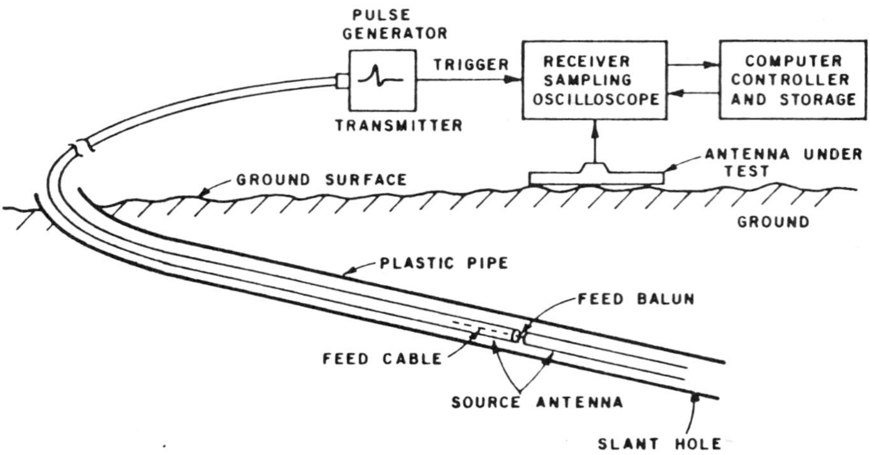

Fig. 9-11. Experiment for evaluating subsurface-sensing radar antennas.

One antenna tested using this apparatus was a prototype for the Terrascan* a buried utility locator system. The design of the antenna was similar to the one shown in Fig. 9-10. Figure 9-12 shows the response of this antenna, with the source antenna positioned at increasing depths. The actual wave shapes are obtained by multiplying the plotted amplitude curves by the amplitude factors shown for each waveform. This series of plots demonstrates the change in signal shape as a function of the propagation path length experienced in the lossy soil of this location. Figure 9-13 shows a set of waveforms recorded with the source antenna at a depth of 64 cm and the sensing antenna location offset from directly above the source in a direction perpendicular to the pipe. Note, again, that both signal amplitude and shape are changing. Thus, a simple amplitude pattern is not sufficient to characterize the performance of this antenna.

Receiver

The receiver used for all transient subsurface probing systems is the sampling oscilloscope, which is synchronized to the transmitting impulse generator and connected to receive the echoes from the receive antenna port. The display waveform is usually recorded on tape or digitized for further processing.

Sampling oscilloscope operation is described elsewhere in this text (see Chaps. 2 and 6). The important parameters to consider for subsurface sensing are its

*Terrascan is a registered trademark of M/A–COM, Inc.

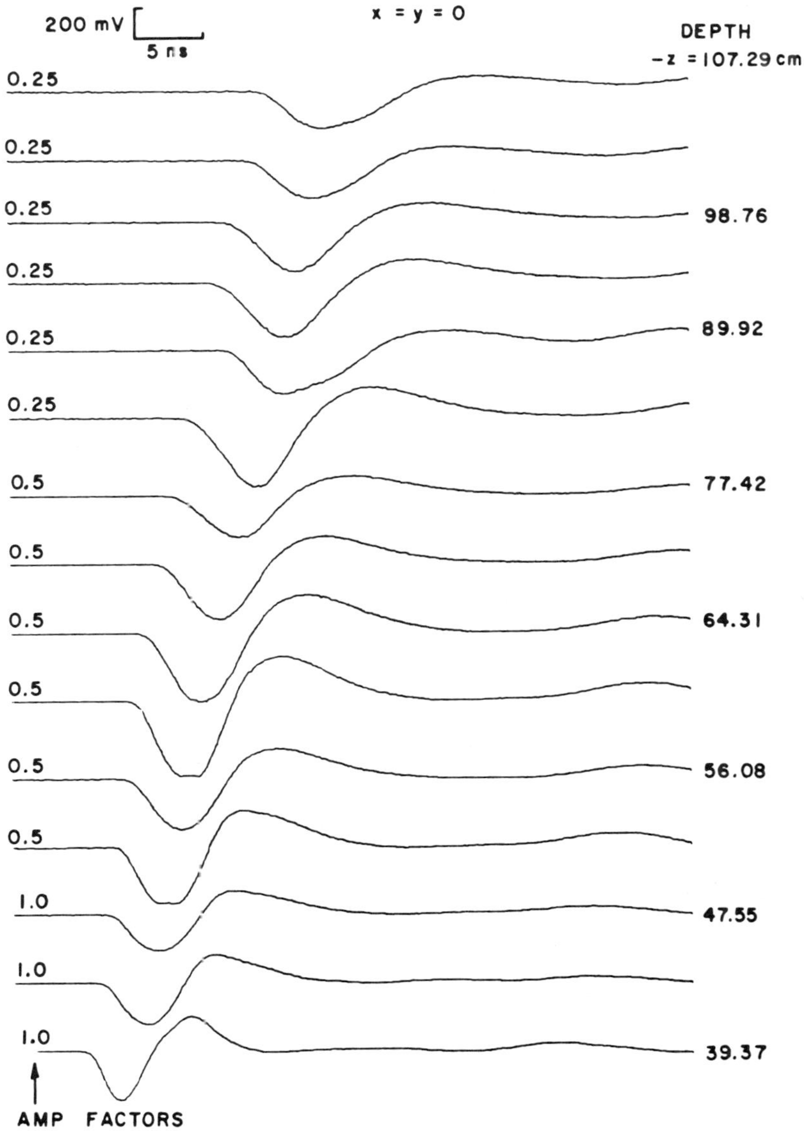

Fig. 9-12. Terrascan antenna response to buried transmitted impulse signal vs depth.

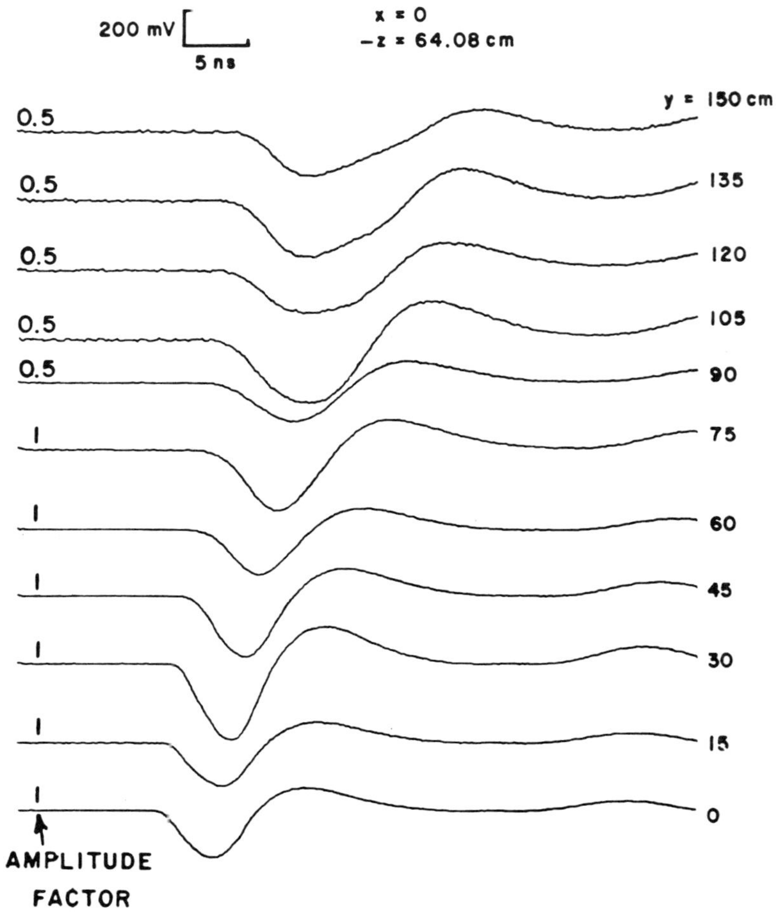

Fig. 9-13. Terrascan antenna signal vs. position offset with respect to a 64-cm-deep pipe.

frequency response (which must span the spectrum of the transmitted and received signals) and its tangential sensitivity (which must be as sensitive as possible).

Processing

The received signal ensemble as seen by the receiver as the subsurface probing radar is moved over the ground surface will contain several identifiable undesired components that are usually stronger than thermal noise. First in the time sequence is the antenna transmit–receive coupling, or ringdown feature. This serves to establish a time-reference point for return-depth determination, but, in general, its amplitude is stronger than target returns and its form changes because of surface roughness. If further delayed in time, returns from surface irregularities and other surface features such as pavement cracks to the side of the antenna will appear. Also, soil nonhomogeneities that are random with respect to position and depth will cause undesired "clutter" echoes.

The target returns that are submerged in the clutter are notable mainly for their spatial behavior. Layers will give an echo at a consistent, possibly slowly varying, depth. Utility lines are recognized as their straight-line paths become evident. Also, targets in general have an impulse signature, one that has been shown to be unique and can be used for target discrimination provided the bandwidth of the transient sensing signal is sufficient.[19]

The factors mentioned here are justification for further processing of the subsurface radar returns. Virtually all modern underground radar systems do processing using either analog circuitry, which is faster but less versatile and less powerful, or digital processors, which range from a few integrated circuit modules to vast array processors. The processing discussed here will be divided into the categories already discussed: (1) clutter removal, and (2) target discrimination.

Clutter Removal. For this discussion, "clutter" is defined as all undesired transient return echo features that are not a thermal noise process. Of course, those returns labelled as "clutter" will depend on system configuration and the purpose of the system. Processing to remove these returns makes it easier to locate and discriminate among targets.

Time Gating. Several undesired returns occur in a time interval before or after the expected target returns. An adjustable time gate that effectively turns off the sensing during these intervals is used to eliminate such returns. A typical application is to "gate out" the antenna coupling return.

Ensemble Averaging. This technique is used to discriminate against those returns that are random with respect to position.

Signal Subtraction. If an undesired echo feature appears consistently, it can be stored in a buffer memory and subtracted from succeeding return echoes. This technique might be used to eliminate the shallow sidewalk return when a utility locator is being moved down a sidewalk to locate utility crossings, for example.

Target Discrimination. This category includes all processing on a single time-domain return waveform to separate expected targets in the medium whose dispersion is known at least approximately.

Fast-Fourier Transform. Utilizing the Fast-Fourier Transform (FFT) to obtain the complex frequency spectrum of the time-gated echo signals is a common processing step. For targets with a strong resonance, this feature may be easily discriminated by observing the frequency spectrum of their return.

Filtering. The first application is made in conjunction with the preceding FFT for discrimination of targets with a strong response in a narrow frequency region. Another application takes advantage of the known frequency propagation characteristics of the soil and is thus a combined target-discrimination and clutter-removal process. It is known that most soils attenuate the high frequencies of the impulse spectrum more rapidly than the low frequencies. Thus, filtering high-frequency returns out of the total echo return for the late-time intervals removes far more surface clutter than buried-target return energy. This is a common step when soil parameters are known.

Difference-Equation Processing. For buried targets having a known ensemble of characteristic resonances, a Corrington difference-equation process has been developed to provide discrimination that is independent of target orientation. This has been shown to be effective for identifying shallow, nonmetallic targets of specific shapes.[16] The processing steps involved are demonstrated and further discussed in those portions of this chapter where specific systems are treated.

9.3 FEATURES OF AN UNDERGROUND UTILITY SENSING TRANSIENT RADAR SYSTEM

A transient radar system has been developed for use like a metal detector to locate nonmetallic as well as metallic utilities to depths up to ten feet (depending on soil properties). This system is one of the simplest and most basic of transient radar sensing systems.

A prototype of the system, which as been given the name "Terrascan," is shown in Fig. 9-14. The impulse generator is housed along with the battery in the lower module on the golf cart. The transmit-receive antenna pair are orthogonal, resistively loaded dipoles. The receiver utilizes sampling circuitry along with rudimentary digital processing and display (in real time) of the echoes. The control panel of the instrument is shown in Fig. 9-15 and a set of typical displays in Fig. 9-16.

Using the Terrascan

The operation of this plastic-pipe locator is quite simple, but some operator training is required. This training involves learning how to recognize typical

Fig. 9-14. Terrascan system.

pipe echoes and the procedure for determining pipe type and direction. The operator first places the antenna on the ground at a place where there are no buried pipelines in order to establish a background or clutter level. The clutter level is less than the return echo from a pipe but varies with soil type, amount of buried rock, and ground cover. When the antenna is not above a pipe, the operator sees a single row of lights across the display, as shown in Fig. 9-16(a). If the clutter is high, the row may have a slight ripple. The top partial row of lights indicates the cursor.

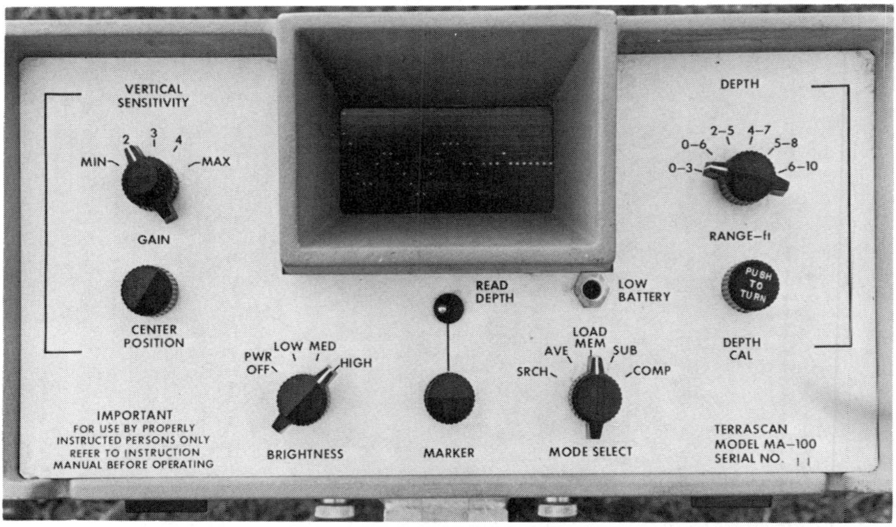

Fig. 9-15. Control panel of the Terrascan.

To find a pipe, the antenna is placed on the ground and the display read. The locator is then moved in 0.3- or 0.6-m (1- or 2-ft) steps until the pipe is found. As the pipe is approached, the line of light becomes a sinusoidal curve, whose maximum amplitude occurs when the antenna is directly above the pipeline, as shown in Fig. 9-16(b). To find the direction of the pipe, the antenna is rotated again to find the point of maximum amplitude. This establishes the pipe as being in one of two orthogonal directions, and moving the antenna in either direction establishes the actual direction. Once the pipe is located, it may be followed. If the pipeline has a tee, a null (straight line) is seen over the tee location. If the line changes from plastic to metal or vice versa, the phase of the waveform on the display reverses. This can be seen by comparing Figs. 9-16(b) and 9-16(c).

The plastic-pipe locator has several display modes. The average mode is provided for use where clutter is high. Averaging greatly reduces the effect of clutter while retaining location accuracy but slightly reduces locating speed. The compare mode is provided to make it possible to store an echo waveform, move the antenna, and then display the stored and the new waveforms alternately. A characteristic of pipe echoes is that their phase reverses if the antenna is rotated through 90 degrees. This feature is useful in distinguishing between clutter and pipe echoes because clutter does not reverse phase.

Another useful feature of the production pipe locator is a cursor that is controlled from the operator's panel. As the control is turned, the eleventh row of

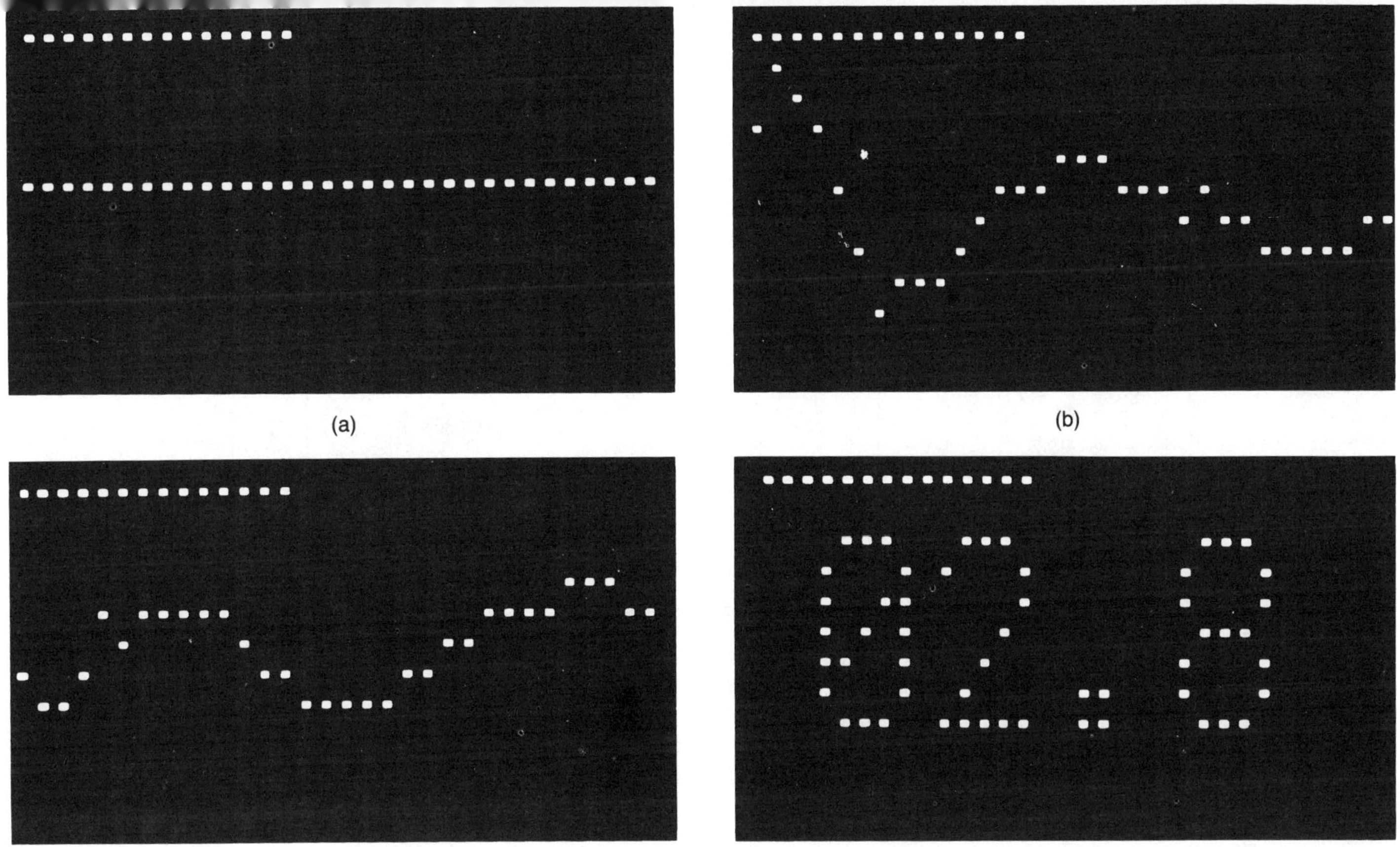

(a) (b) (c) (d)

Fig. 9-16. Typical Terrascan displays: (a) no target—a centered straight line; (b) a metal-pipe return; (c) a plastic-pipe return—upper cursor line moved to center of echo; and (d) depth read-out corresponding to cursor line position.

the display lights up in one-column increments. When the end of the lighted row matches the center of the echo and a button is pushed, the display will digitally show pipe depth, as shown in Fig. 9-16(d).

Terrascan Echo Results.

Some typical echo time-domain response waveforms for the Terrascan system are shown in Figs. 9-17 through 9-21. Figure 9-17 shows a time-domain echo waveform with several important features circled. The first feature, which is always seen, is a portion of the transmitted pulse energy and is used to locate the time reference (zero-depth point). The next feature is a random variation that is due to the ground roughness. The final feature, after a time delay that is proportional to the pipe depth, is the pipe echo.

Figure 9-18 shows the effect of moving the locator from a spot directly above the pipe; Fig. 9-19, of rotating the locator when it is directly above the pipe. (In both figures, the circled portion of the waveform contains the pipe echo.) With this antenna and data-processing technology, the time-domain echo waveform provides information on the presence of buried pipe, its location, its depth, and its direction.

To an experienced observer, the shape and polarity of the echo provide information about the identity of the target as well. The waveforms in Fig. 9-20 show that the echoes of metal and plastic pipes have approximately the same

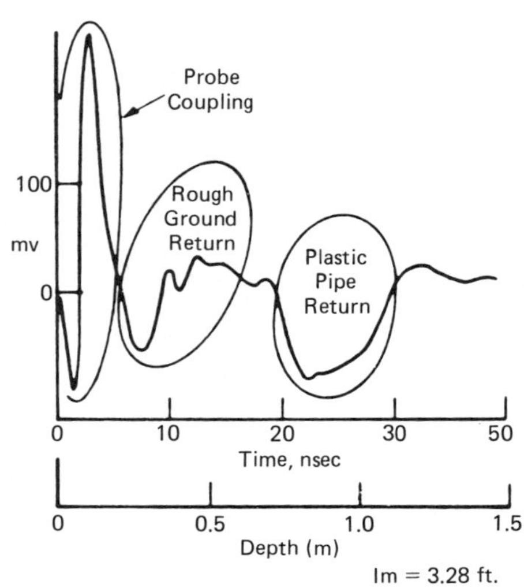

Fig. 9-17. Typical plastic-pipe echo.

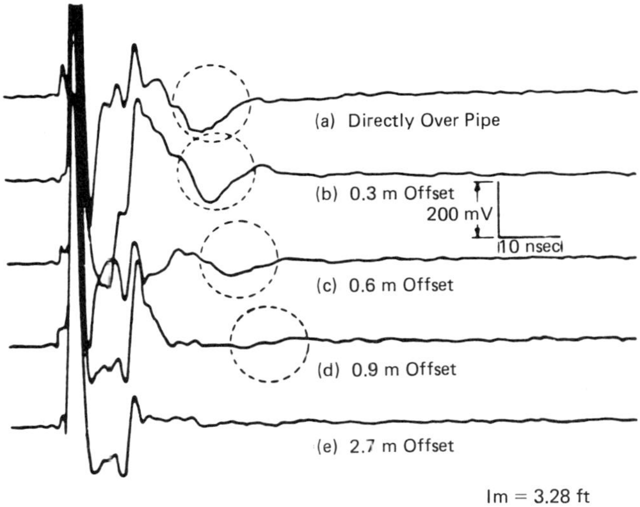

Fig. 9-18. Change in echo as antenna is moved away from target.

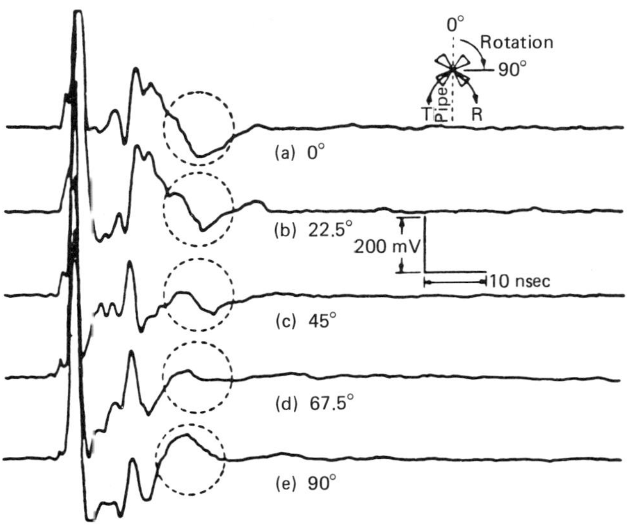

Fig. 9-19. Change in echo as antenna is rotated over target.

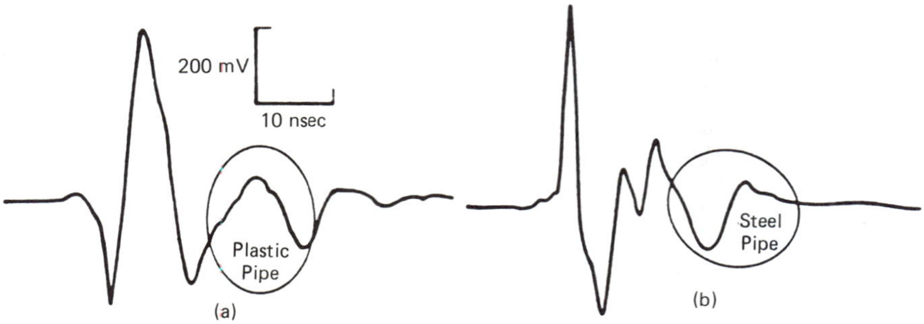

Fig. 9-20. Comparison of plastic and metal-pipe echoes.

shapes but opposite polarities. Figure 9-21 shows the echo of a horizontal 0.3-m-square (1-ft) copper plate buried 0.1 m (4 in.) below the earth surface. The double-bump feature that characterizes this kind of target is circled. The separation of the two peaks is proportional to the size of the plate. The additional bumps in the echo clearly distinguish it from that of a buried pipe.

The Terrascan system has been tested over a wide variety of sites located throughout the United States and England. It has been demonstrated that an experienced operator can locate utility lines to depths of 2 meters in "normal" soil. Perhaps just as important, it has been discovered that the parameters of the soil vary widely, sometimes changing drastically in a distance of only one city block. Relative dielectric-constant values from 3 to 30 have been observed, and propagation loss has varied from less than 3 dB/ft to more than 20 dB/ft. Thus,

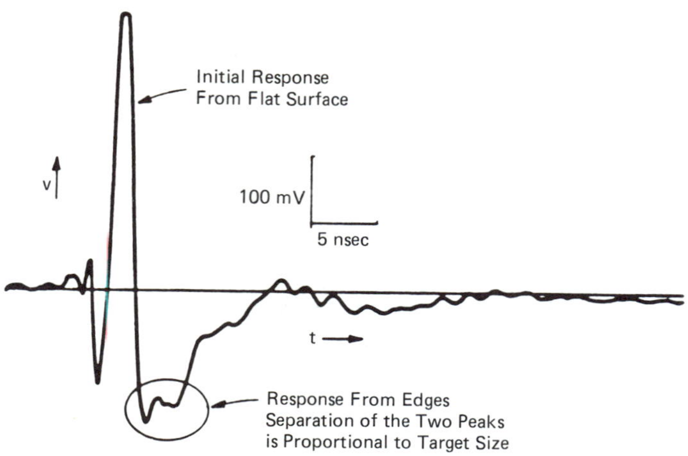

Fig. 9-21. Echo from buried copper plate.

all near-surface-radar sensing systems, including the Terrascan, must be able to adapt to wide variations of propagation conditions.

Figure 9-22 shows a plot of relative echo strength for plastic and metal pipes of 3-in. diameter, buried to a depth of 1 meter in soil of various conductivity and dielectric-constant values. It is seen that the signal strength can vary more than 70 dB, depending on soil propagation parameters.

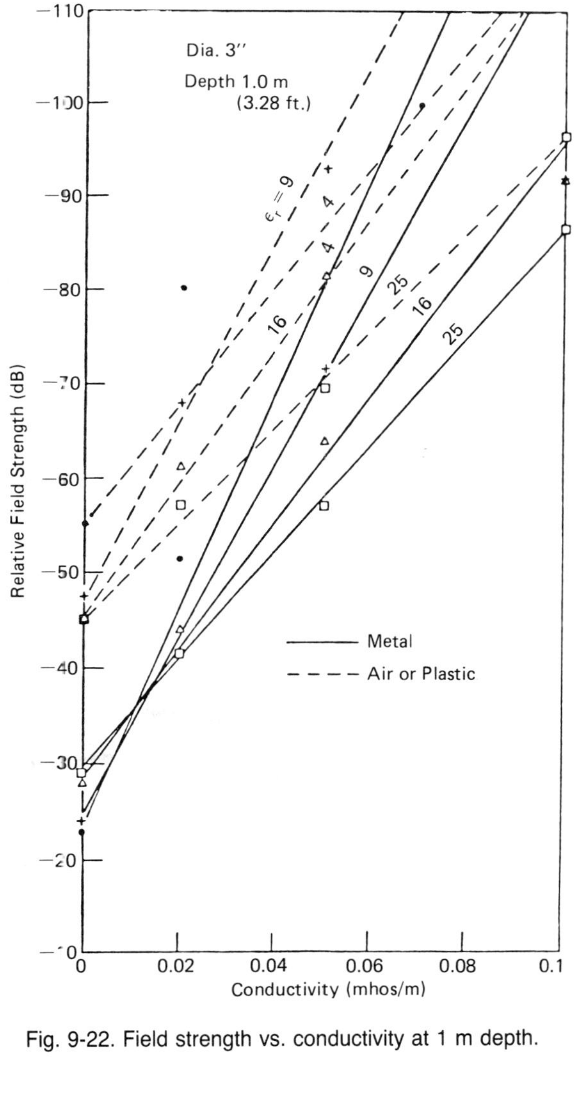

Fig. 9-22. Field strength vs. conductivity at 1 m depth.

9.4 SUBSURFACE RADAR APPLICATIONS

The preceding section of this chapter has focused attention on the operation of the Terrascan system. The remaining sections will discuss the use of it and similar systems for the detection of subterranean targets. There are two major questions involved in the use of video-pulse radar systems. First, is the received signal sufficiently strong to be detected, and, second, once detected, can the target be identified from the data available?

The first is primarily an electromagnetics problem and involves antennas, interface phenomena, propagation, and scattering. Obviously, the means for tackling such a problem would fill a book. Thus, we will discuss results that have already been achieved and only casually describe the technology used to obtain those results.

The second, that of target identification, lies more in the realm of data processing and, the techniques used involve a variety of concepts. The identification of utility lines was outlined earlier; it consists simply of observing the polarization characteristics of the target, i.e., one transmits one linear polarization and observes the measured scattered signal on the cross-polarized receiving antenna. This signal is maximum when both of the crossed antenna pair are at a 45-deg angle with respect to the utility line (provided, of course, that its diameter is not too large in terms of the smallest wavelength being observed). Also, one can "track" the utility line if it follows a straight line, which readily classes it as this type of target. This also makes the concept of grey-level mapping a very attractive one. It has been achieved using a Terrascan type of radar system and follows procedures similar to those used in geophysical exploration using acoustic systems.

The identification of the other classes of targets is somewhat more difficult. It is fortunate that the video-pulse radar system produces a signal at the receiver that is much richer in information content than conventionally pulsed radar. This results from the extreme bandwidth available to the receiver that makes it practical to consider time-domain processing schemes. For example, a difference-equation predictor scheme proposed for the identification of aerospace targets[30] has actually been used for the identification of land-mine targets. The difference-equation approach has also been very useful in the case of the detection of tunnels, for the purpose of enhancing the signal to clutter level. One might also observe that the Tomographic techniques proposed by Lytle[2] represent another mapping concept for tunnel detection, and the Southwest Research Institute[27] has carried this one step further by applying it to the video pulse concept by observing the earliest time of arrival of the signal transmitted between a pair of boreholes. When the propagation time is reduced by an air path, then a tunnel or at least a void has been located.

Finally, it is to be observed that this difference-equation approach makes it practical to process the broad-band, video-pulse radar data received from a dispersive antenna system to obtain the same result achieved by a nondispersive antenna.[27]

The remaining sections of this chapter will focus attention on experimental results (with the exception of the deep probing). There is, however, a substantial body of theoretical work, and the reader is referred to these references for further study.[25,28–30]

9.5 TUNNEL DETECTION

The detection of tunnels and voids in the earth is important from a military as well as a civilian viewpoint. The use of tunnels in Korea and Vietnam has been highly publicized. The hazards created by abandoned mining complexes is, perhaps, not so well known. Horror stories are numerous however, in regions where tunneling systems in the form of abandoned mines exist. These authors have been involved in several such situations in the vicinity of Youngstown, Ohio where the Video Pulse radar system was used to explore a school playground under which such tunnels were known to exist.[31] Subsequent excavations substantiated the measured results.

Tunnel detection using such radars requires operation in the high frequency window (HFW). The reason that practical tunnels are detectable only in the HFW is that their cross-sectional area becomes quite small in terms of λ within the LFW and the scattered field from them is consequently very small. Not only are tunnels resonant, but the antenna used in the HFW video pulse radar system has a resonant behavior as well. This represents a major problem in that the antenna "rings" in such a way that the oscillations exist in the range window occupied by the target.

9.6 COMBINATION OF MAPPING AND SIGNATURE PROCESSING

If a (set of) natural resonance(s) can be attributed to a desired target, then it would appear desirable to focus attention on it (or them). For example, antenna resonances contribute no information concerning the identity of a target. Furthermore, the residue associated with a resonance may be relatively large so that identification algorithms are relatively insensitive to changes in the natural resonances of the target. Recent studies have led to techniques for removing or filtering these unwanted resonances from the data using a difference-equation approach.[32,33] For cases where desired target echoes are submerged in clutter, the combination of filtering with mapping produces quite improved results.

A geometry considered for this investigation was a tunnel in the Rocky Mountains near Gold Hill, Colorado, at a site known as the Hazel A. Mine.[33,34] A modified version of the Terrascan system was used to obtain a sequence of waveforms for positions 20 feet above a known tunnel site. The radar system was moved along a line transverse to the axis of the tunnel, a position represented by the horizontal scale on the maps to follow; the vertical scale will be the arrival time of the signal corresponding to depth-in-ground. The surface over which the radar was operating was rough, and strong clutter signals existed in the time range where the tunnel response would be expected. A grey-level map was created by digitizing the signal magnitude and then assigning darker characters to the stronger signals. No clear evidence of the tunnel could be seen in this map. A most dramatic improvement resulted however, when natural resonances not associated with the tunnel were removed from the raw waveform via a difference-equation approach described in a paper by Volakis and Peters.[33] The result is shown in Fig. 9-23. Clearly the tunnel is properly located. One observes a "premonition" of it in the map of Fig. 9-23, which was created by data processing, but the onset of the signal reflection from the tunnel can be clearly observed in processed waveforms as discussed by Volakis.[35] In these, the tunnel has not only been clearly detected, but identified as an object with a particular natural resonance. This process, in addition to being useful in preparing improved maps, also clearly indicates the time at which the reflected signal first appears, thus fixing the distance to the nearest part of the tunnel with reasonable accuracy.

The removal of the antenna resonance is reasonably straightforward. The equations needed to remove a single pole from the waveform will now be given. If the original waveform is designated as $R(t)$. The waveform after pole removal is given by

$$R_f(t) = R(t) - 2Re(Z_A)R(t - NT_B) + \|Z_A\|^2 R(t - 2NT_B) \quad (9\text{-}8)$$

where

Z_A = the pole to be removed in the Z-domain
T_B = waveform sampling interval (of sampling oscilloscope)
NT_B = interval between samples used in the extraction process

Thus, the new signal $R_f(t)$ (filtered waveform) is constructed from samples of the original waveform spaced at time intervals of T. In general, $2M + 1$ such points are required when M pole pairs (resonances)[42] are concurrently removed. It is apparent that the process can be applied only when $T \geqslant 2MT$ (assuming that the waveform starts at $t = 0$). This limitation introduces errors in the initial portion of $R_f(t)$. This signal section can be recovered by other processes.

It has been proved for a single case,[35] where the scattered signal consists of two resonances—one corresponding to the antenna and the other to the target—

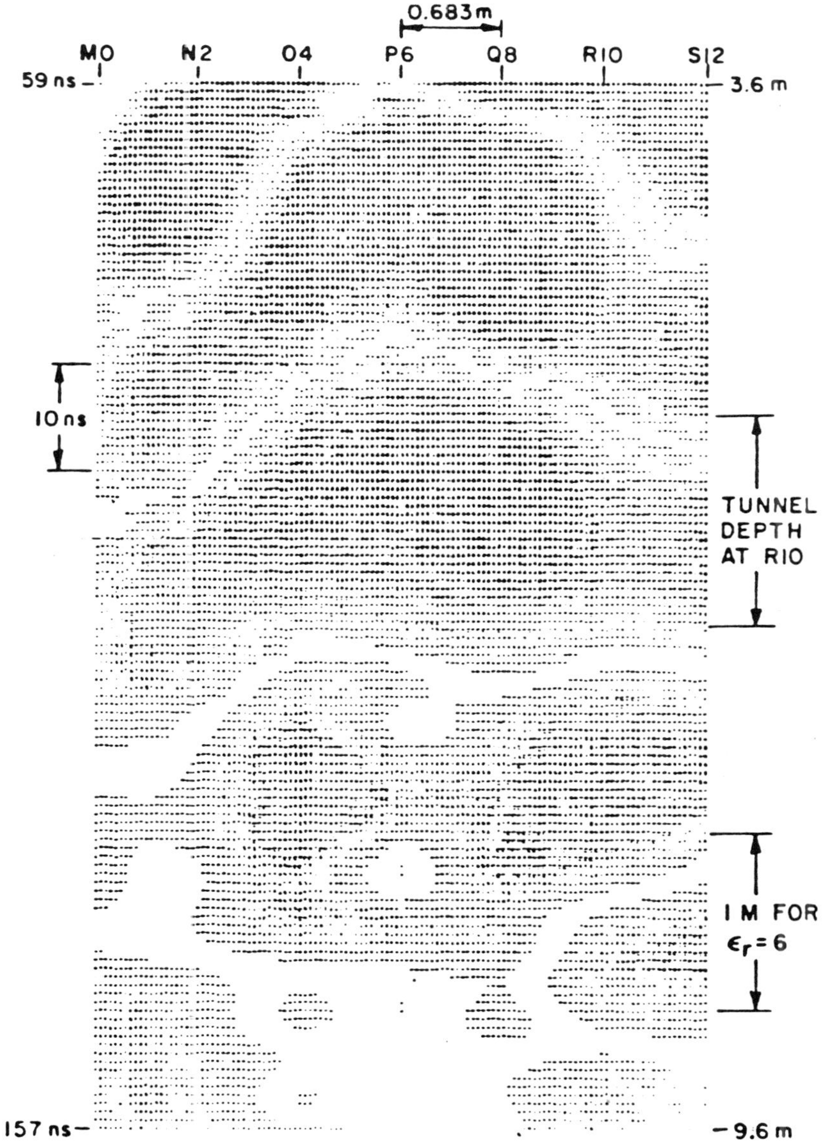

Fig. 9-23. Mapping of top traverse over tunnel. (Raw data has been processed using new clutter-removal techniques based on tunnel resonance.)

that the following distortions are introduced to the remaining signal when the antenna pole pair is removed. There is, first, a phase shift,

$$\Delta = -\omega_T N T_B \tag{9-9}$$

whose amplitude is approximately multiplied by the factor

$$A_f = A sin \frac{\omega_T + \omega_A}{2} NT sin \frac{\omega_T - \omega_A}{2} NT \tag{9-10}$$

where

ω_T = the imaginary part of the target pole pair in the S-domain
ω_A = the imaginary part of the removed antenna pole pair

Furthermore, the condition,

$$e^{\sigma_T T} \approx e^{\sigma_A T} \approx 1 \tag{9-11}$$

where σ_T and σ_A are the real parts of the corresponding S-domain poles, should be imposed.

One may now "clean up" the early part of the waveform on the premise that only the desired resonance remains in the corrected waveform. We do so by marching backward through the waveform starting at a point where the corrected waveform is untainted. The process is given by

$$\|Z_T\|^2 R_f(t - 2NT_B) = 2Re[Z_T]R_f(t - NT_B) - R_f(t) \tag{9-12}$$

where $R_f(t - 2NT_B)$ is the final waveform.

Figure 9-24 is a hypothetical example consisting of two damped sinusoids as defined by

$$R_f(t) = A_1 e^{-\sigma_1 t} cos_1 t + A_2 e^{-\sigma_2 t} cos_2 t \tag{9-13}$$

where A_i represents the values of the residues. Equation 9-13 represents the original waveform of Fig. 9-24. The filtered waveform is obtained when the S_i poles $[S_i = (-6.0 + j2\ 15)10^6]$ have been removed by using Eq. 9-8. This result is then corrected according to Eqs. 9-9 and 9-10.

Figure 9-25 shows the effect of the correction process. It is seen that the corrected waveform is in precise agreement with $A_2 e^{-\sigma_2 t} cos\omega_2 t$. Thus, if this were the signal scattered by the target, the desired signal would have been recovered. Obviously, there is no need for further processing. Measured data,

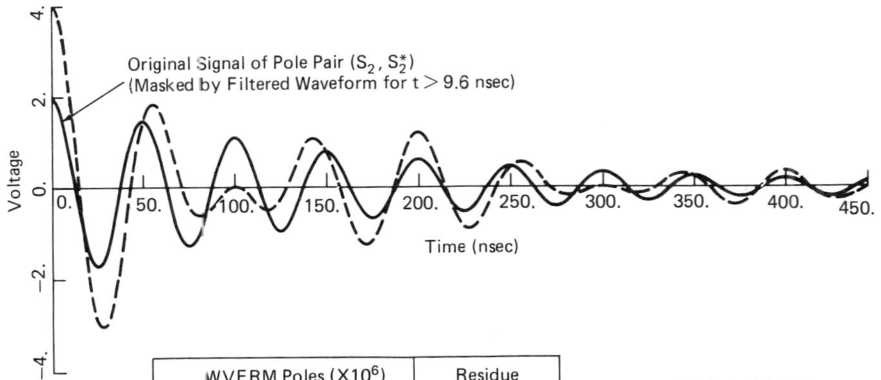

WVFRM Poles (X10^6)		Residue	
σ_i	f_i	Real	IMAG
−6.00	20.00	2.0	0.0
−6.00	15.00	2.0	0.0

- - - - Original WVFRM
——— Filtered (Corrected)
Extraction Interval =
Extracted Poles (X 10^6):

Fig. 9-24. Example of amplitude and phase effects caused by pole-extraction process.

however, will never appear in the nice form given by Eq. 9-13 or Fig. 9-24. Clutter, other poles, possible signal clipping, etc., will usually be present.

Consider now the example of the raw video pulse-radar waveform shown in Fig. 9-26, which was obtained from the tunnel at Gold Hill with the system gain set so that the echo tunnel could be observed. Under these circumstances, the

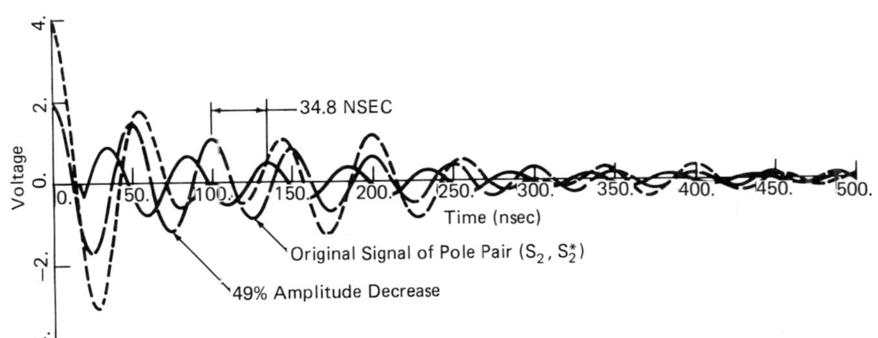

WVFRM Poles (X10^6)		Residue	
σ_i	f_i	Real	IMAG
−6.00	20.00	2.0	0.0
−6.00	15.00	2.0	0.0

- - - Original WVFRM
——— Filtered WVFRM
Extraction Interval = $5T_B$
Extracted Poles (X 10^6): σ_i f_i
−6.0 15.0

Fig. 9-25. Application of the correction process to the results of Fig. 9-14.

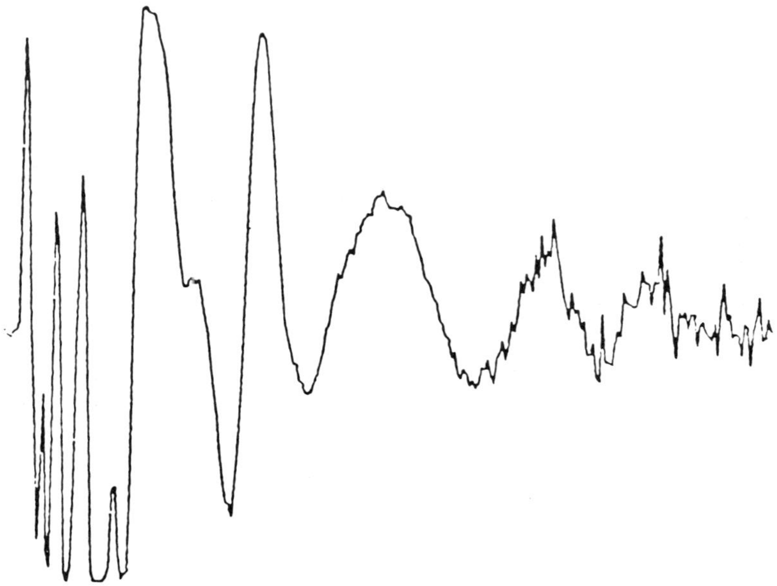

Fig. 9-26. Raw video-pulse waveform obtained over tunnel at Gold Hill.

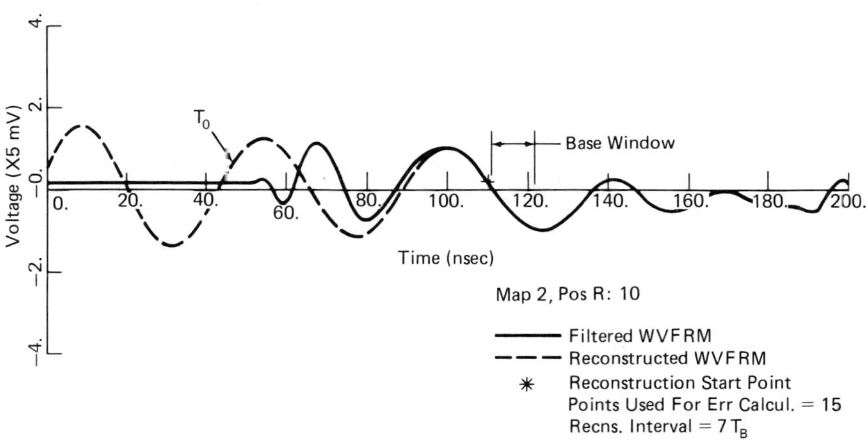

Fig. 9-27. Example of reconstruction process.

peaks in the early part of the waveform were clipped.* A high frequency jitter appeared in the latter portion of the waveform, moreover, as a result of faulty triggering. The waveform with the antenna pole extracted is identified as the filtered waveform in Fig. 9-27. The reconstructed waveform (see Eq. 9-12) is compared with the filtered waveform. The point, T_0, where these two waveforms separate represents the initial point of the signal reflected by the tunnel.

A final word here is that the target identification of the landmine type of target could have been improved if the antenna resonance had been removed before applying the target identification process.

9.7 MINE DETECTION AND IDENTIFICATION

The one area where substantial success has been observed has been in the identification of buried land mines. For some underground radar applications, the targets are not only resonant and can be detected, but can possibly also be identified. Since such objects possess unique transient signatures, identification based solely on the characteristics of a single return echo may be attempted. The target identification techniques have been applied to land-mindlike geometries.[15,16,19] The broad spectrum of the video pulse is necessary to excite different transient signatures or resonances.

The set of targets shown in Fig. 9-28 were buried to a shallow depth in a lossy earth ($\varepsilon_r \simeq 20$; $\sigma \approx 0.03$ s/m). A series of measurements were made over these targets using a radar system with a 500-ps, 1000-V periodic pulser. The received time-domain signals are shown in Fig. 9-29. Clearly, the no-target response of Fig. 9-29 shows a maximum on the order of 100 mV after the initial coupling spike. Thus, it appears that scattered signals are observed at about 80 dB below the transmitted pulse. Cable losses, etc., would reduce this to about 70 dB. Measurements in the vicinity of a single target showed much more variation in waveform shapes than that seen in Fig. 9-29. These waveforms and their received spectrum are dependent on the relative positions of the antenna and the target. The natural resonances (poles) of the target are not dependent on the excitation, however, except for the magnitude of the excitation. A procedure known as “Prony's method” can be used to calculate such poles from the time-domain data. Some typical results are shown in Figs. 9-30 and 9-31. The minelike target is a penetrable body, and we observe that the resonances are independent of ground conditions (or electrical parameters), whereas the brass cylinder is a conducting body, and we observe the resonances to be highly dependent on ground conditions.

*This has since been corrected by incorporating a computer-controlled time-dependent gain in the system.

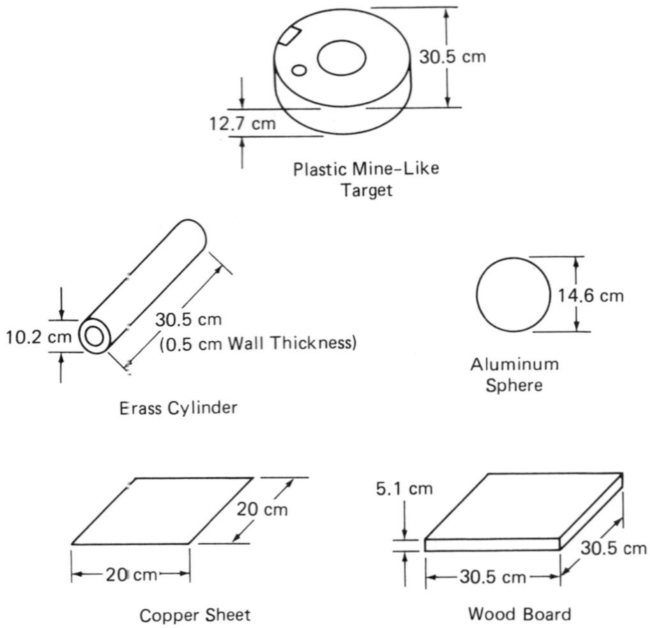

Fig. 9-28. Physical characteristics of subsurface targets.

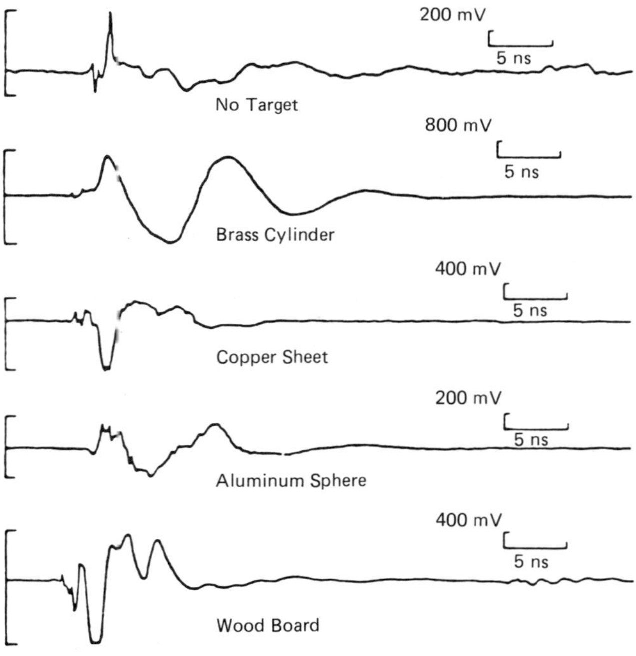

Fig. 9-29. Processed waveforms from the subsurface targets.

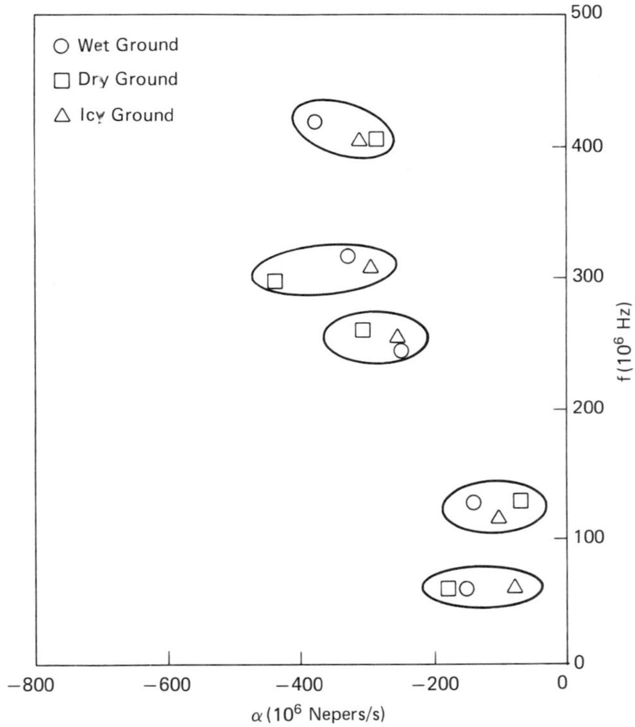

Fig. 9-30. Locations of average extracted resonances of minelike target under different ground conditions.

Even though the waveforms of a given target are quite different as the radar position is changed, the poles associated with these waveforms are clustered in a manner quite similar to those of Fig. 9-30.

The generation of these natural resonances from the raw data via the Prony procedure is a lengthy process and requires substantial computation time on a computer. Once determined, however, these resonances can be used to generate a predictor-correlator equation that is extremely rapid even on a microcomputer.[16] The output of this predictor correlator is shown in Fig. 9-32. Since the abscissa is completely under the control of the operator, the mine-like target can easily be separated from the other targets if one sets a threshold, $\rho(T) \sim 0.3$. This scheme has proven to be extremely successful for a large number of measurements.

To shorten computation time, three other constraints were introduced in the target identification algorithms: (1) maximum and minimum peak signal levels, (2) total energy content, and (3) limits on the arrival time of the maximum. On

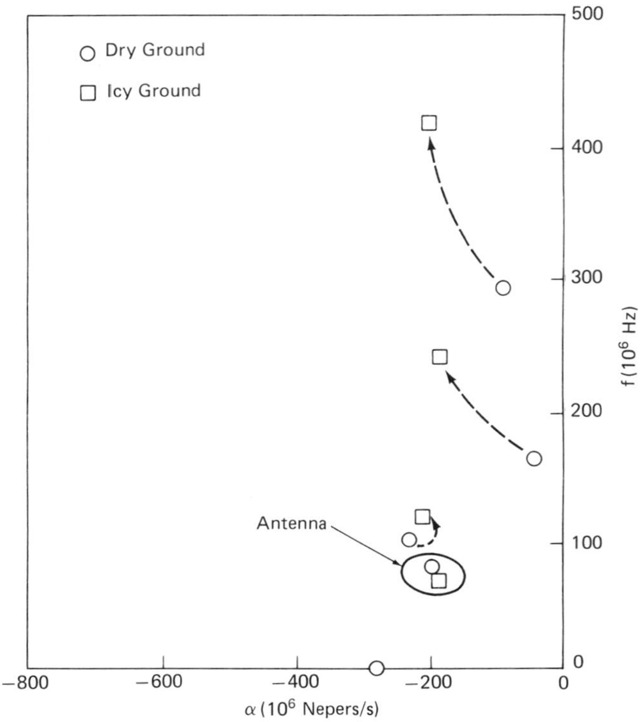

Fig. 9-31. Average extracted resonances of brass cylinder under different ground conditions.

this basis, we achieved 100 percent identification with a single pulse and a false alarm rate on the order of a few percent, using signals that included reflections from a number of unknown false targets in a dirt, debris-filled road outside our laboratory. Fifty-three waveforms whose level exceeded the minimum value were measured over these sites.

9.8 POTENTIAL APPLICATIONS USING THE LOW FREQUENCY WINDOW (LFW)

The systems discussed in the previous sections all make use of the High Frequency Window (HFW). In this section, we will consider the potential of the Low Frequency Window (LFW). Note that there are no actual applications of video pulse radars in this frequency band at the present time. We observe that Barringer did make use of low-frequency radars from an air-borne platform.[36,37] His antenna system was a trailing wire with a radar operating in the HF band, which is not the type of system to be outlined here. We believe that the state-

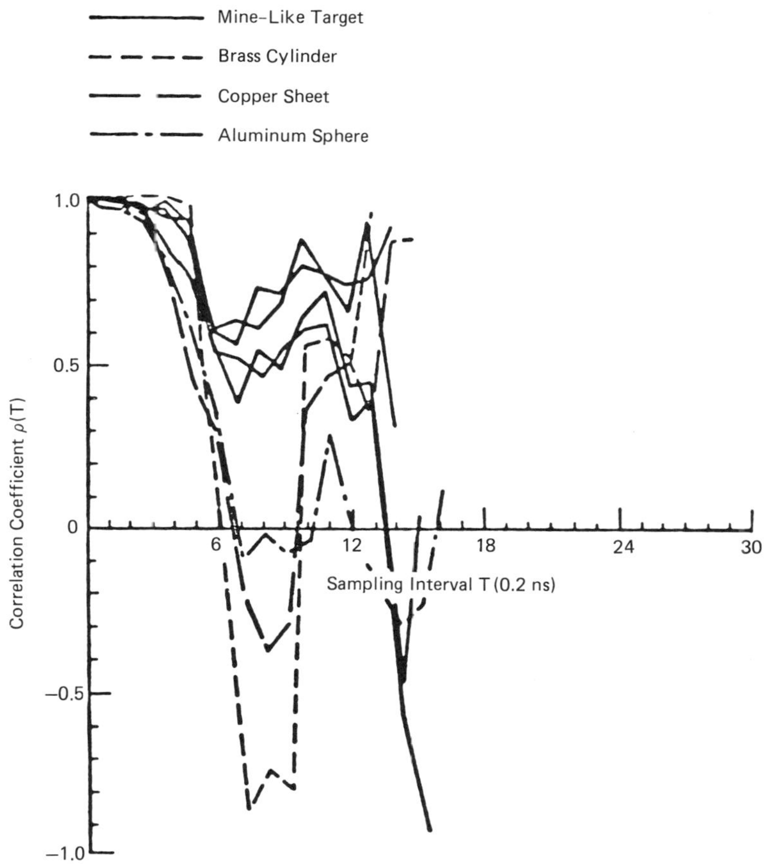

Fig. 9-32. Typical ρ(T) curves for identification of minelike target in dry ground.

of-the-art can be extended beyond the work of Barringer, but doing so would require the use of antennas placed on the surface, thereby increasing the depths and making it possible to detect smaller targets.

Detection of Tunnels with Conducting Wires

The previous discussion involved the detection of tunnels from a change in medium electrical parameters, e.g., a void in the medium. The resonance behavior of the scattered signal involved only the transverse dimensions of the tunnel. If one studies the data given—sensitivity of the ground parameters, etc.—one quickly finds that, for reasons such as these, the practical maximum range at which a tunnel can be detected is on the order of 100 feet. This assumes a weathered-granite medium and substantial improvement in the state-of-the-art of current equipment. This relatively small depth of detection may be increased by several fortuitious factors. The major factor would be attentuation loss in the weathered granite. Less significant factors would include: (1) an improved antenna with reduced ringing, or, conversely, the development of a K-pulse[15] excitation to eliminate such ringing; (2) increased sensitivity of sampling systems; and (3) further development of data-processing steps. If one requires an order-of-magnitude increase in the depth of detection, however, the potential of longitudinal resonances would have to be examined. Furthermore, one would now be attempting to detect any conducting wires, rails, etc., that may be located in the tunnel. The goal of these next few sections is to discuss the detectability of conductors of appropriate length in abandoned tunnels. Since this enterprise involves theory, we will first outline some of the theoretical concepts developed at Ohio State University.

Propagation in the LFW[10,38]

In contrast to the HFW, there are several distinct phenomena related to propagation in the LFW. The first of these, provided the antenna is in electrical contact with the media, is that the input impedance is real and independent of frequency. The second is that the depth of penetration in the ground is restricted to less than 3.85 skin depths at f_m (see Eq. 9-16). Since the medium is dispersionless for signals in this band, no pulse distortion appears as a result of propagation or because of the antenna transfer function.

Figure 9-33 gives the radiated fields at a depth of 30 m in a homogeneous media as a function of frequency for various dipole lengths with a 1-volt excitation.[28] For frequencies less than f_m, the uninsulated dipole radiated field is almost frequency independent. This results in the dispersionless propagation. One interesting result is that the insulated dipole produces nearly the same radiated field in the HFW region as does the uninsulated dipole.

The results of transmission between two dipoles immersed and in electrical contact with a homogeneous conducting media are best summarized in terms of Fig. 9-34. The geometry from which these results were obtained is shown in Fig. 9-35. Figure 9-35(a) shows the propagation effects resulting from propagation over a range $2d$ to an image dipole. Solutions of propagation between the dipoles of Fig. 9-35 would require including the effects of the interface. To do so precisely for a volume of data would be extremely expensive.

The primary requirement of an antenna for a subterranean radar is that it should transmit the applied pulse efficiently into the ground. When the radar is designed to operate in the LFW, the frequency band of operation is low enough for the input pulse duration to become comparable to the propagation time of the pulse to the target and back again. With such a system, any multiple reflections

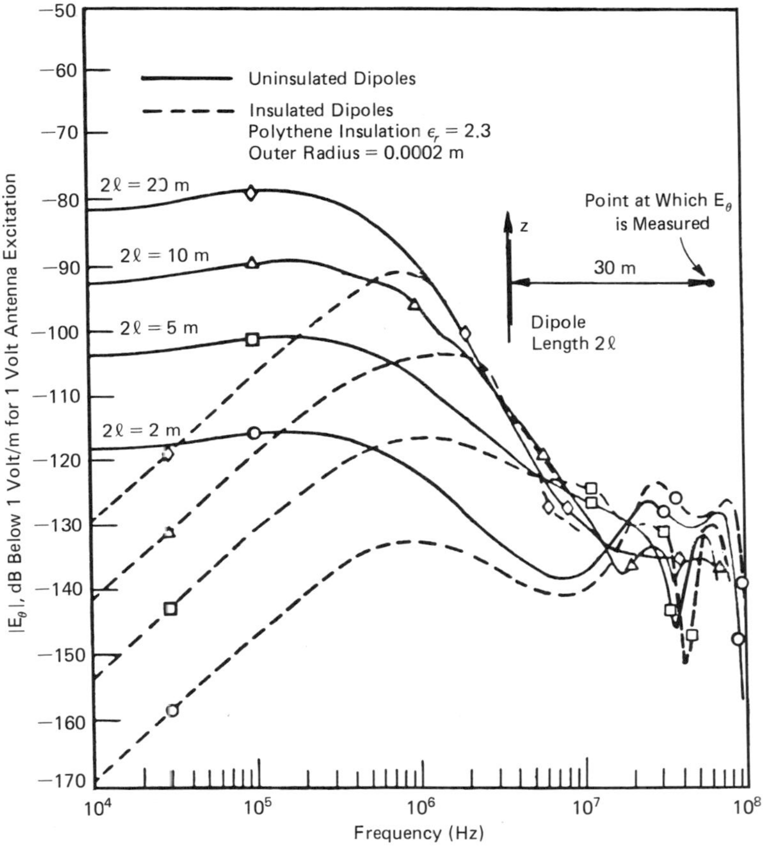

Fig. 9-33. Magnitude of the electric-field strength 30 meters from 0.001-m-radius, perfectly conducting dipoles in a homogeneous medium where $\varepsilon_r = 4$, $\sigma = 0.005$ mhos/m. Results are presented for dipole lengths of 2 m, 5 m, 10 m, and 20 m.

or "ringing" of the input signal on the antenna effectively extends the duration of the transmitted signal so that the low-level reflected signal wanted is masked.

Effective coupling of the transmitted energy into the ground is achieved simply by laying the antenna on the surface of the ground, as shown by the following simple argument. Consider an antenna that is designed to respond to the horizontal electric field (e.g., a dipole). The characteristic impedance of the ground at most frequencies is much lower than the characteristic impedance of free space, and the voltage reflection coefficient of the air-ground interface, ρ, is approximately −1 for a wave normally incident from the air medium. In this case, the electric field of the direct and reflected waves interfere destructively. Since the reciprocity theorem states that the reaction of an electric element is proportional to the

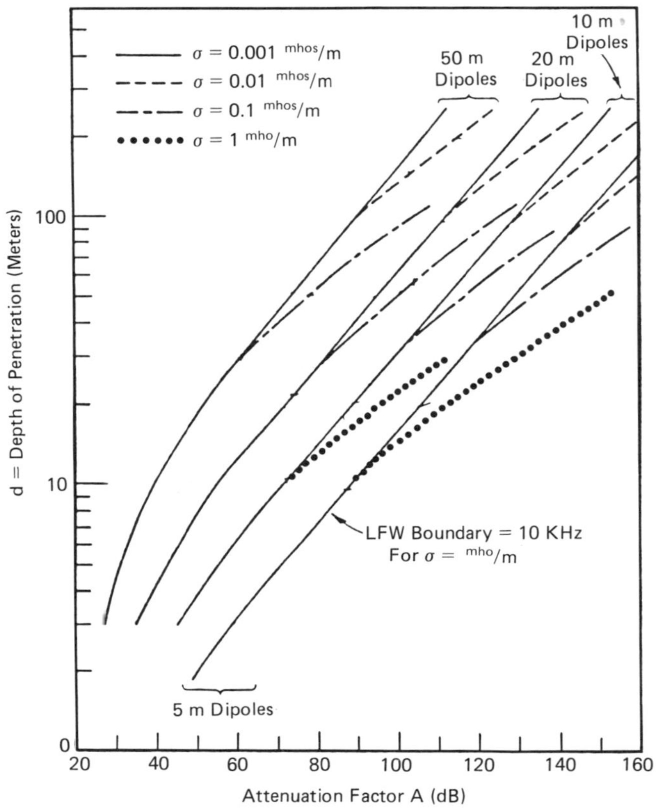

Fig. 9-34. LFW-radar attenuation-factor curves for 5-m, 10-m, 20-m, and 50-m dipoles. The data was computed for a 100-μs Gaussian input pulse bandlimited at about 10 kHz. The LFW breakpoints, therefore, are for a frequency of 10 kHz.

electric field at the current element, the voltage generated at the antenna terminals is small. If the wave incident on the interface comes from below the surface—i.e., from the lossy or conducting medium—the voltage reflection coefficient, ρ, is approximately +1, and thus the response of a linear antenna on the surface of the conducting interface is approximately 1 + ρ times that of the same antenna immersed in a homogeneous conducting medium (in this case, ρ is the voltage reflection coefficient of the air-ground interface for signals arriving from within the ground). It follows that such antennas lying on the surface of the ground respond to signals arriving from within the ground and, by the reciprocity theorem, transmit most of their energy into the earth (in reality, some energy may be supplied to surface waves, which propagate along the earth-air interface).

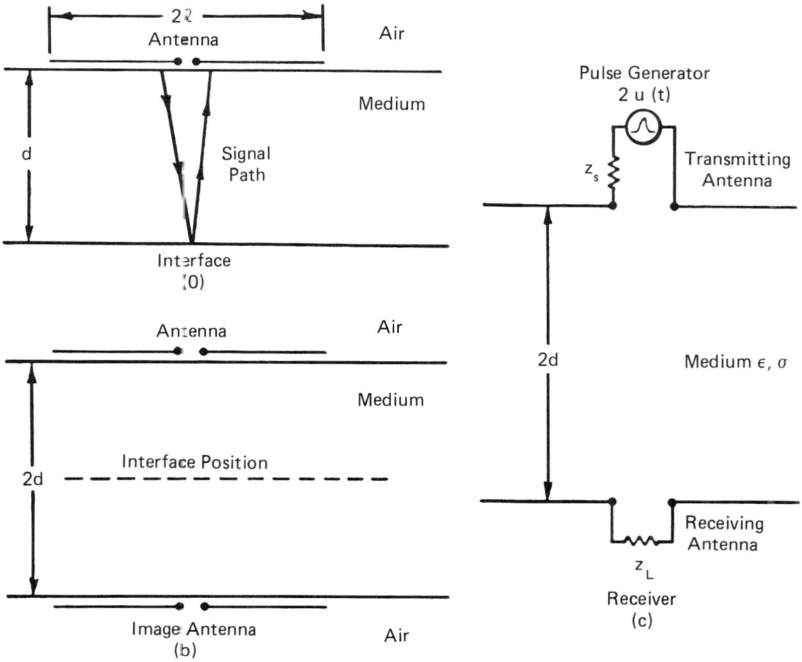

Fig. 9-35. Development of a model for analysis of pulse propagation into the earth: (a) reflection from a plane interface d meters deep; (b) model of reflection from a perfectly reflecting interface; and (c) simplified model of reflection from a perfectly reflecting interface.

This reaction results in what we call an "interface gain," which can be up to two times, or 6 dB, for each dipole of a transmit–receive pair.

There is an additional effect that occurs on transmit. It is convenient to feed the antennas with a voltage pulse, with the source impedance of the pulse generator chosen to give a desirable match to the antenna input. Propagation calculations between antennas placed in a homogeneous medium and driven by voltage pulses will be presented later. For a given input voltage, the current on the transmitting antenna (which determines the fields) is determined by the antenna input impedance. When the antenna is placed at the air-ground interface and in electrical contact with the ground, the input impedance is approximately twice what it would be if the antenna were immersed in the homogeneous medium because of the removal of one conducting half space and because the impedance of the antenna in the homogeneous medium is much less than its impedance in free space.[39,40] The antenna current and the fields are therefore reduced by 6 dB. The "interface gain" of about 6 dB acts to make up this apparent loss, however, and the net effect is that the "interface gain" for a dipole on transmit is about 0 dB. Therefore, because of the voltage pulse input we use in our propagation calculations, the interface gain is about 0 dB for the transmitting dipole and 6 dB for the receiving dipole. These effects have been verified on a limited scale by numerical results obtained when the interface has been included via the Sommerfeld integral.

An important advantage of the effective coupling to the ground when the antenna is lying on the surface is that the antenna is insensitive to signal sources or reflections from objects above the ground. The advantage arises from the fact that most noise sources are above the ground.

We will now assume that the transmitting antenna is fed by a pulser whose voltage output is a Gaussian pulse, as shown in Fig. 9-36. Because the propagation in the LFW is dispersionless, the received pulse is a replica of the trans-

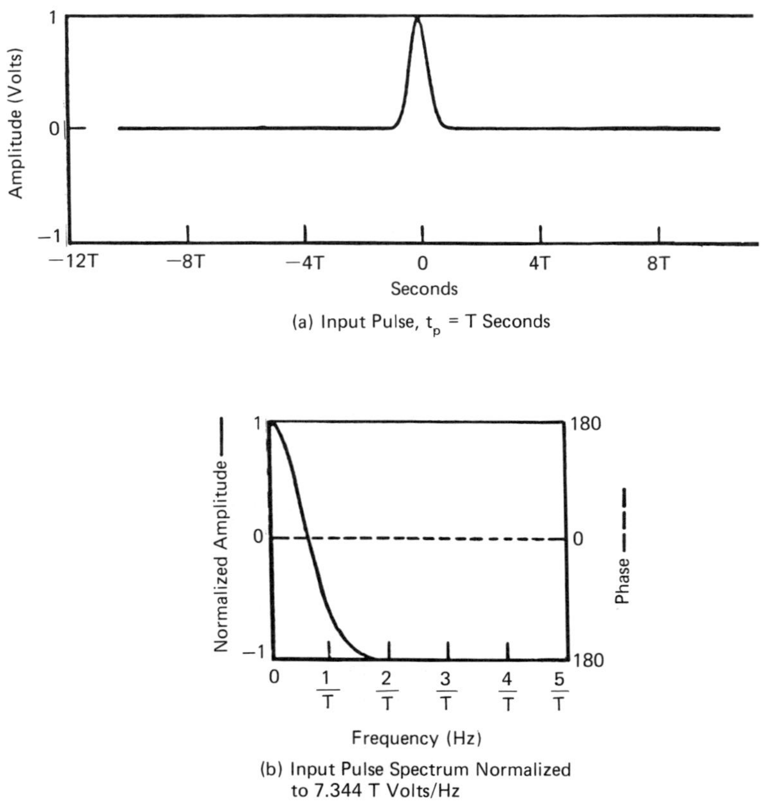

Fig. 9-36. (a) Normalized Gaussian pulse ($t_p = T$ sec), and (b) its spectrum normalized to 7.344T V/Hz.

mitted pulse after it has been passed through a low-pass filter with a critical frequency given by f_m. This makes it practical when the input voltage is a Gaussian pulse to define a pulse attenuation factor as follows:

$$A = \frac{V_r}{V_i} \tag{9-14}$$

where

V_r = peak amplitude of received pulser
V_i = peak amplitude of input pulse

The curves of Fig. 9-34 are for a 100-μsec pulser and give the pulse attenuation factor A for a wide range of conductivities and dipole lengths. These results can be scaled to any other pulse width simply by using Eq. 9-16, which relates f_m, σ, and d. In this case, f_m is taken to be the maximum frequency for the pulse of interest. For Fig. 9-34, this is 10 kHz, the point at which the value of A deviates from that given by the solid lines or the junction of the solid line and the broken lines. One such point is specifically illustrated. If the pulse width were to be shortened to 10 μsec, this maximum frequency would become 100 kHz, and the curves denoted by "..." would represent a conductivity of 0.1 instead of 1.0. Each value of conductivity in the legends would in turn be multiplied by 0.1. This can all be summed up by the following empirical formula:

$$A = 56.5 \log_{10} d - 56.5 \log_{10} (2\ell) + 73 \tag{9-15}$$

where $\ell < d < 970/\sigma f_m$ and where the radar range, d, and the antenna half length, ℓ, are in meters, σ is in mhos/meter, and f_m is in hertz. The latter is given by the equation,

$$f_m = \frac{3.76 \times 10^6}{(2d)^2 \sigma} \tag{9-16}$$

where d is the target depth.

The video-pulse-radar scheme involves using a transmit and receive antenna at the same position. As a result, the reflected pulse of interest may be lost in the clutter created by the coupling of the transmitted pulse directly to the receive antenna.

Figure 9-37 shows the relative location of the received signal and the input signal for 50-m dipoles spaced at 200 m in a medium where $\varepsilon_r = 4$ and $\sigma = 0.01$ mhos/m. Two input signals are shown, one a 100-μsec pulse that is approximately bandlimited at 10 kHz (the LFW cutoff for the system) and the other a 10-μsec

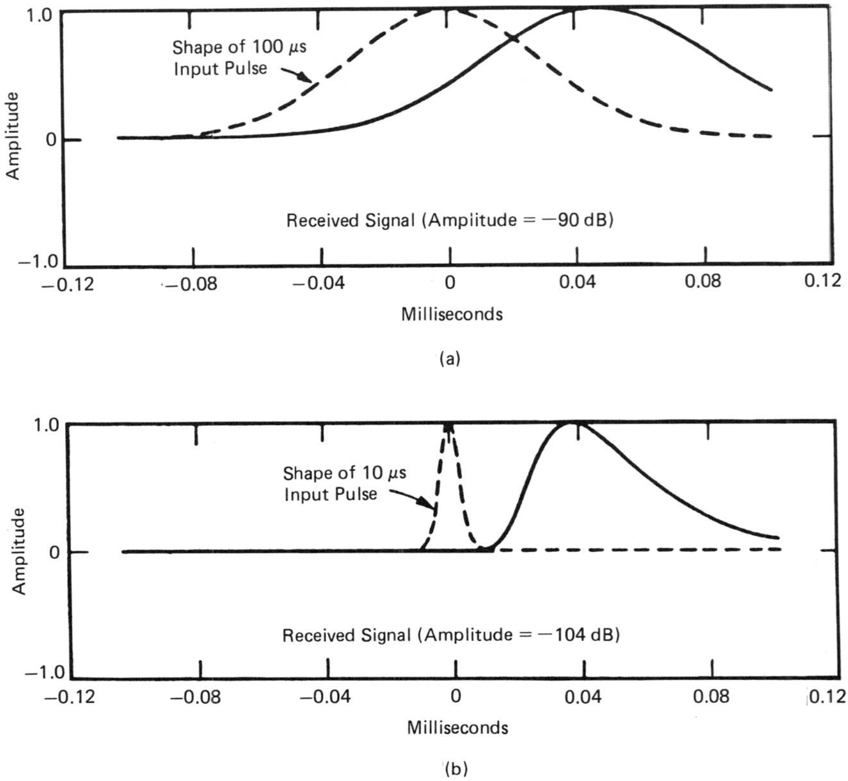

Fig. 9-37. Relative locations of received signal for 50-m dipoles spaced at distance $2d = 200$ m in a medium where $\varepsilon_r = 4$ and $\sigma = 0.01$ mhos/m: (a) 100-μs input pulse, and (b) 10-μs Gaussian input pulse.

pulse that contains significant energy to about 100 kHz. It is seen that, for the shorter input pulse, the range delay is sufficient to separate the transmitted and received signals. The penalty is an increase in attenuation of 14 dB. The isolation between the transmitting and receiving antennas could be as much as 100 dB if a well-designed orthogonal antenna system is used.[38] For a successful design, however, it is desirable to have a total isolation of greater than 120 dB (this is a figure arrived at by considering possible input-pulse amplitudes and receiver sensitivities; rather than being a fixed design parameter, it is merely a guideline). This same system can also be scaled to operate at kilometer depths for geophysical exploration.

It would seem feasible to detect signals at quite low levels using these concepts of LFW radar. In the next section, we will introduce the scattering mechanism for empty tunnels into this basic model.

Scattering Model[41,42]

A means of including target effects is required. This model should make it practical to separate target and propagation properties so that one need not make a separate computation for every target/range combination. This is the goal of the model used in References 41 and 42 to introduce the scattering mechanism for the tunnel as illustrated in Fig. 9-38. Figure 9-38(c) illustrates the propagation model we discussed in the preceding section. To introduce the scattering function for the tunnel, its scattered fields must first be obtained using the model of Fig. 9-38(a), which is then normalized by the direct transmitted field of the line source to the image position shown in Fig. 9-38(b). The case considered in Fig. 9-38 is for the electric line source. The response to a thin, linear, metallic cylindrical scatterer is then given by

$$V_R = \frac{1}{2}\left(\frac{V_R'}{V_T}\right)\left(\frac{E^S}{E^I}\right) \tag{9-17}$$

where

V_R = voltage at terminals of receive antenna
V_T = voltage at transmitting antenna terminals
E^S/E^I = normalized scattering amplitude of cylinder in Fig. 9-38

This model has been compared with results obtained using a precise moment-method solution. The agreement is within 0.8 dB when the signal being computed is 142.8 dB below the transmitted peak.

For penetrable scatterers, such as tunnels, there is a significant response for both polarizations, i.e., for an electric and for a magnetic line source. When both polarizations are taken into account, the net received voltage is

$$V_R' = \frac{1}{2}\frac{V_R}{V_T}\frac{E_{\parallel}^S}{E^I}(1 - a_{\perp}) \tag{9-18}$$

where

$$a_{\perp} = \frac{E_{\perp}^S}{E_{\parallel}^S}$$

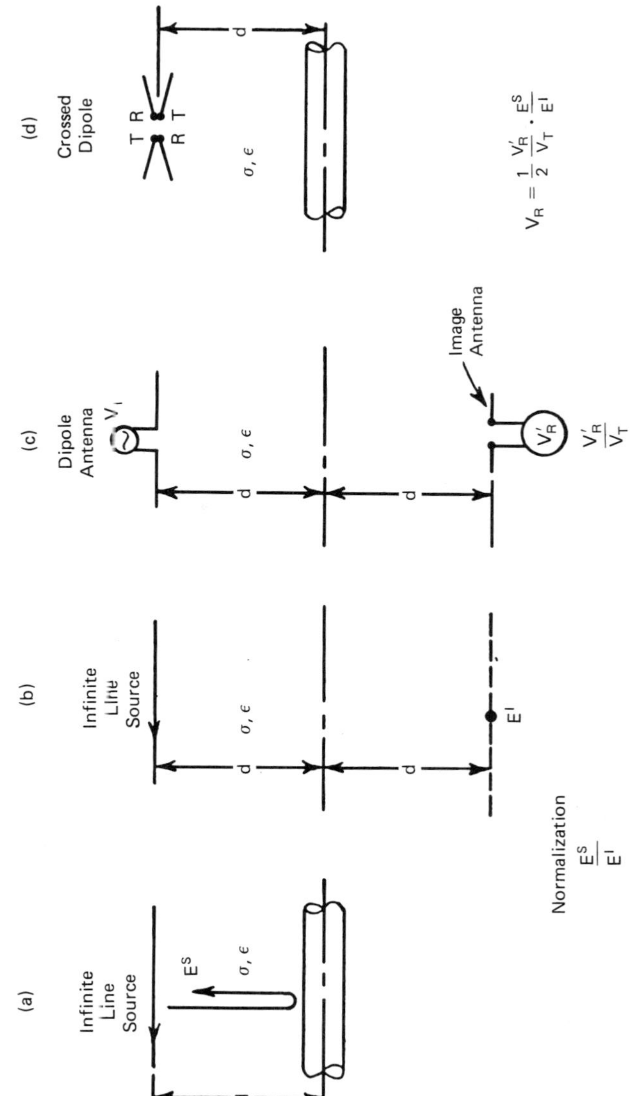

Fig. 9-38. Introducing the scattering attenuation function of a cylinder in the radar model.

Deep Probing Via LFW Video Pulse Radars

In this section we will explore the potential of the video pulse radar to detect large targets—for example, large ore deposits or petroleum-bearing strata—at substantial depths.

The dominant features for operation of the LFW radar system are the propagation losses and the frequency band of operation. The upper frequency for the LFW band is given in Fig. 9-3. The attenuation factor A has been given in the form of the curves of Fig. 9-34 and also Eq. 9-15. This format explicitly includes the range and the antenna length and implicitly includes the pulse width and the conductivity.

The remaining design parameter needed is the scattering function of the target. Figure 9-39 gives the reflection coefficient of various layered media as a function of electrical parameters. Additional curves are easily computed for other combinations of layer and medium parameters. These curves show a decrease in reflection for layers as frequency is decreased. Thus, the critical frequency should exceed the frequency at which the reflection coefficient is, say, −10 dB in order to detect a given layer. Of course, we have little interest in detecting horizontal layers, and, indeed, the cross-dipole antenna system will not detect them. Of

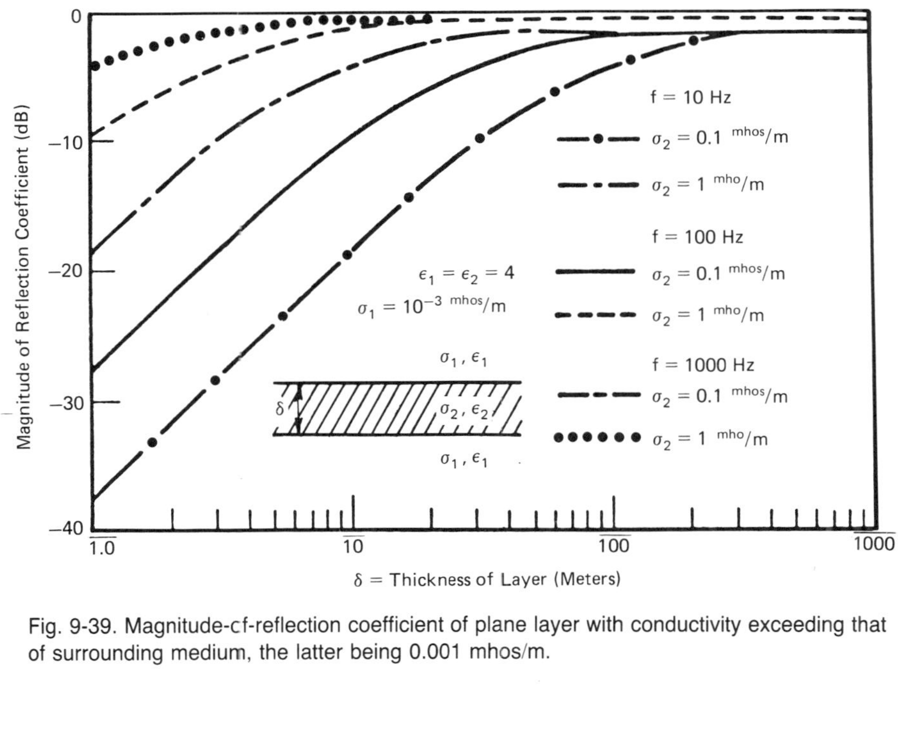

Fig. 9-39. Magnitude-of-reflection coefficient of plane layer with conductivity exceeding that of surrounding medium, the latter being 0.001 mhos/m.

major interest is the detection of anomalies. The essential feature of the anomaly would be that it is of the same thickness as that given in Fig. 9-39. Alternatively, perhaps, the lower edge should not be parallel to the upper edge so that the cancellation causing the low-frequency decay of the reflection coefficient will be minimized.

One substantial difference introduced by the LFW system is the need for the antenna system to be in electrical contact with the conducting ground. Indeed, this is one of the major problems with such a system. When the proper contact is achieved, the antenna impedance becomes essentially a frequency independent resistance. Otherwise, the reactive component of impedance would be dominant, and one then must carefully match impedances to prevent multiple reflection on the antenna that would "ring" into the target range window. Another essential characteristic of the antenna system is that the transmit and receive antennas must be isolated electrically. Isolation as a function of antenna distortion is discussed elsewhere for the crossed dipole, where it is indicated that isolations on the order of 60 to 100 dB may be possible be very careful antenna placement.[38] It may be possible to develop adaptive techniques to improve this isolation since the directly coupled signal occurs in a different time window than does the signal reflected from the target.

9.9 CONCLUSIONS

This chapter demonstrates that the transient radar concept has real advantages for subsurface probing applications. The use of a transient interrogating signal permits target identification and discrimination with a variety of techniques. The transient signal provides the maximum possible resolution of objects submerged in lossy, nonhomogeneous, dispersive soil media and achieves good depth penetration with graceful degradation as soil becomes more lossy for radar signals.

A number of systems have already implemented this concept for a variety of applications. Systems for shallow, buried target identification, for utility line location and/or mapping, and for tunnel detection have been briefly discussed here. Doubtless, various other types will be developed in the future.

Finally, there is a need for more research in this area, especially for soil-radar propagation studies that provide regional average data and relate radar propagation to structural and geological soil parameters. Given the inaccessibility of buried targets, there must be a continuing effort on reliable target-identification processing. Improved components and system approaches will open up more applications for this technology.

Acknowledgments. The authors are greatly indebted to the many researchers at the Ohio State University who have contributed very significantly to the material contained in this chapter. It would be most time-consuming to list them all, and this is left to the references. However, this document would not have been feasible without the efforts of Professor Jack H. Richmond, Dr. Geoffrey Burrell, Dr. C. A. Davis, III, Dr. Luen Chan, Professor David Moffatt, and Mr. Ross Caldecott.

REFERENCES

1. Lerner, R. M. Ground radar system. 1974. U.S. Patent No. 3, 831 173.
2. Dines, Kris A., and Lytle, R. J. 1979. Computerized geophysical tomography. *Proc. of IEEE* 67, No. 7.
3. Lytle, Jeffrey R. 1974. Measurement of earth medium electrical characteristics: Techniques, results and applications. Lawrence Livermore National Laboratory, Livermore, CA.
4. Hoekstra, P., and Delaney, A. 1974. Dielectric properties of soils at UHF and microwave frequencies. *Journal of Geophysical Research* 79, No. 11: 1699–1708.
5. Davis, J. L. 1975. Relative permittivity measurements of a sand and clay soil in situ. *Geological Survey of Canada,* Paper 75-1C, pp. 361–363.
6. Davis, J. L.; Topp, G. C.; and Omran, A. P. 1977. Measuring soil water content in situ using time domain reflectometry techniques. *Geological Survey of Canada,* Paper 77-1B, pp. 33–36.
7. Falls, Bernard A. 1977. Apparatus for measuring the electromagnetic impedance of soils. U.S. Patent No. 4,013,950.
8. Wait, James R. 1955. Mutual electromagnetic coupling of loops over a homogeneous ground. *Geophysics* XX, No. 3: 630–637.
9. Hayes, P. K. 1979. An on-site method for measuring the dielectric constant and conductivity of soils over a one gigahertz bandwidth. *Technical Report 582X-1.* The Ohio State University ElectroScience Lab., Dept. of Electrical Engineering (prepared for Columbia Gas System Service Corporation).
10. Gabillard, R.; Degauge, P.; and Wait, J. R. Subsurface electromagnetic telecommunication—Review. *IEEE Trans on Communication Technology* COM-19: 1217–1228.
11. Burrell, G. A.; Peters, Jr., L.; and Terzuoli, Jr., A. J. 1976. The propagation of electromagnetic video pulses with application to subsurface radar for tunnel application. *Technical Report 4460-2.* The Ohio State University ElectroScience Lab., Dept. of Elecrical Engineering (prepared under Contract DAAG53-76-C-0179 for U.S. Army Mobility Equipment Research and Development Command, Ft. Belvoir, VA. 22060).
12. Tribuzi, C. A. 1977. An antenna for use in an underground (HFW) radar system. *Technical Report 4460-4*. The Ohio State University ElectroScience Lab., Dept. of Electrical Engineering (prepared under Contract DAAG53-76-C-0179 for U.S. Army Mobility Equipment Research and Development Command, Ft. Belvoir, Va. 20060).
13. Wald, L. W. 1979. Modification of the HFW underground antenna based on experimental studies. *Technical Report 4460-6. The Ohio State University ElectroScience Lab., Dept. of Electrical Engineering (prepared under Contract DAAG53-76-C-0179 for U.S. Army Mobility Equipment Research and Development Command, Ft. Belvoir, Va. 22060).*
14. Kennaugh, E. M. 1981. The K-pulse concept. *IEEE Trans. on Antennas and Prop.* AP-29, No. 2.
15. Chan, L. C.; Moffatt, D. L.; and Peters, Jr., L. 1979. A characterization of subsurface radar targets. *Proc. IEEE* 67, No. 7: 991–1001.
16. Chan, L. C.; Moffatt, D. L.; and Peters, Jr., L. 1979. Estimation of the complex natural resonances from a class of sub-surface targets. In *Acoustic electromagnetic and elastic wave scattering—Focus on T-matrix approach.* New York: Pergamon Press, pp. 453–463.

17. Chan, L. C.; Peters, Jr., L.; and Moffatt, D. L. 1981. Improved performance of a subsurface radar target identification system through antenna design. *IEEE Trans. on Antennas and Prop.* AP-29, No. 2.: 307–312.
18. Chan, L. C.; Moffatt, D. L.; and Peters, Jr., L. 1981. Subsurface radar target imaging estimates. *IEEE Trans. on Antennas and Prop.* AP-29, No. 2: 413–418.
19. Chan, L. C. 1979. Subsurface electromagnetic target characterization and identification. Ph.D. dissertation, The Ohio State University.
20. Smith, G. 1983. Antenna pattern performance near the interface of a dielectric medium. Paper submitted to *IEEE APS.*
21. McWane, P. D. 1972. A look at the antenna radiation problem in the time domain. Ph.D. dissertation, Ohio State University.
22. Morey, 1974. U.S. Patent No. 3,806,795.
23. Young, J. D., and Caldecott, R. 1976. U.S. Patent No. 3,967,282.
24. Young, J. D., and Caldecott, R. 1977. U.S. Patent No. 4,062,010.
25. Davis, III, C. W. 1979. A computational model for subsurface propagation and scattering for antennas in the presence of a conducting half space. *Technical Report* 479X-7. The Ohio State University ElectroScience Lab., Dept. of Electrical Engineering (prepared for Columbia Gas System Service Corporation).
26. Izadian, J. 1980. An underground near-field antenna pattern range. *Quarterly Report 529-81-3.* The Ohio State University ElectroScience Lab., Dept. of Electrical Engineering (prepared under Contract 5014-352-0234 for Gas Research Institute).
27. Owens, T. E. 1981. Cavity detection using VHF hole-to-hole electromagnetic techniques. Paper presented at Symposium on Tunnel Detection, Colorado School of Mines, Golden, Col.
28. Peters, Jr., L., and Richmond, J. H. 1982. Scattering from cylindrical nonhomogeneities in a lossy medium. *Radio Science* 17, No. 5: 973–987.
29. Izadian, J. S.; Peters, Jr., L.; and Richmond, J. H. 1984. Computation of scattering from penetrable cylinders with improved numerical efficiency. Paper accepted for publication in *IEEE Trans. on Geoscience.*
30. Izadian, J. S. 1982. Two dimensional EM-scattering by buried penetrable noncircular cylinders using the method of moments. Ph.D. dissertation, The Ohio State University.
31. Davis, C. W., and Peters, L. 1979. Application of a video pulse radar system to detect Tunnels at the Curtis School Yard in Trumbull County, Ohio. Technical Report 784460-7. The Ohio State University ElectroScience Lab., Dept. of Electrical Engineering (prepared under contract DAAG53-76-C-0179 for U.S. Army Mobility Equipment Research and Development Command).
32. Volakis, J., and Peters, L., Jr. 1980. Elimination of undesired natural resonances for improved target identification. Paper presented at International Union of Radio Science, Quebec.
33. Volakis, J., and Peters, Jr., L. 1983. Improved identification of underground targets using video pulse radars by elimination of undesired natural resonances. *IEEE Transactions on Antennas and Prop.* AP-31, No. 2.
34. Davis, C. W., and Peters, Jr., L. 1979. Characteristics of a video pulse radar system operating in the high frequency window at the Hazel A. Mine near Gold Hill, Colorado. *Technical Report* 784460-8. The Ohio State University ElectroScience Lab., Dept. of Electrical Engineering (prepared under Contract No. DAAG53-76-C-0179 for U.S. Army Mobility Equipment Research and Development Command, Ft. Belvoir, Va. 22060).
35. Volakis, J. 1979. Improved identification of underground targets using video-pulse radars by elimination of undesired natural resonances. M.Sc. thesis, The Ohio State University.
36. Barringer, A. R. 1963. The use of audio and radio frequency pulses for terrain sensing. *Proc., 2nd Symposium on Remote Sensing of Environment,* Center of Remote Sensing Information and Analysis, Willow Run Laboratories, University of Michigan, pp. 201–214.

37. Gelengnse, M., and Barringer, A. R. 1964. Recent progress in remote sensing with audio and radio frequency pulses. *Proc., 3rd Annual Symposium on Remote Sensing of Environment,* pp. 469–494, (see Ref. 36.)
38. Burrell, G. A., and Peters, Jr., L. 1979. Pulse propagation in lossy media using the low-frequency window for video pulse radar application. *Proceedings of the IEEE* 67, No. 7.
39. Burrell, G. A., and Munk, B. A. 1976. The array scanning method and applying it to determine the impedance of linear antennas in a lossy half space. *Technical Report 4460-1.* The Ohio State University ElectroScience Lab., Dept. of Electrical Engineering (prepared under Contract DAAG53-76-C-0179 for U.S. Army Mobility Equipment Research and Development Command, Ft. Belvoir, Va. 22060).
40. Burrell, G. A., and Munk, B. A. IEEE AP/S paper.
41. Peters, Jr., L.; Burrell, G. A.; and Tran, H. B. 1977. A scattering model for detection of tunnels using video pulse systems. *Technical Report 4460-3.* The Ohio State University ElectroScience Lab., Dept. of Electrical Engineering (prepared under contract DAAG53-76-C-0179 for U.S. Army Mobility Equipment Research and Development Command, Ft. Belvoir, Va. 22060).
42. Burrell, G. A.; Richmond, J. H.; Peters, Jr., L.; and Tran, H. B. 1979. A scattering model for detection of tunnels using video pulse radar systems. In *Acoustic, electromagnetic and elastic wave scattering—Focus on T-matrix approach.* New York: Pergamon Press, pp. 667–683.
43. Kouyoumjian, R. G.; Peters, Jr., L.; and Thomas, D. T. 1963. A modified geometrical optics method for scattering by dielectric bodies, *IEEE Trans.* LAP-11: 690–703.

10
Time-Domain Measurements of Components and Materials

Harry M. Cronson

The MITRE Corporation

10.0 INTRODUCTION

Scope

The chief focus of this chapter is time-domain measurements of components and materials where the end results are the microwave frequency-scattering parameters of the device under test. The measurement system is referred to as a time-domain automatic-network analyzer (TDANA), which usually consists of a pulse generator, delay-line network, sampling system, analog-digital interface, and computer. The distinction between components and materials measurement is somewhat artificial, since the TDANA is similar in both cases; it is maintained in this discussion, however, because components and materials usually remain distinct in the literature. The TDANA systems described are primarily those built at Sperry Research Center (Sudbury, MA) because these were often the ones developed first[1,2] and also because of the author's familiarity with them during his 12 years there.* Other excellent TDANA systems have been built by the National Bureau of Standards (NBS), Boulder, CO.,[3,4] and other organizations.[5,6]

An Elementary Time-Domain System

The fundamentals of time-domain techniques for an insertion-loss measurement are illustrated in Fig. 10-1(a), which shows a basic configuration consisting of

*In particular, the author would like to acknowledge his debt to Dr. G. F. Ross and Dr. A. M. Nicolson for their guidance, encouragement, and pioneering contributions, and to express his appreciation to his co-worker, P. G. Mitchell.

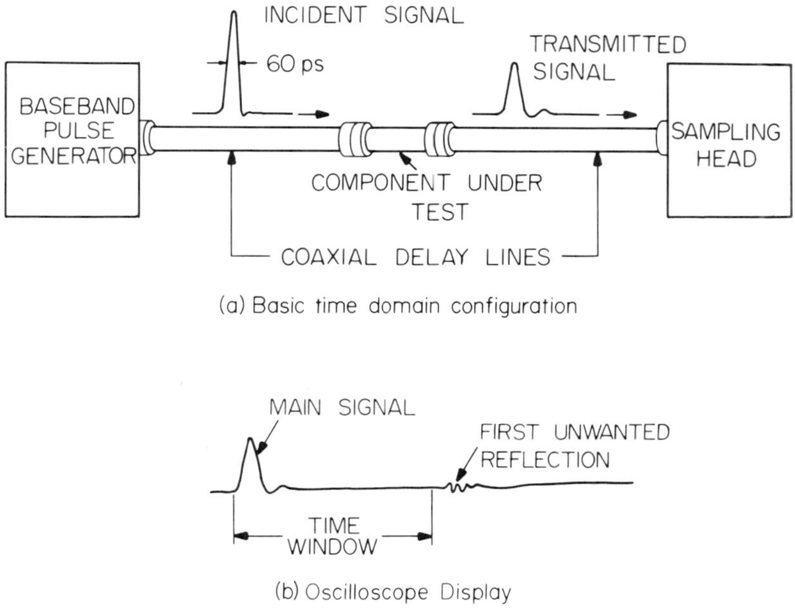

Fig. 10-1. An elementary time-domain system: (a) basic configuration, and (b) oscilloscope display.

a baseband pulse generator, two sections of delay line, the component under test, and a sampling head. The generator produces an impulse (which in practice may have a half-width of about 60 ps) and a synchronized trigger pulse for the oscilloscope. This transmitted signal, $v_{tr}(t)$, travels through the delay lines and component-under-test into the sampling head, where it is sampled, digitized, and stored. To complete the measurement procedure, the component under test is removed, and the incident impulse passing through the two delay lines, $v_i(t)$, is acquired. These two time-domain waveforms may then be translated into the frequency domain using the fast Fourier transform (FFT). Their ratio, $F[v_i(t)]/F[v_{tr}(t)]$—where $F[v(t)]$ represents the *FFT* of $v(t)$—is the insertion loss of the component being tested.

One of the principal advantages of this time-domain method is the ability to "window out" extraneous system reflections. Only a selected time window of the entire signal incident on the sampling head is acquired, as shown in Fig. 10-1(b), so that unwanted reflections from the generator and sampling head cannot distort the desired data. The window width must be chosen to be less than the two-way transit time in the individual delay lines.

The use of a step or impulse waveform is usually satisfactory for transformed time-domain measurements up to about 10 GHz. Above this frequency the spectral amplitude of available step and impulse generators decreases rapidly, and it is very difficult to obtain accurate results.[7] To overcome this limitation, RF burst generators with appreciable spectral amplitude between 9 to 16 GHz have been developed.[8,9]

For some applications, the time-domain waveforms themselves may be used directly without frequency translation. For example, certain flaws in passive components can cause decreased risetime, undershoot, and/or ringing in the transmitted pulse. These component problems may be detected easily on a sampling oscilloscope display by semiskilled production line personnel.

The Time-Domain Automatic-Network Analyzer

The function of the basic components in a time-domain measurement system can perhaps best be described with reference to a specific system. An often-used configuration for insertion-loss measurements containing the five essential parts of a pulse generator, delay-line network, sampling system, analog-digital interface, and computer is shown in Fig. 10-2. To determine the insertion loss, the processed waveforms with and without the device under test (DUT) are compared.

The generator supplies a subnanosecond pulse that propagates through the delay lines and DUT to the sampling head. Most workers use a high-repetition-

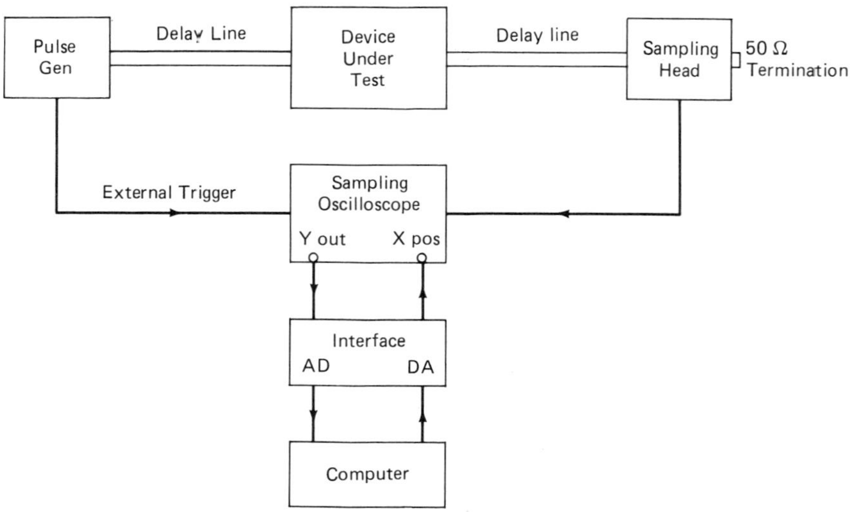

Fig. 10-2. Insertion-loss measurement system.

rate, solid-state source with a pretrigger and average the data from many pulses to reduce noise. The pulse shape may be either a step, as supplied by a tunnel-diode generator, or a smoothed impulse obtained from combination avalanche-transistor, step-recovery-diode generators.[8,10] Since the sampling diodes are voltage limited, it becomes important to select the pulse shape most appropriate for the measurement. If higher frequency results are required, the impulse waveform is chosen, since for the same voltage amplitude, the impulse has a greater high-frequency spectral amplitude than the step. Note that if information is required over only octave or smaller bandwidths, other pulse shapes, like the doublet or several RF cycles, will provide larger spectral amplitudes over a limited range.

The delay lines provide a sufficient time window for the transmitted pulse to be observed free of unwanted reflections. In Fig. 10-2, the lines on both sides of the DUT must be long enough for the reflections between the DUT and pulse generator, and between the DUT and sampling head, to be outside the measurement window. If T_w is the time-window interval and l is the line length, then $T_w < 2l/v$, where v is the velocity of propagation in the lines. For accurate measurements, the lines themselves, and the connectors adjacent to the DUT, must be precisely matched. The sampling head and oscilloscope are used as a sample-and-hold circuit with a very narrow sample gate. The interface averages and converts the analog scope Y output to digital information for the computer. To minimize noise, the computer controls the horizontal scan of the scope through the interface. Once the computer acquires the sampled time-domain waveform, it performs an FFT to extract the desired frequency information.

Although we could measure the insertion loss of the DUT with the arrangement shown by acquiring the waveforms with and without the DUT in the circuit, several additional components are usually included to improve measurement accuracy. To reduce noise and jitter with "3-point" scanning,[11] a reference waveform derived from the pulse generator may be introduced outside the measurement window through the other channel of a dual-channel sampler. In some cases a separate pulse is not necessary, for the pulse may serve as its own reference. To maintain maximum amplitude waveforms at the sampling head, a calibrated attenuator may be inserted immediately after the pulse generator. Since the signal-to-noise ratio (SNR) is usually proportional to the input-signal amplitude, SNR is optimized when the input-signal amplitude is a maximum. The input-signal amplitude is limited, however, by the 1-V burnout level of the sampling diodes. Present pulse generators furnishing about an 8-V amplitude pulse with 60-ps half-width must be attenuated before entering the sampling head. When an attenuator with an insertion loss of 20 dB or greater is measured, however, removing the initial pulse-generator attenuation is desirable since it is no longer needed to protect the sampling diodes with the test attenuator in place. In practice, the calibrated attenuator is used in a substitution method to limit the voltage entering the sampling head and to serve as a comparison standard.

Sampling and Frequency Translation

The desired frequency response must be determined from the sampled waveforms rather than from the original continuous transients. For a continuous signal to be represented by values of equi-spaced samples, the sampling theorem must be satisfied. The theorem states that a function of time $v(t)$ that contains no frequencies greater than f_0 is uniquely determined by the values of $v(t)$ at any set of sampling points spaced no more than $1/2f_0$ apart. In practice, this means the sampling interval must be small enough to meet the above constraint. Frequency conversion for the properly sampled time functions is given by the discrete Fourier transform,

$$v(f) = T \sum_{n=0}^{N-1} v(nT)e^{-jn2\pi fT} \tag{10-1}$$

for $f < 1/2T$, where T is the sampling interval and the $v(nT)$ are N samples of the waveform. Although the above equation can be used to determine f at any frequency between 0 and $1/2T$, FFT techniques that restrict f to discrete frequencies—$f = k/NT$, where $k = 0,1 \ldots . N/2$—are usually employed to save computation time when N is large.[12] Other conversion techniques, such as Samulon's method[13] and modified FFT techniques for a small N,[14] have also been used. Further discussion on signal processing is provided in Chap. 12.

The choice of N and T is determined by the highest frequency present in the waveform, the time duration of the waveform, and the computer capacity for an N-array FFT. The spectrum of the sampled waveform is limited by the pulse width of the input signal, the bandwidth of the oscilloscope, and the measurement network. If T is too large, aliasing errors will contaminate the computed spectrum.[15] The time-window width NT should be large enough to encompass the entire nonzero portion of the response, or else truncation errors will appear. For a fixed computer-storage capacity of N numbers, reducing the aliasing error depends on reducing T, whereas reducing the truncation error requires that NT, and hence T, be as large as possible. A compromise must be made in the choice of T.

10.1 MEASUREMENT OF COMPONENTS

Evolution from Reflectometry to Computerized Measurement Systems

Between about 1964 and 1972, time-domain techniques evolved from qualitative observations of step waveforms reflected from discontinuities in transmission lines[16] to complete scattering parameter characterizations of two-port components as a function of frequency. Time-domain reflectometry (TDR) has long been used to locate faults in telephone lines, but TDR techniques were applied to microwave components only after the introduction of sampling oscilloscopes and pulse generators with combined risetimes less than 100 ps.[17] Figure 10-3 illustrates a typical TDR configuration. In TDR, reflections due to mismatches in the line under test are clearly discernible as separate echoes if the transmit time between discontinuities is greater than about half the risetime of the incident pulse. In addition, the shape of the reflection can provide valuable information about the nature of the discontinuity. Thus, the oscilloscope display can be directly related to line mismatch.

These early techniques provided the impetus for subsequent advances because they demonstrated the power and simplicity of time-domain methods for wideband applications. Original TDR methods were used to give only qualitative information about the frequency response of the discontinuity. In England in 1965, Davis and Loeb were among the first to demonstrate experimentally that transient waveforms could be Fourier-transformed to obtain the frequency response of the network under test.[18] By analyzing the step-response waveforms reflected and transmitted through a transistor embedded in a 50-Ω line, they calculated scattering parameters of the transistor from 0.125 to 1 GHz. The amplitude of each waveform as plotted on an X-Y recorder was measured at time intervals of 100 ps, and the data were used in a computer program to calculate the transform. Mismatches in the system were separated by suitable lengths of coaxial line, and signal-degradation effects were eliminated since both reference and test waveforms traversed the same signal path. Good agreement was obtained between the frequency response obtained by time-domain techniques and conventional bridge measurements.

In the United States, the basic concept for time- to frequency-domain translations using a computer-controlled sampling oscilloscope was developed by Nicolson and Ross.[19] In 1968, Nicolson presented a refined comprehensive treatment of their method to measure the broadband, microwave-transmission

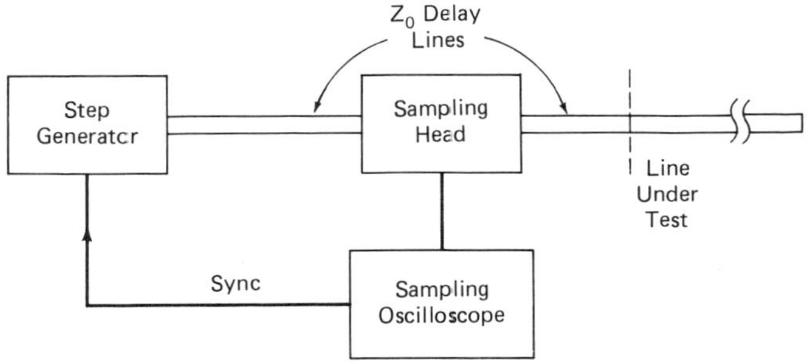

Fig. 10-3. Time-domain reflectometry configuration.

characteristics from a single measurement of the transient response.[1] Results up to 5 GHz on short-circuited stubs, couplers, and filters were reported. The experimental configuration was similar to that of Fig. 10-2, with the response taken directly from the vertical output of the 12-GHz sampling oscilloscope and subsequently filtered and digitized. (Included in Reference 1 is an extensive discussion of measurement error caused by noise, nonlinearities, and frequency translation.)

This analysis established the error bounds on time-domain metrology techniques, with the prime source of error being low-frequency timing jitter. Measurements made on known components showed good agreement with predicted errors. Later, in 1969, Nicolson described a method of closed-loop, computer-controlled scanning for dynamic correction of errors due to vertical amplifier drift and timing shifts.[11] This "3-point" scanning procedure greatly reduced lower-frequency noise components and permitted longer term averaging to obtain higher SNR.

These techniques provided new precision and enabled time-domain metrology to proceed from testing to calibrating microwave components. Another system, independent of Nicolson's and reported in 1969 by Farber and Ho for both passive and active devices,[20] utilized an on-line computer connected directly to a sampling oscilloscope to facilitate time-to-frequency translation. By 1972, time-domain systems covering a wide range of applications were reported.[2,21] Some of these systems with their later improvements and refinements will now be described.

Measurement Results

Some representative time-domain measurement results are given in this section. The measurement configuration is basically that of Fig. 10-2 but with the addition of a calibrated step attenuator that is separated from the pulse generator and component under test by delay lines.[22] A substitution procedure is used here. With the component under test in place, the insertion loss of the calibrated attenuator is reduced until the transmitted signal amplitude approaches the 1-V limitation of the sampling head. The waveform is scanned and digitized with a limited-variance, nonsequential scanning procedure.[2] This 256-point scan over a 2.5-ns window takes about 1 minute. The component under test is removed and the calibrated attenuation increased until the incident signal at the sampling head is again a maximum. After this waveform is acquired, the FFT of both signals is computed. The ratio of the transmitted-to-incident signal transforms, along with the calibrated attenuator values, provides the insertion loss of the network.

Results of measurements on 10-dB and 40-dB wideband attenuators are illustrated as points in Fig. 10-4, together with frequency-calibration values in-

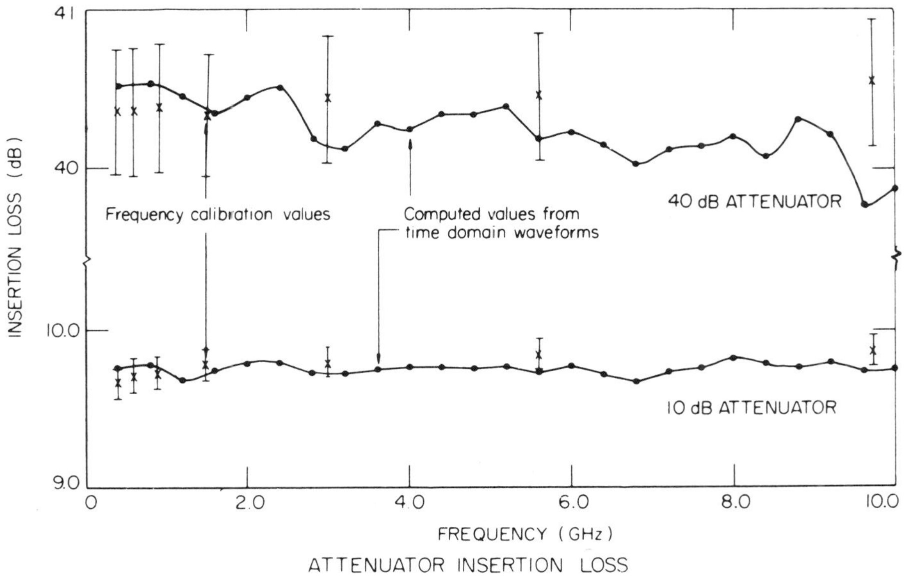

Fig. 10-4. Attenuator insertion loss.

dicated by crosses with ±1-percent error bounds. These results are a part of measurements taken on six attenuators ranging in value from 10 to 60 dB. When compared with frequency-calibration data for the same attenuators, the tests showed better than 1-percent agreement for insertion losses of 10 to 50 dB in the 0.4- to 8-GHZ frequency range.

The accuracy of this method depends on the calibration of the substitution attenuator. A step attenuator was calibrated over insertion values of 10 to 60 dB in 10-dB steps, using a previously calibrated 10-dB attenuator as a standard. The standard attenuator was calibrated with a Hewlett-Packard 8542A, computer-controlled, phase-locked network analyzer with an estimated accuracy of ±0.05 dB. In the calibration, a substitution procedure was followed in which the averaged waveforms through the standard attenuator were compared with averaged waveforms through the step attenuator at similar deflections. The step attenuator was first set to 0 dB, with the 10-dB standard in the device-under-test position. After the waveform was scanned four times and the results averaged, the attenuator was stepped to the 10-dB position, the 10-dB standard was removed, and the resultant waveform was scanned four times and averaged. The ratio of the FFTs of the averages, together with the 10-dB standard values, provided the insertion loss difference between the 0-dB and 10-dB settings. To determine the insertion-loss difference between the 10-dB and 20-dB settings, the procedure was repeated with the step attenuator at 10 dB (with the standard in line) and the attentuator stepped to 20 dB (with the standard removed). In this way, the step attenuator was calibrated through the 60-dB setting.

Another configuration to measure S_{11} as well as S_{21} in a 2.5-ns window is illustrated in Fig. 10-5.[23] With the switches in the T (transmit) position, the insertion loss of the component under test is measured, whereas for return-loss measurements, the R position is used. For this application, a directional coupler was designed to have an approximate smoothed impulse response in a 2.5-ns time window for both the transmitted and coupled pulse.[24] This was accomplished by extending the length of the uniform coupling region so that the two-way travel time between the inevitable coupler discontinuities was greater than 2.5 ns.

Figure 10-6 illustrates S_{21} and S_{11} measurement results for a 40-dB attenuator. Each graph shows three sets of data points: two time-domain measurements to assess repeatability and an automatic, frequency-domain, network-analyzer result to serve as an accurate, independent reference. The discrepancies in the S_{21} transformed time-domain data above 6 GHz are due to the lower available spectral amplitude of the impulse generator. Although a reasonable agreement between the transformed S_{11} time-domain runs and network-analyzer values exists, the spreads are much larger than the S_{21} results. The discrepancy is thought to be caused by residual ripples in the directional coupler.

Fortunately, accurate measurements of S_{11} are usually not required. One need only show that S_{11} is below some minimum value; for this situation, present time-domain techniques should be satisfactory.

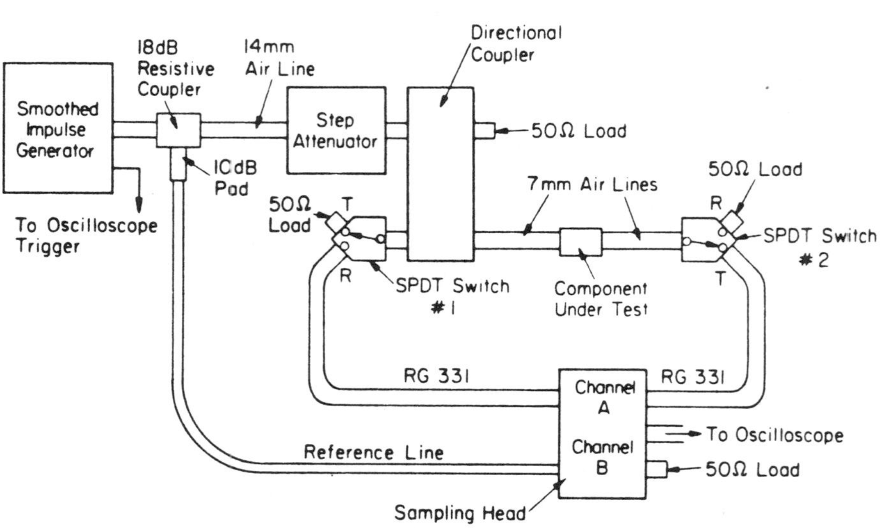

Fig. 10-5. 0.4- to 10-GHz measurement configuration.

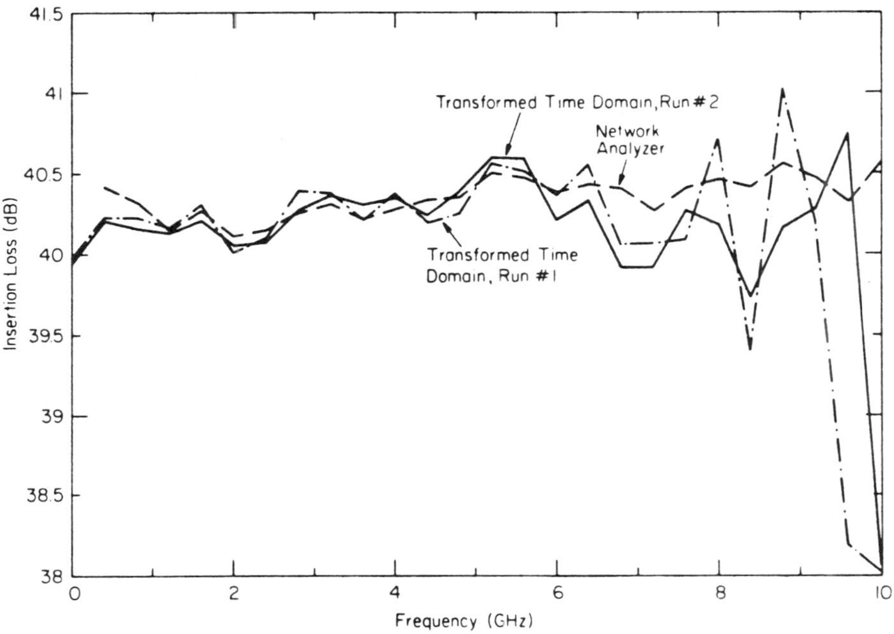

(a) insertion loss

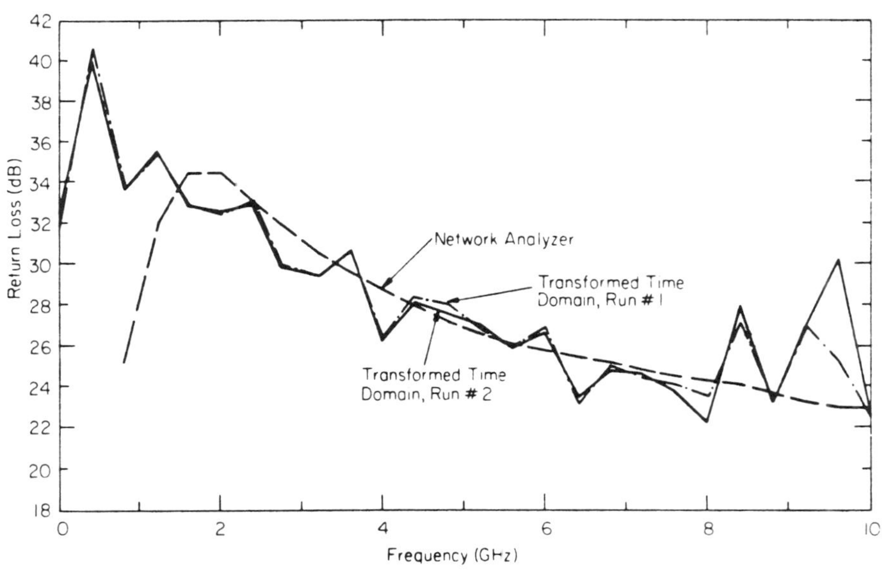

(b) return loss

Fig. 10-6. S_{21} and S_{11} measurement of 40-dB attenuator: (a) insertion loss, and (b) return loss.

Insertion-loss measurements can also be taken using an RF pulse waveform to overcome the high-frequency deficiencies of the impulse waveform. Some of these results for a 30-dB attenuator are illustrated in Fig. 10-7. Note, in general, the good agreement among the three sets of data. The divergencies seen below 10 and above 16 GHz are due to the decreased spectral amplitude of the pulse in this range.

To overcome this inconvenience and the inaccuracies of using two pulse generators to cover a very large frequency range, a composite generator was developed that simultaneously applies both an impulse and RF waveform.[9] As an example of the results obtainable with the composite generator, Fig. 10-8 shows the comparison between a 20-dB attenuation measurement taken on an automatic, frequency-domain network analyzer and the transformed time-domain data. The time-domain measurement was taken with two 46-cm sections of precision 7-mm air line separating the attenuator from the composite generator on one side and the sampling head on the other.

Error Sources

A fairly comprehensive error investigation was reported by Cronson and Mitchell.[22] Their experiments were designed to isolate and evaluate various sources of error, among them: random errors due to noise and drifts; systematic errors caused by substitution-attenuator inaccuracies, line mismatch, deflection nonlinearities, and

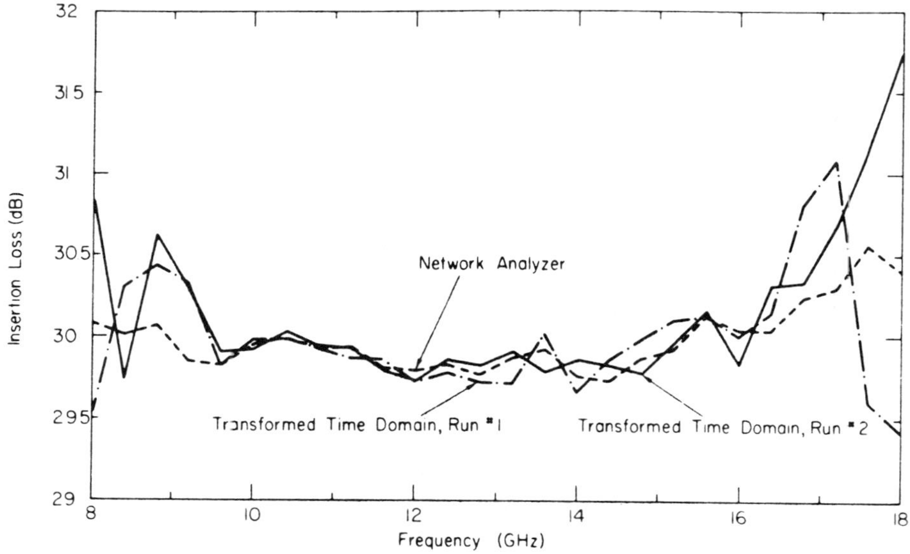

Fig. 10-7. Insertion-loss measurement of 30-dB attenuator.

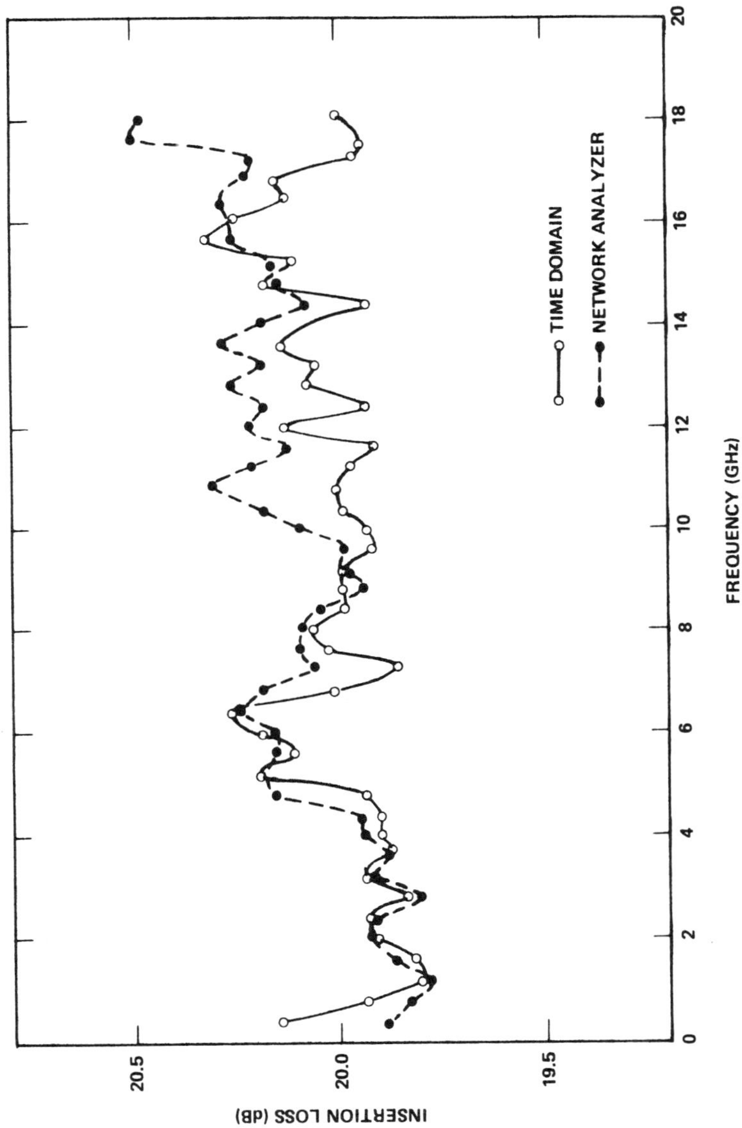

Fig. 10-8. Insertion-loss measurement of 20-dB attenuator using the composite generator.

inaccurate time-window widths; time-to-frequency translation errors of aliasing and truncation; and mechanical errors due to connect-disconnect cycles. An operational approach was adopted so that the individual errors could be expressed in decibels and directly related to the measured insertion loss. The salient facts of the investigation are summarized in Table 10-1, which contains a general classification of the four main types of error. These types are further divided into known types within each group. For each of these types, the causes are identified; a procedure to evaluate the error magnitude is given; the error magnitude is estimated from the results of analyses and/or experiments; and methods to reduce the error are suggested. Random noise proved to be the major contributor to measurement error. Reference 22 and the work of Gans and Andrews[3] should be consulted for more detail.

Comparison of Time-Domain and Frequency-Domain Automatic Network Analyzers

Most readers are more familiar with swept or stepped-frequency automatic network analyzers (SFANA) than with the TDANA. The most widely used systems are vector ANAs based on heterodyne principles[25] and a variety of scalar ANAs using diode detection.[26] There are also SFANAs based on homodyne principles[27] and six-port techniques.[28] Over the years, commercially available SFANAs have become less expensive and easier to use because of rapid advances in desk-top computers and microprocessors. One of the major advantages of the TDANA is the much lower cost of the pulse generator, made up of a tunnel diode or an avalanche transistor with step-recovery diodes, compared with that of a computer-controlled swept-frequency source, to cover the 10-MHz to 18-GHz frequency range. A disadvantage of the pulse generator is its considerably lower signal-to-noise ratio per frequency point, making narrow-band and very-wide-dynamic range measurements difficult. The TDANA can be easily used for direct time-domain measurements, however, whereas the SFANA requires considerable software[29] or sophisticated microprocessor techniques[30] to present a similar display.

10.2 MEASUREMENT OF MATERIALS

Material Measurement Basics

As mentioned, the distinction between component and material measurement is somewhat artificial since similar techniques are used in both applications. In this section, material measurements are subdivided into two categories. First, and most closely related to component measurements, is the method introduced by

Table 10-1. System Errors.

CLASSIFICATION	TYPE	CAUSES	EVALUATION	MAGNITUDE	METHODS OF REDUCTION
Random	Measured with 1-min. scan	Amplifier noise and drift, timing jitter and shifts, quantization	Successive scans of the waveform in an undisturbed configuration	<0.13 db in 10 db to 10 GHz	Increase averaging
Systematic	Substitution attenuator	Error in values of calibrated attenuator	Knowledge of error bounds of standard attenuator	<0.05 dB in 10 dB	Smaller error bounds for standard attenuators
	Mismatch	Mismatches in connectors or lines in the vicinity of the component under test	Analysis of reflections with SWR data of connectors and components	0.01 dB	Care in selecting and assembling lines; calculated correction from known line imperfections
	Nonlinearity deflections	Nonlinearity of timing ramp and errors in horizontal drive voltages	Measure waveforms with simulated nonuniform sampling intervals	0.02 dB	Waveforms occupy same relative time position; computer linearization
	Frequency displacement	Error in absolute value of time window	Measure time window	0.05% in frequency	Precise time window calibration
Time-to-frequency conversion	Aliasing	Spectral overlap from sampling interval	Calculate spectrum for various sampling intervals	0.025 dB	Decrease sampling interval
	Truncation	Loss of waveform data from finite time window	Compare spectrum for different truncations	0.02 dB	Waveforms in same relative time positions
Mechanical motion	Connect, disconnect, and line displacement	Changes of connector SWR due to mechanical displacement	Compare data over many connect-disconnect cycles	0.02 dB	Minimize mechanical motion

Nicolson and Ross for measuring radar absorbing materials.[31] Second are methods used by physical chemists to study the behavior of polar molecules and biological substances, methods referred to as *time-domain spectroscopy* (TDS).

In material measurement, the test sample is usually prepared to fill the complete volume between the inner and outer conductors of a section of coaxial line. Subsequent measurements are based on the differences between the characteristic impedance and propagation constant of the filled and unfilled lines. If an air coaxial line with characteristic impedance, Z_0, and propagation constant, γ_0, is filled with a material having a relative complex permittivity, $\varepsilon = \varepsilon' - j\varepsilon''$; a dc conductivity, σ; and a relative complex permeability, $\mu = \mu' - j\mu''$, then the new characteristic impedance is

$$Z_1 = Z_0[\mu/(\varepsilon - j\sigma/\omega)]^{1/2} \tag{10-2}$$

and the new propagation constant is

$$\gamma_1 = \gamma_0[\mu(\varepsilon - j\sigma)\omega)]^{1/2} \tag{10-3}$$

The constitutive parameters of the material are then deduced from the scattering parameters of the altered lines.

The Nicolson-Ross Method

The basic elements of the Nicolson-Ross method are covered by Ross in Chap. 1 of this volume.[32] This section provides a more complete description of their system and others built at Sperry Research Center. The actual delay-line circuits are more complicated than the simple schematics shown by Ross because of the need to introduce reference waveforms that reduce waveform drift and jitter. Figure 10-9(a) shows the complete coaxial system; Fig. 10-9(b) the sampling oscilloscope waveforms. The pulse partially reflected back to the sampling head is designated V_R. Another part of the pulse, V_T, arrives at the sampling head after a round trip through the material via reflection at a short circuit. Timing markers V_X and V_Y are introduced into the time window to provide stability in measuring the desired waveforms, V_R and V_T. The measurement procedure consists of scanning various reference waveforms and V_R and V_T for the sample under test.

Several of these systems built by Nicolson are described by Cronson and Ross.[21] One employs General Radio's 14-mm precision air lines and provides two 2.5-ns time windows, each containing 256 sampling points. The other unit is a larger and more automated system. Constructed from semiflexible RG-/U 331 cable, it has two 10-ns windows, each with up to 1024 points. The latter

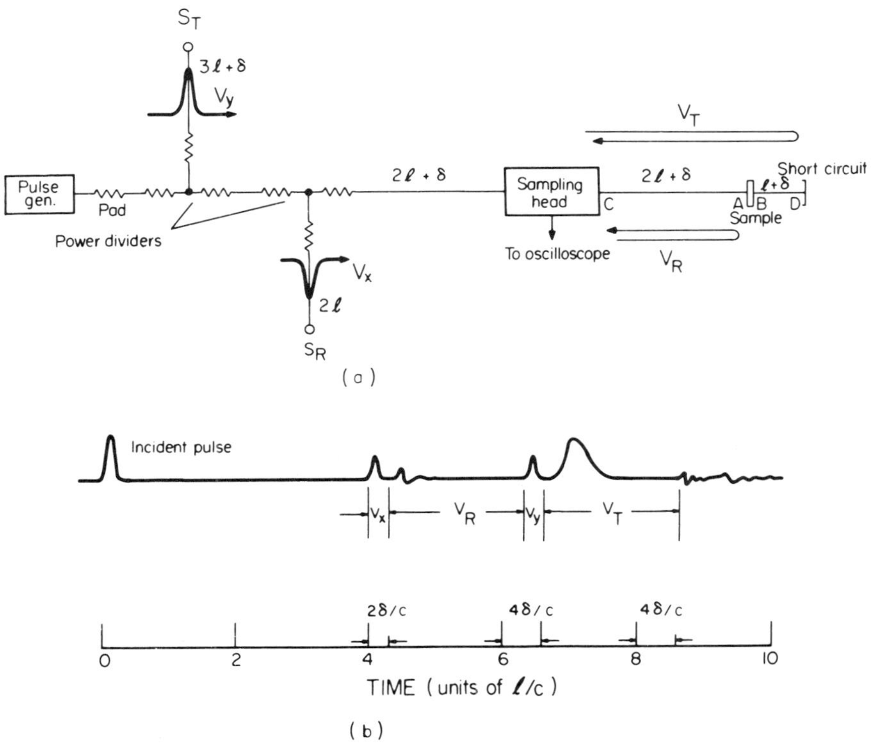

Fig. 10-9. Delay-line configuration and waveforms for material measurement: (a) coaxial line system, and (b) sampling oscilloscope waveforms.

system, delivered to the U.S. Air Force in 1971, measures ε and μ over a frequency range of 0.1-10 GHz in about 10 min. Both systems utilize a 5-V step-recovery diode pulse generator with a −6-dB width of 90 ps and a variable-dwell-time scanning procedure for more efficient waveform acquisition.

The 10-ns system has an automated design with coded lighted pushbuttons on the control panel to direct the measurement sequence. In addition, a special sample holder for measurements up to 1000°F is provided. The instrument rack, containing a sampling oscilloscope, operation control panel, controller, and digital voltmeter, is shown in Fig. 10-10. A typical result obtained with the 2.5-ns system is shown in Fig. 10-11.

One unique application of the 10-ns system is measurements at very high temperatures.[33] The high-temperature measurements employ a sample holder made from Kovar with gold-plated conducting surfaces that is contained in an

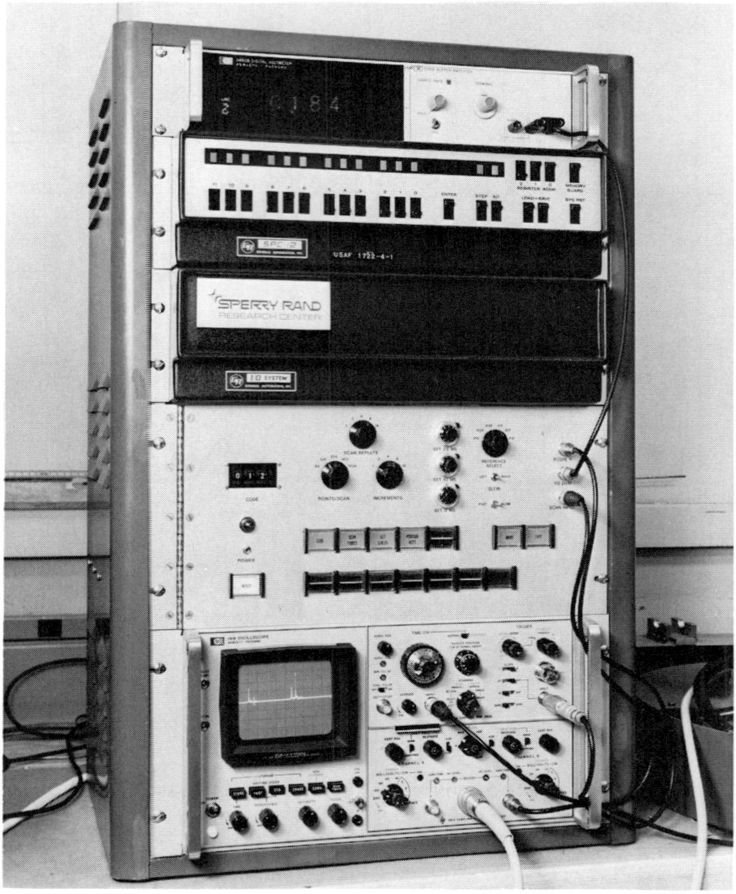

Fig. 10-10. Material-measurement-system instrument rack.

oven and capable of being cycled between room temperature and 1000°F. A picture of the high-temperature equipment with the sample holder enclosed and supported in a quartz tube with ends covered by asbestos blocks is shown in Fig. 10-12. Here, the 14-mm air lines have reduced wall thickness over most of their length to reduce heat loss. Water-cooling jackets for the sample holder outside the oven are provided, but experience has shown that they are not needed below 1000°F.

In 1974, the frequency range of these time-domain systems was extended to 16 GHz with a specially designed RF pulse generator.[7] This system was built to share the same signal-storage and processing equipment as the 0.1- to 10-GHz unit but had a different pulse-generator and delay-line configuration. The result of a measurement taken on both systems is shown in Fig. 10-13, along

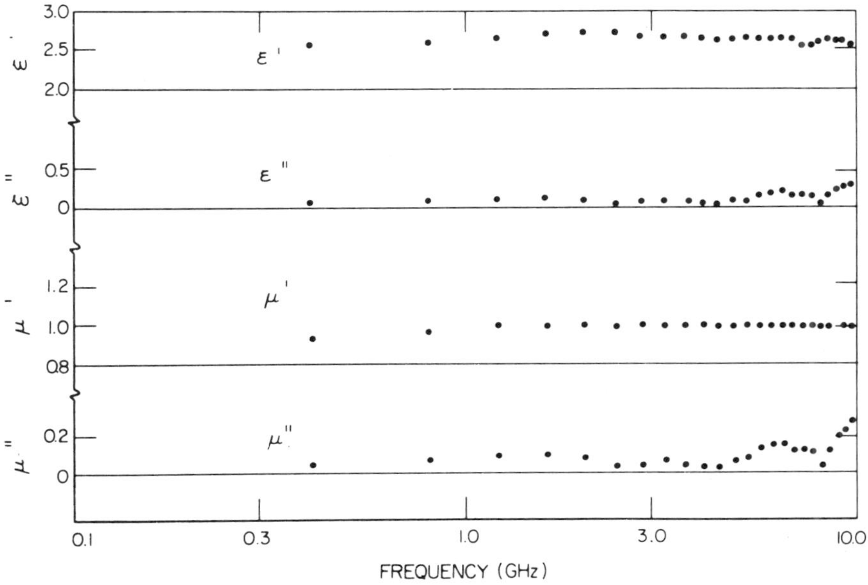

Fig. 10-11. Constitutive parameters for 0.25-in.-thick Plexiglas.

with swept-frequency analyzer results. Note the good agreement between the two time-domain analyzers and the swept-frequency-domain analyzer.

It should also be mentioned that the sample thickness of the material should be chosen to avoid an integral number of half-wavelengths in the material over the frequency range of interest. From the point of view of signal-to-noise ratio, the thickness of the sample is optimum when the signals in the reflection and transmission windows are about the same amplitude. Reference 7 treats these considerations.

As of this writing (1983), these systems with modified data acquisition hardware have been in almost daily use at the U.S. Air Force Avionics Laboratory, U.S. Air Force Systems Command, Dayton, Ohio, since their delivery in 1971 and 1974.

Time-Domain Spectroscopy

Almost independently of the work of Nicolson, Ross, and their colleagues in electrical applications, physical chemists and biologists began to use TDR and later time-domain spectroscopy (TDS) to study liquids and various biological specimens. Their work has generated an extensive literature with at least several reviews.[34–36] As among electrical engineers, the methods evolved from TDR.

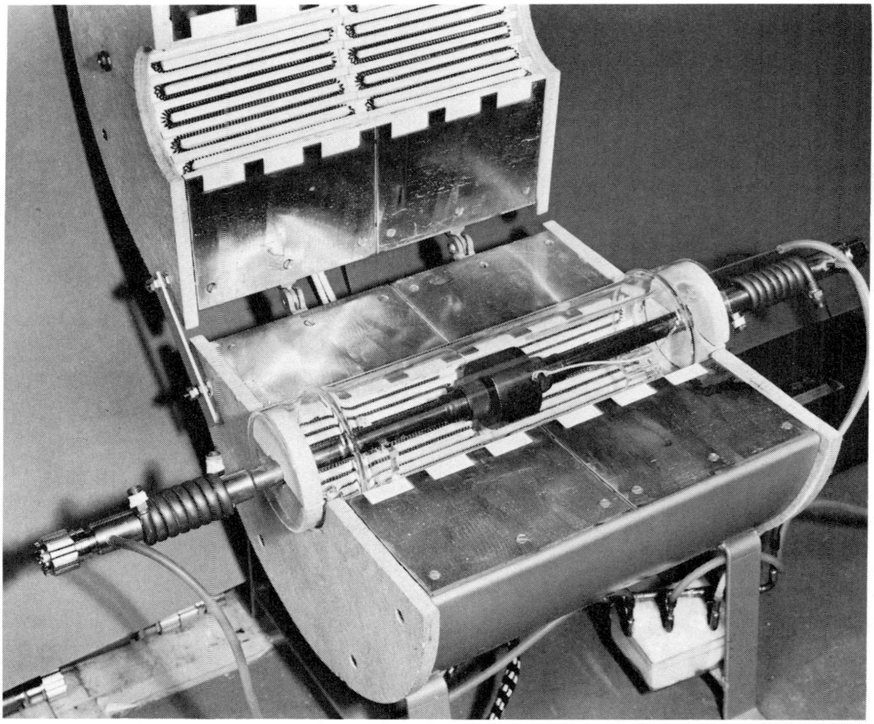

Fig. 10-12. High-temperature sample holder.

Fellner-Feldagg was the first to indicate to physical chemists how TDR could replace tedious point-by-point measurements to obtain the dc and high-frequency permittivity and relaxation time of polar liquids.[37] Fellner-Feldagg showed how the dc and high-frequency permittivity for a non-conductive dielectric could be calculated from the initial and final values of the normalized reflected step response. A theory developed independently by Bagozzi[38] for the pulse response of polar, liquid-filled coaxial lines, including the effects of skin-effect losses, was applied by Bagozzi, Ives, and Nahman[39] to determine very small dielectric-relaxation times in a Debye binary liquid using a time-domain transmission method. Work on TDR of polar liquids was extended and generalized by van Gemert and de Graan[40] and van Gemert,[41] who treated Debye, Cole-Cole, and Davidson-Cole dielectrics with dc conductivity. They concluded that standard TDR methods do not have sufficient accuracy to distinguish the type of dielectric model being tested from the shape of the step response. Therefore, unless the type of liquid is known *a priori*, it is impossible to use the shape of the response to derive the relaxation time, but the desired dielectric properties can be found by transforming the time-domain data into the frequency domain.

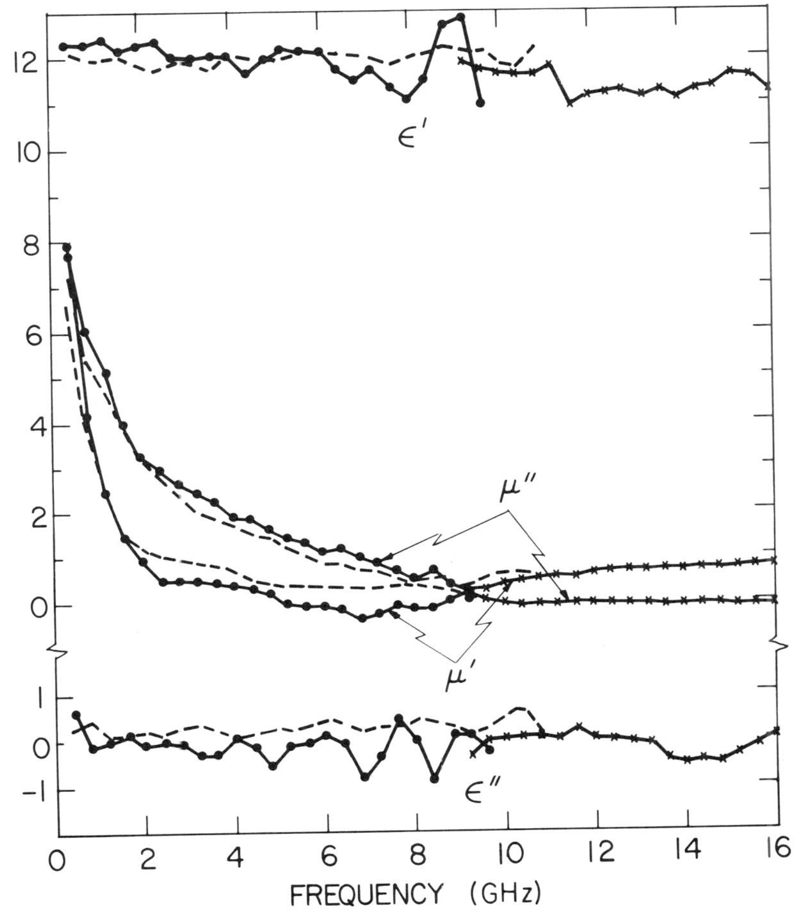

Fig. 10-13. Relative permittivity and permeability for a 0.05-in.-thick ferrite. Dashed lines represent stepped-frequency, automatic-network-analyzer results.

A frequency-conversion approach to long-sample TDR, based on methods originally developed for network analysis, was applied in 1970 by Suggett et al[42] and resulted in a significant improvement in accuracy. Their method, designated as time-domain spectroscopy (TDS), avoided any *a priori* assumptions about the nature of the test liquid and provided much more consistent semicircular Cole-Cole plots than frequency-domain methods. In 1971, Loeb et al.[43] provided a more detailed description of TDS techniques developed to obtain reasonably precise measurements over a 10-MHz to 13-GHz frequency range. Methods to extend this work to lower frequencies were devised by Fellner-Feldagg[44] and DeLoor, van Gemert, and Gravesteyn.[45] Another technique based on a small shunt capacitor terminating a transmission line was developed by Iskander and Stuchly[46] and later reviewed by the Stuchlys.[47] TDS techniques are extensively used today by physical chemists for dielectric relaxation studies.

10.3 CURRENT RESEARCH AND FUTURE DIRECTIONS

Time-domain research on measurements of components and materials was most active in the 1970s. Although a time-domain measurement system has advantages over a frequency-domain system, a commercial turnkey system was never produced. Thus, the TDANA was used only in a few laboratories and could not gain widespread acceptance among engineers and scientists.

Current research and application areas include optical TDR for testing optical fibers,[48,49] Josephson junction samplers,[50,51] and lower cost TDANA.[6] There has been some effort to extend time-domain techniques above 20 GHz using existing sampling heads and heterodyne methods.[52] Computer programs are available to translate swept-frequency ANA results into the time domain and translate a windowed portion of the time-domain response back into the frequency domain.[53,29] Recently, microprocessor based instrumentation has been introduced to perform these translations in real time.[30]

Not to appreciate the information available directly in the time domain or transformed from the frequency domain is like covering one eye and losing perspective. When engineers realize the advantages of time-domain techniques in testing very-high-speed integrated circuits, designing complex optical components, or investigating electromagnetic interference problems, the TDANA will find many more applications.

REFERENCES

1. Nicolson, A. M. 1968. Broadband microwave transmission characteristics from a single measurement of the transient response. *IEEE Trans. Instrum. Meas.* IM-17:395–402.
2. Nicolson, A. M.; Bennett, C. L.; Lamensdorf, D.; and Susman, L. 1972. Applications of time domain metrology to the automation of broadband microwave measurement. *IEEE Trans. Microwave Theory Tech.* MTT-20:3–9.
3. Gans, W. L., and Andrews, J. R. 1975. Time domain network analyzer for measurement of RF and microwave components. *NBS Tech. Note 672,* NBS Boulder, CO.
4. Andrews, J. R. 1978. Automatic network measurements in the time domain. *Proc. IEEE* 66:414–423.
5. Hannaford, D. 1974. The application of picosecond pulses in time domain metrology. *Proc. 1974 Joint Measurement Conf.,* NBS, Gaithersburg, MD.
6. Rigg, P. R., and Carroll, J. E. 1980. Low-cost computer-based time-domain microwave network analyzer. *IEE Proc.* 127, Pt. H, No. 2:107–111.
7. Cronson, H. M.; Nicolson, A. M.; and Mitchell, P. G. 1974. Extensions of time domain metrology above 10 GHz to materials measurements. *IEEE Trans. Instrum. Meas.* IM-23:463–468.

8. Nicolson, A. M.; Cronson, H. M.; and Mitchell, P. G. 1976. Subnanosecond risetime pulse generators. *IEEE Trans. Instrum. Meas.* IM-25:104–107.
9. Cronson, H. M. 1977. Generation and use of composite subnanosecond waveforms for measurements from 0.4 to 18 GHz. 7th European Microwave Conference, Copenhagen, pp. 566–568.
10. Andrews, J. R. Fast pulsers—small and large. Chapter 4 of this volume.
11. Nicolson, A. M. 1969. Wideband system function analyzer employing time to frequency domain translation. WESCON Convention, session 22, San Francisco, CA.
12. Brigham, E. O. 1974. *The Fast Fourier Transform.* Englewood Cliffs, NJ: Prentice-Hall.
13. Samulon, H. A. 1951. Spectrum analysis of transient response curves. *Proc. IRE* 39:175–186.
14. Loeb, H. W. 1972. Efficient data transformation in time-domain spectrometry. *IEEE Trans. Instrum. Meas.* IM-21:166–168.
15. Bergland, G. D. 1969. A guided tour of the fast Fourier transform. *IEEE Spectrum,* July, pp. 41–52. (Also see other articles in the same issue.)
16. Oliver, B. M. 1964. Time domain reflectometry. *Hewlett-Packard Journal* 15.
17. Time domain reflectometry. Hewlett-Packard. 1964. *Application Note 62.*
18. Davis, F., and Loeb, H. W. 1965. Time-domain measurements for transistor and network characterization up to 1 Gc. *Proc. IEEE* 53:1649–1650.
19. Nicolson, A. M., and G. F. Ross. 1972. System to measure the frequency-domain response of a radar component. U.S. Patent No. 3,634,755.
20. Farber, A. S., and Ho, C. W. 1969. Wide-band network characterization by Fourier transformation of time-domain measurements. *IEEE J. of Solid-State Circuits* SC-4:231–235.
21. Cronson, H. M., and Ross, G. F. 1972. Current status of time-domain metrology in material and distributed network research. *IEEE Trans. Instrum. Meas.* IM-21:495–500.
22. Cronson, H. M., and Mitchell, P. G. 1973. Time domain measurements of microwave components. *IEEE Trans. Instrum. Meas.* IM-22:320–325.
23. Cronson, H. M. 1975. TDM: An alternate approach to microwave measurements. *Microwaves,* Dec., pp. 40–45.
24. Cronson, H. M. 1976 A wideband directional coupler for time-domain measurements. *IEEE Trans. Instrum. Meas.* IM-25:15–17.
25. Adam, S. F. 1968. A new precision automatic-microwave measurement system. *IEEE Trans. Instrum. Meas.* IM-17:308–313.
26. Mendel, E. 1983. The systems approach to scalar automatic network analyzers. *Microwave J.,* April, pp. 49–70.
27. King, R. J. 1978. *Microwave homodyne systems.* Peter Peregrinus Ltd., England.
28. Hoer, C. A. 1979. Performance of a dual six-port automatic network analyzer. *IEEE Trans. Microwave Theory Tech.* MTT-27:993–998.
29. Stinehelfer, Sr., H. E. 1982. Deembedding the capacitance of a resonant circuit using time-domain reversal and subtraction. *1982 IEEE MTT-S International Microwave Symposium Digest,* pp. 354–356.
30. ———. 1984. Real-time measurements in a wideband network analyzer. *Microwave J.* 27:138–142.
31. Nicolson, A. M. and Ross, G. F. 1970. The measurement of the intrinsic properties of materials by time domain techniques. *IEEE Trans. Instrum. Meas.* IM-19: pp. 377–382.
32. Ross, G. F. Early developments and motivations for time-domain measurements, Chapter 1 of this volume.
33. Krueger, C. H., and Carter, L. E. 1974. Measurement of the high temperature electrical properties of ferrites using time-domain techniques. Conf. Precision Electromagnetic Meas., London.
34. van Gemert, M. J. C. 1973. High frequency time-domain methods in dielectric spectroscopy. *Philips Res. Repts.* 28: pp. 530–572.

35. Fel'dman, Yu. D.; Zuev, Yu. F.; and Valetov, V. M. 1979. Time domain spectroscopy of dielectrics. Translated in *Instrum. and Exp. Tech.* 22, No. 3, Pt. 1:611–29.
36. Kaatze, V., and Geise, K. 1980. Dielectric relaxation spectroscopy of liquids: Frequency domain and time domain experimental methods. *J. Phys. E: Sci. Instrum.* 13:133–141.
37. Fellner-Feldegg, H. 1969. The measurement of dielectrics in the time domain. *J. Phys. Chem.* 73:616–623.
38. Bagozzi, R. P. 1969. Pulse response of polar-liquid filled coaxial lines. M. S. Thesis, Dept. Elec. Eng., Univ. of Colorado, Boulder, CO.
39. Bagozzi, R. P.; Ives, W. R.; and Nahman, N. S. 1969. Determination of the dielectric relaxation time in a Debye binary liquid by pulse measurements. Int. Union Radio Sci. (URSI) 16th Gen. Assembly (Sessions 1–4: New Appl. Audio Meas.), Ottawa, Canada; also, *Prog. Radio Sci.* 2:257–265, 1966–1969; Int. Union Radio Sci., Brussels, Belgium, 1971.
40. van Gemert, M. J. C., and De Graan, J. G. 1972. Calculations on the time domain reflection behavior of a transmission line partly filled with a polar dielectric. *Appl. Sci. Res.* 26:1–17.
41. van Gemert, M. J. C. 1972. Time domain reflectometry as a method for the examination of dielectric relaxation phenomena in polar liquids. Doctoral thesis, University of Leiden.
42. Suggett, A.; Mackness, P. A.; and Tait, M. J. 1970. Dielectric relaxation studies by time domain spectroscopy. *Nature* 228:456–457.
43. Loeb, H. W.; Young, G. M.; Quickenden, P. A.; and Suggett, A. 1971. New methods for measurement of complex permittivity up to 13 GHz and their application to the study of dielectric relaxation of polar liquids including aqueous solutions. *Ber. Bunsen Ges.* 75:1155–1165.
44. Fellner-Feldegg, H. 1972. A thin-sample method for the measurement of permeability, permittivity, and conductivity in the frequency and time domain. *J. Phys. Chem.* 76:2116–2112.
45. DeLoor, G. P.; van Gemert, M. J. C.; and Gravesteyn, H. 1973. Measurement of dielectric permittivity with time domain reflectometry: Extension of the method to lower frequencies. *Chem. Phys. Lett.* 18:295–299.
46. Iskander, M. F., and Stuchly, S. S. 1972. A time domain technique for measurement of the dielectric properties of biological substances. *IEEE Trans. Instrum. Meas.* IM-21:425–429.
47. Stuchly, M. A., and Stuchly, S. S. 1980. Coaxial line reflection methods for measuring dielectric properties of biological substances at radio and microwave frequencies—A review. *IEEE Trans. Instrum. Meas.* IM-29:176–183.
48. Personick, S. D. 1977. Photon probe—An optical-fiber time-domain reflectometer. *Bell Syst. Tech. J.* 56:355–366.
49. Fontaine, J. J.; Diels, J.; Wang, C.; and Sallaba, H. 1981. Subpicosecond time domain reflectometry. *Opt. Lett.* 6:405–407.
50. Hamilton, C. A.; Lloyd, F. L.; Peterson, R. L.; and Andrews, J. R. 1979. A super-conducting sampler for Josephson logic circuits. *Appl. Phys. Lett.* 35:718–719.
51. Sampler resolves 8.5-ps rise times. 1983. *Electronics* January 27, pp. 49–50.
52. Nahman, N. S.; Allred, C. M.; Andrews, J. R.; Hoer, C. A.; and Lawton, R. A. 1979. Radar absorber measurement techniques at frequencies above 20 GHz. *NBS IR 79-1613*.
53. Hines, M. E., and Stinehelfer, Jr., H. E. 1974. Time domain oscillographic microwave network analysis using frequency domain data. *IEEE Trans. Microwave Theory Tech.* MTT-22:276–282.

11 Measurement of Lightning Waveforms

J. E. Nanevicz and E. F. Vance

SRI International

11.0 BACKGROUND

Natural lightning produces one of nature's largest electromagnetic transients. Although its electromagnetic properties have been an academic curiosity since Franklin's time, it took the development of the electric power, telephone, and telegraph industries for engineering interest in measuring them to grow. Protection of the large overhead line networks of these systems required a knowledge of the surge delivered by a direct lightning strike.

The need for information about lightning motivated specialists in natural atmospheric electrical processes. These atmospheric electricians pondered the processes that lead to charge separation and buildup in the atmosphere and studied the details of lightning initiation and discharge processes. Throughout the history of lightning research, the measurement of electromagnetic transients produced by these processes has set the pace of scientific development in the discipline.

Early experience with the damage that lightning could do to electrical and communication equipment motivated the development of protective devices, such as lightning arresters and guard wires. It was thus necessary to know the energy, charge transfer, and other basic properties of the lightning discharge that would affect the design of these devices. This interest was reflected in the emphasis put on the high-current return stroke and follow-on currents that deliver large energies and transfer great quantities of charge. Much less attention was given to the fast processes associated with lightning.

The advent of wireless communications sparked interest in the effects of lightning processes on the radio-frequency spectrum. Atmospheric noise and whistlers produced by lightning discharges impose a limit on communication in parts of the spectrum and are a significant element in the background noise in much of the radio spectrum from VLF to UHF. Although this interference is mostly associated with the fast processes in the discharge, almost all of the data on it have been obtained with narrowband receivers instead of wideband transient measurements. This occasionally leads to some confusion about the true nature of the lightning transients.

Current Interest in Lightning Transients

The coming-of-age of digital electronics, sophisticated integrated circuits, and the advances in nonmetallic materials for use in aircraft and rockets have been factors in a heightened concern about the hazards of lightning. In aviation, an increased dependence on avionics for flight control and navigation has been accompanied by a reduction in electromagnetic protection as the all-metal construction is succeeded by insulating or poorly conducting composites. The avionics, which once contained only narrowband analog circuits, now use mostly digital circuits, which are more susceptible to transients than to the spectral amplitude.

Likewise, the increasing use of microprocessors in industry and in the home have pressed the makers of this equipment—and, hence, the atmospheric electricians—to develop a better understanding of lightning as a transient interference source. Whereas past concern has been for the direct damage that a stroke can produce in the form of melting metal and exploding timbers, etc., the present concern is for the transient that is induced deep inside structures that are electromagnetically very complicated. Because digital circuits are inherently broadband and because coupling to them frequently depends on mutual inductance and capacitance terms such as $M di/dt$ and $C dv/dt$, the early-time or fast-changing part of the lightning-produced transient is of primary interest.

In addition, for airborne systems, the properties of lightning at flight altitudes are important. Almost all of the direct measurements of lightning have been done on the ground and on cloud-to-ground strokes. Most of the strikes on aircraft occur at altitudes of several thousand feet, however, and are believed to be intracloud strokes. There is also speculation that the presence of the aircraft may affect the properties of the lightning stroke. Thus, there is a desire to make direct measurements of the intracloud stroke properties and to acquire a better understanding of the stroke initiation processes at flight altitudes.

Finally, there is a growing interest in developing effective lightning location and warning systems that can be used to guide aircraft away from active lightning threats, give accurate data on the progress of lightning activity in the vicinity of fuel and explosive operations, and locate lightning-initiated forest fires before they develop into major fires. Thus, there is a need for better association between lightning activity and the remote observations of that activity. Much progress in the area of locating and tracking lightning activity has been made in the last few years.

Special Problems in Lightning Measurements

The measurement of the lightning-caused transients poses some unique challenges to the experimenter. A lightning strike is a rare event even in lightning-prone areas; and about 0.1 strokes per year occur inside a 100-m diameter circle in the vicinity of Orlando, Florida. Thus, the instruments for measuring lightning directly may have to wait years to capture an event that lasts less than a second and has important details lasting only a few nanoseconds.

In the typical laboratory experiment, a trigger signal is provided to gate the recorders on prior to the arrival of the transient. Natural lightning recorders, however, must be self-triggered or continuously recording.

Because of the rarity of direct strikes to average terrain, most lightning data are obtained from remote measurements or from direct strikes to tall structures. Many early measurements were made on the Empire State Building in New York, whose height causes it to be struck several times per year. Recent experiments have triggered lightning by launching rockets that reel out a grounded wire into promising thunderstorm areas.[1–6] Although the data-acquisition rate is much greater in these experiments, it has not been determined whether the characteristics of strokes to tall structures are the same as those of strokes to average terrain; there is speculation that there may be differences, particularly in the fast and early-time properties.

The data-acquisition rate can also be enhanced by remotely observing all strokes within a radius of, say, 10 kilometers of the sensor. Remote measurements of lightning properties usually depend on measuring the electric or magnetic fields (or both) and inferring the stroke current and other properties from these. To interpret remote measurements, it is necessary to know the location of the sensor relative to the stroke. In early experiments, photographic and acoustic data were used to help determine the location of the stroke. More sophisticated lightning location systems are now available that use time-of-arrival of key parts of the signal at two or more receivers to locate the source.

The interpretation of remote measurements also requires an understanding of the effects of propagation between the source and the sensor, as well as the ability to associate the observed response with the part of the source producing it. Thus, for example, the effect on the waveshape of propagating several kilometers over soil of varying (and often unknown) electromagnetic properties may be important to the interpretation of measurements made near the ground. The effects of the airframe on the waveform must be determined to interpret airborne measurements of lightning effects. Many of the techniques described in other chapters may be used when adequate data on the terrain or airframe are available.

Associating an observed effect with a particular part of the lightning discharge is perhaps the most confusing aspect of lightning research. Lightning is a very complicated phenomenon, parts of which occur over a very large volume of the thunderstorm region. The spatial separation is illustrated in Fig. 11-1, where the locations of different sources of VHF energy in a three-stroke flash have been plotted.[7] Each such source is separated in both space and time in the lightning flash; this fact is illustrated in Fig. 11-2, where the individual responses in receivers of various frequencies are plotted.[8] Hence, great care is required to scale responses to a common distance from the stroke and to relate measurements with narrowband receivers to time-domain waveforms.

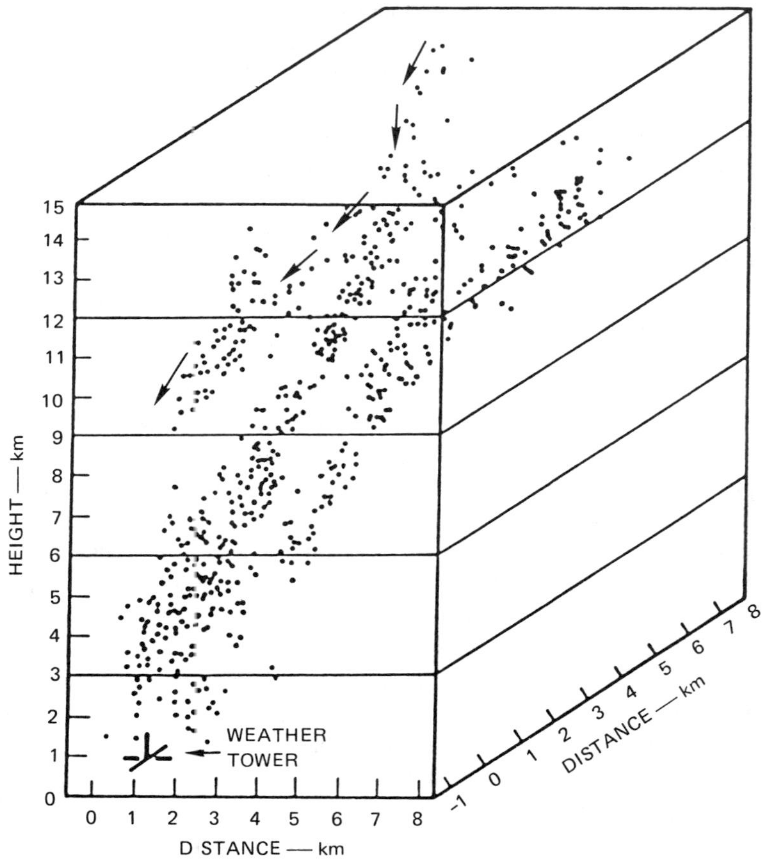

NOTE: Arrows indicate direction of propagation of groups of sources which occured in bursts

SOURCE: Reference 7

Fig. 11-1. VHF-noise-source locations during the three-stroke lightning flash to the 150-m weather tower at KSC on July 19, 1976 at 16:59:59.

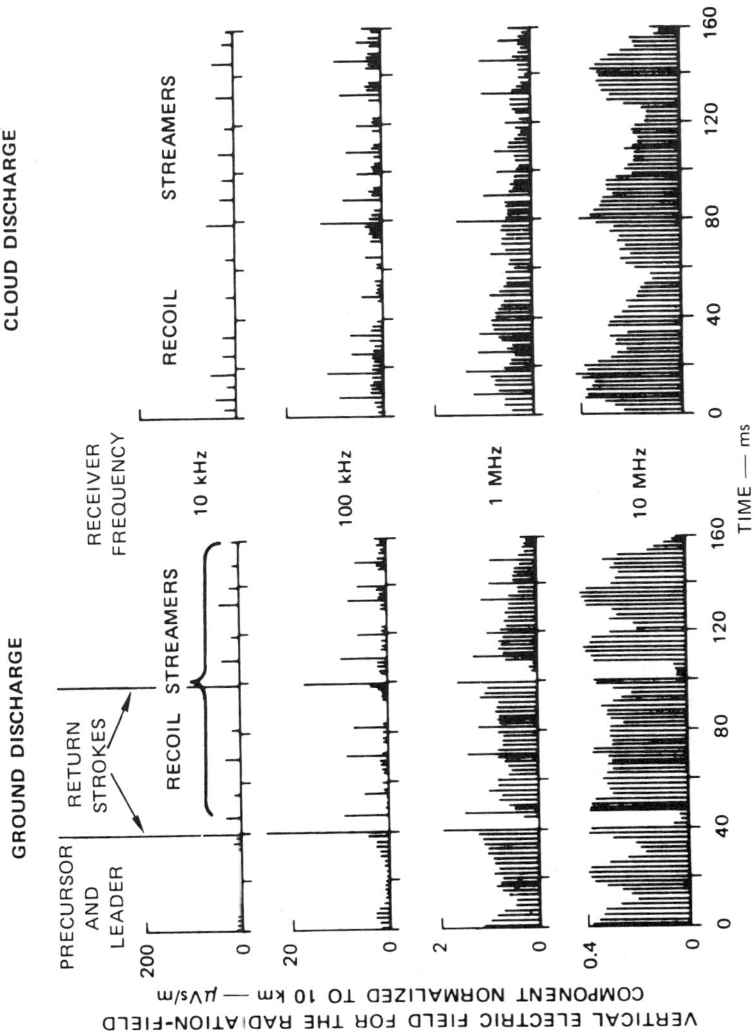

SOURCE: Reference 8

Fig. 11-2. Structure (illustrative) of the fields radiated by lightning as a function of time and frequency.

Since a single lightning flash appears to consist of many discrete discharges, each of which may contain rapidly changing currents and fields and since perhaps as many as 10,000 of them occur in less than one second, the recording task is challenging. Given writing speeds fast enough to record a 100-ns change on 0.25 cm of film, the record of a one-second flash would occupy 25 km of film! Since even raster scopes or multiple traces on slowly moving film cannot provide a record of this effective length, an entire flash has never been recorded. Moreover, whenever less than an entire flash is recorded, there is frequently some uncertainty about what part has been observed.

Obviously, the recording system must be triggered on the lightning transient to capture even part of the flash; continuous recording at the rate suggested above is not feasible. To avoid too many false triggers on locally produced transients, the trigger level must be set above the ambient transient level. For ordinary oscilloscopes, the leading edge of the transient will always be lost unless a large delay is made in the signal line.

For some digitizing oscilloscopes, the trigger freezes the information that is in storage regardless of whether it arrived before or after the trigger; these instruments can be used to capture the leading edge of the transient even though the trigger level is reached well after the beginning of the transient.

Careful experimenters use several different sensing and recording schemes with overlapping capabilities to avoid some of the uncertainties inherent in individual techniques. Thus, for example, electrostatic fields and narrowband RF measurements may accompany broadband transient measurements so that the observed responses may be related to each other and to the events in the flash.

Measurement of Lightning Fields

The fields produced by lightning and associated thunderstorm processes vary temporally from the near-static fields of the separated charge to nanosecond changes associated with short sparks and leaders that occur in and about the charged cloud and stepped leaders. The peak amplitude of the stroke current and the electric and magnetic fields can vary over two decades from flash to flash. Thus, important characteristics of sensors for measuring lightning fields are their bandwidth and dynamic range. Small electric-dipole antennas are commonly used to measure the electric fields of lightning, as subsequently discussed. See Chap. 5 for a more general discussion on transient sensors.

11.1 MEASUREMENT OF FIELDS

For the measurement of transient fields produced by lightning, a sensor that does not display resonances or other irregular behavior in the lightning spectrum is desired. Electric and magnetic dipoles that are small compared to the shortest wavelength in the lightning spectrum of interest are usually used for these measurements. These simple antennas have no natural resonances in that part of the frequency spectrum where they are electrically small. Because they are small, however, these antennas are also insensitive. Fortunately, the fields produced by nearby lightning are quite large so that the limited antenna sensitivity can often be overcome.

As we shall discuss in greater detail later, the small electric dipole has an open-circuit terminal voltage that is proportional to the electric field impressed on the antenna and independent of frequency. Thus, in principle, this antenna could be used throughout the spectrum from the highest frequency at which it is electrically small down to dc. At low frequencies, however, it is very difficult to measure the open-circuit voltage induced in a small dipole because the impedance of the dipole becomes very large. Thus, almost any instrument for measuring voltage will load the antenna so much that the open-circuit conditions are no longer approximated. Hence, for static and very low-frequency electric fields, an electric field mill is used. The field mill, which mechanically chops the electrostatic field to convert it to an alternating signal, will also be described later.

Small Electric Dipoles

The small electric dipole employed for lightning measurements is almost always used with a ground plane, as illustrated in Fig. 11-3. The shape of the antenna element is somewhat arbitrary insofar as the antenna's response to the electric field is concerned, but the shape does affect the tractability of antenna analysis, the mechanical design, and the source impedance of the antenna.[9] The linear antenna on the left of Fig. 11-3 has the smallest source capacitance. Since it also protrudes from the surface, it would be unsuitable for use in airborne experiments where aerodynamic drag is a consideration. On the other hand, the flush-plate antenna on the right of Fig. 11-3 can be built into the skin of an aircraft or rocket in such a way that it adds no drag.

The conical dipole is shaped so that it resembles a constant-impedance, conical

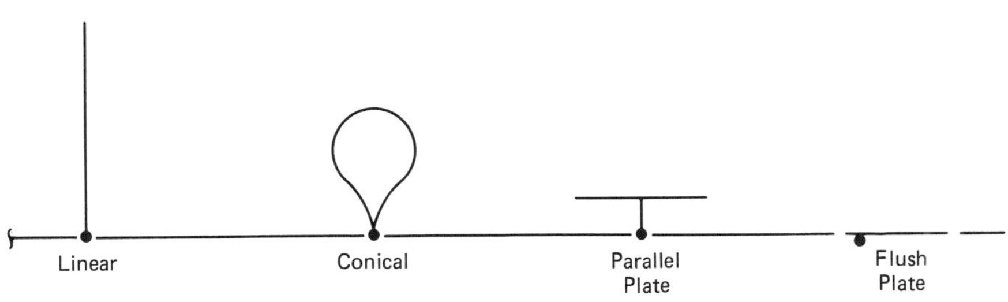

Fig. 11-3. Small electric dipoles.

transmission line near the feedpoint, and its overall contour is that of an equipotential surface about an elementary linear dipole.[9–10] It can be used in high fields without inducing corona from its upper end. The parallel plate dipole has the largest source capacitance of the antennas shown, but this has been achieved with a sacrifice in sensitivity.

The open-circuit voltage at the terminals of a linear dipole immersed in a uniform electric field is given by

$$V = h_e E \tag{11-1}$$

where h_e is the effective height of the antenna and E is the electric field strength (in the absence of the antenna). This equation may be taken as the definition of effective height. As illustrated in Fig. 11-4, the effective height is the height of the equipotential containing the dipole surface when the terminals are open. Equation 11-1 is applicable to either the frequency or time domain, subject to the size restrictions on the antenna.

The antenna impedance is an almost pure capacitive-reactance at frequencies where the antenna is electrically small. Hence, its short-circuit current is

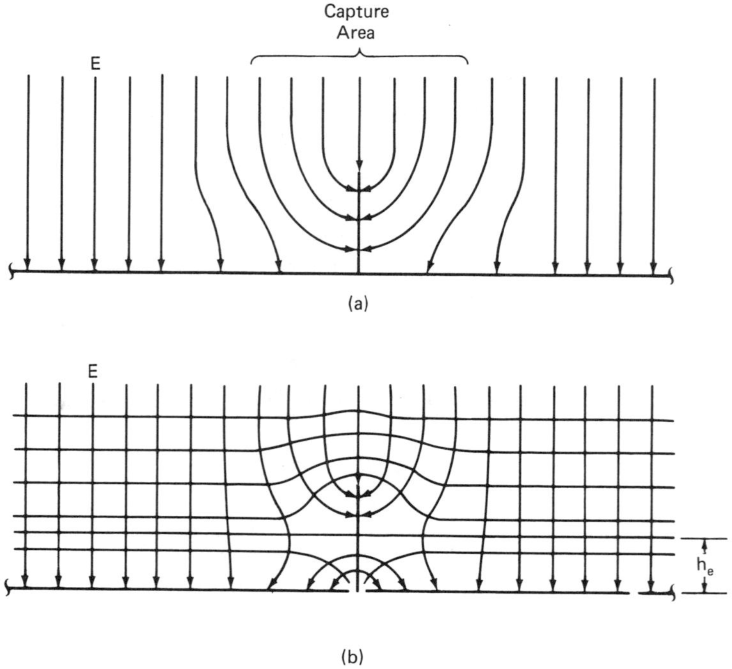

Fig. 11-4. Effective height and capture area for a linear dipole: (a) terminals shorted, and (b) terminals open.

$$I = Ch_e \frac{dE}{dt} \tag{11-2}$$

where C is the antenna capacitance. An alternate expression for the short-circuit current is suggested by Fig. 11-4(a). The electric field within the capture area depicted terminates on the antenna element and induces a charge, $\varepsilon_0 EA$, on the element, where A is the capture area and ε_0 is the permittivity of the medium. The rate of change of this charge is the short-circuit current,

$$I = \varepsilon_0 A \frac{dE}{dt} \tag{11-3}$$

From a comparison of Eqs. 11-2 and 11-3, it is apparent that

$$\varepsilon_0 A = h_e C \tag{11-4}$$

For linear antennas, h_e is about half the actual height of the antenna, and, for flush-plate antennas with small isolation gaps, the capture area is about equal to the area of the plate.

It is apparent from Eqs. 11-1 to 11-3 that the response of the small dipole can be proportional to the electric field, the derivative of the field, or perhaps neither. To measure electric field directly, the open-circuit voltage must be measured; that is, the measuring instrument must not significantly discharge the antenna capacitance during the lightning transient, and the time constant of the antenna capacitance with the instrument load must be longer than the duration of the transient field.

If a coaxial cable is used for carrying the signal from the antenna to the recording instruments, as illustrated in Fig. 11-5(a), the time constant of the measurement system is Z_cC, where Z_c is the characteristic impedance of the cable; it is assumed that the cable is terminated in its characteristic impedance at the recorder. For 50-ohm cable and an antenna capacitance of 50 pF, the antenna terminal voltage will be proportional to the field for times short in comparison to 2.5 ns; it will be proportional to the derivative of the field for times long in comparison to 2.5 ns. Thus, it seems that the simplest measurement to make with a small dipole and laboratory instruments is the time derivative of the electric field.

To increase the time constant so the electric field can be recorded directly, it will be necessary either to increase the capacitance of the antenna, as illustrated in Fig. 11-5(b), or the resistance of the load, as illustrated in Fig. 11-5(c), or both. Increasing the capacitance by loading the terminals with additional capacitance reduces the sensitivity of the antenna in the same proportion as the time

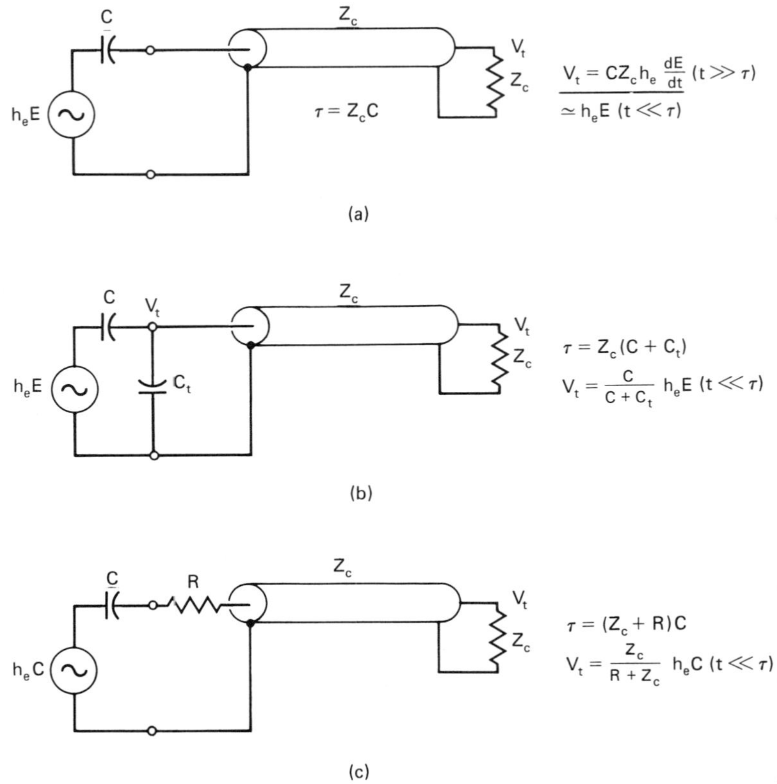

Fig. 11-5. Antennas with passive loading: (a) use of coaxial cable for carrying signal, (b) capacitance loading, and (c) resistance loading.

constant is increased, provided the same cable impedance is retained (the range of cable impedance available is quite limited). Increasing the resistance of the load reduces the sensitivity of the antenna in a similar way, as illustrated in Fig. 11-5(c).

If we wished to measure the fields of a lightning transient lasting 100 μs, we would need to increase the time constant by about 500,000, which would reduce the signal by 500,000 if either of these methods were used. This sacrifice in sensitivity may be acceptable if the fields being measured are very large, as they are near a lightning flash. For remote measurements, however, the loss in sensitivity may not be acceptable. For these measurements, a high-input-impedance amplifier, such as a FET follower, can be used to achieve a large time constant without a corresponding loss in sensitivity, as shown in Fig. 11-6. Time constants of several ms can be achieved with high-feedback amplifiers and some shunt capacitance across the antenna terminals.

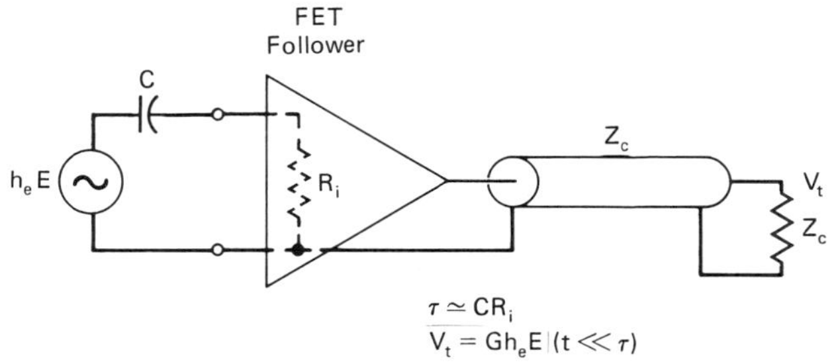

Fig. 11-6. Active amplifier at antenna terminals for direct measurement of fields.

An inexpensive direct measurement of fields can be made using a plate antenna and a passive oscilloscope probe, as illustrated in Fig. 11-7. The plate antenna can be fabricated from sheet metal, with plastic film serving as the dielectric that separates the antenna from the ground plane. With a passive 10-Mohm probe attached directly to the plate, the transient can be recorded directly on the oscilloscope. A one-square-meter antenna with 10 mils (0.25 mm) of solid insulation would have a capacitance of about 80 nF and also a time constant of about 80 ms when loaded with the 10-Mohm probe. Although the 10-Mohm probe typically attenuates the signal by a factor of 10 and the capacitance of the antenna plate reduces the terminal voltage, this method is nevertheless very useful for the measurement of large fields.

Small Magnetic Dipoles

The small magnetic dipole is a dual of the electric dipole; the short-circuit current induced in a perfectly conducting, electrically small loop is proportional to the magnetic-field strength. There are practical differences that make the design of

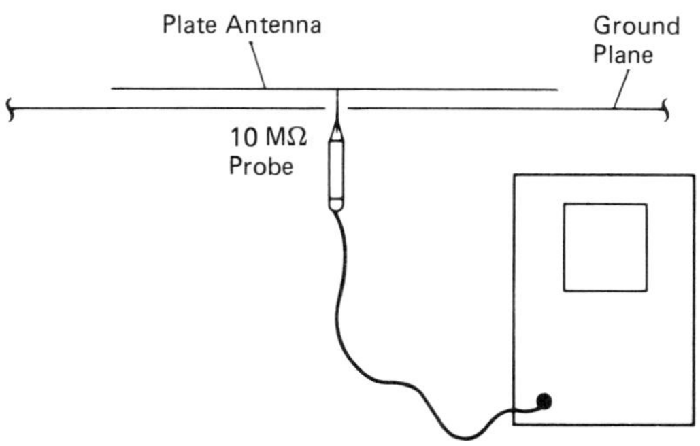

Fig. 11-7. Easy-to-make antenna for direct measurement of electric fields.

wideband loops somewhat more difficult, however, than the design of wideband electric field antennas.

Consider first a single-term loop of area A that is immersed in a uniform magnetic field H, as illustrated in Fig. 11-8. The open-circuit voltage induced in the loop is

$$V = \mu_0 A \frac{dH}{dt} \tag{11-5}$$

and the short-circuit current of a perfectly conducting loop is

$$I = \frac{\mu_0 A}{L} H \tag{11-6}$$

where L is the inductance of the loop given by

$$L \simeq \mu_0 r \left[\ln\left(\frac{8r}{a}\right) - 2 \right] \tag{11-7}$$

where r is the loop radius, and a is the conductor radius, when $r/a \gg 1$. Thus, the loop response can be proportional to either the magnetic field or its derivative.

It is often difficult to measure the magnetic field directly, however, because of the resistance that must be included in most practical loop designs. If the loop is used to drive a 50 ohm-cable, as illustrated in Fig. 11-9, the current through the 50 ohms will be proportional to the magnetic field only for times that are short compared to the time constant L/R. For single-turn loops suitable for measuring the fast-changing magnetic fields, this time constant may be only

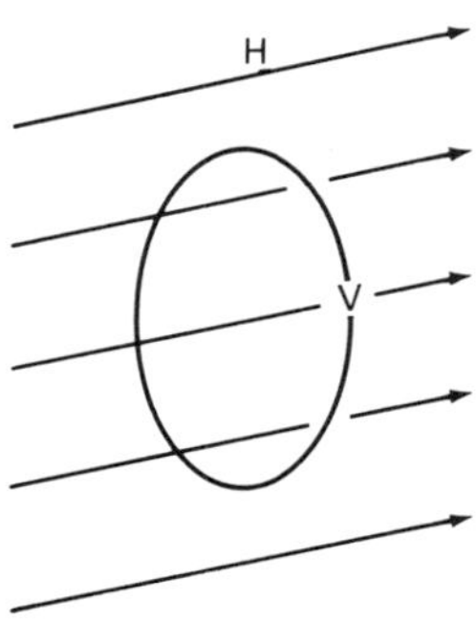

Fig. 11-8. Single-turn loop in magnetic field.

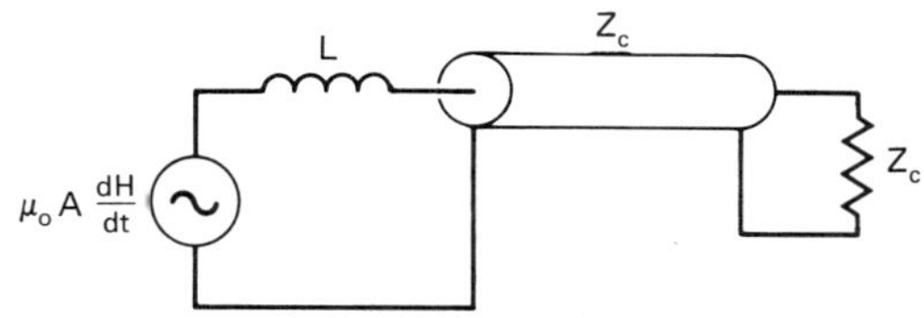

Fig. 11-9. Loop antenna loaded with a transmission line.

a few ns; for times longer than a few ns, the current and voltage on the 50-ohm load are proportional to the rate of change of the field.

To increase the time constant, L must be increased, R must be reduced, or both. L can be increased by making the loop larger and by adding more turns, both of which reduce the highest frequency at which it is electrically small. R can be reduced by using current probes with a low insertion resistance; these probes produce a voltage (across 50 ohms, typically) that is proportional to the current through them. The insertion resistance may be 1 ohm or less in commercially available RF probes. Hence, the time constant of single-turn loops can be increased to 1 to 100 μs by using current probes with a 1-m diameter loop. Very long time constants can be obtained with many turn loops, but these multiturn loops tend to have self-resonances that limit their usefulness at high frequencies.

Loop antennas also have current induced in them by the electric field. When the terminals are in the side of the loop that is perpendicular to the electric field, this current is a common-mode current and can be rejected by balanced circuits. Another option is to use shielded loops so that the current induced by the electric field flows on the shield, rather than in the internal loop. A half-loop version of the shielded loop is illustrated in Fig. 11-10.

The discussion of magnetic dipoles should also mention slot antennas. The electrically small slot, illustrated in Fig. 11-11(a), is particularly useful where the antenna must be flush-mounted to avoid aerodynamic drag on aircraft and rockets or to avoid damage and trip hazards in high traffic areas. The slot antenna is an interruption in the otherwise uniform current density induced on a con-

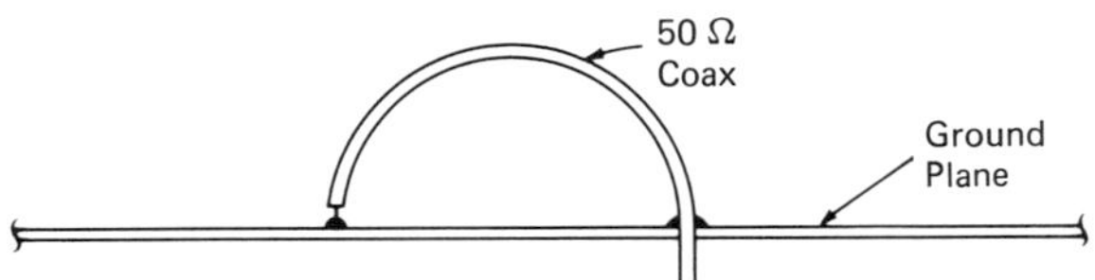

Fig. 11-10. Shielded half loop on ground plane.

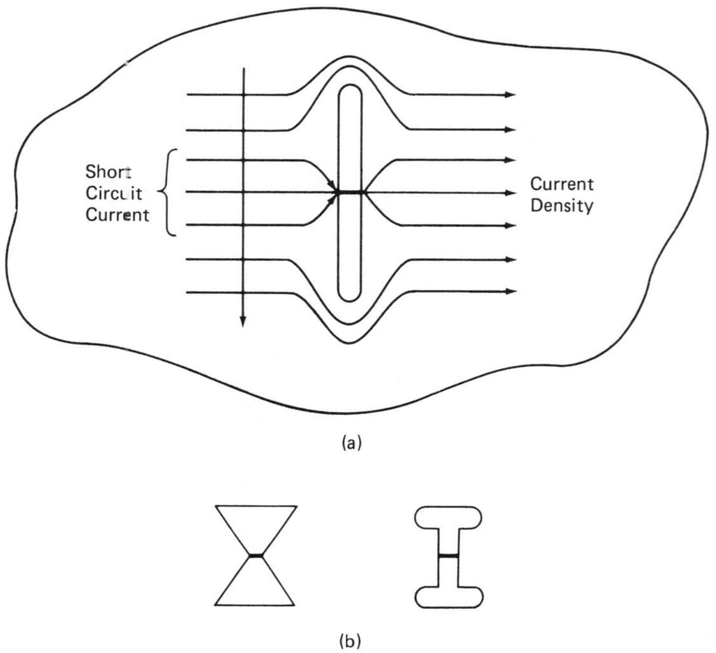

Fig. 11-11. Slot antennas in conducting sheet: (a) simple slot, and (b) top-loaded slot.

ducting plane by the magnetic field. The current flowing through a short circuit at the midpoint of the slot is

$$I \simeq l_e H \tag{11-8}$$

where l_e is the effective length of the slot. The open circuit voltage across the slot at this point is

$$V = L l_e \frac{dH}{dt} \tag{11-9}$$

The effective length of the long narrow slot is about half its actual length; that is, as illustrated in Fig. 11-11(a), about half the sheet current interrupted by the slot flows through the shorted terminals at the midpoint. The effective length can be increased by "top-loading" the slot as illustrated in Fig. 11-11(b).

Some experimenters prefer to measure the rate-of-change of the field rather than the field itself. The sensor design effort may then be concentrated on making the loop or slot maintain its derivative response at high-frequencies where ωL

is large. The high frequency performance is enhanced by making L small and the load resistance large. Baum and his associates[10] have developed an array of multigap loops that reduce the inductance by using a wide strip instead of a round conductor and that increase the load by using several terminals (gaps) about the circumference of the loop, each loaded with a transmission line. Thus, for example, the four-gap loop illustrated in Fig. 11-12 is loaded with four 100-ohm cables that are combined to give 100-ohm-balanced output terminals. Thus, a very low inductance and fairly large load resistance are achieved.

Electrostatic Fields

Because of the complexity of the lightning flash, it is useful to have multiple measurements to aid in interpreting the data. A very useful instrument for this is the electric-field mill. This device measures the electrostatic field and slowly varying electric fields. Typical risetimes for field mills are tenths to hundredths

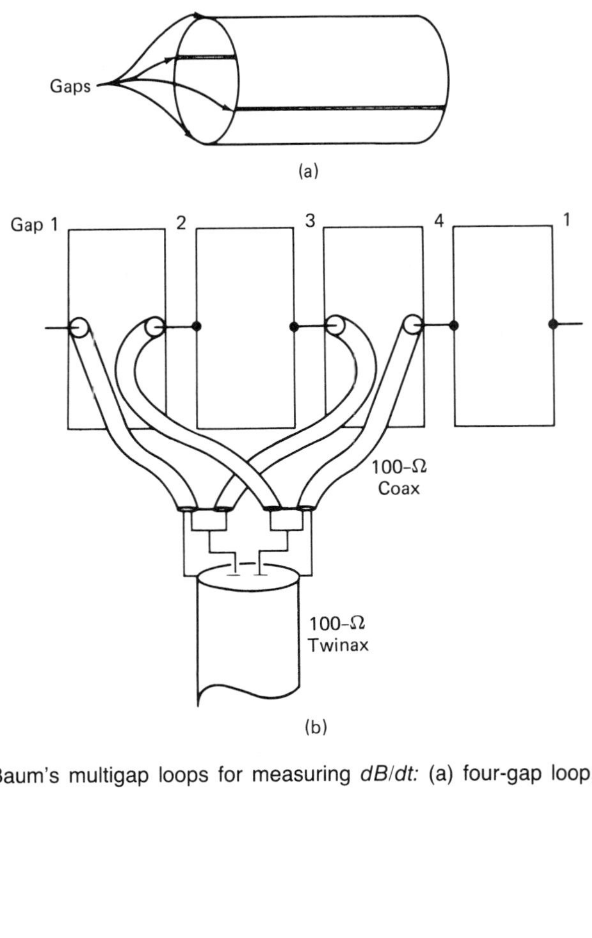

Fig. 11-12. Baum's multigap loops for measuring *dB/dt:* (a) four-gap loop, and (b) cable connectivity.

of a second. Nevertheless, because lightning is an electrostatic discharge, the field mill's indication of the buildup and collapse of the static field is very useful in interpreting the transient measurements obtained from electric and magnetic dipoles. On airborne measurements, an array of field mills can be used to determine the aircraft potential and the direction of the external static field before and after the flash. On the ground, the field mill is often used to measure the large field strengths that indicate a lightning flash is imminent or can be easily triggered.

The field mill is an electromechanical chopper that converts the electrostatic field into an alternating current. The most common type uses a rotating vane alternately to cover and uncover a stationary sensor plate, as illustrated in Fig. 11-13. Thus, charge is alternately induced on the sensor plate and on the grounded vane in accordance with the following formulas:

$$q_{pk} = \varepsilon_0 EA \qquad \text{(sensor uncovered)} \tag{11-10}$$

$$q = 0 \qquad \text{(sensor covered)} \tag{11-11}$$

where A is the area of the exposed sensor plate. For the rotating-vane instrument illustrated, $q(t)$ is a triangular wave alternating between 0 and q_{pk}, as illustrated

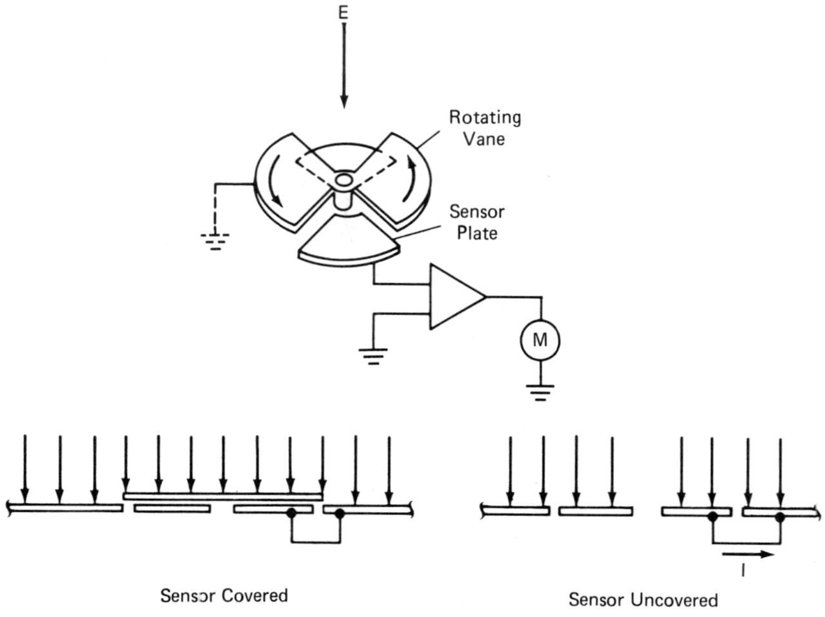

Fig. 11-13. Principle of operation of a field mill.

in Fig. 11-14(a). The short circuit current to the sensor plate is a square wave [see Fig. 11-14(b)] of amplitude

$$I = \frac{dq}{dt} = \pm 4\varepsilon_0 EAS \tag{11-12}$$

where S is the speed of rotation in revolutions per second. This current is amplified and detected to obtain a voltage or current proportional to the electrostatic field. Other chopper mechanisms used include oscillating shutters, rotating cylinders, and vibrating wires and reeds.

With synchronous detectors, the field mill can be used in an ionized medium to measure both static field and current density, since the current density is also chopped. The ion current variation is 90 degrees out of phase with the field-induced current variation so that they can be separated in the synchronous detection. Rotating-vane mills have been used to measure fields and ion-current density on space probes and ionospheric sounding rockets.

Direct Measurement vs. Data Processing

Since the electric and magnetic fields can be obtained from their derivatives and vice versa, one might question the necessity for measuring the field directly

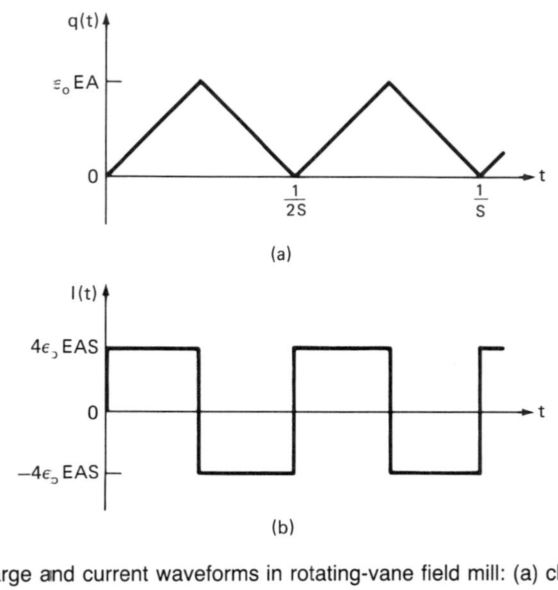

Fig. 11-14. Charge and current waveforms in rotating-vane field mill: (a) charge on sensor plate, and (b) short-circuit current.

when its derivative can be easily measured and integrated to obtain the field. It is important to recognize that any processing of measured data includes processing of the noise and uncertainty in the data. This is particularly important when integrating measured data; when the value of dE/dt or dH/dt is smaller than the ambient noise, the integration process merely integrates the noise. This effect is illustrated in Fig. 11-15.

Effects of Surroundings

Structures between the lightning discharge and the sensor also affect the sensor response. Imperfectly conducting soil causes attenuation and dispersion of the wave as it propagates between the source and the sensor. As shown in Fig. 11-16, the wave propagating over finitely conducting soil has a tangential electric field which is, in part, responsible for energy dissipated in the ground. The tangential electric field is given approximately by

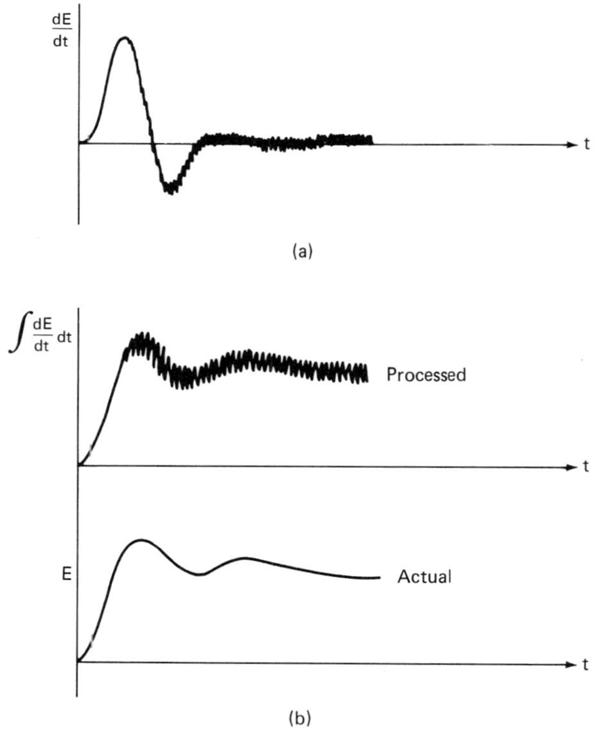

Fig. 11-15. Effect of processing data: (a) derivative, and (b) field.

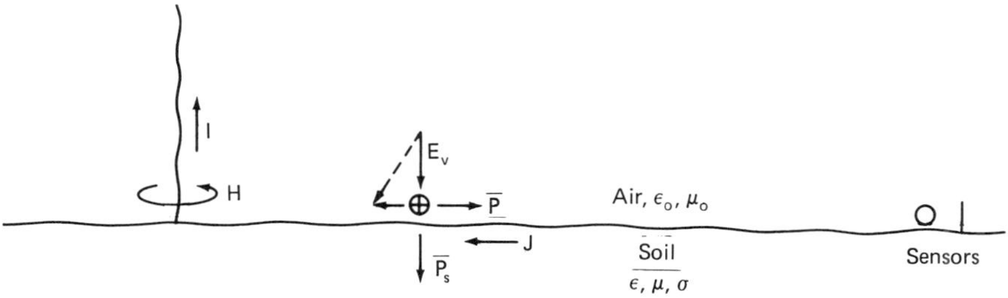

Fig. 11-16. Propagation of lightning transient over soil.

$$E_t \simeq H_t \frac{1 + i}{\sigma\delta} \tag{11-13}$$

where σ is the soil conductivity and δ is the skin depth in the soil given by the equation, $\delta = (\omega\mu\sigma/2)^{-1/2}$. This expression for the tangential electric field is accurate for a distance from the source that is large compared with δ.

The tangential electric field at the surface varies as $\omega^{1/2}$ through the skin depth, δ; the higher the frequency, the greater are E_t and the losses to the soil. Hence, a sensor at the surface some distance from the lightning channel receives a distorted pulse with the high-frequency spectrum suppressed. This may be one reason the risetime of lightning generated fields has recently been observed to be much faster than was earlier believed. Weidman and Krider have measured the fields near shore from lightning strokes offshore and found significantly faster risetimes than were observed on land.[11] Note that the conductivity of seawater is 100 to 1000 times that of soil.

The other major source of anomalies in lightning measurements is local structure near the sensors. Sensors mounted near tall structures such as towers or buildings will be affected by scattering from them. Similarly, sensors mounted on such structures may be affected by the currents and fields induced on the latter. This problem is particularly acute for airborne measurements since all of the antennas must respond to the surface fields induced on the airframe on which they are mounted. Isolation of the loop antennas from local currents by placing them at the ends of structural elements (e.g., wing tips) or on booms or drogues has been proposed as a means of obtaining undistorted magnetic field measurements.[12]

11.2 CURRENTS

Stroke-current measurements were among the earliest systematic measures of lightning properties. The current measurements were almost always made on tall structures such as skyscrapers, radio towers, and masts, most of them being made under the auspices of power and telecommunication groups. As a result, the use of current shunts and related techniques was common, and the emphasis was on energy and charge transfer rather than high-fidelity reproduction of the complete waveform. Much of the early lightning current data was obtained from magnetic links that record only the peak current.

Shunts are popular because they can be used to measure very large currents and because they can be constructed *in situ* on tower legs, building frames, etc. For example, the voltage drop between two points on a tower leg can be used as a measure of the current in the leg. Such shunts can be calibrated by discharging a known current through the leg. The difficulty with these shunts is that they do not accurately measure the fast changing parts of the current. The voltage induced in the measuring circuit by large, fast-changing currents depends strongly on the wire dress and routing. Relating the measured voltage to the current in the shunt is thus difficult, if not impossible.

To obtain high-fidelity measurements of currents in structures, measurement techniques similar to those used for electromagnetic fields are required. One approach is to use a loop antenna adjacent to the tower leg to measure the magnetic fields of the current. There are commercial current probes available that concentrate this field in a ferromagnetic core and have a sensor winding calibrated for current through the core and a standard transmission line impedance. These ferromagnetic core devices saturate with moderate charge transfers, however, so that they are not suitable for large-amplitude, long-lasting currents. Nevertheless, they are suitable for the fast-rate of change currents that are difficult to interpret from measurements using current shunts.

11.3 NARROWBAND MEASUREMENTS

Instrumentation and Rationale

Historically, atmospheric electricians have had to exercise ingenuity in devising schemes that would permit them to apply available but severely limited instrumentation to the study of lightning parameters. For example, until very recently, no instruments were available to record the high-frequency processes in the time waveform of the lightning current or its radiated signal. Accordingly, many experimenters were led to carry out measurements to study the high-frequency properties of lightning using narrowband measuring setups of the sort illustrated in Fig. 11-17. With this arrangement, it is only necessary to have receivers that can cover the frequency range included in the lightning spectrum. Limitations of receiver bandwidth and recorder response restrict the ability to identify single

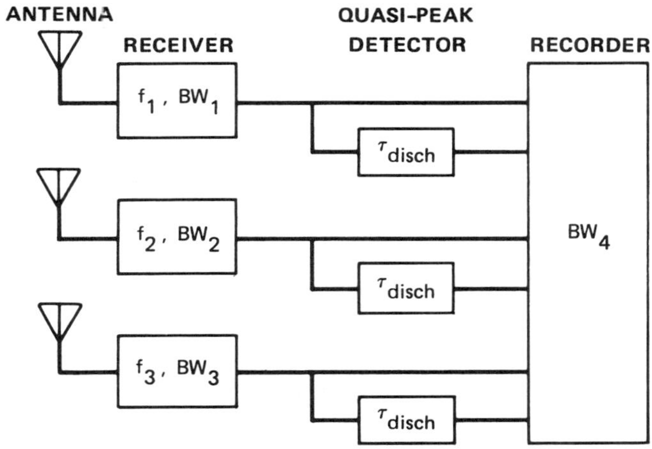

Fig. 11-17. Experimental arrangement used for measuring spectral properties of lightning.

events, but it is possible to make inferences regarding the existence of high-frequency processes (fast risetimes) and other important characteristics of the lightning flash.

A further practical consideration for early experimenters was that a need existed for data regarding lightning noise levels throughout the RF spectrum. Since narrowband communication receivers were the principal devices of concern, a narrowband measurement of noise spectral density was adequate.

Thus, over the years, measurements of the sort illustrated in Fig. 11-17 have been made by a number of experimenters using equipment with a variety of characteristics.[12–20] Varying degrees of attention were paid to locating the storm or the lightning events. In some cases, the center frequency of one or more of the receivers was switched during the course of a storm to permit more frequencies to be covered. The receiver output bandwidth varied with the experimenter, and some used quasi-peak detectors with a discharge time constant as long as 0.6 s.[14]

Characteristics of Data

Is is important to observe that narrowband measurements tend to yield a spectrum that is a composite of the processes in the lightning flash. If the frequency spectrum of a single input pulse is essentially flat over the passband (BW) of a narrowband receiver, the output signal will correspond to the receiver's impulse response. The impulse response of a tuned narrowband system is an exponentially decaying sinusoid (or its exponentially decaying envelope if an appropriate output envelope detector is included). The output decay time constant, τ, is given approximately by the equation,

$$\tau = Q/\pi f_0 \tag{11-14}$$

where $Q = f_0/BW$. Therefore, τ is given approximately by $1/\pi BW$. As an example, Horner's measurements[13,15] were made with receiver bandwidths of 250 Hz. Thus, the characteristic output decay time of his system, $\tau = 1/250\pi$, is given as approximately 1.3 ms. For an average lightning interpulse spacing of 50 μs, this means that 28 additional pulses have arrived and contributed energy (or been "stacked") before the effect of the first one has died out.

Frequently, lightning data must be applied to the study of the effects of lightning on wideband systems such as aircraft wiring. Since the "stacking" of pulses does not occur in such wideband systems, the fact that narrowband spectral data include the contributions from successive pulses injects complications into its application and interpretation for these purposes.

Oetzel collected the available narrowband data, adjusted as well as possible to a common bandwidth of 1 kHz and a lightning distance of 10 km, and plotted the resulting individual spectra.[21] Later, Cianos and Pierce added the results of additional measurements, but, instead of presenting the data as a collection of spectral curves generated by a variety of experimenters, they simply plotted the data points as shown in Fig. 11-18.[22] To unify the presentation, they plotted a $1/f$ line representing an empirical relationship between peak field strength and frequency derived earlier for the VLF range.[23]

It is evident from Fig. 11-18 that at frequencies above about 1 MHz there is a great spread in the data and that the empirical $1/f$ line generally lies along the upper bound of the narrowband data, whereas a $1/f^2$ line beginning at 1 MHz would lie along the lower bound. Cianos and Pierce, who clearly understood the complexities of lightning processes, cautioned the reader that the $1/f$ line ". . . is only an analytical tool and does not imply any physical justifications." Unfortunately, this line is now frequently taken, without the precaution, to represent the Cianos and Pierce perception of the lightning spectrum.

It is interesting to consider the data above 1 MHz in Fig. 11-18 more carefully to determine if there is a reason for the great data spread observed. Iwata's data,[16] which lie along the dashed $1/f^2$ line, were generated using receivers with bandwidths of 10 and 80 kHz (which imply characteristic times of 32 and 4 μs, respectively). Thus, his system response was sufficiently fast for significant pulse "stacking" to be precluded. As indicated earlier, Horner's system had a bandwidth of 250 Hz so that considerable pulse stacking would be expected and could drive his data upward to several times the single-pulse levels. Since Takagi's system[14] used a quasi-peak detector with a decay time of 0.6 s, his measurements constitute a composite of the entire flash. The great variability in Takagi's data probably

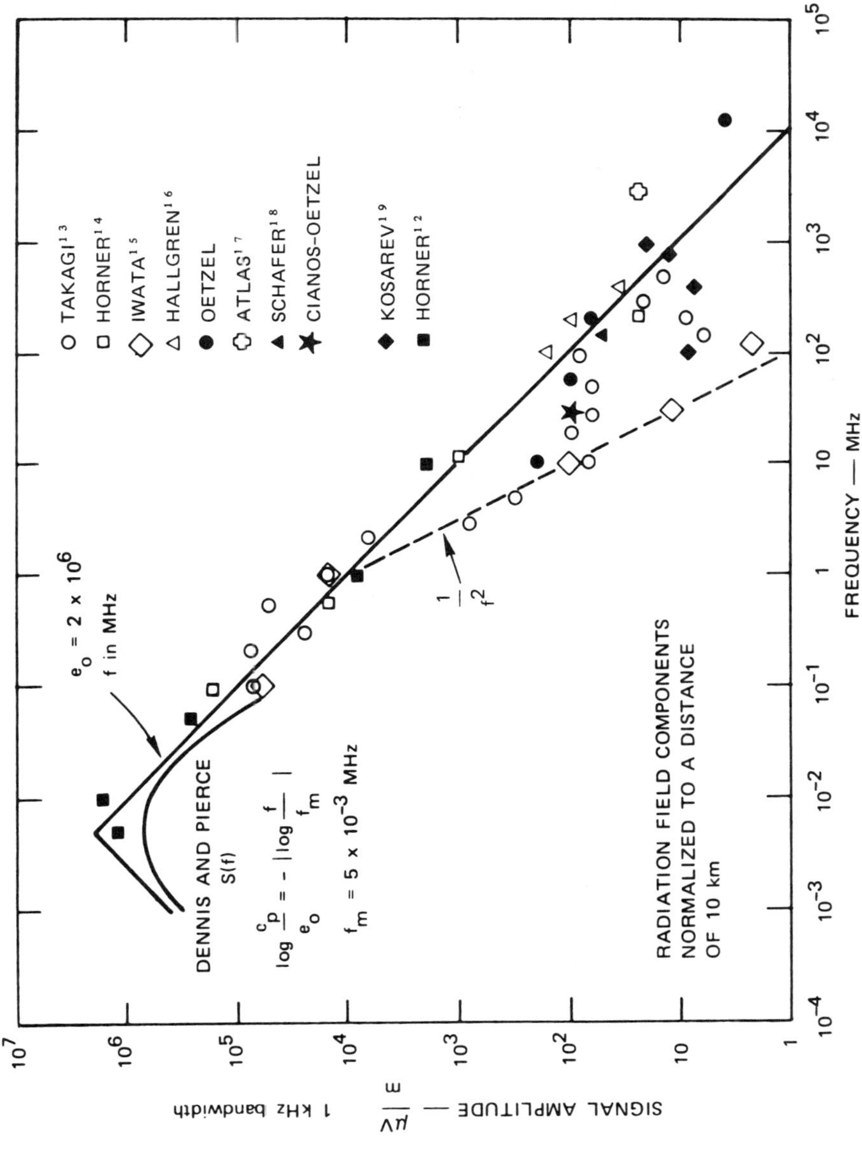

SOURCE: Reference 10

Fig. 11-18. Peak received amplitude for signals radiated by lightning.

stems from the fact that he had only two receivers available, which were shifted from frequency to frequency as the storm progressed.

Application of Narrowband Data to Wideband Systems

In conclusion, it appears that direct measurement of spectral density using multiple narrowband receivers can yield data valid for single-pulse analysis provided the receiver bandwidth is sufficiently great to prevent stacking. This approach was taken by Iwata, whose data in Fig. 11-18 lie along the dashed $1/f^2$ line and are also in excellent agreement with the recent spectral data of Weidman, Krider, and Uman, which were generated by taking Fast Fourier Transforms (FFT) of time waveforms of lightning-produced electromagnetic events.[24] If the bandwidth of the system of Fig. 11-17 is increased sufficiently, it becomes possible to distinguish many of the individual events in a lightning flash,[21] and the data assume the form shown earlier in Fig. 11-2.

11.4 LIGHTNING LOCATION AND WARNING

General

A lightning flash produces a wide variety of effects—electromagnetic, optical, acoustic, chemical, and physical. All of these have been used individually and in combination for the purpose of lightning location and warning. For each general class of effect, furthermore, special features may be identified and used for lightning study. For example, electromagnetic sensors have been designed to detect features ranging from the gross dc-field change accompanying a flash to the VHF/UHF radiation occurring during the flash. Each feature provides information about a different process in the overall flash. The one chosen for a given measurement is determined by the overall objectives of the measurement program.

Often, several effects are measured and used to gain insights not readily achievable from the study of one of them alone. For example, acoustic measurements were used by Few[25,26] to develop a picture of the spatial configuration of a lightning current channel inside a cloud not achievable at that time from either optical or electromagnetic techniques. More recently, Mazur[27] has used the scattering of 200-MHz radar signals from the free electrons in the lightning channel to determine that the NASA F-106 lightning-test aircraft triggered virtually all of the lightning strokes it experienced during the time its flights were observed with the radar.

Field-Change Measurements

One of the earliest electrical parameters measured in connection with lightning location studies is the electrostatic field change accompanying a lightning flash. (During the flash, a portion of the charge in the thunderstorm cell is neutralized by the current flowing in the strokes comprising the flash. This charge neutralization, in turn, causes the electrostatic field in the vicinity of the cell to be diminished.) Although a measurement of the truly static electric field either prior to or after the flash would require a rotating-vane field meter or a corona discharge probe, the field change may be measured using a simple electric dipole antenna and an appropriate amplifier. An indication of the approach of active storm cells may be obtained by noting the increase in the number and amplitude of field changes registered by the field-change detector.

In particular, flash (field-change) counters are usually designed to operate at frequencies below 1 kHz to take advantage of the fact that the electrostatic component of the field change varies with distance approximately as d^{-3}; for the radiation component, the variation is as $1/d$. As discussed by Pierce, this choice of response function has the desirable result of minimizing the effect of source variability in the uncertainties of distance estimation.[28]

The basic circuit first used by Pierce for counting and warning is shown in Figure 11-19. Simple solid-state versions are possible. In operation, capacitor C_1 is charged through R_1 with a time constant of about a second; this is ap-

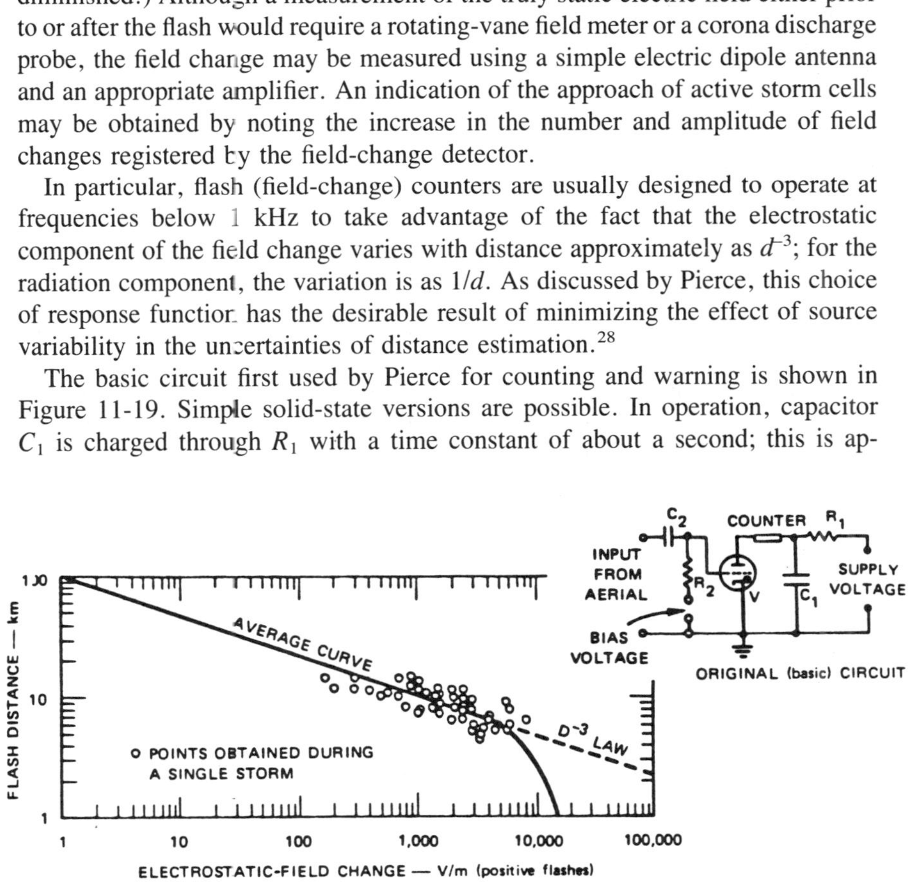

NOTES:

1. SENSITIVITY (range) CONTROLLED BY BIAS VOLTAGE.
2. DEAD TIME CONTROLLED BY R_1C_1.
3. FREQUENCY BAND (~100–1000 Hz) OF ANTENNA CIRCUIT ADJUSTED TO PRESERVE ELECTROSTATIC (D^{-3}) RESPONSE WHILE MINIMIZING SPURIOUS FUNCTIONING.

Fig. 11-19. Electrostatic field-change lightning-warning device.

proximately the upper limit of flash duration. The bias voltage is set so that the discharge device V—a gas discharge tube, originally—is nonconductive. If an impulse field change of sufficient magnitude is incident on the antenna, V becomes conducting, C_1 is then discharged through V, and the counting or recording device operates. The circuit then resets with the time constant R_1C_1.

Figure 11-19 also illustrates some actual electrostatic field-change measurements for which the flash distance was determined from the lightning–thunder interval. The clustering around the $1/d^3$ relationship is quite marked. Note, finally, that the electrostatic field changes are large in magnitude and, even when $d = 100$ km, are typically 1 V/m. Thus, the signal strength is more than adequate for the operation of electrostatic field-change circuitry up to considerable ranges.

Crossed-Loop Direction Finders

Crossed loops can be used as the basis of a lightning direction-finding system that functions as follows: A time-varying magnetic field penetrating the area enclosed by a wire loop induces in the wire a voltage and current proportional both to the time derivative and strength of the magnetic field and to the area of the loop. If the normal vector of the loop surface and the magnetic field vector are parallel, the induced voltage is a maximum. If the two vectors are orthogonal, the induced voltage is zero. At an arbitrary angle of incidence, α, the induced voltage depends on the cosine of this angle. If there are two loops whose planes are at right angles to one another (crossed loops), the induced voltages in the two loops are proportional to $\cos \alpha$ for one loop and to $\sin \alpha$ for the other. The ratio of the two loop voltages is a function of the incidence angle but not of the magnetic field strength since this cancels out. From this ratio, therefore, the azimuthal direction of the lightning discharge can be determined.

The simple, crossed-loop direction-finding technique, of course, yields no information on distance to the channel. The coordinates of the lightning channel can be determined by the use of a pair of crossed loop systems time-tied to make certain that the same flash is being observed.

An alternative scheme that allows range to be estimated from a single crossed-loop system involves using the amplitude of the received field change signal to generate a "pseudo range." With this method, it is assumed that a constant amount of charge is neutralized in a stroke and that the amplitude of the received signal is inversely proportional to the distance to the lightning channel. This concept is the basis of the Ryan "Stormscope" employed on small aircraft for lighting detection and ranging. "The range data generated by such a system, of course, suffers from the fact that all lightning discharges are *not* indeed alike. However, for the pilot of a light aircraft trying to avoid severe weather, it is of great help to know that storm conditions exit along a certain heading roughly so many miles away."[29]

A problem with crossed-loop systems is that the location algorithm requires that assumptions be made about the polarization of the source. Generally, it is assumed that the RF signal is generated by a vertically oriented current channel. In actuality, the lightning channel is extremely tortuous and may actually be horizontal over a distance of several kilometers in the region of the cloud base. Thus, the azimuthal direction inferred for the lightning current channel will vary with time after the initiation of the stroke as the channel orientation deviates from the vertical. These effects are familiar to modern researchers and have been taken into account in the development of their location systems. In particular, the system developed and marketed by Krider, Noggle, and Uman[30] uses only the first few microseconds of the signal radiated by the lightning return stroke. Their argument for adopting this operating principle is that at the time the stepped leader is contacted by the upcoming streamer from the ground, the channel is largely vertical. Thus, if attention is restricted to the early phases of the return stroke, there is very little error in assuming that the lightning signal is being radiated by a purely vertical channel. Restricting attention to ground strokes in this way is ideal for applications involving damage to objects on the ground. These include forest-fire initiation and damage to utility lines. Once the position of the channel is defined, the amplitude of the measured pulse can be used to infer the current and charge flowing in the return stroke channel. This information is also useful to forestry departments and utilities since it is indicative of the likelihood of forest-fire initiation or the probable severity of damage to the utility system.

Time-of-Arrival Measurements

Recordings of received signals from lightning flashes indicate that noise pulses are radiated at frequencies extending to microwaves.[20] These observations led Cianos, Oetzel, and Pierce[31] to suggest that VHF time-of-arrival measurements could be used to locate the positions of the short, fast processes within a thunderstorm cell responsible for the generation of these signals. The concept was later demonstrated by Cianos, Oetzel, and Pierce,[32] and was adopted by NASA as the basis of the LDAR lightning location system at Kennedy Space Center.[33]

Much of the electrical activity responsible for the VHF radiation from a lightning flash is associated with the probing processes accompanying the development of the leader and the redistribution of charge preceding the successive return strokes. Thus, the sources located by VHF time-of-arrival systems are spread throughout the thunderstorm cell, as indicated in Fig. 11-1. The location of the VHF sources and their intensity is not closely related to the location for the return stroke or its intensity. Consequently, this location system is not suitable for use by foresters or utilities, these being the most interested parties in the precise location and characterization of the ground return stroke. This system is very useful, however, in studying the development of active regions within a thunderstorm cell and the way in which a lightning flash evolves.[7,34]

Interferometer System

Like the time-of-arrival systems, interferometric systems rely on the short, fast, electromagnetic process in the lightning discharge for their functioning. The relative phases of the signals arriving at a pair of omnidirectional antennas contain the desired information regarding the direction of arrival of the signals. By using two pairs of such antennas, it is possible to locate the source.

Such a system operating at 34.3 mHz with a 2λ baseline was used by Warwick, et al., for the study of the steps of a lightning stepped leader.[35] Richard, et al., have continued the development and application of interferometric systems for lightning studies.[36] Their system, which operates at 300 mHz, is arranged to determine angular location (elevation and azimuth) by means of two interferometric pairs formed by a three-antenna system. Each interferometer station is built with two complimentary systems: a large one with a 10-λ baseline for accurate but ambiguous location of sources (since phase measurement is obtained with a 2 π ambiguity) and a small one with a λ baseline for ambiguity removal.

Lightning Channel Location by Radar

Although, strictly speaking, the radar detection of a lightning channel does not in itself provide information about lightning electromagnetic waveforms, it can be used for location and to provide very important auxiliary information regarding lightning characteristics. In particular, the use of a UHF radar by Mazur, et al., has provided substantial insight into the problem of lightning strikes to aircraft.[27] This information, in turn, will be of use in defining the electromagnetic signals experienced by aircraft.

At UHF, the electron density in the lightning channel is sufficiently high to produce substantial electromagnetic scattering with an acceptably slow decay as the channel cools. Also, at UHF, the scattering from precipitation particles is very small so that the lightning channel data are not contaminated by returns from the cloud itself. Mazur's UHF radar was slaved to an S-band radar system capable of tracking an aircraft in flight. By observing the development of the lightning channel in the vicinity of an F-106B lightning test aircraft, he has determined that, for every lightning strike to the aircraft, the channel echo began at the aircraft and propagated outward. This implies that, in the F-106 experiments, the aircraft triggered the strike.

11.5 AIRBORNE LIGHTNING MEASUREMENTS

Early Programs

Reporting. As aircraft began to be operated in the proximity of thunderstorm cells, it was found that lightning could inflict damage to the aircraft and its associated equipment. In order to design aircraft properly for safe operation in the presence of lightning, it was necessary to study its properties and develop simulation techniques. Initially, much of this necessary information was generated by means of lightning-strike reporting programs.[37,38] In the normal course of operation, following a lightning strike, the pilot of the aircraft filled out a report indicating the circumstances of the lightning encounter and describing the damage experienced.

Initially, attention was directed at controlling the damage caused by lightning. Accordingly, simulators were developed for duplicating the damage observed in flight. In effect, the earliest airborne lightning measurements were the pits, burns, and other damage observed following a direct strike to the aircraft. These simple measurements were remarkably informative. They permitted the identification of critical zones on the aircraft where lightning strikes and attachments were most likely to occur. They also allowed the evaluation of simulators of varying degrees of complexity to test evolving aircraft designs.

Project Roughrider Flight Tests. The first major effort to measure directly the currents associated with lightning strikes to aircraft was carried out on Project Roughrider.[39] This was a program in which an F-4 aircraft, specially hardened for thunderstorm cell penetration, was flown into the lightning-producing regions of active cells in an effort to attract direct lightning strikes to the aircraft. Instrumentation on the aircraft included oscilloscopes to record lightning currents flowing in special electrodes mounted at the extremities of the aircraft where direct lightning attachment was likely to occur.

During the course of the program, 24 strikes were recorded. The lightning data accumulated were published largely in the form of oscillograms. Since this was a pioneering effort at storm-cell penetration and airborne lightning-triggering/recording, the experience and results have been of great interest to later programs.

More Recent Programs

General. In recent years, aircraft designers and operators have had to give progressively more attention to the electromagnetic signals lightning induces in aircraft avionic systems. More and more, these systems are being based on digital technology, which is inherently more susceptible to transient impulses than the old analog circuitry. Also, new aircraft structural materials, such as fiberglass and other composites, provide substantially less shielding than all-metal skins. In keeping with this change in problem areas, lightning-flight-test objectives have also changed to stress measurements that help to define the electromagnetic properties and effects of lightning.

SRI/Air Force/NASA Learjet Tests. The first of the recent airborne lightning-measurement programs accenting electromagnetic effects used a Learjet test aircraft and was conducted by SRI, the Air Force, and NASA in the vicinity of Kennedy Space Center (KSC) and Patrick Air Force Base, Florida.[40,41] These tests were designed to take advantage of the availability of an aircraft instrumented by SRI for launch-support activity in the vicinity of thunderstorms at KSC. The lightning measurements were intended to investigate the feasibility of using modern instrumentation and sensors to make airborne measurements of lightning-related fields and to generate data on the transients generated on the aircraft and its interior wiring. No attempt was to be made to penetrate thunderstorm cells deliberately or to strive for direct lightning strikes to the aircraft.

The Learjet instrumentation system was designed to accommodate the characteristics of lightning. Since a lightning flash generates a complicated electromagnetic signal with significant components ranging from virtually dc to tens of MHz, it is necessary that all measured data be recorded in such a way that accurate time correlation of all components is possible. In particular, there are sufficient differences from flash to flash, and even on the same components of a given stroke within a flash, that a meaningful comparison of the outputs of different measuring instruments is possible only if the data are obtained for the same flash. Accordingly, all data were recorded either on an analog tape recorder or on floppy-disk storage along with the output of an accurate clock to permit time correlation of the various events. Provisions were made for continuous recording of aircraft parameters and a voice log was provided. Sensors were installed on the exterior and interior of the test aircraft to assist in following the lightning transient through the various states from propagating field to induced aircraft skin currents to noise signals on internal wiring.

The instrumentation system is shown in block form in Fig. 11-20. The spectrum analyzer systems were made up of individual fixed-frequency receiver channels. Provisions were included in the overall instrumentation to allow any of the spectrum analyzer systems to be used to monitor any transient sensor. Spectrum analyzer systems No. 1 and No. 2 each incorporated four receiver channels covering the range of 10 kHz to 30 MHz, whereas systems No. 3 and No. 4 each consisted of only two channels. This was a compromise arrangement necessitated by the fact that the analog tape recorder available for the flight tests could accommodate only 14 data channels.

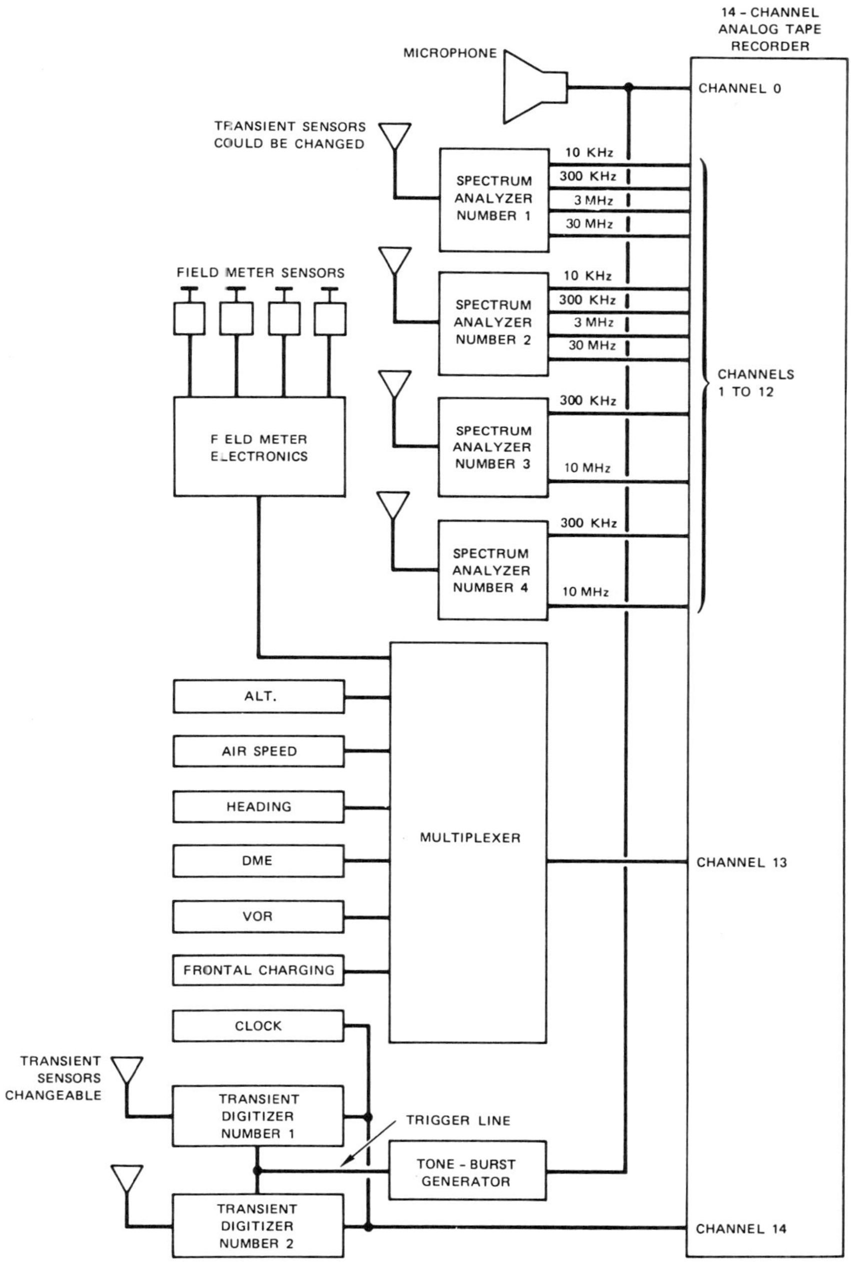

Fig. 11-20. Instrumentation for 1977 lightning-transient study.

As indicated in Fig. 11-2, each stroke within a lightning flash is composed of a large number of components of widely varying spectral nature. To be able to define the nature of the lightning electromagnetic source, it is necessary to be able to resolve these individual components. For this reason, each of the spectrum analyzer channels was designed to have a 20-kHz bandwidth to take full advantage of the bandwidth capabilities of the analog tape recorder. Post-detection filter characteristics were also designed to be compatible with the available 20-kHz bandwidth.

The frequency range covered by the spectrum analyzers was chosen to accommodate several considerations. A substantial fraction of the spectral energy in a flash is contained in the low frequencies (10 kHz), and, accordingly, much ground-based data has been generated using low-frequency monitors. Therefore, the flight system included provisions for measurements at 10 kHz for comparison wih published ground-based data. Experience gained in nuclear electromagnetic-pulse (EMP) analysis indicated that signals coupling through apertures in a shielded system are proportional to *di/dt* where *i* is the current flowing on the surface of the shield.[42] This means that the high-frequency processes in the stroke are important in determining the signals induced in interior wiring. Consequently, provisions were included for measuring signal components up to 30 MHz.

In general, the spectrum analyzer provides a means for breaking the lightning pulse down according to activity in the various frequency regimes. To do so is important because the time waveform of the lightning signal is dominated by the high-level, low-frequency signal associated with the return strokes. Accordingly, high-frequency activity of great interest in determining aircraft system response is not prominently displayed on normal oscillograms.

Also shown in Fig. 11-20 are two transient digitizers that could be connected to members of the same set of electromagnetic sensors used with the spectrum-analyzer system. Each digitizer channel is essentially a wideband (100 MHz or more) oscilloscope that is triggered by the leading edge of the transient and writes the transient signal as a pattern on an array of diode cells. The stored patterns are subsequently read, processed, and stored in the floppy-disk memory of a computer making up part of the digitizer system.

The provisions used for time correlation of the digitizer and spectrum analyzer data are shown in the lower part of Fig. 11-20. A binary-coded-decimal (BCD) time signal was recorded by the PDP-11 computer for each digitizer system trigger, whereas a continuous IRIG B time-code signal was recorded on one of the analog tape-recorder channels used in conjunction with the spectrum analyzer. In addition, digitizer firings were marked on the spectrum analyzer record by initiating a tone-burst generator using the digitizer trigger signal. The tone burst was recorded on the voice log channel of the analog tape recorder.

All of the low bandwidth data were multiplexed onto a signal channel of the tape recorder, including outputs from the field-meter system; aircraft parameters such as altitude, heading, etc.; and the frontal charging current generated by flight through precipitation. The field-meter system includes an analog data processor that yields the free-space field components E_x, E_y, and E_z. Aircraft parameters were derived largely from existing systems on the aircraft. Synchro-to-dc converters were used to convert aircraft signals into analog dc signals for inclusion in the multiplexed data.

The Learjet test aircraft carried a wide variety and large number of sensors. (The design philosophies for the electromagnetic sensors have been discussed earlier in Sec. 11.2). The sensors that protruded through the aircraft skin are shown in Fig. 11-21. Five field meters were installed as shown. The two wing-tip units and the two fuselage units were connected to the analog data-processing system. The field-meter system measured the ambient static-electric field at the aircraft location. In addition, this system provided an indication of the static field change resulting from the neutralization of charge during a flash. This change takes place over a period on the order of a second and can be used to obtain a rough estimate of the distance to a flash.

The propagating field generated at the aircraft location by a lightning flash could be measured by two separate sensors on the exterior of the aircraft—an *E* field antenna located on the top of the fuselage and an *H*-field loop antenna mounted in the nose radome.

In considering the response of an airborne system to lightning transients, it is convenient to break the problem up into several steps. First, the propagating field induces currents and charge displacements on the airframe. These airframe currents excite apertures or diffuse through the skin to generate fields that induce currents in wiring on the interior of the aircraft. To assist in the analysis of this problem, the Learjet was equipped with a set of skin current sensors, as shown in Fig. 11-21, to help define the way in which the airframe is excited by the lightning fields. The sensors were half loops and slots, as illustrated in Figs. 11-10 and 11-11. Photographs of the actual sensors used are shown in Fig. 11-22. The current arriving on the aircraft as the result of fictional charging by impinging precipitation particles was measured by means of an electrically isolated metal patch mounted on the nose of the right-wing-tip tank.

In addition to the external sensors discussed in connection with Fig. 11-21, the Learjet incorporated a number of instrumented wires on its interior, as shown in Fig. 11-23, that could be used to measure the signals induced on typical cabling on the interior of the aircraft.* In general, the pickup wires were installed along the major structural members of the airframe. Because each wire couples electromagnetically to currents flowing on the member along which it is installed,

*Ordinary No. 20 hook-up wire was used for this purpose, e.g., Alpha No. 1553-20-10/30 PVC type MW-V (Mil. No. W-76B).

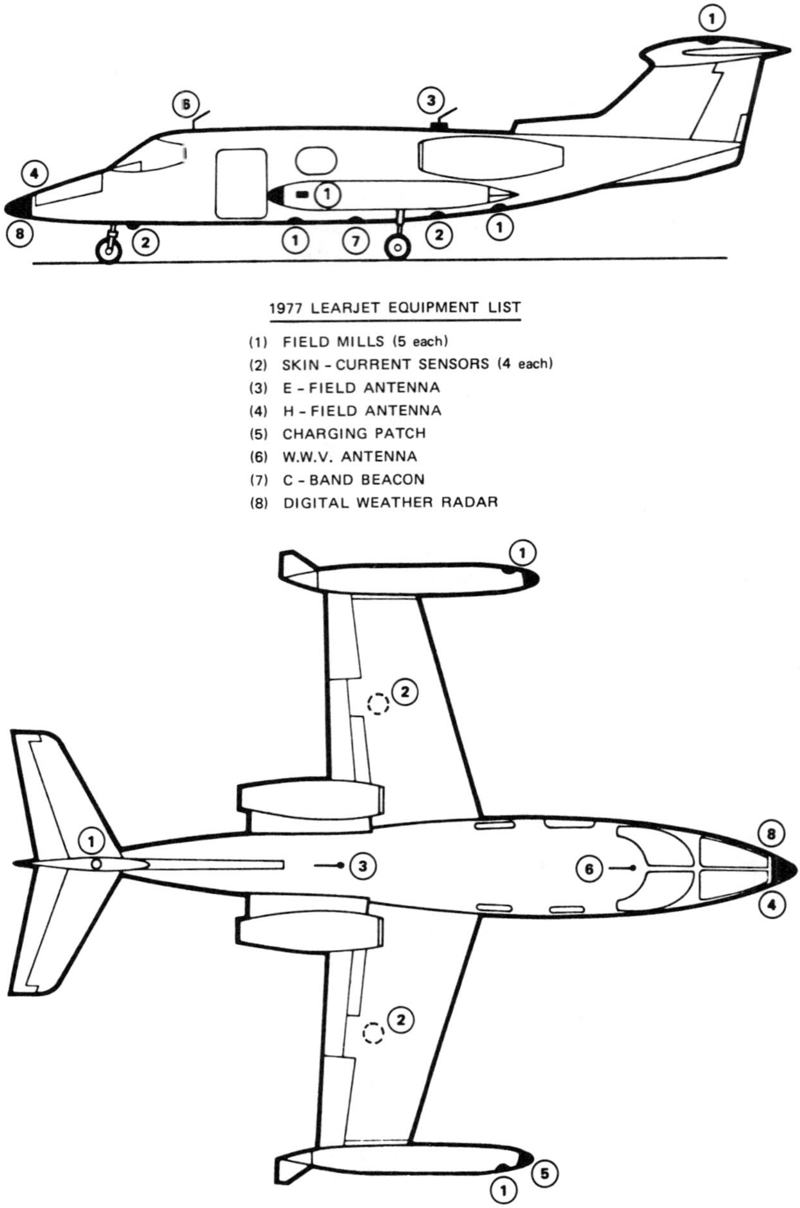

Fig. 11-21. Sensors mounted in skin of aircraft.

(a)

(b)

Fig. 11-22. Skin-current sensors: (a) half-loop, and (b) slot.

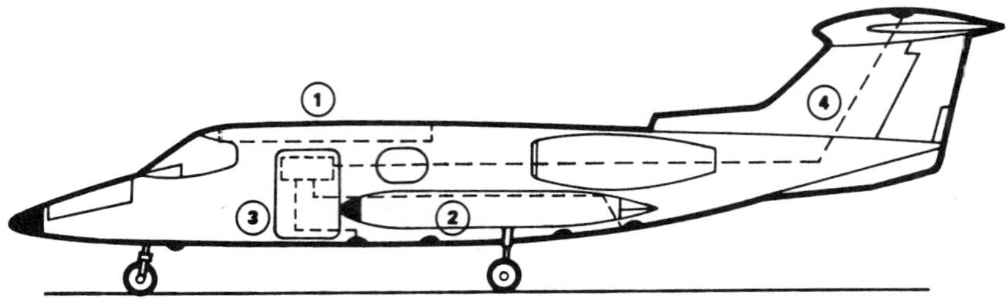

1977 LEARJET EQUIPMENT LIST

(1) CABIN WIRE
(2) AFT FUSELAGE
(3) FORWARD FUSELAGE
(4) FIN CAP
(5) RIGHT WING
(5A) RIGHT WING (after 8/18/77)
(6) LEFT WING

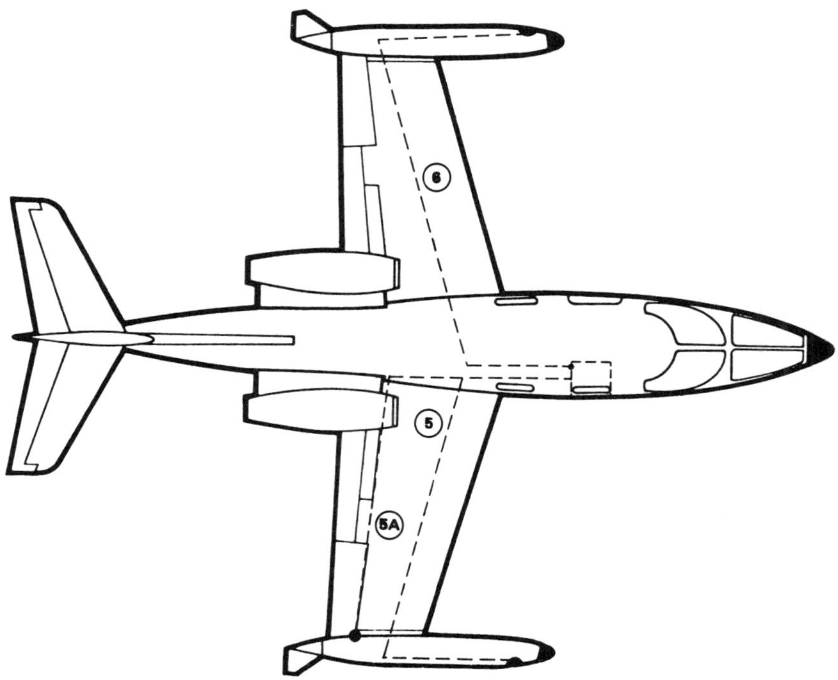

Fig. 11-23. Wire pickup sensors on interior of Learjet.

these current measurements provide a technique for investigating the coupling between airframe skin currents and wiring on the interior. The locations of the wires and their numbers were determined by the simple expedient of running two extra wires in each bundle leading from the cabin to each of the field-meter sensors. One of the wires in each pair was shorted to the airframe at its field-meter sensor location. The second wire of each pair was terminated with a 68-ohm resistor (which was a rough estimate of the wire-to-structure characteristic impedance) at its field-meter sensor location.

A patch panel inside the cabin permitted the special sensor wires to be terminated in either a short circuit or the characteristic impedance of 68 ohms. Tektronix CT-2 current transformers (Z_T = 1 mV/mA) were used to measure the current in the wires of interest.

In general, the specrum analyzers and transient digitizers could be connected to any of the external or internal sensors. Power splitters mounted in a patch panel in the cabin allowed selected sensors to be monitored simultaneously by both a spectrum analyzer and a digitizer.

NASA F-106. Since 1980, Pitts and his coworkers have operated an instrumented F-106 in active thunderstorm cells in an effort to gather data on direct strikes to the aircraft.[43,44] Unlike the Learjet discussed earlier, the F-106 is deliberately operated inside active thunderstorm cells in an effort to induce such strikes. The external sensors used on the F-106 are conceptually similar to the field and skin-current sensors used on the Learjet program. In addition, the F-106 carries an instrumented pitot tube to measure directly the currents flowing in strikes to the nose. Movie cameras pointed fore and aft are used to locate lightning attachment points on the aircraft.

Since the space available on the inside of the F-106 is much smaller than that on the Learjet, severe limitations are imposed on the design of the data-acquisition system. As the testing using this aircraft has progressed, the instrumentation system has systematically been upgraded, principally by increasing the channel capacity of the recording system. The transient recording system used on the F-106 is one based on a commercial transient digitizer (Biomation Model 6500) with the memory increased by over two orders of magnitude to allow recording of 1.3 milliseconds of data at 10-ns resolution.

The UHF radar lightning-location system discussed in the previous section has been used successfully on the F-106 program to vector the test aircraft into electrically active regions of the storm cell to maximize the probability of lightning strikes and to provide important insights regarding lightning triggering and interception.

Air Force WC-130. Between 1979 and 1981, the U.S. Air Force operated an instrumented WC-130 aircraft in the vicinity of thunderstorm cells as part of a study of lightning characteristics.[45,46] Since the purpose of these tests was not to study direct strikes to the aircraft but to investigate the characteristics and effects of nearby lightning, the test aircraft, instrumented for lightning interaction studies, was flown in the vicinity of thunderstorms but not through active cells. The aircraft contained a variety of electromagnetic sensors located at various positions on the fuselage (conceptually similar to the Learjet sensor arrangement shown in Figs. 11-21 and 11-23). The onboard instrumentation system gradually evolved and ultimately included 10 channels with 20-MHz bandwidth (using transient digitizers). Simultaneous ground-based measurements were obtained, including a VLF system to locate the lightning relative to the aircraft.

Transall (French). The French Office National d'Etudes et de Recherches Aerospatiales (ONERA) has assembled and flight-tested an instrumentation system for lightning testing using the Transall aircraft. The system includes the following sensors and instruments:

1. Current probes on front and rear booms responding to amplitudes of ± 150 kA with a bandwidth of about 3 MHz
2. Four field mills to provide measurements of the atmospheric electric field and charge on the aircraft. A real-time display of the field and its direction is used to vector the plane into high-field regions where the probability that it will be struck is high.
3. Electric and magnetic field sensors inside and outside a window on the aircraft surface. Induced voltages on three wires inside the window (one short-circuited, two terminated in 50 Ω) can also be measured. The signals are sampled by four Tektronix transient analyzers with 10-ns resolution. Most signals are recorded on tape by means of two 3-MHz video recorders, two 14-channel 400-kHz tape recorders, and two 10-MHz magnetic-tape recorders. The digital records are registered with 1,500 pretrigger points at 10-ns intervals and 500 posttrigger points at 50 ns. Several additional magnetic field measurements can be made on the aircraft surface and transcribed on the 400-kHz tape recorder.
4. Sensor for measurements of precipitation charging on a special insulated windshield
5. Film and TV camera photographs of the lightning strikes
6. Photoelectric sensors for measurements of the light impulses from the tip and front of the fuselage

Most analog data are transmitted from the current or field sensors through fiber-optic links that have a dynamic range of 41 dB and a bandwidth from 500 Hz to 180 MHz.

Mockup testing and instrument calibration has been completed, and a shakedown test of the instrumentation on the Transall aircraft has been carried out. Some test flights were conducted in the southwestern part of France during the summer of 1984.

REFERENCES

1. Brook, M. G.; Armstrong, G.; Winder, R. P. H.; Vonnegut, B.; and Moore, C. B. 1961. Artificial initiation of lightning discharges. *J. Geophys. Res.* 66:2967–2969.
2. Newman, M. M. 1975. Use of triggered lightning to study the discharge process in the channel. *Problems of atmospheric and space electricity.* New York: American Elsevier, pp. 482–490.
3. Fieux, R. C.; Gary, C. H.; and Hubert, P. 1975. Artificially triggered lightnings above land, *Nature 257,* pp. 212–214.
4. Fieux, R. P.; Gary, C. H.; Hutzler, B. P.; Eybert-Berard, A. R.; Hubert, P. L.; Meesters, A. C.; Perroud, P. H.; Hamelin, J. H.; and Person, J. M. 1978. Research on artificially triggered lightning in France. *IEEE Trans. On Power Apparatus and Systems* PAS-97, No. 3: 725–733.
5. Eybert-Berard, A. P.; Hubert, A. P.; and Laroche, P. 1982. Triggered flashes during TRIP-81. *Proc. International Aerospace Conference on Lightning and Static Electricity.* St. Catherine's College, Oxford, England. (Published by Public Relations Department, Culham Laboratory, Abingdon, Oxon.)
6. Laroche, P. 1983. A contribution to the analysis of triggered lightning: First results obtained during the TRIP-82 experiment. *Proc. International Aerospace and Ground Conference on Lightning and Static Electricity.* DOT/FAA/CRT-83/25, Fort Worth, Tex.
7. Rustan, P. L.; Uman, M. A.; Childers, D. G.; Beasley, W. H.; and Lennon, C. L. 1978. Lightning source locations from VHF radiation data for a flash at Kennedy Space Center. *J. of Geophysical Research* 85, No. C9:4893–4903.
8. Cianos, N.; Oetzel, G. N.; and Pierce, E. T. 1972. Structure of lightning noise—especially above HF. Lightning and Static Electricity Conference Papers, AFAL-TR-72-325, Wright Patterson AFB, Ohio.
9. Wheeler, H. A. 1947. Fundamental limitations of small antennas. *Proc. IRE,* December:1479–1484.
10. Baum, C. E.; Breen, E. L.; Giles, J. C.; O'Neill, J.; and Sower, G. D. 1978. Sensors for electromagnetic pulse measurements both inside and away from nuclear source regions. *IEEE Trans. EMC* 20:22–35. See also Baum, C. E. 1969. Parameters for electrically-small loops and dipoles expressed in terms of current and charge distributions, *AFWL Sensor and Simulation Note* 74.
11. Weidman, C. D.; and Krider, E. P. 1980. Submicrosecond risetimes in lightning return stroke fields. *Geophys. Res. Lett.* 7:955–958.
12. Nanevicz, J. E.; and Vance, E. F. 1982. Instrumentation design tradeoffs for the airborne characterization of lightning. International Conference on Lightning and Static Electricity, St. Catherine College, Oxford, England.
13. Horner, F. 1964. Radio noise from thunderstorms. In *Advances in radio research,* ed. by J. A. Saxton, vol. 2, pp. 122–215. New York: Academic Press.
14. Takagi, M.; and Takeuti, T.. 1963. Atmospherics radiation from lightning discharge. *Proc. Res. Inst. Atmos.* 10:1. Nagoya University, Japan.
15. Horner, F.; and Bradley, P. A. 1964. The spectra of atmospherics from near lightning discharges. *J. Atmos. Terrest. Phys.* 26:1155–1166.
16. Iwata, A.; and Kanada, M. 1967. On the nature of the frequency spectrum of atmospheric source signals. *Proc. Res. Inst. Atmos.* 14:1–6. Nagoya University, Japan.

17. Hallgren, R. E.; and MacDonald, R. B. 1963. *Atmospherics from Lightning 100 to 5600 MHz.* Report No. 63-538-89, IBM Federal Systems Division.
18. Atlas, D. 1959. Radar lightning echoes and atmospherics in vertical cross-section. *Recent advances in atmospheric electricity* ed. by L. G. Smith. London: Pergamon Press, pp. 441–459.
19. Schafer, J. P.; and Goodall, W. M. 1939. Peak field strengths of atmospherics due to local thunderstorms at 150 Megacycles. *Proc. IRE.* 27:202–207.
20. Kosarev, E. L.; Zatsepin, V. G.; and Metrofanov, A. V. 1970. Ultrahigh frequency radiation from lightnings. *J. Geophys. Res.* 75:7524–7530.
21. Oetzel, G. N.; and Pierce, E. T. 1969. The radio emissions from close lightning. In *Planetary Electrodynamics,* vol. 1, ed. by S. C. Coroniti and J. Hughes. New York: Gordon and Breach, pp. 543–571.
22. Cianos, N.; and Pierce, E. T. 1972. A ground-lightning environment for engineering usage. *Technical Report 1,* SRI Project 1834, SRI International, Menlo Park, Calif.
23. Dennis, A. S., and Pierce, E. T. 1964. The return stroke of the lightning flash to earth as a source of VLF atmospherics. *Radio Science* 68D:777–794.
24. Weidman, C. D.; Krider, E. P.; and Uman, M. A. 1981. Lightning amplitude spectra in the interval from 100 kHz to 20 MHz. *Geophys. Res. Lettrs.* 8, No. 8:931–934.
25. Few, A. A. 1970. Lightning channel reconstruction from thunder measurements. *J. Geophys. Res.* 75:7517.
26. Few, A. A.; and Teer, T. L. 1974. The accuracy of acoustic reconstructions of lightning channels. *J. Geophys. Res.* 79, No. 33:5007.
27. Mazur, V.; Fisher, B. D.; and Gerlach, J. C. 1984. Conditions for lightning strikes to an airplane in a thunderstorm. AIAA 22nd Aerospace Sciences Meeting.
28. Pierce, E. T. 1956. The influence of individual variations in the field changes due to lightning discharges on the design and performance of lightning flash counters. *Arch. fur Met. Geophys. und Bioklimat.* 9A:73–88.
29. Ryan, P. A.; and Spitzer, N. Stormscope, U.S. Patent No. 4,023,408.
30. Krider, E. P.; Noggle, R. C.; and Uman, M. A. 1976. A gated wide-band magnetic direction finder for lightning return strokes. *J. of Appl Meteorology* 15:301–306.
31. Oetzel. G. N.; and Pierce, E. T. 1969. VHF technique for locating lightning. *Radio Sci.* 4, No. 3:199–202.
32. Cianos, N.; Oetzel, G. N.; and Pierce, E. T. 1972. A technique for accurately locating lightning at close ranges. *J. of Appl. Meteorology* 11, No. 7:1120–1127.
33. Lennon, C. L. 1975. LDAR lightning detection and ranging presentation. RF Systems Branch of J.F.K. Space Center, NASA, IN-TEL-32.
34. Uman, Martin A.; Beasley, W. H.; Tiller, S. A.; Lin, Y. T.; Krider, E. P.; Weidman, C. D.; Krehbil, P. R.; Brooks, M.; Few, A. A.; Bohannon, J. L.; Lennon, C. L.; Poehler, H. A.; Jaffens, W.; Gulick, J. R.; and Nicholson, J. P. 1978. An unusual lightning flash at Kennedy Space Center. *Science* 201, No. 4350.
35. Warwick, J. W.; Hayenga, C. O.; and Brosnahan, J. W. 1979. Interferometric directions of lightning sources at 34 MHz. *J. Geophys. Res.* 84:2457–2468.
36. Richard, P.; Delannoy, A.; Labaune, G.; and Laroche, P. 1983. UHF interferometric imaging of lightning. 1983 International Aerospace and Ground Conference on Lightning and Static Electricity, Fort Worth, Tex.
37. General Electric Company. 1975. Data from the airlines lightning strike reporting project summary report (Report period: 1 June 1971 to 1 November 1974), compiled by B. I. Hourihan.
38. Perry, B. L. 1969. Lightning and static electricity conference. AFAL-TR-68-290, Part II. Wright-Patterson Air Force Base, Ohio.
39. Petterson, B. J.; and Wood, W. R. 1968. Measurements of lightning strikes to aircraft. Final report to Federal Aviation Agency, contracts FA65WA1-94 and FA66NF-AP-12.

40. Nanevicz, J. E.; Adamo, R. C.; and Bly, R. T. 1977. Airborne measurement of electromagnetic environment near thunderstorm cells (TRIP-76). Technical report AFFDLTR-77-62, SRI project 5536, SRI International, Menlo Park, Calif.
41. Dijak, J. T. 1977. Airborne measurements of transient electric fields and induced transients in aircraft due to close lightning. Technical Report AFFDLTR-77-64.
42. Lee, K. S. H., ed. 1980. EMP interaction: Principles, techniques, and reference data. AFWL-TR-80-402, Air Force Weapons Laboratory.
43. Pitts, F. L.; Thomas, M. E.; Campbell, R. E.; Thomas, R. M.; and Zaepfel, K. P. In flight lightning characteristics measurements system. FAA-RD-79-6, FAA-Florida Institute of Technology Workshop on Grounding and Lightning Technology, 105–111.
44. Trost, T. F., and Zoepfel, K. P. Broadband electromagnetic sensors of lightning research. *Lightning Technology,* NASA CP-2128, FAA-RD-80-30, NASA Langley Research Center, pp. 131–152.
45. Baum, R. K. 1980. Airborne lightning characterization. *Lightning Technology,* NASA Technical Symposium, NASA Conference Publication 2128, Report No. FAA-RD-80-30.
46. Rustan, P. L.; Kuhlman, B. P.; DuBro, G. A.; Reazer, M. J.; Risley, M. D.; and Serrano, A. V. 1982. Correlated airborne and ground measurement of lightning. Proceedings of the International Aerospace Conference on Lightning and Static Electricity, St. Catherine's College, Oxford.

12 Transient Electromagnetic Signal Processing: An Overview of Techniques

J. V. Candy, G. A. Clark, and D. M. Goodman

Lawrence Livermore National Laboratory

12.0 INTRODUCTION

The acquisition of high-frequency electromagnetic data poses difficulties from both measurement and signal processing viewpoints. Because of the high frequencies required and the possibility of reflections from objects other than the one being tested, it is often necessary to conduct transient, rather than steady-state, experiments. Consequently, several of the assumptions common to many signal-processing techniques are violated. Two such assumptions that clearly do not apply to transient signal experiments are that the signals are stationary (i.e., the statistics of the signals do not change with time) and that there exists a "persistently exciting" input (roughly, the input signal excites all modes of the object being tested, and this excitation does not die out with increasing time). In this chapter, we discuss typical signals encountered in transient electromagnetic measurements and analyze the most common signal-processing approaches for extracting the desired information. We do not attempt to develop the theory behind these methods, but we do provide references for those who wish to read further. We also discuss briefly some of the more common applications of the techniques we present.

Background

A nuclear device detonated at high altitudes generates an intense, transient, electromagnetic pulse (EMP) that typically has a very short risetime (a few nanoseconds) and a high peak field amplitude (tens of kilovolts).[1] A common approach in EMP analysis (and in countless other areas as well) is to use a given set of excitation-response measurements of an object being investigated (aircraft, ship, antenna, waveguide, etc.) to characterize somehow its response to an arbitrary excitation. In other words, we require an input–output model of the object. Typical time- or frequency-domain experiments used to obtain this model include the following:

1. Scattering measurements
2. Field measurements
3. Reflectometry measurements
4. Scale-model measurements
5. Full-scale EMP simulator measurements

The development of models that allow extrapolations from transient measurements has long been a concern of EMP researchers. Initial attempts to extrapolate from scale-model measurements were made with Fourier transform techniques.[2] Unfortunately, these attempts often failed because the data were noisy, and the extrapolated signals were not consistent with data from full-scale measurements. Time-domain nonparametric estimates of the model were often more successful.[3,4] Parametric (pole-zero) modeling of electromagnetic object responses[5–7] was first motivated by the singularity-expansion method,[8] which also gave a theoretical justification for such an approach. Initial attempts to determine parametric models used Prony's method; later, other techniques were borrowed from the system-identification literature.

We want to emphasize that merely applying system-identification techniques to EMP extrapolation will not solve the problem. Most parametric-identification methods assume that input and output data are available simultaneously (rather than collected from separate records as are data in many electromagnetic experiments) and that the input signal is persistently exciting. Furthermore, most of what is known about the quality of the parameter estimates obtained with these techniques is asymptotic; e.g., many estimators can be shown to be consistent (i.e., to converge to the correct values as the amount of data increases) and asymptotically efficient (i.e., to converge to the minimum variance estimate as the amount of data increases). The quality of the estimates for short data lengths is not well understood. The data quality is also limited by the state-of-the-art in high-frequency digitizing systems and, in many cases, by the damping in the object being excited. Although satisfactory results are often obtained with experimental data, sometimes the results are not satisfactory, and no single technique has proved to be sufficiently robust to solve the extrapolation problem without fail. Even when a model appears to have been successfully identified from measured data, often the resonances in the model do not coincide with the ones predicted by theory, suggesting inadequacies in either the model or the theory, or both. This issue will be discussed further in Chap. 13.

In this section, we have described the source of the problem we discuss in the remainder of the chapter; in the next, we will define this problem more carefully.

Problem Definition and General Considerations

Simply stated, the problem we study in the remainder of this chapter is

> *Given* a set of input (excitation) and output (response) measurements, *find* the best model of the electromagnetic object.

We call the above statement the *electromagnetic estimation problem*. It raises two issues that require further discussion. Note first that we start with a given set of input and output measurements. In fact, the selection of a reasonable input signal is in itself an important and difficult problem (i.e., one of experiment design). Certainly, one cannot estimate all of the modes of an object if all of them are not excited. Unfortunately, the equipment used to perform transient experiments for the electromagnetic estimation problem usually limits the choice of an input signal. Therefore, we limit our discussion of the input signal to how it can complicate the problem if it is not satisfactory. A detailed discussion of optimal input selection is given elsewhere.[9,10]

The second issue we need to consider is what we mean by "best model." Actually, this phrase can have several interpretations. For example, we might want our model to produce the best possible fit to our observed output when it is driven by our observed input. On the other hand, it is more likely that we wish our model to be such that, when we drive it with an arbitrary input, its output is a good estimate of the output that would have been produced by the actual object being tested. As we shall see, these two objectives are quite different and can sometimes be at odds with each other. A third interpretation occurs if we already have a mathematical model for the test object but do not know the parameters in the model. In this case we probably will consider the best model to be the one with parameters that are as close as possible to those of the actual object.

We may think of the search for the best model as a process involving three steps. First, because it is impractical to consider all possible models, we must restrict our search to a particular class or set. We discuss several classes of impulse response and difference-equation models in this chapter, and another class of models is presented in the next chapter. Usually the selection of the model set is based on the issues of the previous paragraph and on the nature of the object and the experiment. The next step is to select a reasonable criterion to measure how well a given model in the model set matches the observed data;

typically, the criterion is based upon squared error. The third step is to use a minimization algorithm to find the model in the model set that minimizes the criterion function.

Because this chapter is concerned with transient experiments, most of the emphasis is on time-domain models; there is only a brief discussion of frequency-domain approaches. For most classes of time-domain models, the process of selecting the best model can be put in more concrete terms by expressing it as a prediction-error problem: The quality of a model is measured by its ability to predict future behavior based on past observations. The criterion in this case is a function that measures the size of the prediction error when the model is applied to the measured data. The best model is, of course, the one that minimizes this function.

An important consideration in finding the best model is *order determination*. The order of a model is a measure of model size (number of model parameters).[3,11] This issue could be folded into the first step above, but most order-selection techniques involve minimizing the criterion function for several different orders and then estimating the correct order based on the results. Therefore, we treat this problem as a separate issue. A common mistake in attempting to identify the best model is overmodeling. At first glance, a high-order model seems attractive because the higher the model order, the lower the minimum value of the criterion function (e.g., if all of the parameters can be varied, a third-order difference equation will always give a better fit to the data than a second-order one) What has been overlooked is something we mentioned earlier: We usually want our model to be a good estimator of the object's response to arbitrary inputs. It turns out that, if the criterion is mean-squared error, then, for a large class of models, once the true model order has been exceeded, the variance of the error in estimating the behavior of the object for an input other than the measured one increases linearly with model order.

This phenomenon has a nice intuitive interpretation: When the model becomes larger than necessary, the extra degrees of freedom are used to model the noise. On the other hand, of course, the bias will increase as the model order is decreased from its true value. Because the input-output characteristics of electromagnetic objects are typically described by very high (infinite) order differential equations, we usually face a bias vs. variance tradeoff in selecting the model order if we are using a differential or difference equation to describe our object. [Recall that mean-squared error $=$ (bias)^2 $+$ variance.] The concept of using just enough parameters in the model, but no more, is called the *parsimony principle*.

Because the purpose of this section is to define the problem and to give the reader an overview of the issues involved in solving it, we avoided discussing specific models and the tradeoffs among them. More detailed and mathematical treatments of these issues can be found elsewhere.[12–15]

12.1 SIGNAL-PROCESSING AND ESTIMATION TECHNIQUES

A vital first step in attacking the electromagnetic estimation problem is analysis, and, if necessary, preprocessing of the measured input-output records. Only after this step should a modeling technique be applied.

The models we consider in this chapter can be divided into two basic classes: nonparametric and parametric. In both cases, we assume that the object is a linear time-invariant system so that its input-output behavior is completely described by its impulse response. With a nonparametric model, we estimate the impulse response directly with no a priori assumptions about its structure; on the other hand, with a parametric model, we assume that the impulse response satisfies some mathematical relationship and try to estimate the parameters in that relationship. This separation of model classes is somewhat artificial; with digital processing we can estimate the value of the impulse response at only a finite number of points, and the values of these points may be thought of as parameters. We consider the use of both frequency- and time-domain methods for the nonparametric modeling problem. The only parametric models we consider in this chapter are difference equations; another class of models will be discussed in the next chapter.

Preliminary Analysis and Data Preprocessing

Most modern data-acquisition schemes used to measure electromagnetic responses of objects record the signals digitally. Therefore, the data consist of discrete representations of continuous-time (analog) signals. There are several ways the measuring equipment can distort the data. Distortions should be analyzed carefully and removed whenever possible because they can bias estimates. Perhaps the easiest distortion to analyze and correct is a nonlinearity in measuring the sampled values. If the nonlinearity can be determined, then its inverse can be applied digitally to the samples to give the correct values. A far more serious problem is a nonlinearity in the time base; i.e., the length of the sampling period is not constant with time. Correcting for nonuniform sampling usually involves a complex interpolation scheme that is used to compute the values of the analog signal at the desired uniformly spaced sample points. This issue is too involved to treat here but is discussed elsewhere.[16,17] Another situation in which distortion can be introduced is if the measuring equipment itself has dynamics (nonunity transfer function). Compensating for this distortion is a deconvolution problem and involves obtaining a model of the dynamics and then filtering with the inverse transform of the model. This filtering is often restricted to a band of frequencies to prevent ill-conditioning.[18] The time-domain techniques for nonparametric modeling, which we discuss in the next section, are also very popular deconvolution methods.

Computing the Fourier transforms of data records is an important first step in preliminary data analysis, and understanding this process also allows us to explain distortion caused by undersampling the analog signal. Suppose that the analog signal, $x_a(t)$, has been uniformly sampled with sampling period T to give the discrete sequence, $\{x_s(k)\} = \{x_a(kT)\}$. (We assume that k ranges from 0 to $N - 1$ in the remainder of this chapter.) The Fourier transform of the discrete signal is given by

$$X_s(e^{j2\pi fT}) = \sum_{k=0}^{N-1} x_s(k)e^{-j2\pi fTk} \tag{12-1}$$

If $X_a(j2\pi f)$ is the Fourier transform of the original analog signal, then its relationship to Eq. 12-1 is given by

$$X_s(e^{j2\pi fT}) = \frac{1}{T}\sum_{l=-\infty}^{\infty} X_a\,[j2\pi(f + lf_s)] \tag{12-2}$$

where f_s is the sampling frequency ($f_s = 1/T$).[19] If $X_a(j2\pi f)$ is zero when $|f| > f_s/2$, then the terms on the right-hand side of Eq. 12-2 do not overlap, and the Fourier transform of the sequence is just a periodic version of the transform of the analog signal. It follows, therefore, that no information was lost in sampling and that the analog signal can be reconstructed from its discrete version. On the other hand, if the analog signal is not bandlimited to the region, $|f| < f_s/2$, then the terms in Eq. 12-2 overlap, with the result that high-frequency components in the analog signal can appear as lower frequency ones in the discrete signal. This phenomenon is called "aliasing." The only way to prevent aliasing is to use a low-pass filter with a cutoff frequency less than $f_s/2$ to bandlimit the analog signal before it is sampled. The fact that to be able to compute $\{x_s(k)\}$ it is necessary to know only the value of Eq. 12-1 at N uniformly spaced frequencies between 0 and f_s is shown by the following discrete Fourier transform relationships:

Discrete Fourier Transform (DFT):

$$\tilde{X}(n) = \sum_{k=0}^{N-1} x_s\,(k)e^{-j(2\pi/N)nk} \tag{12-3}$$

Inverse Discrete Fourier Transform (IDFT):

$$x_s(k) = \frac{1}{N}\sum_{n=0}^{N-1} \tilde{X}(n)e^{j(2\pi/N)nk} \tag{12-4}$$

It is these transforms that are evaluated with the computer, usually via an algorithm called the Fast Fourier Transform (FFT), which reduces the order of the computational effort from N^2 to $N \log N$ operations.

Space limitations prevent a thorough discussion of the issues involved in using Eq. 12-3 rather than Eq. 12-1, but we wish to warn the reader of a few traps that await the unsuspecting. First, compare Eqs. 12-1 and 12-2 and note that the DFT must be multiplied by a factor of T when it is used to calculate the frequency components of the original analog signal. Second, observe that the convolution of two sequences each having finite length N yields a sequence of length $2N - 1$, and that, for Eq. 12-1, convolution in the time domain is equivalent to multiplication in the frequency domain. Now observe that, if we compute the N-point DFTs of the two sequences, multiply transforms, and compute the IDFT, the result is a sequence of length N. This discrepancy occurs because the IDFT of the product of two DFTs must be interpreted as a circular convolution rather than the usual linear convolution. Linear convolutions can be calculated with the DFT by adding the appropriate number of zero-valued samples to the ends of the original sequences and then computing DFTs of length $2N - 1$ or greater. Increasing the length of the DFT is equivalent to sampling more densely in the Fourier domain, and the problems we encountered with convolution can be explained in terms of time-domain aliasing by a time-domain equation similar to Eq. 12-2. Time-domain aliasing also causes problems if we use the DFT to compute the inverse Fourier transform of $1/X_s(e^{j2\pi fT})$ because the actual time sequence will usually have an infinite number of nonzero values. The required length of the DFT depends on how fast the actual sequence goes to zero. Detailed treatments of the DFT are contained elsewhere.[20,21]

At this point, the reader may have the impression that the continuous signal should be sampled as finely as possible, but, in fact, oversampling can also cause difficulties. The most obvious problem is computational load. If the input sequence, output sequence, and impulse-response sequence all are assumed to have length N, then the number of floating-point operations increases as N^3 for certain nonparametric time-domain methods. Oversampling also causes the Fourier transforms of the discrete signals to be nearly zero at high frequencies, and, as we show in the next section, ill-conditioning may result. (To avoid confusion, we should point out that the spectrum of a discrete sequence is taken to be the region between $-f_s/2$ and $f_s/2$, so that by "high frequency" we mean a frequency near, but less than, half the sampling rate.) Other issues should also be considered when selecting the sampling rate.[22–24]

If the sampling rate of the measuring equipment cannot be changed, using digital techniques to convert the sampling rate of the discrete signals will give satisfactory results in some, but not all, cases. We consider the usual situation in which the sampling rate is changed by an integer factor, but it is also possible to use rational factors. Decimating (reducing the sampling rate) by a factor M involves two steps:

1. To avoid aliasing, apply a low-pass digital filter to the original sequence. The cutoff frequency should be less than $f_s'/2$, where f_s' is the new sampling rate, f_s/M, and the pass-band gain should be 1.
2. Obtain the decimated sequence by retaining only every Mth sample of the filtered sequence.

Interpolating (increasing the sampling rate) by a factor L also involves two steps:

1. Between each sample of the original sequence, place $(L - 1)$ zero-valued samples. The new sampling rate is then: $f_s' = Lf_s$.
2. Apply a low-pass digital filter to the zero-interpolated sequence. The cutoff frequency should be less than $f_s/2$, and the pass-band gain should be L.

Note that thus far in this chapter we have used only the analog domain-frequency variable, f. Often, the normalized variable, $\omega = 2\pi fT$, where T is the sampling period, is used in the discrete domain. Neither decimation nor interpolation can correct for aliasing introduced when the analog signal was first sampled. On the other hand, if the Fourier transform of the analog signal is assumed to be zero for frequencies greater than f_s/M, then nothing is lost when the signal is decimated. More details on decimation and interpolation issues will be found in References 25 through 27.

In many applications, it is necessary to impose a window function, $\{w(k)\}$, on the data sequence, $\{x(k)\}$, before its Fourier transform is computed; i.e., we compute the transform of $\{w(k)x(k)\}$, not $\{x(k)\}$. The purpose of windowing is to prevent spectral leakage by bringing both ends of $\{x(k)\}$ smoothly to zero.[19,21] Windowing is particularly important for data records that are truncated versions of some other record. On the other hand, because a transient signal is one that starts out with zero value, increases in magnitude, and then dies back to zero, the application of a window is pointless if the entire record is included. In fact, windowing can severely distort the transform if, as is often the case, the first part of the signal is a sharp transition from zero. One special case often occurs with measurements of transient electromagnetic signals: The record must be truncated after a certain time, T_0, because it has been distorted by reflections from objects other than the one being tested. Because the data record has been truncated at only one end, only a one-sided window should be used, i.e., only the last half of any of the conventional windows so that the data is brought smoothly to zero at T_0 but not at the beginning of the record.

Because multiplication in the time domain is equivalent to convolution in the frequency domain, windowing is equivalent to convolving the transforms of the data and the window function. The transform of a typical window function consists of a main lobe at zero frequency and lower magnitude sidelobes at other frequencies. Truncating the data is equivalent to applying a rectangular window to it. The rectangular window has particularly large sidelobes, and it is these that produce the spectral leakage (i.e., energy at one frequency in the unwindowed record "leaks" to other frequencies in the truncated record).

Because the main lobe of the transform of the window function has finite width, another consequence of windowing is "smearing" (i.e., local averaging) of the transform of the original data record. For example, if $\{x(k)\}$ is a truncated record, T_t seconds long, of a signal consisting of two sinusoids separated in frequency by Δf, then, because the window function smears the two spectral lines of the original signal, the transform of $\{x(k)\}$ will have two distinct peaks only if (approximately) $\Delta f > 1/T_t$. Otherwise, the two spectral components will merge into a single peak. This limit on the resolving ability of the Fourier transform is a consequence of the uncertainty principle.[28] Even though the rectangular window has the largest sidelobes of any of the common data windows, its main lobe is the narrowest, so that resolution cannot be improved by using a different window. For a given window shape, the width of the main lobe (hence also the amount of smearing) is inversely proportional to the window length. The choice of window shape usually boils down to a tradeoff between main-lobe width and sidelobe magnitude.[29]

The smearing and resolution issues are crucial to an experimenter who is trying to obtain information about the resonances of an object with transient experiments. Estimating the Q of an object is often a goal of an experiment, and, from our discussion, it should be obvious that working with truncated-data records will cause peaks to be widened by smearing, thereby giving Q estimates that are too low. Similarly, if the data record is short, it may not be possible to determine, with transform techniques, whether a given spectral peak was caused by a single resonator or by two resonators whose frequencies are only slightly different. If the resonators are highly damped, the problem is even worse because the peaks are wide in any case. We should point out that nonlinearities in the time base may also smear transforms. One solution to the problems outlined in this paragraph is the use of parametric techniques. Even if the impulse response has already been estimated by means of a nonparametric technique, a parametric technique can be applied to the estimate. Variables like Q or resonant frequency can then be computed from the estimated parameters. Examples of using a parametric technique to estimate Q and resonant frequency from input-output records are contained in Reference 30. It is exactly such an approach that permits parametric techniques (e.g., maximum-entropy spectral analysis) to resolve spectral peaks that, because of the uncertainty principle, could not be resolved with conventional transform methods. An overview of parametric spectral-analysis techniques is given in Reference 31.

Models for the Electromagnetic Estimation Problem

This section is divided into three parts according to type of model being discussed. Some of the issues involved in selecting a particular model class were discussed in the Introduction; we will mention others in what follows. We denote input, output, and impulse response sequences by $\{x(k)\}$, $\{y(k)\}$, and $\{h(k)\}$, respectively, but the expressions for input and output can mean either the actual signals or ones corrupted with noise. In one special case we use $\{u(k)\}$ to represent a noise-corrupted output record.

Nonparametric Frequency-Domain Models. Because the Fourier transform of the output of a linear, time-invariant system is the product of the transforms of the input and the impulse response, the most obvious way to estimate the impulse-response sequence is to divide the transform of the output record by that of the input record and then compute the inverse transform:

$$\{\hat{h}(k)\} = \mathcal{F}^{-1}\left\{\frac{Y(e^{j2\pi fT})}{X(e^{j2\pi fT})}\right\} \tag{12-5}$$

[We denote estimates with a circumflex (ˆ).] If the Fourier transform of the input signal is small in some region of the spectrum, then the transform of the output will be small in the same region. In addition, if the input and output records are corrupted by noise, then the frequency response estimates in this region will have extremely high variances because they are computed by dividing small noise-corrupted numbers by other small noise-corrupted numbers. The result is often that the impulse-response estimate given by Eq. 12-5 behaves very erratically. What we have observed is the obvious fact that the response of an object at a given frequency cannot be determined unless the object is excited at that frequency. The best solution, of course, is to use an input signal that has energy at all frequencies, but physical limitations usually prevent doing this. Often the problem of having to divide by small numbers is a consequence of oversampling. It can be solved by decimating the data until the Fourier transform of the input record becomes sufficiently large at all frequencies. A third solution is discussed in the section on nonparametric time-domain models below.

Another problem with Eq. 12-5 occurs if it is necessary to truncate the data records. The records must then be windowed to prevent spectral leakage from biasing the frequency estimates, but the windowing process itself introduces another form of bias by smearing the transforms. The reader should also recall that DFT techniques are used to evaluate Eq. 12-5 on the computer, and time-domain aliasing effects may cause the shape of the estimated impulse response to vary with the order of the DFT.

Another frequency-domain approach is the use of power spectral-density techniques. Suppose that the input signal is a stationary random process and that the only measurement noise, $\{n(k)\}$, appears at the output and is uncorrelated with the input. Then, the observed output signal is

$$u(k) = y(k) + n(k) \tag{12-6}$$

and, because the noise is uncorrelated with the input, $S_{xu}(f) = S_{xy}(f)$. Consequently, the Fourier transform of the impulse response is given by

$$H(e^{j2\pi fT}) = \frac{S_{xu}(f)}{S_{xx}(f)} \tag{12-7}$$

where $S_{xx}(f)$ is the power-spectral-density function of $\{x(k)\}$ and $S_{xu}(f)$ is the cross-spectral density function of $\{x(k)\}$ and $\{u(k)\}$. From Eq. 12-7 it appears that the effects of the noise term have been completely eliminated, but, of course, the spectral-density functions are unknown and must be replaced by estimates obtained from the data. The usual approach is to estimate the autocorrelation of the input, $\{R_{xx}(k)\}$, and the cross-correlation of the input and output, $\{R_{xu}(k)\}$, and then compute Fourier transforms to get estimates of the corresponding spectral-density functions. It can be shown that the variance of the spectral-density estimates does not decrease with increasing data length unless windows are applied to the estimated correlation sequences so that relatively few of their terms are retained. On the other hand, such windowing introduces bias because of smearing. Therefore, in minimizing the mean-square error of the spectral estimates, we face a bias-vs.-variance tradeoff because reducing the length of the window lowers variance but increases bias. The selection of a window of reasonable length is a trial-and-error process. The method we have outlined in this paragraph is discussed in greater detail elsewhere.[2,14,32–34]

Of course, the method described in the last paragraph is not directly applicable to transient signals because such signals are very different from stationary processes, and the arguments used to justify this method lean heavily on the assumption that the signals are stationary. In fact, windowing the correlation functions of transient signals usually introduces an unacceptable amount of bias. On the other hand, suppose that the correlation estimates are the usual "deterministic" correlations obtained from the data records, i.e.,

$$\{\hat{R}_{xx}(k)\} = \{x(k)\} * \{x(-k)\} \quad \text{and} \quad \{\hat{R}_{xu}(k)\} = \{x(k)\} * \{u(-k)\} \tag{12-8}$$

where $*$ denotes convolution and the sequences are assumed to be zero outside the interval of observation. Then, if the spectral densities are calculated by computing transforms of Eq. 12-8 without windowing, the estimate of the frequency response in Eq. 12-7 is the same as would be obtained by dividing the transform of the output record by that of the input record, and we are back to the estimate shown in Eq. 12-5. Furthermore, to correct truncation effects (particularly of the one-sided variety) in transient signals, windows should be applied directly to the data records, not their correlation sequences. Power-spectral-density methods for stationary processes obtain low variance estimates by time averaging; the problem with applying such methods to transients is that there are not enough data over which to average. On the other hand, if the experiments are repeatable, ensemble averaging is a good way to obtain low variance estimates from transient signals. Chapter 10 of Reference 33 discusses spectral-analysis techniques for transients. We discussed the spectral-density method primarily to point out the pitfalls of applying it blindly to the transient problem. Another reason, however, is that the time-domain method of the next section has a frequency-domain interpretation as an estimator of a form similar to Eq. 12-7.

Nonparametric Time-Domain Models. If we assume that the impulse response has length M and that the input and output records are zero outside the observation interval (a reasonable assumption for untruncated transient signals), then the convolution relationship can be written in matrix form as follows:

$\boldsymbol{y} = \boldsymbol{Xh}$, where $\boldsymbol{y} = [y(0), y(1), \ldots, y(N-1); 0 \ldots, 0]^T$

$\boldsymbol{h} = [h(0), h(1), \ldots, h(M-1)]^T$, and

$$\boldsymbol{X} = \begin{bmatrix} x(0) & 0 & & 0 & 0 \\ x(1) & x(0) & & 0 & 0 \\ \vdots & \vdots & & \vdots & \vdots \\ x(N-1) & x(N-2) & \cdots & x(0) & 0 \\ 0 & x(N-1) & & x(1) & x(0) \\ \vdots & \vdots & & \vdots & \vdots \\ 0 & 0 & & x(N-1) & x(N-2) \\ 0 & 0 & & 0 & x(N-1) \end{bmatrix} \tag{12-9}$$

If the input-output records were completely noise free, then $\boldsymbol{h}$ could be calculated by inverting an $M \times M$ submatrix of $\boldsymbol{X}$. If, however, as in the last section, the noise appears at the output according to Eq. 12-6, then the output record, $\{u(k)\}$, in vector form is expressed as $\boldsymbol{u} = \boldsymbol{y} + \boldsymbol{n}$, and, because $\boldsymbol{y}$ is no longer available, $\boldsymbol{h}$ cannot be computed exactly but must be estimated from the observed input-output records, $\{x(k)\}$ and $\{u(k)\}$. Because $\boldsymbol{Xh} = \boldsymbol{u} - \boldsymbol{n}$, a seemingly reasonable approach is to treat Eq. 12-9 as a linear regression problem,[35] i.e., $\hat{\boldsymbol{h}}$ is chosen to be the vector that minimizes

$$(\boldsymbol{X}\hat{\boldsymbol{h}} - \boldsymbol{u})^T (\boldsymbol{X}\hat{\boldsymbol{h}} - \boldsymbol{u}) \tag{12-10}$$

This approach has a prediction error interpretation since $\hat{\boldsymbol{u}} = \boldsymbol{X}\,\hat{\boldsymbol{h}}$ is the output predicted by the model and $\hat{\boldsymbol{h}}$ is selected to minimize the prediction error, $\boldsymbol{u} - \hat{\boldsymbol{u}}$, with the criterion being least-squares. The solution to Eq. 12-10 is found by solving the normal equation,

$$\boldsymbol{X}^T\boldsymbol{X}\hat{\boldsymbol{h}} = \boldsymbol{X}^T\boldsymbol{u} \tag{12-11}$$

If we define $\hat{\boldsymbol{R}}_{xx}$ as $\boldsymbol{X}^T\boldsymbol{X}$ and $\hat{\boldsymbol{r}}_{xu}$ as $\boldsymbol{X}^T\boldsymbol{u}$, then their entries can be obtained from Eq. 12-8, i.e., $[\hat{\boldsymbol{R}}_{xx}]_{ij} = \hat{R}_{xx}(i - j)$ and $[\hat{\boldsymbol{r}}_{xu}]_i = \hat{R}_{xu}(i)$. One might expect, therefore, that the estimate computed using Eq. 12-11 and the estimate computed by substituting the transforms of Eq. 12-8 in Eq. 12-7 would be very similar. In fact, Hunt[36] has shown that, if the impulse response is assumed to have a length, $M = 2N - 1$, then the DFT of the linear-regression estimate is

$$\{\tilde{H}(n)\} = \frac{\tilde{S}_{xu}(n)}{\tilde{S}_{xx}(n)} \tag{12-12}$$

where $\{\tilde{S}_{xx}(n)\}$ and $\{\tilde{S}_{xu}(n)\}$ are the (length M) DFTs of the correlation functions in Eq. 12-7.

If the noise signal is zero-mean, then the linear-regression estimate is an unbiased linear (linear with respect to $\boldsymbol{u}$) estimate of $\boldsymbol{h}$, and, if the noise is white (the covariance matrix of $\boldsymbol{n}$ is $\sigma^2\boldsymbol{I}$, where σ^2 is the noise variance), this estimate is, in the mean-square sense, the Best Linear Unbiased Estimate (BLUE) of $\boldsymbol{h}$. If the noise is not white, the BLUE estimator is obtained by minimizing the weighted least-squares criterion function, $(\boldsymbol{X}\boldsymbol{h} - \boldsymbol{u})^T\boldsymbol{W}^{-1}(\boldsymbol{X}\boldsymbol{h} - \boldsymbol{u})$ where $\boldsymbol{W}$ is the covariance matrix of $\boldsymbol{n}$.[23,34] We consider only the white-noise case, but the arguments extend to the non-white case. The total variance of the linear-regression estimate can be shown to be

$$E[(\hat{\boldsymbol{h}} - \boldsymbol{h})^T (\hat{\boldsymbol{h}} - \boldsymbol{h})] = \sigma^2 \sum_{m=1}^{M} \frac{1}{\lambda_m} \tag{12-13}$$

where E is the expectation operator, $\lambda_1, \lambda_2 \ldots, \lambda_m$ are the eigenvalues of $\boldsymbol{X}^T\boldsymbol{X}$, and σ^2 is the variance of the white noise. In our special case, $\hat{\boldsymbol{R}}_{xx}$ is a symmetric Toeplitz autocorrelation matrix, and

$$\frac{\lambda_{\max}}{\lambda_{\min}} \approx \frac{\max_f \tilde{S}_{xx}(f)}{\min_f \hat{S}_{xx}(f)} \tag{12-14}$$

where $\hat{S}_{xx}(f)$ is the Fourier transform of $\{\hat{\boldsymbol{R}}_{xx}(k)\}$.[37] Therefore, if the spectrum of the input signal is nearly zero at some frequency, the variance of the impulse-response estimate may be very large, even in the low noise case. This is the same problem we encountered with frequency-domain models. By rewriting Eq. 12-13 in the form,

$$E\{\hat{\boldsymbol{h}}^T\hat{\boldsymbol{h}}\} = \boldsymbol{h}^T\boldsymbol{h} + \sigma^2 \sum_{m=1}^{M} \frac{1}{\lambda_m} \tag{12-15}$$

we see that the energy in the estimated impulse response, $\hat{\boldsymbol{h}}$, will usually be much larger than the energy in the actual impulse response, $\boldsymbol{h}$. If Eq. 12-10 is minimized subject to the condition, $\hat{\boldsymbol{h}}^T\hat{\boldsymbol{h}} < C$, an energy constraint on $\hat{\boldsymbol{h}}$, then the form of the estimate is

$$\hat{\boldsymbol{h}}(\delta) = (\hat{\boldsymbol{R}}_{xx} + \delta \boldsymbol{I})^{-1}\boldsymbol{r}_{xu} \tag{12-16}$$

where $\delta > 0$ is a nonlinear, monotonic decreasing function of C and $\hat{\boldsymbol{h}}(\delta)$ is a biased estimate of $\boldsymbol{h}$. Here, as in the frequency-domain case, however, we face a bias-vs.-variance tradeoff, except that this time the tradeoff involves selecting a good value for δ (equivalently, C). Often, if δ is selected properly, $\hat{\boldsymbol{h}}(\delta)$ will have a mean-square error that is much lower than that of $\hat{\boldsymbol{h}}(0)$, the best unbiased estimate. One method for selecting δ is to increase it until the solution "stabilizes." Another is to select δ in such a way[38] that the prediction error is what would have been expected given that the noise variance was σ^2, i.e., pick δ to satisfy

$$[\boldsymbol{u} - \boldsymbol{X}\hat{\boldsymbol{h}}(\delta)]^T [\boldsymbol{u} - \boldsymbol{X}\hat{\boldsymbol{h}}(\delta)] = \sigma^2 (N + M - 1) \tag{12-17}$$

where $N + M - 1$ is the length of **n**. Several other methods for selecting δ that are based on statistical considerations are presented in Reference 4.

Since the criterion function (Eq. 12-10) increases with δ, any positive value of δ will increase the fitting error for the observed data. On the other hand, a good positive value gives a better estimate of the impulse response and, thereby, better estimates of the responses to arbitrary inputs. Roughly speaking, the bias in Eq. 12-16 is that the Fourier transform of $\hat{\boldsymbol{h}}(\delta)$ is forced to be small at frequencies where $\{x(k)\}$ has little energy. When $M = 2N - 1$, there is an exact frequency-domain interpretation. As shown in Reference 36, the DFT of $\hat{\boldsymbol{h}}(\delta)$ is

$$\tilde{H}_{\delta}(n) = \frac{\tilde{S}_{xu}(n)}{\tilde{S}_{xx}(n) + \delta} \tag{12-18}$$

This kind of bias is unimportant if the "arbitrary" inputs of interest also have little energy at the same frequencies. If the response at these frequencies is of interest, the only option is to excite the object with a better test input; signal processing will not help.

Thus far, we have discussed only the untruncated case, but this method is particularly useful if the records must be truncated after a certain time, T_0. In this case, the linear regression estimator is obtained by replacing $\boldsymbol{X}$ and $\boldsymbol{u}$ with truncated versions from which all bottom rows involving samples occurring after time T_0 have been removed. The estimate obtained from Eq. 12-11 will still be unbiased (albeit high-variance), and, unlike in the frequency domain, no windows need be applied to handle spectral leakage. The variance can, as before, be lowered by using Eq. 12-16. However, $\boldsymbol{X}^T\boldsymbol{X}$ will no longer be Toeplitz, so all of the frequency-domain interpretations no longer apply. In particular, there are examples in which $\{x(k)\}$ has a white spectrum, and yet the minimum $\boldsymbol{X}^T\boldsymbol{X}$ is arbitrarily small.

Another modification to the technique we have presented is the replacement of the identity matrix by some other matrix. Depending on the matrix selected, the simple energy constraint we used can be changed to weighted constraints in either the time domain or the frequency domain, or both. Such modifications are discussed in several of the references. Computational issues are important because the dimensions of $\hat{\boldsymbol{R}}_{xx}$ may be large. Fortunately, in the untruncated case, $(\hat{\boldsymbol{R}}_{xx} + \delta \boldsymbol{I})$ *is* Toeplitz, and Eq. 12-16 can be solved using the Levinson algorithm.[39] An even faster frequency-domain approach is possible if the value, $M = 2N - 1$, is acceptable.[36] References 35–44 treat various aspects of the method we have outlined in this section; Reference 4 gives a more detailed overview with emphasis on the truncated case.

Because the input and output signals are measured in the same way in many experiments, the assumption that noise affects only the output, as in Eq. 12-6, may not seem reasonable. Usually, however, the excitation signal is much larger than any measurement noise; furthermore, because the same excitation signal is typically used for many experiments, the experimenter should be willing, if necessary, to use ensemble averaging to obtain a low-variance estimate of $\{x(k)\}$. The estimation problem becomes much more complex if noise on the input record must be considered, because the $\boldsymbol{X}$ matrix is no longer deterministic and the linear regression model no longer applies.

Parametric Time-Domain Models. In this section, we outline the issues involved in estimating the parameters of a difference equation in order to describe the input–output behavior of our test object, i.e., we model this behavior with the equation,

$$\sum_{i=0}^{M_a} a(i)y(k - i) = \sum_{j=1}^{M_b} b(j)x(k - j) \tag{12-19}$$

If the object satisfies such a relationship exactly, then its impulse response can be determined by replacing $\{x(k)\}$ with the unit impulse sequence; the resulting output will be $\{h(k)\}$. The sum over the input starts at $j = 1$ because the object is not assumed to respond instantaneously.

Parametric models have several advantages over nonparametric ones. If the object's input–output behavior satisfies a finite-order difference equation, then estimating its parameters, instead of estimating $\{h(k)\}$ directly, will often give a better mean-square estimate of the impulse response. Furthermore, the poles of Eq. 12-19—i.e., the roots of $\Sigma_{i=0}^{M_a} a(i)z^{-1}$—may have physical significance[8] and may indeed be the goal of the experiment. Even if a finite-order difference equation is not an appropriate model for the test object, often such an equation of sufficiently high order can be used to give a good estimate of the impulse response or to improve resolution and obtain better estimates of Q.

Because the data records, $\{x(k)\}$ and $\{y(k)\}$, are noisy (and because the model does not describe the object exactly), there will usually be no values of the parameters, $\{a(i)\}$ and $\{b(j)\}$, such that Eq. 12-19 is satisfied exactly. The most obvious solution is to add an error term, $\{v(k)\}$, to Eq. 12-19 and then select the parameters to make the error small. By rewriting Eq. 12-19, we can show that $\{v(k)\}$ has a prediction error interpretation [that $a(0) = 1$ is assumed]:

$$y(k) = \left[-\sum_{i=1}^{M_a} a(i)y(k - i) + \sum_{j=1}^{M_b} b(j)x(k - j) \right] + v(k) \tag{12-20}$$

where the term in square brackets uses previous input and output values to predict the next value of the output, $y(k)$. Dudley[7] gives a detailed treatment of this approach to the electromagnetic estimation problem; we give a brief summary of his discussion in the next two paragraphs.

Because $\{v(k)\}$ is a linear function of the parameters, the optimal parameters can be obtained, if the usual sum-of-squares criterion is applied to the prediction error, by solving a linear least-squares problem. For convenience, we assume that $M_a = M_b = M$ and define Eq. 12-21 as shown on the following page.

$$\boldsymbol{\theta}^T = [-a(1), \ldots, -a(M); b(1), \ldots, b(M)],$$
$$\mathbf{y}^T = [\mathbf{y}_u^T \vdots \mathbf{y}_l^T] = [y(1), \ldots, y(M - 1) \vdots y(M), \ldots, y(N - 1)],$$
$$\mathbf{v}^T = [\mathbf{v}_u^T \vdots \mathbf{v}_l^T] = [v(1), \ldots, v(M - 1) \vdots v(M), \ldots, v(N - 1)],$$

$$\mathbf{W} = \begin{bmatrix} \mathbf{W}_u \\ \cdots \\ \mathbf{W}_l \end{bmatrix} = \left[\begin{array}{cccccc|cccccc} y(0) & 0 & 0 & \ldots & 0 & 0 & x(0) & 0 & 0 & \ldots & 0 & 0 \\ y(1) & y(0) & 0 & \ldots & 0 & 0 & x(1) & x(0) & 0 & \ldots & 0 & 0 \\ \vdots & \vdots & \vdots & \ldots & \vdots & \vdots & \vdots & \vdots & \vdots & \ldots & \vdots & \vdots \\ y(M-2) & y(M-3) & y(M-4) & \ldots & y(0) & 0 & x(M-2) & x(M-3) & x(M-4) & \ldots & x(0) & 0 \\ \hline y(M-1) & y(M-2) & y(M-3) & \ldots & y(1) & y(0) & x(M-1) & x(M-2) & x(M-3) & \ldots & x(1) & x(0) \\ \vdots & \vdots & \vdots & \vdots & \vdots & \vdots & \vdots & \vdots & \ldots & \ldots & \vdots & \vdots \\ y(N-2) & y(N-3) & y(N-4) & \ldots & y(N-M-2) & y(N-M-1) & x(N-2) & x(N-3) & x(N-4) & \ldots & x(N-M-2) & x(N-M-1) \end{array}\right] \tag{12-21}$$

In terms of this expression, Eq. 12-20 becomes

$$\mathbf{W}\theta = \mathbf{y} - \mathbf{v} \tag{12-22}$$

and, of course, the value of θ that minimizes $v^T v$ is the solution to the normal equation:

$$\mathbf{W}^T \mathbf{W}\theta = \mathbf{W}^T \mathbf{y} \tag{12-23}$$

Note that, if $M_a = 0$, then Eq. 12-21 is the same linear-regression problem we solved in the truncated-data case of the previous section. Unfortunately, however, if $M_a = 0$, then $\{y(i)\}$ appears in both $\mathbf{y}$ and $\mathbf{W}$, and, because $\{y(i)\}$ cannot be considered deterministic, Eq. 12-21 is not a standard linear-regression problem, although it has the appearance of one. In fact, unless $\{v(k)\}$ is white, the solution to Eq. 12-23 is inconsistent (i.e., it is biased even as N gets large). This bias can be severe even at only moderate noise levels. It is important to understand that $\{v(k)\}$ is not output noise. In particular, if the only noise in the system does enter at the output according to Eq. 12-6 and $\{u(k)\}$ is substituted for $\{y(k)\}$ in Eq. 12-20, then

$$v(k) = \sum_{i=0}^{M_a} a(i)n(k - i)$$

Therefore, even if the only error is white measurement noise on the output, the parameter estimates will be biased.

In Eq. 12-21, the signals were assumed to be zero before the observation interval but not after. It is also possible to assume that the signals are zero after the observation interval, but this is usually not done with parametric modeling. The zero-signal assumption before the observation interval is made because, roughly speaking, it forces the estimator to take a closer look at the (often important) first part of the signals. If this assumption cannot be made because the first few samples of the observed data are not close to zero, then $\mathbf{W}$, $\mathbf{y}$, and $\mathbf{v}$ should be replaced with W_l, y_l, and v_l in Eqs. 12-22 and 12-23. The first attempts at parametric modeling of transient electromagnetic systems used Prony's method[5,6] to fit an exponential series to a single data record. If W_l, y_l, and v_l are used, if $\{x(k)\}$ represents the unit-impulse sequence, and if $2M = N$ so that there are as many parameters as samples, then the error term is zero and Eq. 12-22 is the "curve-fitting" form of Prony's method. If $2M < N$, then solving Eq. 12-23 yields the $\{a(i)\}$ parameters that would result from the least-squares form of Prony's method, but the $\{b(j)\}$ parameters are different because they are chosen to match the first few data points exactly. Because the columns of W are also shifted versions of the data records, oversampling causes the same conditioning problems with inverting $W^T W$ as it did with $X^T X$ of the last section. There is a recursive algorithm (one in which the data records are used one sample at a time) for solving Eq. 12-23 that makes this method particularly attractive for small computers.

One way to reduce bias in the solution to Eq. 12-23 is to introduce a noise model. Several different models are available[45,46]; the most useful for our purposes results from assuming that the error term is produced by filtering a white-noise sequence, $\{e(k)\}$, through a moving average filter, i.e.,

$$\sum_{i=0}^{M_a} a(i)y(k - i) = \sum_{j=1}^{M_b} b(j)x(k - j) + \sum_{l=0}^{M_c} c(l)e(k - l) \qquad (12\text{-}24)$$

That both $a(0) = 1$ and $c(0) = 1$ is assumed. Equation 12-24 is the so-called autoregressive moving-average model with exogenous input, and it is often denoted by ARMAX(M_a,M_b,M_c). This equation can be rewritten in terms of polynomials in the backward-shift operator q^{-1} [where $q^{-1}y(k) = y(k - 1)$], as follows:

$$A(q^{-1})y(k) = B(q^{-1})x(k) + C(q^{-1})e(k) \qquad (12\text{-}25)$$

In turn, Eq. 12-25 can be rewritten in prediction-error form, as follows:

$$\frac{1}{C(q^{-1})}[A(q^{-1})y(k) - B(q^{-1})x(k)] = e(k) \qquad (12\text{-}26)$$

That $e(k)$ is the error in predicting $y(k)$ from past values of the sequences $\{y(k)\}$ and $\{x(k)\}$ is shown in Reference 46. If the data records are generated by an ARMAX system, then a reasonable estimate of the model parameters can be obtained by minimizing the sum-of-squares of the sequence $\{e(k)\}$ that results from substituting the data records into Eq. 12-26. This minimization is computed over $\{a(i)\}$, $\{b(j)\}$, and $\{c(l)\}$. If the order of the model is correct, then under reasonable assumptions, this estimate will be consistent and asymptotically efficient. Of course, such asymptotic criteria may not be useful for the transient case, but simulations have shown that this model often works well. Unfortunately, however, although the prediction error is still a linear function of $\{a(i)\}$ and $\{b(j)\}$, it is nonlinear with respect to $\{c(l)\}$, and thus iterative nonlinear minimization techniques must be applied. Often the solution to Eq. 12-23 is used as a starting point. The estimator for the ARMAX model also has a recursive implementation, but it may be necessary to run the same data through the recursion several times before the solution converges because the recursive algorithm is just a special version of an iterative minimization algorithm.

The two techniques we have considered thus far are called equation-error methods because the input–output data are substituted into a difference equation whose parameters are then adjusted to minimize the error resulting from this substitution. The third technique we consider is an output-error method in which a difference equation is actually driven by the observed input signal, i.e.,

$$\sum_{i=0}^{M_a} a(i)\hat{y}(k - i) = \sum_{j=1}^{M_b} b(j)x(k - j) \tag{12-27}$$

The output error, $\{e(k)\}$, is the difference between the output of the model and the output data record, as follows:

$$e(k) = y(k) - \hat{y}(k) \tag{12-28}$$

The parameter estimates are obtained by computing the $\{a(i)\}$ and $\{b(j)\}$, which minimize the sum-of-squares of the output error sequence. This minimization problem is linear in the $\{b(j)\}$ but not the $\{a(i)\}$. Although $\{e(k)\}$ is still the error in predicting $\{y(k)\}$ from past data, now only the input sequence, $\{x(k)\}$, is used in the prediction. This technique has a "robustness" advantage over the previous ones because, if noise enters the system only as in Eq. 12-6, then the parameter estimates are consistent even if the noise is correlated and no noise model is used. This method is considered in more detail in the next chapter; a particular implementation is described in Reference 30.

The models we have discussed in this section appear to be adequate for simple electromagnetic objects like spheres, cylinders, etc., but may not always be satisfactory for highly coupled, complex structures such as scale models of ships or aircraft. More complicated algorithms based on model-based techniques and the Kalman filter are capable of handling multiple-input and multiple-output data records as well as nonstationary statistics.[47,48] Some initial research has been conducted with these algorithms, but more work is needed to determine their suitability for the transient problem.[49]

12.2 APPLICATIONS

This section treats the application of the techniques previously described to some typical electromagnetic estimation problems. First, we consider the application of parametric methods to characterize an electromagnetic transient facility. Second, we show how signal-processing techniques can be used to estimate the impedance and effective height of an antenna from noisy time-domain reflectometry measurements. Third, we discuss EMP calculations for a scale model F-14 fighter aircraft, and, finally, we estimate the parameters of a simple dipole using noisy transient measurements.

Characterization of an Electromagnetic Test Facility

Identification techniques have been used to improve signal quality at the Lawrence Livermore National Laboratory's Electromagnetic Test Facility (EMTF) (see Chap. 7). The EMTF was developed primarily to investigate the responses of objects subjected to transient electromagnetic fields. The facility has been used (1) to characterize the EMP response of scale models (e.g., tanks, aircraft, ships); (2) to validate experimentally theoretical and computer model predictions for simple structures (e.g., dipoles, cylinders, wires); and (3) to characterize antennas by means of reflectometry and scattering measurements. The facility is versatile and can be reconfigured easily to meet changing project needs. Signal-processing and estimation techniques have been applied to improve the processing of noisy data and to develop models to characterize the facility for calibration and experiment-design purposes.

The EMTF can be divided into three major subsystems: (1) the incident-field simulator, (2) the test object, and (3) the measurement system. The field simulator is designed to produce an approximation to a propagating plane wave and consists of a short-duration (200-psec), high-intensity (1-KV) pulser; a conical antenna with a 45-deg half angle; and a ground plane. The antenna, when excited by the pulser, produces a transient field that excites the test object and produces measurable responses over a band 3-GHz wide. The incident energy is scattered from the test object and is then sampled by voltage or current probes whose outputs are fed to either a transient digitizer or a sampling oscilloscope. These measurements are corrupted by noise and distortion caused by the recording equipment, the source antenna, cables, probes, reflections from other objects, and environmental disturbances. To improve signal-to-noise ratio and compensate for distortion, signal-processing techniques must be employed. Figure 12-1 is a diagram of the EMTF. For a more detailed discussion than we present here, see References 50 through 52.

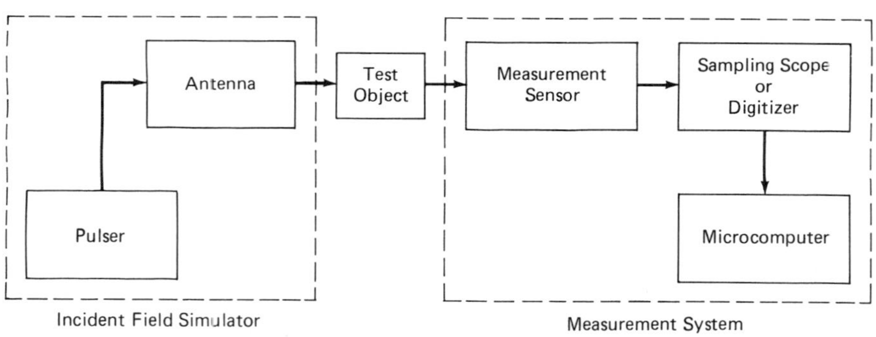

Fig. 12-1. EMTF system diagram.

A black-box modeling approach to characterizing the EMTF (i.e., to obtaining models of the required instrumentation) has been taken. "Black box" means that the models are fitted to the input–output data records with no attempt to relate the fitted or identified model to the physics of the problem. The identified models may, of course, be thought of as digital filters, and these filters (or their inverses) can be used to reduce noise and distortion in the data records. Therefore, we may view the identification of the various components of the EMTFs instrumentation as a means for designing noise-reduction filters.

We now describe an identification we performed on the pulse generator. Because no input was available and because we needed a model for simulation purposes, we assumed that the input was the impulse sequence, as shown in Fig. 12-2. Before proceeding with the identification, it is first necessary to consider some of the factors we outlined in the Introduction. We decided to use an ARMAX model and minimize the square of the prediction error via a recursive algorithm. The particular algorithm we used is the recursive maximum-likelihood method.[46] For a starting point, we used the estimate given by the linear least-squares estimator of Eq. 12-23.

In performing the actual identification, we took the following steps:

1. Data preprocessing
2. Order testing

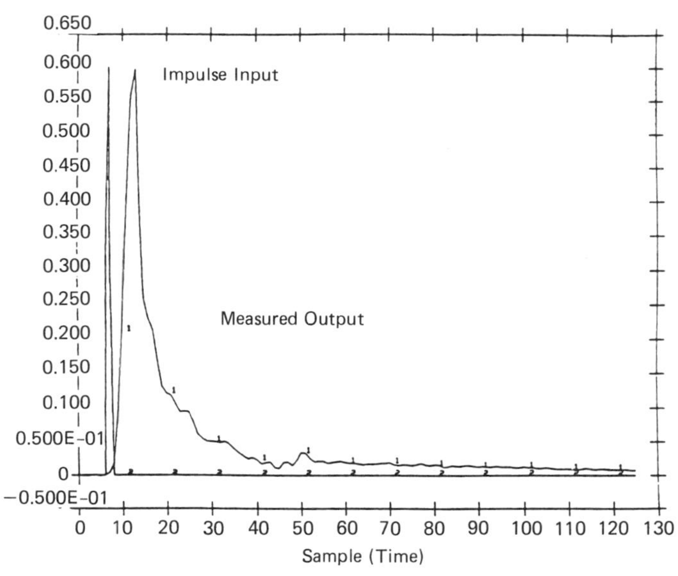

Fig. 12-2. Preprocessed pulser measurement data.

3. Parameter estimation
4. Model validation

We have discussed all of these steps previously except for model validation, which we will treat in this section. To get a better estimate of the output of the pulser, an ensemble of ten data records was collected, and these records were averaged. The resulting signal is shown in Fig. 12-2. The Fourier transform of this signal showed it to be highly oversampled; therefore, to reduce the computational load and improve conditioning, the signal was filtered at 500 MHz and decimated by a factor of ten.

The appropriate model order can be estimated in a variety of ways. One of the methods suggested in Reference 11 is to estimate ARMAX models of several different orders, excite each of the models with the observed input, and overlay the resulting outputs on the observed output. The results for the pulser are shown in Fig. 12-3. This method is called the *signal error test*.

Once the order and the parameters have been estimated, the final step is validation; that is, we must verify that the estimated model makes sense. Before proceeding to more complicated tests, it is often useful to overlay the one-step-ahead predicted response, $\{\hat{y}(k)\}$, on the actual response, $\{y(k)\}$, to see how well the model behaves as a predictor. The results for a sixth order model are

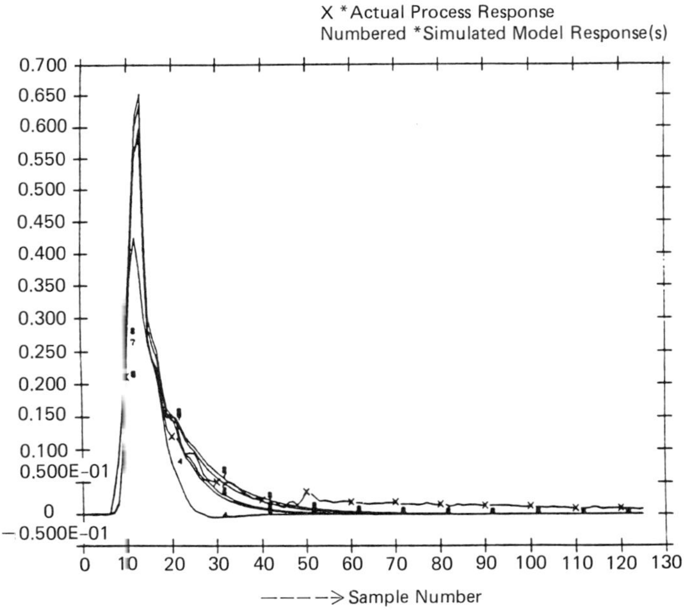

Fig. 12-3. Signal error order test.

shown in Fig. 12-4. If our model is actually the same as the system that generated the data records, then the prediction error sequence, $\{e(k)\} = \{y(k)\} - \{\hat{y}(k)\}$, will be zero-mean and white.[11] Therefore, another way to validate the model is to perform a whiteness test on the prediction-error sequence. One such test concludes that the whiteness hypothesis is true if 95 percent of the sample autocorrelation values of $\{e(k)\}$ lie within the two-sigma bounds.[32,48] The results for our sixth order model are shown in Fig. 12-5; the dotted lines are the bounds. The prediction error is almost white because only 10 percent of the samples exceed the bounds and the mean is almost zero.

A third method for validating the model is to perform statistical tests on the ensemble of data records. One such technique is to subtract the estimated response sequence generated by the model, $\{\hat{y}(k)\}$, from each of the output records in the ensemble to form an ensemble of model errors, i.e.,

$$\{e_j(k)\} = \{y_j(k)\} - \{\hat{y}(k)\} \tag{12-29}$$

where j corresponds to the jth record in the ensemble. The error ensemble is then evaluated statistically by computing the ensemble mean and standard deviation. These statistics can then be used to give a measure of model performance. We know that the mean should be small and that the variance should be close to that of the assumed white noise input.[3] The filtering of noisy data—collected

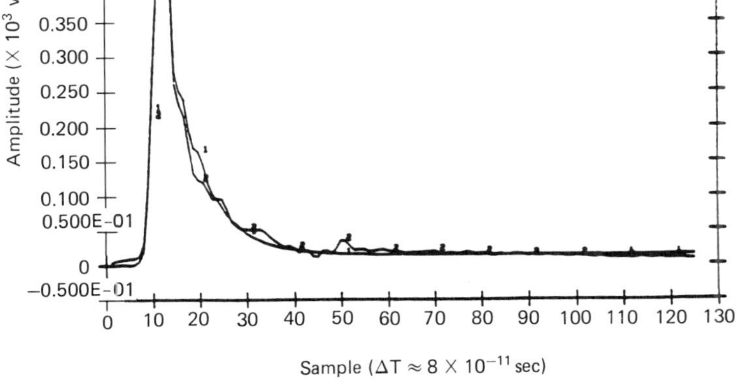

Fig. 12-4. Estimated and actual pulser response at the EMTF.

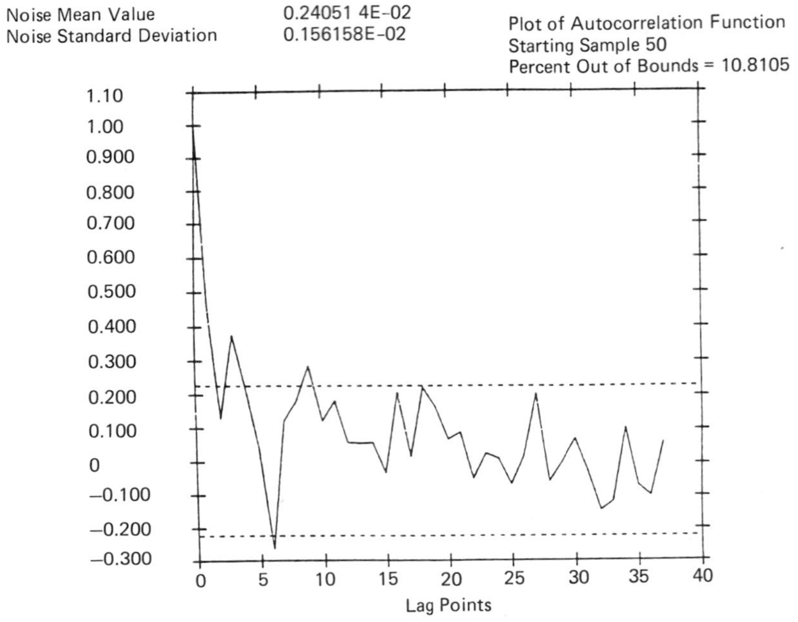

Fig. 12-5. Whiteness tests of estimated pulser data.

by a free-field probe at the EMTF—is depicted in Fig. 12-6(a). Note that the identified model for the given input has filtered or reduced the noise significantly. Details of the simulation can be found in References 51 and 52.

A simple method for removing instrumentation effects from the measured data requires the solution of the *deconvolution problem,* which can simply be stated as follows:

> *Given* the measured data $\{y(k)\}$ and impulse response $\{h(k)\}$ of a linear system, *find* the unknown input sequence $\{u(k)\}$ driving the system.

Our problem is more complex because the measured data are noisy. We can use the identified model to perform the deconvolution simply by constructing the corresponding inverse filter. We use the following *deconvolution procedures:*

1. Identify the ARMAX model (impulse response) from an independent set of input-output data.
2. From the identified *model,* $[B(z)/A(z)]$ construct the inverse model, $[A(z)/B(z)]$.
3. Select a set of measured outputs, $\{y(k)\}$, that are different than the set used to identify the model in step 1.

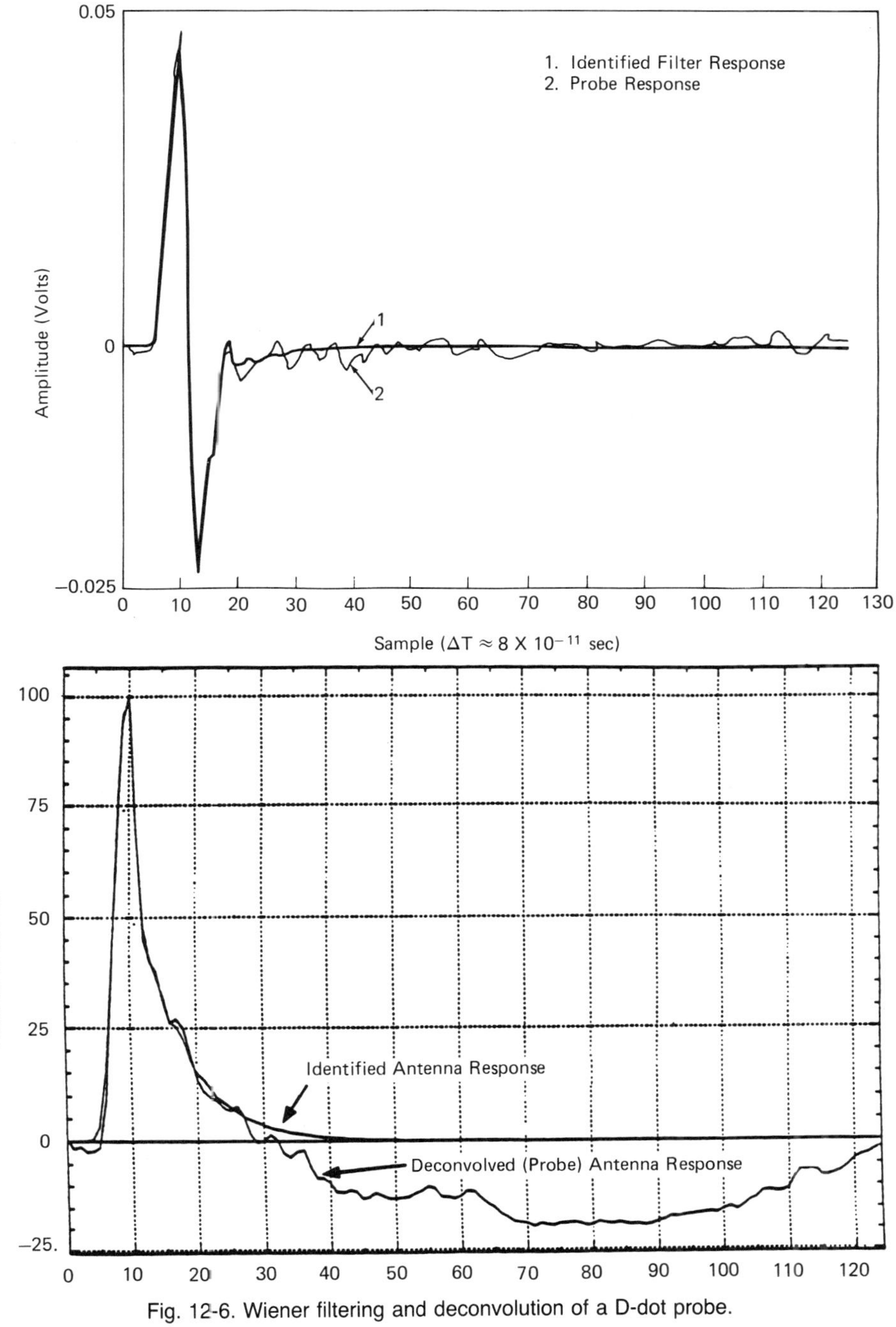

Fig. 12-6. Wiener filtering and deconvolution of a D-dot probe.

4. Apply the $\{y(k)\}$ from step 3 to drive the inverse model $[A(z)/B(z)]$ to generate an estimate of the input sequence $\{u(k)\}$.

In the case of the D-dot probe, we identified the process model as shown in Fig. 12-6(a). The probe was deconvolved from the free-space response data using the above procedure. The estimated input is overlayed with the simulated antenna response for comparison in Fig. 12-6(b). Here we see that over approximately the first 30 samples of the response (the high signal-to-noise ratio region, attributable to the antenna only), the deconvolution procedure works quite well; however, when the signal-to-noise ratio deteriorates (past 30 samples), the deconvolved input is not very good. This deterioration is to be expected, since, when the transient signal (antenna response) dies out, the deconvolver merely integrates the noise. More sophisticated deconvolution techniques for transient signals have become available.[4,18,48] We mention in passing that, if the probe model is not desired for simulation purposes, then, in step 1 of the procedure, the deconvolver could be identified directly by letting the measured output be the input to the identification algorithm and the measured input be the output. In this way, we can identify the deconvolver directly and merely apply it as a digital filter to the measured response in order to obtain the deconvolved input. This concludes our discussion of the signal processing at the EMTF. In the next section, we will use the signal-processing approach to estimate the impedance and effective height of an antenna.

Antenna Impedance and Effective Height

Scale-model testing and analysis is often a useful tool in designing electronic equipment that is hardened against the effects of EMP. Time-domain-reflectometry (TDR) measurements are particularly useful for antenna testing and modeling. In this section we show how signal processing techniques can be used to estimate the impedance and effective height of a test antenna from noise-corrupted TDR measurements.

The load voltage $v_l(t)$, of an antenna being tested is measured by exciting the antenna with a transient pulse [see Fig. 12-7(a)]. Next the open-circuit voltage, $v_0(t)$, is measured by replacing the test antenna with a probe and conducting the same experiment. These two measurements permit the calculation of the antenna input impedance, $Z_a(f)$. The relationship between the two measured voltages is given by the reflection coefficient,

$$R(f) = V_l(f)/V_0(f) \tag{12-30}$$

The input impedance is in turn given by

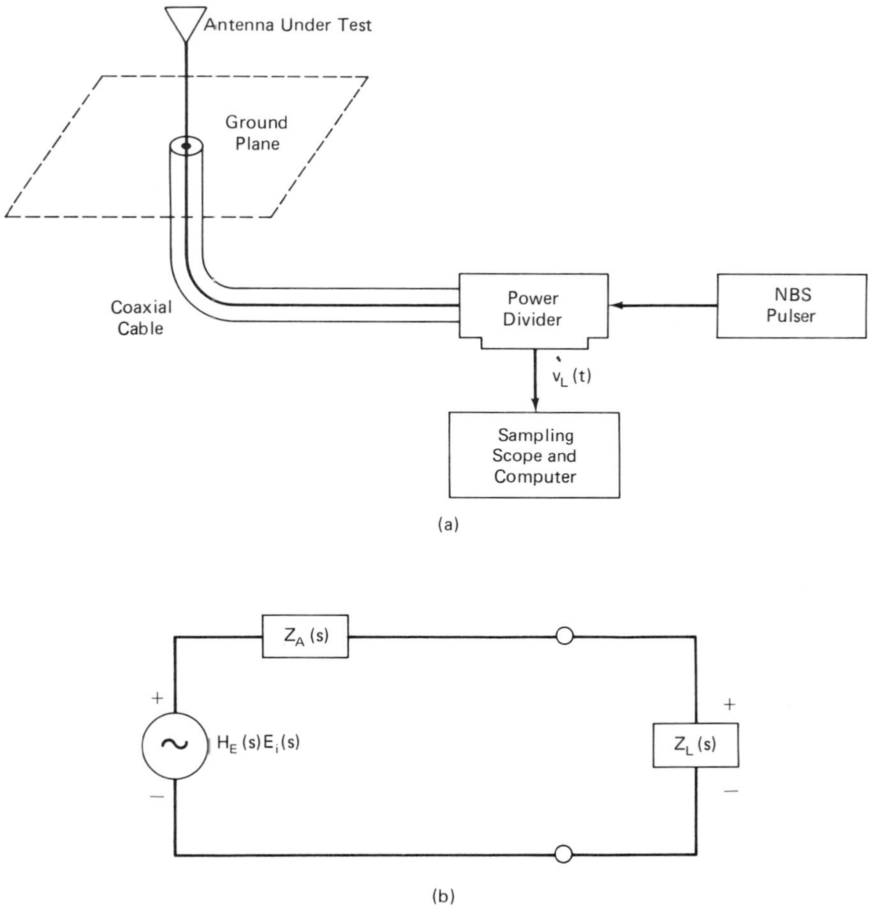

Fig. 12-7. (a) Antenna impedance measurement, and (b) effective height determination using Thevenin equivalent circuit [$Z_A(S)$ = antenna impedance; $Z_L(S)$ = load-circuit impedance; $H_E(S)$ = antenna effective height; and $E_i(S)$ = incident field].

$$Z_a(f) = \frac{[1 - R(f)]}{[1 + R(f)]} Z_0 = \frac{[V_0(f) + V_1(f)]}{[V_0(f) - V_1(f)]} Z_0 \qquad (12\text{-}31)$$

where Z_0 is the impedance of the transmission line used in the TDR experiment. One way to estimate the input impedance would have been to apply an identification technique to Eq. 12-30 and substitute the result in Eq. 12-31, but in Reference 53, the rightmost expression in Eq. 12-31 was used to estimate the impedance directly. The difference between $v_0(t)$ and $v_l(t)$ was taken to be the measured input to the unknown system, and their sum was taken to be the output. The estimated model is, of course, an estimate of $Z_a(f)$. The open-circuit-voltage records are shown in Fig. 12-8(a). These records were preprocessed and their sum and difference were computed. Impedance estimates were calculated using both frequency-domain and time-domain nonparametric methods and a parametric-output error method. The frequency responses of the three estimated models are shown in Fig. 12-8(b).

If the antenna is excited by an incident field, $e_i(t)$, and its output is the scattering voltage $v_s(t)$, then these two signals must satisfy the equivalent circuit shown in Fig. 12-7(b). From this figure it follows that the effective height can be calculated by using the equation,

$$H_e(f) = [1 + Z_n(f)]T(f) \tag{12-32a}$$

where

$$Z_n(f) = Z_a(f)/Z_1(f) \tag{12-32b}$$

and

$$T(f) = V_s(f)/E_i(f) \tag{12-32c}$$

Because $Z_a(f)$ has already been estimated and $Z_l(f)$ is known, an estimate of $Z_n(f)$ is available, but additional measurements must be made to estimate $T(f)$. A transient scattering experiment was conducted to obtain records of $e_i(t)$ and $v_s(t)$. A probe placed in the ground plane of the EMTF was used to measure the incident field at that point. Next, the probe was replaced with the test antenna and the output from the antenna measured. The three procedures we used to identify the impedance are now applied to estimate the transfer function, $T(f)$. These estimates, along with the estimates for the input impedance, are substituted into Eq. 12-32 to obtain estimates of the effective height, the results being shown in Fig. 12-9. Clearly, the three estimates are consistent with each other and can also be shown to be consistent with theory.[53]

By the use of frequency scaling, scale-model results are easily extrapolated to predict the responses of full-scale devices to EMP. If K is the scale factor (units of full-scale length per unit model length), then the frequencies of the full scale model will be lowered by a factor of $1/K$, and, of course, the lengths will be increased by a factor of K. In particular,

$$Z_a^{\text{full}}(f) = Z_a^{\text{scale}}(f/K) \tag{12-33a}$$

and

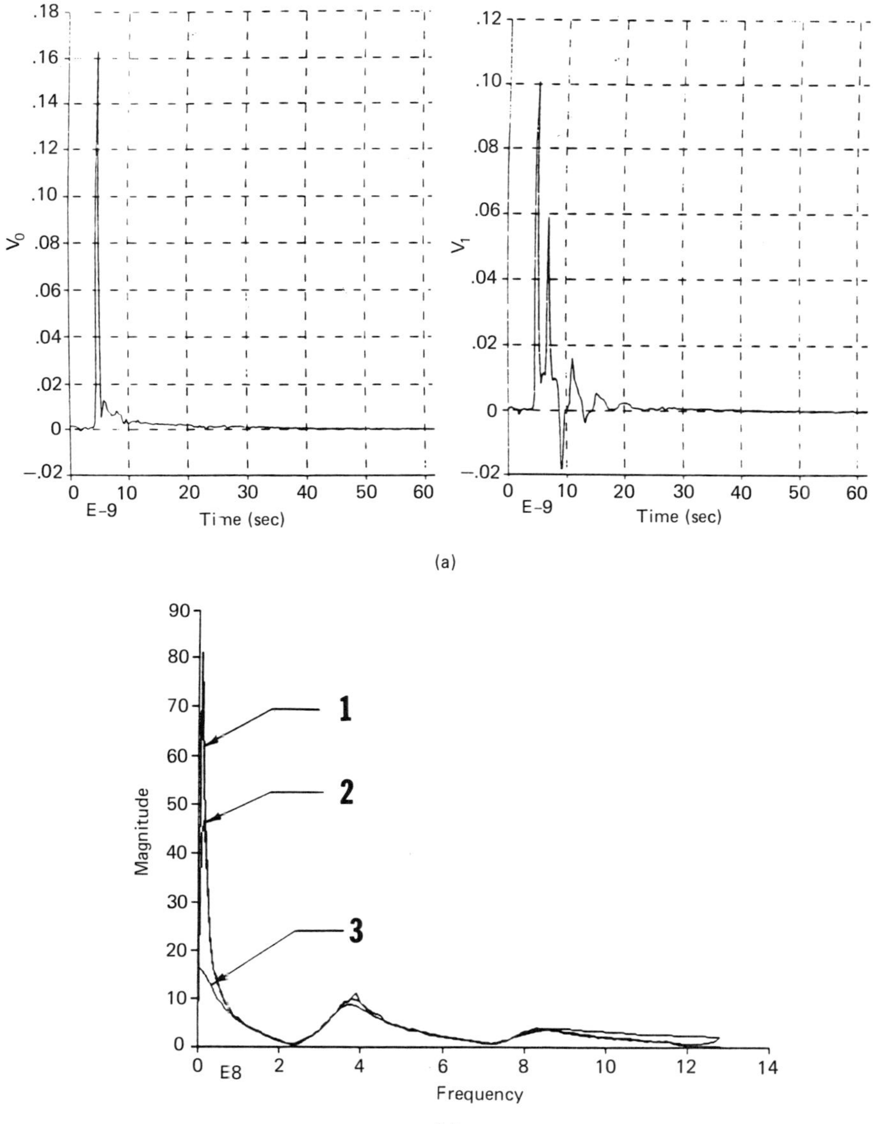

Fig. 12-8. Antenna impedance: (a) measurements, and (b) estimated models [(1) frequency-domain division, (2) FIR identification, and (3) sixth-order IIR identification].

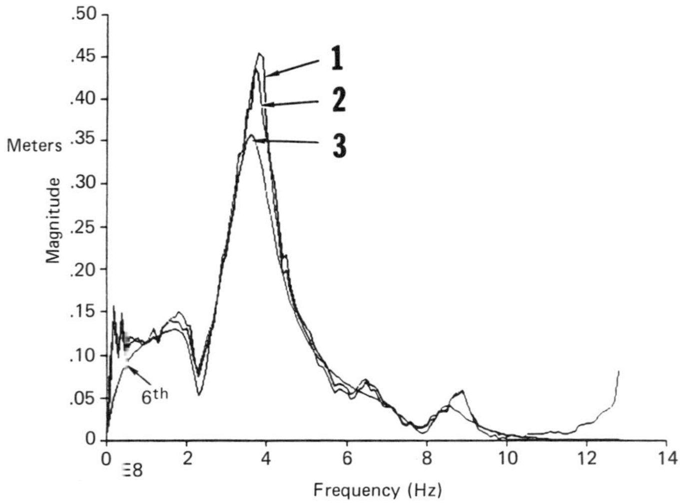

Fig. 12-9. Antenna effective height using three methods: (1) frequency-domain division, (2) FIR identification, and (3) sixth-order IIR identification.

$$H_e^{\text{full}}(f) = K\, H_e^{\text{scale}}(f/K) \qquad \text{(12-33b)}$$

The procedures we outlined in this section are covered in much greater detail in Reference 53; for a comparison of scale model estimates to full-scale test results for a ship, see Reference 54.

EMP Scale Model Testing and Estimation

In this section, we will discuss the application of identification techniques to the problem of extrapolating an F-14 aircraft's response to EMP from experiments performed on a scale model. The main objective of this effort is to compare the predicted responses with measurements made on an actual aircraft. The test performed on the scale-model F-14 duplicated the measurements of external surface current and charge made on a full-scale F-14 at the Air Force Weapons Laboratory's horizontally and vertically polarized dipole facilities. The identification techniques described in Sec. 12.1 were used to obtain transfer-function models of the scale model's behavior.[55] The objective is to scale these transfer functions and drive them with records of the excitations at the full-scale facility. We limit our discussion to the estimation of a single transfer function.

The tests on scale models were performed at the EMTF described in Sec. 12.2. In these tests, the aircraft was placed in a particular flight configuration and instrumented with probes located at points of interest. It was then excited with a wave of the appropriate polarization and its response sampled and recorded. After the data was preprocessed, an identification technique was applied. The frequency response of the model for one such configuration is shown in Fig. 12-10. As in the tests described in the last section, a parametric method as well as frequency-domain and time-domain nonparametric methods were applied. This time, however, the results, although similar to each other, are not as consistent as they were in the other tests. This is because an aircraft is a highly complex structure and is much more difficult to model. Furthermore, the data records were noisier.

Electromagnetic Object Identification

In this section, we will discuss the identification of the parameters of a simple electromagnetic object—a dipole. Our purpose is to estimate the modes ("poles") of the dipole and compare them with the ones predicted by the singularity expansion method.[8,56]

A 0.4-m, thin-wire dipole [Fig. 12-11(a)] was mounted on the EMTF ground plane and excited with a transient electromagnetic field. Its response was measured by means of an electric-field probe mounted at its center and connected

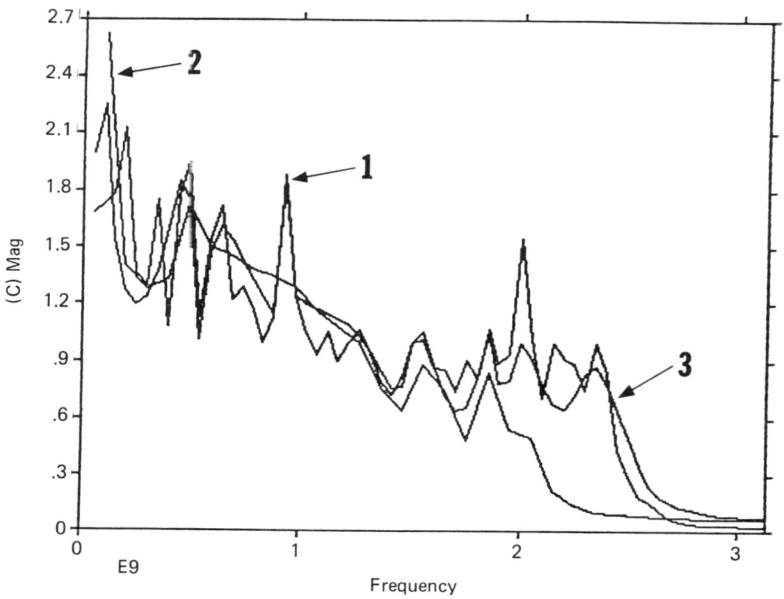

Fig. 12-10. Estimated EMP response of F-14 aircraft using scale-model tests: (1) frequency-domain division, (2) FIR identification, and (3) IIR identification.

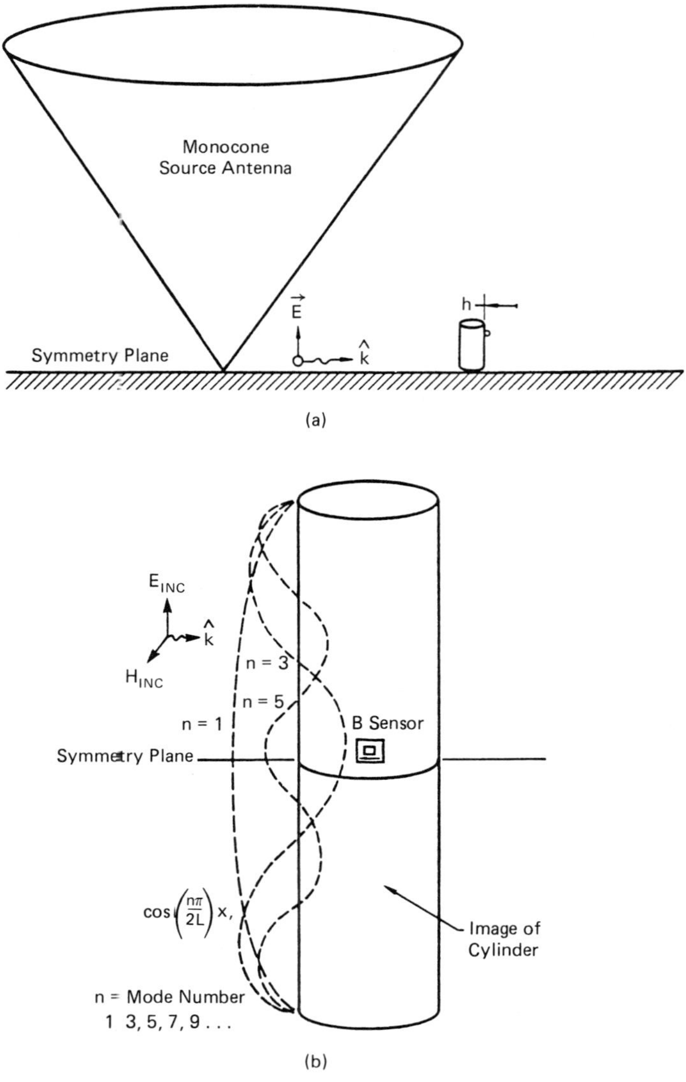

Fig. 12-11. Dipole response: (a) experimental setup, and (b) modes.

directly to a sampling scope. Preprocessed versions of the excitation (scaled by 0.01) and the dipole response are shown in Fig. 12-12(a). These data were low-pass filtered by an antialiasing filter and decimated by a factor of two. Furthermore, they were obtained from an ensemble average of ten records. The spectrum of the response is shown in Fig. 12-12(b). The first two resonances appear to

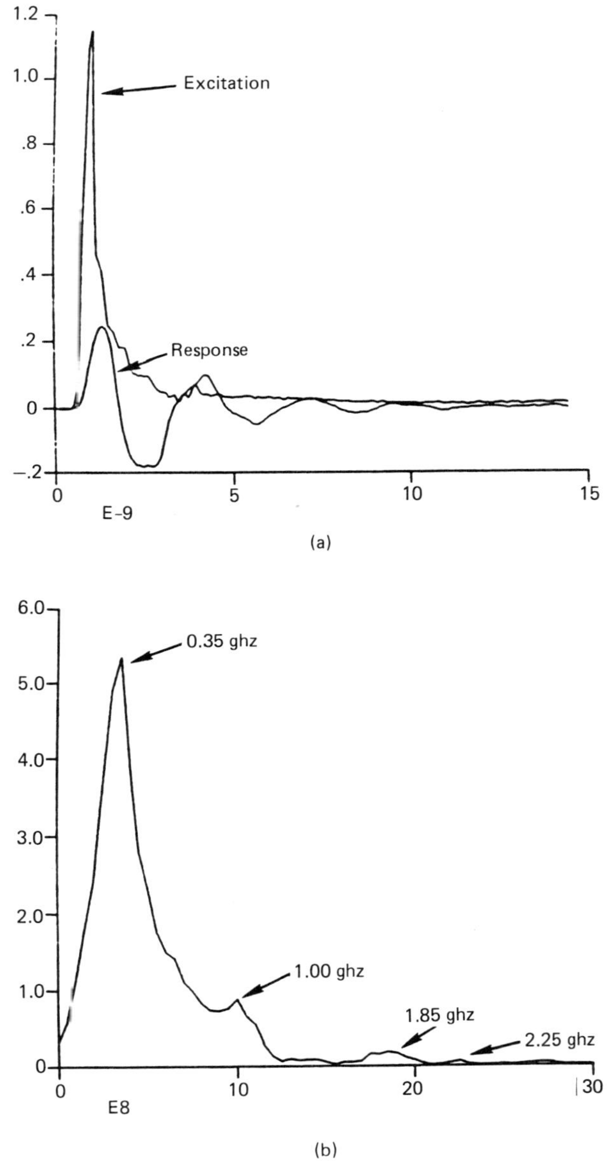

Fig. 12-12. Measured dipole response: (a) temporal data, and (b) frequency spectrum.

be adequately excited, but the others are not and are probably badly contaminated with noise. We should expect to be able to identify the first two easily, therefore, but to have trouble with the others because of the low signal levels.

The two algorithms we used to estimate the parameters were the recursive extended least-squares (RELS) algorithm for ARMAX-equation-error modeling[46] and the nonlinear least-squares (NLS) algorithm for output-error modeling.[30] Models were tried for several different orders and for several different starting points. Many local minima were observed for orders ranging from 5 to 15. To obtain better starting points and, one hopes, to avoid converging to a nonglobal minimum, the least-squares estimate given by Eq. 12-23 was used as a starting point. The recursive least-squares (RLS) algorithm was used to compute this estimate. This algorithm was easily programmed to run on a small minicomputer at the EMTF, but the more complicated RELS and NLS algorithms required a larger machine.

Order tests performed with the RLS algorithm revealed that either a twelfth or fifteenth order model contained approximations to most of the theoretically predicted resonances and provided reasonable fits to the data. Furthermore, the prediction-error sequence satisfied a statistical test for whiteness. We decided that an ARMAX(15,12,0) model estimated with RLS was a reasonable starting point for the RELS and NLS algorithms.

The RELS and NLS algorithms were applied to the data records with the starting point provided by RLS. The RELS algorithm was used to estimate an ARMAX(15,12,2) model, and the result, according to the final prediction error criterion, was found to be superior to the model estimated with RLS. Similarly, an output error model, with $M_a = 15$ and $M_b = 12$, was estimated with NLS. The identified response and spectrum are shown in Fig. 12-13; the fits are reasonable. The identified parameters were transformed to the continuous domain, and estimates of the poles were calculated by the impulse-invariant transformation method.[19]

Tesche[56] has calculated theoretical values for the poles of a dipole; these results are dependent on both its length and its diameter and are typically normalized to remove the effects of the former. Table 12-1 shows the theoretical and the estimated poles. Since the response is measured at the center of the dipole, only the odd numbered poles in References 51 and 52 are excited. It is clear that neither algorithm gives good estimates of the damping factors (real part of the poles). This result is not unexpected because the amount of data in a transient signal is limited. Both algorithms perform much better on the frequency estimates (imaginary parts) and are even able to extract some of the higher frequency resonances. We conclude that, for a simple object such as a dipole, the singularity-expansion method coupled with the appropriate identification technique can be a useful tool for evaluating transient responses.

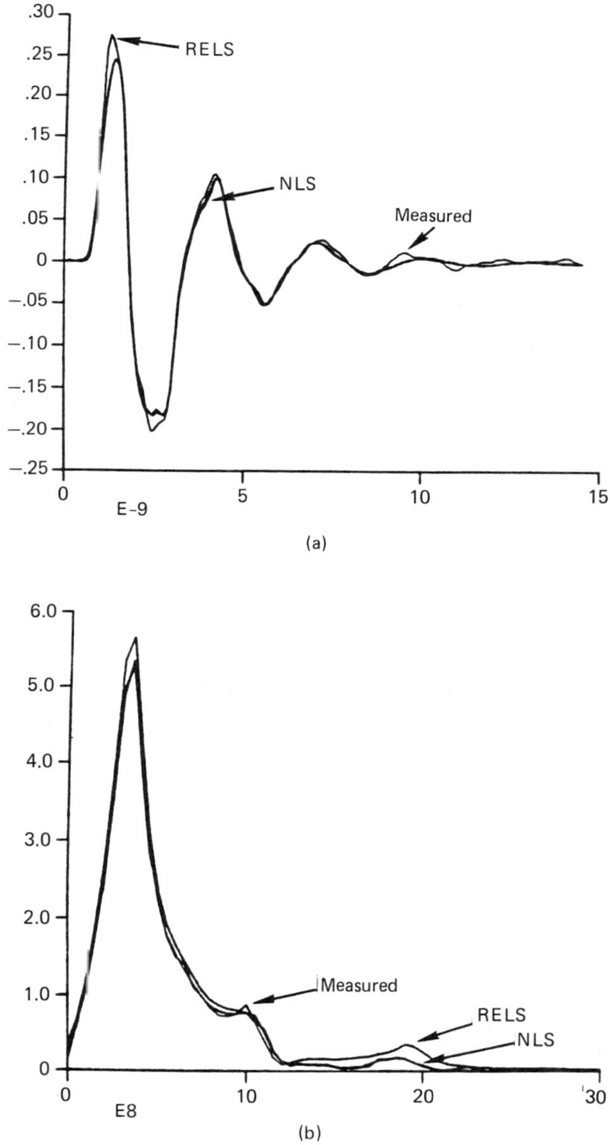

Fig. 12-13. Identified dipole response: (a) temporal, and (b) spectra.

Table 12-1. Dipole Resonance Identification.

POLE	THEORETICAL		RELS (on-line)—ARMAX (15,12,2)				NLS (off-line)—ARMAX (15,12,0)			
	$-d(\times 10^8)$	f(GHz)	$-\hat{d}(\times 10^8)$	$\hat{d}_e$*	$\hat{f}$(GHz)	$\hat{f}_e$(GHz)†	$-\hat{d}(\times 10^8)$	$\hat{d}_e$*	$\hat{f}$(GHz)	$\hat{f}_e$(GHz)†
(1)	0.30	0.35	0.78	1.57	0.33	0.04	0.79	1.59	0.33	0.04
(3)	0.53	10.87	1.05	0.99	10.77	0.01	0.84	0.59	10.87	0.
(5)	0.66	18.25	0.40	0.30	19.35	0.06	1.43	1.17	17.17	0.06
(7)	0.79	25.65	2.38	2.01	28.19	0.10‡	1.03	0.31	18.99	0.26
(9)	0.99	32.76	1.75	0.76	37.83	0.16‡	1.02	0.02	35.9	0.10
(11)	1.32	40.04	1.77	0.34	48.13	0.20	0.50	0.62	49.15	0.23

$*\hat{d}_e = \dfrac{d_{\text{TRUE}} - \hat{d}}{d_{\text{TRUE}}}$

$\dagger\hat{f}_e = \dfrac{f_{\text{TRUE}} - \hat{f}}{f_{\text{TRUE}}}$

‡Residue order of magnitude low.

12.3 CONCLUSIONS

We began this chapter with a discussion of several practical problems in interpreting data from transient electromagnetic experiments. This discussion motivated our definition of the electromagnetic estimation problem. We then showed that this problem could be placed in a system identification context, and we did this following the same general structure used by Ljung.[13,14,46] Next we presented some of the most useful techniques for solving the electromagnetic-estimation problem. We argued that preprocessing the data to prevent aliasing, to remove distortions, and to provide an optimal-sampling rate was a critical first step before applying any technique to model the test object. We then presented some of the most popular modeling approaches, including both frequency-domain and time-domain nonparametric methods as well as parametric methods for difference equation models.

From the results presented in this and the subsequent chapter, we conclude that simple ordinary difference equation models appear to be adequate for simple electromagnetic structures such as cylinders, spheres, slabs, or dipoles, but not for highly coupled, complex structures such as ships or aircraft. Very little work has been done to understand EMP behavior by identifying more sophisticated models (such as ones using partial differential equations). In the ideal situation, the researcher would develop a model using Maxwell's equations, design an experiment to excite the significant modes of the test object and then use a model-based identification algorithm to estimate the parameters in his model.[48,51–61] Depending on the outcome, the model is then changed until an adequate one evolves. Such an approach has been taken for complex mechanical parts subjected to vibration tests.[12]

One problem with model-based estimation techniques is that the numerical methods used to minimize whatever criterion function is selected may need to be very sophisticated and the computational load may be quite high. In the case of electromagnetic systems, the complexity of the minimization problem depends, of course, on the model set that was selected. For example, the singularity-expansion theory leads to identification methods for ordinary difference equations, but one approach to modeling the test object as a distributed system (i.e., as on object that satisfies a partial differential equation) leads to algorithms based on state-space models.[3,48] Obviously, the more complex the model, the more complex the algorithm to estimate its parameters. In principle, algorithms already exist to solve the electromagnetic-estimation problem for several classes of complex models. Unfortunately, however, many of the algorithms will prove to be insufficiently robust in practice to converge to a physically meaningful solution. Parameter estimation for more complex classes of models is currently an active area of research. In the future, as more computer power becomes available and the models are better understood, it is possible that robust, large-scale, model-based estimation algorithms that are applicable to the electromagnetic-estimation problem will be developed.

REFERENCES

1. Longmire, C. L. 1978. On the electromagnetic pulse produced by nuclear explosions. *IEEE Trans. Antennas Propag.* AP-26.
2. Wellstead, P. E. 1981. Nonparametric methods of system identification. *Automatica* 17.
3. Eykhoff, P. 1974. *System Identification*. New York: Wiley.
4. Bhargava, U. K.; Kashyap, R. L.; and Goodman, D. 1985. Impulse response identification with deterministic inputs using non-parametric methods. *Proceedings on 23rd Allerton Conference on Communications, Control and Computing*. Also available as LLNL Report UCID–
5. VanBlaricum, M. L , and Mittra, R. 1975. A technique for extracting poles and residues of a system directly from its transient responses. *IEEE Trans. Antennas Propag.* AP-23.
6. Poggio, A. J.; VanBlaricum, M. L.; Miller, E. K.; and Mittra, R. 1978. Evaluation of a processing technique for transient data. *IEEE Trans. Antennas Propag.* AP-26.
7. Dudley, D. G. 1979. Parametric modeling of transient electromagnetic systems. *Radio Science* 14.
8. Baum, C. E. 1971. On the singularity expansion method for the solution of EM interaction problems. *AFWL Interaction Note 88*.
9. Goodwin, G. C., and Payne, R. L. 1977. *Dynamic system identification experiment design and data analysis*. New York: Academic Press.
10. Zarrop, M. B. 1979. *Optimal experiment design for dynamic system identification*. Berlin: Springer-Verlag.
11. van den Boom, A., and van den Enden, 1974. The determination of the orders of process and noise dynamics. *Automatica* 10.
12. Lager, D.; Azevedo, S.; and Candy, J. 1982. The application of large-scale estimation to vibrating structures *Proceedings of the Sixth IFAC Symposium on Identification and System Parameter Estimation*.
13. Ljung, L. 1982. Aspects on the system identification problem. *Signal Processing* 4: 445–460.

14. Ljung, L., and Glover, K. 1981. Frequency-domain versus time-domain methods in system identification. *Automatica* 17.
15. *Proceedings of Sixth IFAC Symposium on Identification and System Parameter Estimation,* Arlington, Va., June 1982. (See Workshop Sessions 00 and 01.)
16. Papoulis, A. 1966. Error analysis in sampling theory. *Proceedings of the IEEE* 54, No. 7: 947–955.
17. Jerri, A. J. 1977. The Shannon sampling theorem—Its various extensions and applications, a tutorial review. *Proceedings of the IEEE* 65, No. 11: 1565–1596.
18. Candy, J. V., and Zicker, J. E. 1982. Deconvolution of noisy transient signals: A Kalman filtering approach. *LLNL Report* UCRL-87432 or *Proc. IEEE Confr. Decision Contr.*, Washington.
19. Oppenheim, A. V., and Schafer, R. W. 1975. *Digital signal processing*. Prentice-Hall: New Jersey.
20. Brigham, E. O. 1974. *The Fast Fourier Transform*. Prentice-Hall: New Jersey.
21. Rabiner, L. R., and Gold, B. 1975. *Theory and application of digital signal processing*. Prentice-Hall: New Jersey.
22. Astrom, K. J. 1969. On the choice of sampling rates in parameter identification of time series. *Information Sciences,* pp. 273–287.
23. Kay, S. M. 1981. The effect of sampling rate on autocorrelation estimation. *IEEE Transactions on Acoustics, Speech and Signal Processing* ASSP-29, No. 4: 859–867.
24. Crittenden, J. L.; Mulholland, R. J.; Hill, IV, J.; and Martinez, E. F. 1983. Sampling and data analysis properties of Prony's method applied to model identification. *International Journal of System Sciences* 14, No. 5: 571–584.
25. Crochiere, R. E., and Rabiner, L. R. 1975. Optimum fir digital filter implementations for decimation, interpolation, and narrow-band filtering. *IEEE Acoustics, Speech, and Signal Proc.* ASSP-23.
26. Crochiere, R. E., and Rabiner, L. R. 1981. Interpolation and decimation of digital signals—A tutorial review. *Proc. IEEE* 69.
27. Clark, G. A., and Bogdan, E.J. 1982. Summary of rules for sampling, decimation, and interpolation of experimental data. *LLNL Report* UCID-19411.
28. Marple, S. L. 1982. Frequency resolution of Fourier and maximum entropy spectral estimates. *Geophysics* 47.
29. Harris, F. J. 1978. On the use of windows for harmonicanalysis with the Discrete Fourier Transform. *Proc. IEEE* 66, No. 1: 51–83.
30. Goodman, D. M. 1983. NLS: a system identification package for transient signals. *LLNL Report* UCID-19767.
31. Kay, S. M., and Marple, Jr., S. L. 1981. Spectrum analysis, a modern perspective. *Proc. IEEE* 69, No. 11: 1380–1419.
32. Jenkins, G. M., and Watts, D. G. 1968. *Spectral analysis and its applications*. San Francisco: Holden-Day.
33. Bendat, J., and Piersol, A. 1971. *Random data: Analysis and measurement procedures*. New York: Wiley.
34. Bendat, J., and Piersol, A. 1980. *Engineering applications of correlation and spectral analysis*. New York: Wiley.
35. Draper, N. R., and Smith, H. R. 1966. *Applied regression analysis*. New York: Wiley.
36. Hunt, B. R. 1971. Biased estimation for nonparametric identification of linear systems. *Mathematical Biosciences* 10: 215–237.
37. Gray, R. M. 1972. On the asymptotic Eigenvalue distribution of Toeplitz matrices. *IEEE Trans. Info. Theory* IT-18, No. 6: 725–730.

38. Twomey, S. 1965. The application of numerical filtering to the solution of integral equations encountered in indirect sensing measurements. *J. Franklin Inst.* 279: 95–109.
39. Treitel, S., and Robinson, E. A. 1966. The design of high resolution digital filters. *IEEE Trans. Geoscience Electronics* GE-4, No. 1: 25–38.
40. Ekstrom, M. P. 1973. A spectral characterization of the Ill-conditioning in numerical deconvolution. *IEEE Trans. Audio and Electroacoustics* AU-21, No. 4: 344–348.
41. Ekstrom, M. P., and Rhodes, R. L. 1974. On the application of Eigenvector expansions to numerical deconvolution. J. Computational Physics 14, No. 4: 319–340.
42. Hoerl, A. E., and Kennard, R. W. 1970. Ridge Regression: Biased estimation for nonorthogonal problems. *Technometrics* 12, No. 1: 55–67.
43. Nashed, M. Z. 1981. Operator-theoretic and computational approaches to Ill-posed problems with applications to antenna theory. *IEEE Trans. Antennas and Progagation* AP-29, No. 2: 220–231.
44. Sarkar, T. K.; Weiner, D. D.; and Jain, V. K. 1981. Some mathematical considerations in dealing with the inverse problem. *IEEE Trans Antennas and Propagation* AP-29, No. 2: 373–379.
45. Astrom, K., and Eykhoff, P. 1971. System identification—a survey. *Automatica* 7, pp 123–162.
46. Ljung, L., and Soderstrom, T. 1983. *Recursive identification*. Boston: MIT Press.
47. Anderson, B. D., and Moore, J. 1979. *Optimal filtering*. Englewood Cliffs, N.J.: Prentice-Hall.
48. Candy, J. V. 1986. *Signal Processing: The model-based approach*. New York: McGraw-Hill.
49. Gavel, D. T.; Candy, J. V.; and Lager, D. L. 1980. Parameter estimation from noisy EM measurements. *Nuclear EMP Mtg* and *LLNL Report* UCRL-84437.
50. Candy, J. V., and Zicker, J. E. 1982. Identification of electromagnetic object response from noisy transient measurements. *LLNL Report* UCRL-86135 or *IFAC Symposium on System Ident.*, Washington, D.C.
51. Zicker, J. E., and Candy, J. V. 1982. Electromagnetic test facility characterization: An identification approach. *LLNL Report* UCID-19620.
52. McGevena, V. G. 1985. Signal processing for the analysis of transient electromagentic data. *LLNL Report* UCID-20304.
53. Clark, G. A.; Martin, L. C.; and Bogdan, E. J. 1983. Identification of antenna parameters from time-domain pulse response data. *LLNL Report* UCID-19770.
54. Deadrick, F. J., and Cabayan, H. S. 1980. EMP coupling to ships. *LLNL Report* UCRL-52803.
55. Clark, G. A., and Bogdan, E. J. 1983. Identification of electromagnetic response models for aircraft using transient time domain measurements. *LLNL Report* UCID-.
56. Tesche, F. M. 1973. On the analysis of scattering and antenna problems using the singularity expansion technique. *IEEE Trans. Antennas Propag.* AP-21.
57. Ljung, L. 1981. Analysis of a general recursive prediction error identification algorithm. *Automatica* 17.
58. Kayshap, R. L., and Rao, A. R. 1976. *Dynamic stochastic models from empirical data*. New York: Academic Press.
59. Box, G. E. P., and Jenkins, G. M. 1976. *Time series analysis forecasting and control*. San Francisco: Holden Day.
60. Sage, A. P., and Melsa, J. L. 1971. *System Identification*. New York: Academic Press.
61. Rosenfeld, A., and Kak, A. C. 1982. *Digital picture processing*. New York: Academic Press. (See Chapter 7.)

13 Transient Identification and Object Classification

D. G. Dudley

Professor of Electrical Engineering University of Arizona

D. M. Goodman

Lawrence Livermore National Laboratory

13.0 INTRODUCTION

In Chap. 12, the authors presented an overview of signal processing and identification techniques applicable to transient electromagnetic data. We call such considerations the "identification problem." Their procedure makes use of a model and a strategy whereby the parameters of the model were adjusted to make it "close," in some sense, to the electromagnetic process. One strategy, for example, is to adjust the model parameters to make the output, $\tilde{y}(t)$, from the model close to the output, $y(t)$, from the process in the sense that the Euclidean norm of the error, $y(t) - \tilde{y}(t)$, is minimized.

In the present chapter, we will consider the possibility of determining certain physical characteristics of the electromagnetic process from transient data. Specifically, we ask the following question: Can the model parameters identified during solution to the identification problem be transformed into parameters that classify the electromagnetic process? We call these considerations the "classification problem." Solution to this problem has important application in fields as diverse as radar-target determination and geophysical exploration. Indeed, the basics transcend electromagnetics and can be considered in a variety of disciplines dealing with experiments producing transient data.

As a simple example of the concepts, consider the transient backscatter from a perfectly conducting sphere of diameter D. Suppose the observables in an experiment involving the sphere are a component, $x(t)$, of the incoming electric field and a component, $y(t)$, of the backscattered electric field. We preprocess $x(t)$ and $y(t)$, using the concepts in Chap. 12, to produce the input and output sequences, $x(k)$ and $y(k)$, respectively, where $k = 1,2 \ldots, K$. As a model of this example electromagnetic process, suppose we adopt the pole-zero (PZ) model in Eq. 12-19. Define a model parameter vector **c** that contains the model parameters by

$$\mathbf{c}^{\mathrm{T}} = [a_0\, a_1 \ldots a_N\, b_0\, b_1 \ldots b_M] \tag{13-1}$$

where T denotes "transpose." Next, suppose that we identify the parameters in **c** by one of the techniques described in Chap. 12. The classification problem for this example can now be stated as follows: Given the identified model parameter vector **c**, find a transformation L such that

$$L\mathbf{c} = D \tag{13-2}$$

where D is the sphere diameter. In this example, if the transformation L can be found, the classification is termed "complete" in the sense that the parameter D produced by the transformation gives all the information necessary to describe the object.

A critical examination of the classification problem reveals many traps for the unwary. In the above example, we must recognize that certain a priori information has been assumed, namely, that the scattering object is spherical and perfectly conducting. Without this a priori information, classification with the parameter D loses its meaning. In addition, all experiments, real or computer-simulated, are carried out in the presence of noise, a fact that can have a serious effect on the accuracy of the identification of **c**, the existence of the transformation L, and the accuracy of the determination of D. Furthermore, the above example is simplistic. In actual application, most objects of practical interest are complex in both shape and constitution to a degree that complete classification is beyond the present state of both science and computation.

In this chapter, we begin with a discussion of object features, emphasizing the difference between local and global descriptions. We follow with an in-depth treatment of complex exponential series identification, emphasizing identification with a difference equation model. We discuss in particular a nonlinear least-squares algorithm (NLS) designed for transient data. Next, we turn to ray-optic identification. Finally, we end with a discussion of the basic issues and some conclusions.

13.1 OBJECT DESCRIPTIONS

Electromagnetic researchers studying transient phenomena are interested in interactions with a wide variety of objects. To describe a particular object, we might consider such physical features as size, local curvature, permittivity and conductivity. An electromagnetic transient interacting with the object cannot, however, measure most physical features directly. We therefore seek divisions of object descriptions into classes compatible with specific electromagnetic problem-solution methods. One such division, described by Heyman and Felsen,[1] distinguishes between "local" and "global" object features. The local features are described by a ray-optic solution, whereas the global features are described by a complex resonance solution. Heyman and Felsen[1] discuss this subject in quite general terms. Here, we shall be content to develop these ideas in a canonical example considered by many authors[2,3,4]—the backscatter from a lossless dielectric slab. We begin by developing both the ray-optic and complex resonance solutions, follow this with a discussion of how the two solutions exhibit object descriptions, and conclude with a short discussion of the relevance of the solution methods to identification and classification.

Example: Normal Backscatter From a Dielectric Slab

Consider a plane wave traveling in free space (Region 1). The wave encounters a dielectric slab (Region 2) of thickness d and dielectric constant ε_r (Fig. 13-1), where incidence is perpendicular to the slab surface. The slab is infinite in extent and backed by a perfectly conducting half-space. This problem has variations only in the z-coordinate direction. Putting the relationships, $\partial/\partial x = \partial/\partial y = 0$, in Maxwell's equations and assuming that the plane wave is linearly polarized with the electric field in the x-direction, we obtain, with the $\exp(i\omega t)$ time convention, the following well-known description[5] of the total fields in the two regions:

$$E_{x1} = e^{ikz} + \rho e^{-ikz} \tag{13-3}$$

$$H_{y1} = -\frac{1}{ik\eta_0}\frac{dE_{x1}}{dz} \tag{13-4}$$

$$E_{x2} = A_2^+ e^{i\beta z} + A_2^- e^{-i\beta z} \tag{13-5}$$

$$H_{y2} = -\frac{1}{ik\eta}\frac{dE_{x2}}{dz} \tag{13-6}$$

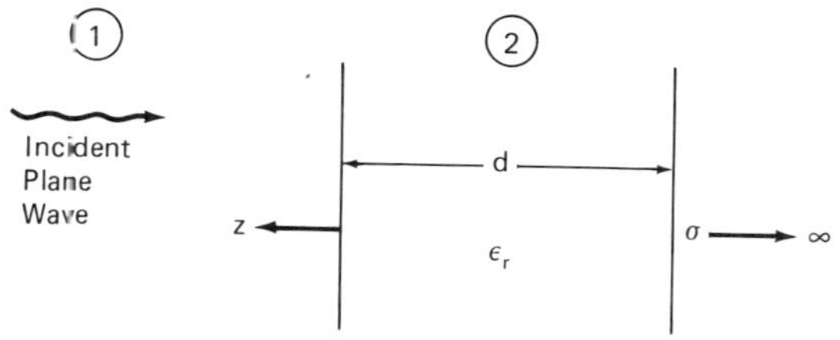

Fig. 13-1. Dielectric slab.

where E_{xi} and H_{yi} (i = 1 or 2) are the x-component of the E-field and the y-component of the H-field, respectively, in Regions 1 and 2. In addition, the wave numbers, k and β, and the intrinsic impedances, η_0 and η, are defined by

$$k = \frac{\omega}{c} \tag{13-7}$$

$$\beta = \frac{\omega}{v} = k\varepsilon_r^{1/2} \tag{13-8}$$

$$\eta_0 = 120\pi \tag{13-9}$$

$$\eta = \frac{\eta_0}{\varepsilon_r^{1/2}} \tag{13-10}$$

where k is the wave number in free space; β, the wave number in the dielectric slab; c, the phase velocity in free space; v, the phase velocity in the slab; ε_r, the dielectric constant of the slab; η_0, the intrinsic impedance of free space; and η, the intrinsic impedance of the slab. The coefficients ρ, A_2^+, and A_2^- are determined by application of the boundary conditions, namely, the continuity of tangential electric and magnetic fields at $z = 0$ and the vanishing of the tangential electric field at $z = -d$. The result for the reflection coefficient, ρ, is as follows:

$$\rho(\omega) = -\frac{R + e^{-i2\beta d}}{1 + Re^{-i2\beta d}} \tag{13-11}$$

where

$$R = \frac{\varepsilon_r^{1/2} - 1}{\varepsilon_r^{1/2} + 1} \tag{13-12}$$

Equation 13-3, when used in conjunction with Eqs. 13-7 through 13-12, gives the total electric field, $E_{x1}(z,\omega)$, in the frequency domain exterior to the dielectric slab. This formulation can be transformed to the time domain as follows: Let $f(t)$ be the input transient signal and $F(\omega)$ its Fourier transform. The electric field, $e_{x1}(z,t)$, in the time domain is then given through multiplication of Eq. 13-3 by the input spectrum, $F(\omega)$, followed by an inverse Fourier transformation, viz:

$$e_{x1}(z,t) = \frac{1}{2\pi}\int_{-\infty}^{\infty} F(\omega)E_{x1}(z,\omega)e^{i\omega t}d\omega \tag{13-13}$$

Substitution of Eq. 13-3 into 13-13 gives

$$e_{x1}(z,t) = f\left(t + \frac{z}{c}\right) + g\left(t - \frac{z}{c}\right) \tag{13-14}$$

where

$$g(t) = \frac{1}{2\pi}\int_{-\infty}^{\infty} F(\omega)\rho(\omega)e^{i\omega t}d\omega \tag{13-15}$$

The first term on the right of Eq. 13-14 is the incoming wave, which is the total field that would be present in the absence of the slab. The second term contains the scattering effects caused by the slab. These effects can be determined by performing the inverse transform indicated in Eq. 13-15. This transform can be accomplished by two different methods. In the first method, the reflection coefficient is expanded in a geometric series and the inverse transform taken term by term. The result is the ray-optic form of solution. In the second method, the complex ω-plane singularities of the reflection coefficient and the input spectrum are determined and the inverse transform performed by contour-integration techniques. The result is the complex-resonance form of solution. The details follow.

The Ray-Optic Solution. Consider the reflection coefficient, $\rho(\omega)$, given by Eq. 13-11. Note that since $R < 1$, we must have $|R\exp(-i2\beta d| < 1$. We therefore can expand Eq. 13-11 in a geometric series, viz:

$$\rho(\omega) = -\left[R + (1 - R^2)\sum_{n=1}^{\infty} R^{n-1}e^{-in2\beta d}\right] \tag{13-16}$$

We substitute Eq. 13-16 into Eq. 13-15 and note that the inverse transform can now be performed immediately term-by-term in the summation. Substitution of the result into Eq. 13-14 gives the electric field as follows:

$$e_{x1}(z,t) = f\left(t + \frac{z}{c}\right) - Rf\left(t - \frac{z}{c}\right) - T_1T_2 \sum_{n=1}^{\infty} (-R)^{n-1} f\left(t - \frac{z}{c} - 2n\frac{d}{v}\right) \qquad (13\text{-}17)$$

where

$$T_1 = 1 + R \qquad (13\text{-}18)$$

$$T_2 = 1 - R \qquad (13\text{-}19)$$

Equation 13-17 is the ray-optic solution for the dielectric slab. The terms on the right-hand side can be described as follows: The first term is the incoming pulse; the second, the pulse reflected from the front face of the dielectric ($z = 0$). The third term consists of an infinite summation representing successive reflections from the back face of the dielectric ($z = -d$). A significant feature of this solution form, provided that $f(t) = 0$ for $t < 0$, is that the series can be truncated after N terms and still give the exact answer for $t < z/c + 2Nd/v$.

The Complex Resonance Solution. To obtain the complex-resonance solution, we again consider the inverse transform in Eq. 13-15. Rather than expand the reflection coefficient in a series as was done in the ray-optic solution, we seek the solution by the calculus of residues.

To begin the procedure, we first note that it is not possible to obtain information on the complex resonances at the observation point until after the arrival of the transient signal from the back face of the slab ($z = -d$). This fundamental fact leads us to seek the complex resonance counterpart to the ray-optic solution only for the time, $t > z/c + 2d/v$. For earlier times, the two solutions will be identical. Indeed, in Eq. 13-17, it is the third term that will be expressed in complex-resonance form.

We multiply Eq. 13-3 by the spectrum $F(\omega)$ of the input pulse and obtain the following:

$$E_{x1}(z,\omega)F(\omega) = F(\omega)e^{ikz} + F(\omega)\rho(\omega)e^{-ikz} \qquad (13\text{-}20)$$

where $\rho(\omega)$ is given by Eq. 13-11. We separate the reflection coefficient in Eq. 13-11 into two terms as follows:

$$\rho(\omega) = -R + \tilde{\rho}(\omega) \tag{13-21}$$

where

$$\tilde{\rho}(\omega) = \tilde{\rho}_0(\omega)e^{-i2\beta d} \tag{13-22}$$

$$\tilde{\rho}_0(\omega) = -\frac{1 - R^2}{1 + Re^{-i2\beta d}} \tag{13-23}$$

Substitution of these results into Eq. 13-20 yields the following:

$$E_{x1}(z,\omega)F(\omega) = F(\omega)e^{ikz} - RF(\omega)e^{-ikz} + \tilde{\rho}_0(\omega)F(\omega)e^{-i(kz+2\beta d)} \tag{13-24}$$

The inverse Fourier transform of Eq. 13-24 gives the following:

$$e_{x1}(z,t) = f\left(t + \frac{z}{c}\right) - Rf\left(t - \frac{z}{c}\right) + \tilde{g}\left(t - \frac{z}{c} - 2\frac{d}{v}\right) \tag{13-25}$$

where

$$\tilde{g}(t) = \frac{1}{2\pi}\int_{-\infty}^{\infty} F(\omega)\tilde{\rho}_0(\omega)e^{i\omega t}d\omega \tag{13-26}$$

Consider the integral in Eq. 13-26 around a closed contour C (see Fig. 13-2), which consists of the interval $(-a,a)$ along the real axis and a semicircle, C_a, of radius a through the upper half of the ω-plane. By the basic theorem of the calculus of residues,[6] if $F(\omega)\rho(\omega)$ is analytic within and on the closed contour, except at a finite number of singular interior points, then

$$\oint_C F(\omega)\tilde{\rho}_0(\omega)e^{i\omega t}d\omega = 2\pi i \sum_{n=1}^{N} \text{Res}[F(\omega)\tilde{\rho}_0(\omega)\, e^{i\omega t},\omega_n] \tag{13-27}$$

$$= \int_{-a}^{a} F(\omega)\tilde{\rho}_0(\omega)e^{i\omega t}d\omega + \int_{C_a} F(\omega)\tilde{\rho}_0(\omega)e^{i\omega t}d\omega \tag{13-28}$$

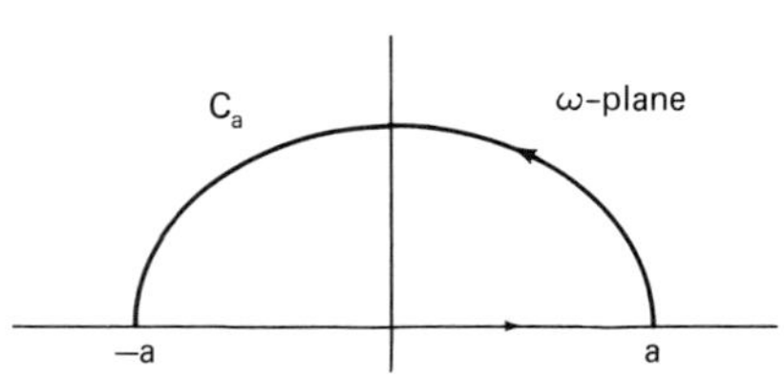

Fig. 13-2. Contour to evaluate Eq. 13.26.

where Res$[f(z),a]$ symbolizes the residue of $f(z)$ at $z = a$ and where the sum is over the residues at all poles enclosed by contour C. Equation 13-28 indicates the division of closed contour C into the real-axis and semicircular segments, respectively.

Consider the singularities of $\tilde{\rho}_0(\omega)$. By equating the denominator equal to zero, we find that the singularities occur as follows:

$$\omega_m = \frac{v}{2d}[(2m + 1)\pi - i(\ln R)] \qquad (13\text{-}29)$$

where $m = \ldots, -2, -1, 0, 1, 2, \ldots$. Since $R < 1$, $\ln R$ is negative. Thus, all the singularities of $\tilde{\rho}_0(\omega)$ occur in the upper half-plane along a line parallel to the real axis and symmetric about the imaginary axis. The contour in Fig. 13-2, being of finite radius a, encloses a finite number of singularities, as required by the theorem. In addition, we may show that the singularities are all simple poles.[7] The proof, being straightforward, is omitted.

Consider the singularities of $F(\omega)$. For definiteness, we assume for the forcing function the well-known double exponential pulse,

$$f_D(t) = u(t)(e^{-\alpha_1 t} - e^{-\alpha_2 t}) \qquad (13\text{-}30)$$

where α_1 and α_2 are both positive. To simplify the analysis here, we investigate the single exponential as follows:

$$f(t) = u(t)e^{-\alpha t} \qquad (13\text{-}31)$$

where $\alpha > 0$. The response to the double exponential can then be obtained by linearity. The Fourier transform of Eq. 13-31 is

$$F(\omega) = \frac{1}{i(\omega - i\alpha)} \qquad (13\text{-}32)$$

We observe that $F(\omega)$ has a simple pole on the positive imaginary axis.

We may now produce the complex-resonance form of solution by the following procedure. Note that in the limit as $a \to \infty$, the first integral in Eq. 13-28 is $2\pi\tilde{g}(t)$. Taking the limit in Eq. 13-26, we have

$$\tilde{g}(t) = i\text{Res}[F(\omega)\tilde{\rho}_0(\omega)e^{i\omega t}, i\alpha] + i\sum_{n=1}^{\infty} \text{Res}[F(\omega)\tilde{\rho}_0(\omega)e^{i\omega t}, \omega_n] \qquad (13\text{-}33)$$

where the summation is over the residues at all the poles of $\tilde{\rho}_0(\omega)$. Equation 13-33 is true provided that

$$\lim_{a\to\infty} \int_{C_a} F(\omega)\tilde{\rho}_0(\omega)e^{i\omega t}d\omega = 0 \tag{13-34}$$

Note in Eq. 13-21 that as $a \to \infty$, $|\rho(\omega)|$ is bounded. Also, in Eq. 13-33, $|F(\omega)| < 1/|\omega|$. Equation 13-34 therefore meets the requirements of a well-known theorem,[8] and the integral vanishes. (We notice that the magnitude of the reflection coefficient, $\rho(\omega)$, does not vanish on the semicircle as $a \to \infty$, and therefore the vanishing of the integral along C_a depends totally on the behavior of the forcing function. If, for example, the forcing function were $\delta(t)$, $F(\omega)$ would equal 1 and the integral would not vanish. Indeed, it can be shown that the inverse transform does not exist for $F(\omega) = 1$ except in the distributional sense. This result is expected, since it is apparent from Eq. 13-17 that the ray-optic solution for the delta function forcing function is a train of delta functions.)

The final steps in the complex-resonance solution are the evaluation of the residues in Eq. 13-33, followed by substitution of Eq. 13-33 into Eq. 13-25. The result is as follows:

$$\begin{aligned} e_{x1}(z,t) &= e^{-\alpha(t+z/c)}\, u\left(t + \frac{z}{c}\right) - Re^{-\alpha(t-z/c)}\, u\left(t - \frac{z}{c}\right) \\ &\quad - (1 - R^2)u\left(t - \frac{z}{c} - \frac{2d}{v}\right)s\left(t - \frac{z}{c} - \frac{2d}{v}\right) \end{aligned} \tag{13-35}$$

where $s[t - (z/c) - (2d/v)]$ is the complex-resonance contribution given by the following:

$$s(t) = \frac{v}{2d}\sum_{n=-\infty}^{\infty} \frac{e^{i\omega_n t}}{\alpha + i\omega_n} + \frac{e^{-\alpha t}}{1 + Re^{2\alpha d/v}} \tag{13-36}$$

Equation 13-35 is the complex-resonance solution for the dielectric slab. The terms on the right side of the equation can be interpreted as follows: The first term, nonzero for $t > -z/c$, is the incoming wave. The second term, nonzero for $t > z/c$, is the wave reflected from the front face of the slab. The third term, nonzero for $t > z/c + 2d/v$, contains two pieces, given by Eq. 13-36. The first piece is an infinite sum of complex exponentials containing the complex resonances of the slab. The second piece is a decaying exponential representing the contribution of the forcing function pole.

Comparison of Solutions. We now have a description of the ray-optic solution (Eq. 13-17) and the complex-resonance solution (Eq. 13-35). The solutions are identical until $t = z/c + 2d/v$, at which time the observer first senses the response from the back face ($z = -d$) of the slab. For $t > z/c + 2d/v$, the two solutions, although yielding the identical answer, are dramatically different in form. In the ray-optic form (Eq. 13-17), each successive term in the summation adds a response containing a particular reflection from the back face. For example, if we specialize Eq. 13-17 to the exponential input form (Eq. 13-31) and truncate the series after the term, $n = 1$, we have

$$e_{x1} = e^{-\alpha(t+z/c)} u\left(t + \frac{z}{c}\right) - Re^{-\alpha(t-z/c)} u\left(t - \frac{z}{c}\right) - T_1 T_2 e^{-\alpha(t-z/c-2d/v)} u\left(t - \frac{z}{c} - 2\frac{d}{v}\right) \tag{13-37}$$

This expression is exact for $t < z/c + 4d/v$. With each successive reflection from the back face, there is one more term added from the general summation. There is no corresponding behavior in the complex-resonance form (Eq. 13-35). Indeed, for any time, $t > z/c + 2d/v$, the entire series is required for an exact solution.

We point out that the complex resonance solution in Eq. 13-35 is, in fact, a hybrid solution, where the terminology is that of Heyman and Felsen.[1] Note that for early times, the solution is ray-optic, and for later times, complex-resonance. This distinction becomes crucial in the transient identification to be considered later in this chapter.

We shall next discuss how object features emerge differently in different solution forms. We shall use the ray-optic and complex-resonance forms for the dielectric slab to illustrate the concepts.

Object Features in the Scattered Field Solutions

Consider the following realization of the dielectric slab: Let the input pulse, $f(t)$, be the double exponential defined in Eq. 13-30, with $\alpha_1 = 10^7$ and $\alpha_2 = 10^8$. This choice of parameters produces an input (see Fig. 13-3) with a 13.4-ns risetime and a 135-ns falltime (measured at the point where the signal has decreased to $1/e$ compared to the peak). Let the slab be 5 m thick with a dielectric constant, ε_r, of 5.5. Using the ray-optic formulation (Eq. 13-17), we obtain the total field when $z = 12$ m (see Fig. 13-4), a result that graphically displays the multiple reflections from the slab. In particular, note that we have referenced time so that the value, $t = 0$, corresponds to the time the initial pulse passes

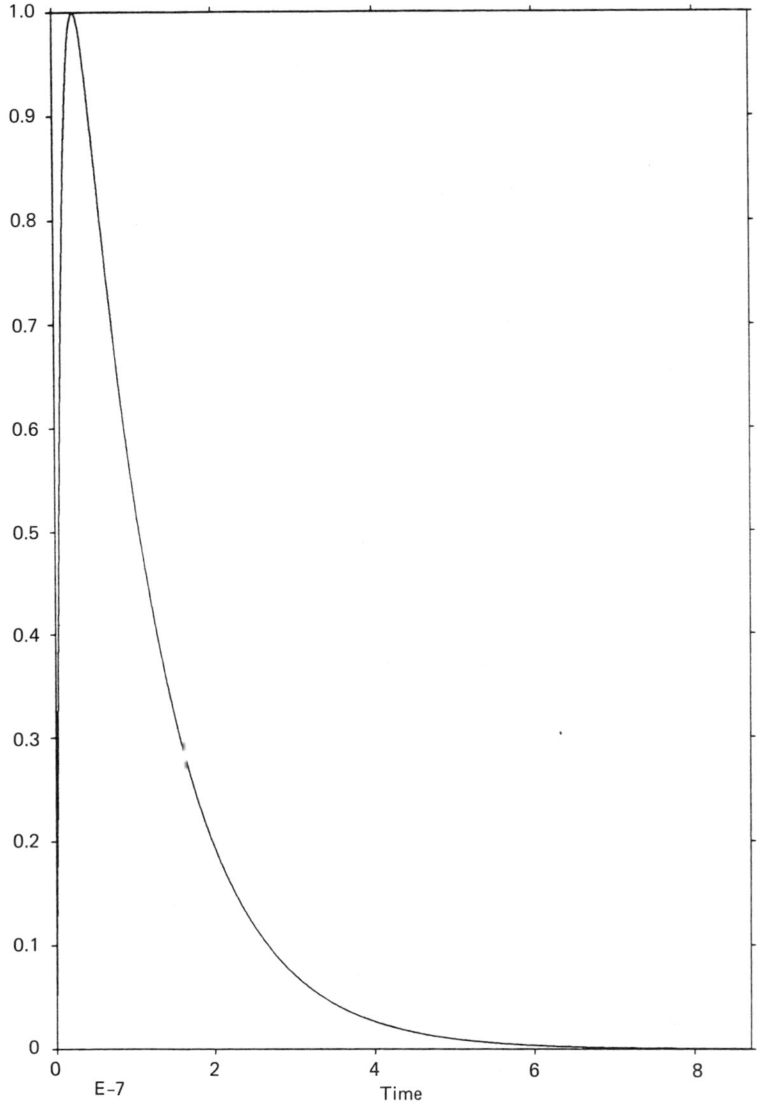

Fig. 13-3. Input double exponential pulse; $\alpha_1 = 10^7$, $\alpha_2 = 10^8$; 1024 samples; $\Delta t = 8.494 \times 10^{-10}$ sec.

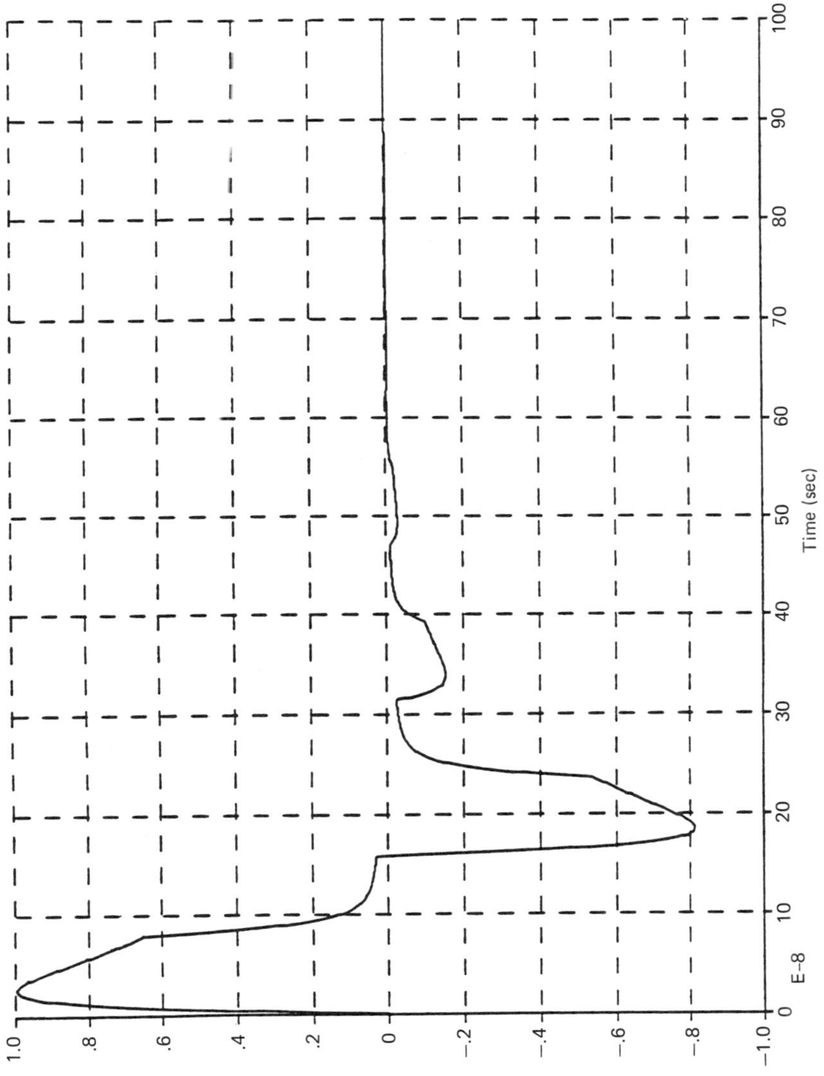

Fig. 13-4. Total field; $z = 12$m, $d = 5$m, $\varepsilon_r = 5.5$; 1024 samples; $\Delta t = 8.494 \times 10^{-10}$ sec.

the observer. The first discontinuity in slope occurs at the value, $t_1 = 80$ ns, when the reflection from the front face ($z = 0$) arrives back at the observer. The time t_1 is calculated by dividing the round trip distance (24 m) by the phase velocity in free space (3×10^8 m/s). The second discontinuity in slope occurs at the value, $t_2 = 158$ ns, when the reflection from the back face ($z = -5$ m) arrives back at the observer. The time t_2 is calculated by dividing the round trip distance in the slab (10 m) by the phase velocity in the dielectric and adding t_1. Subsequent reflections arrive at the observer every 78 ns, or

$$t_n = 80 + (n - 1) \times 78 \qquad (13\text{-}38)$$

where $n = 1,2, \ldots$ and t_n is measured in nanoseconds.

Although the preceding calculations are straightforward, we now take the inverse point of view. Indeed, we now inquire if we can reconstruct the object features from the input and output waveforms as given in Figs. 13-3 and 13-4, respectively. In particular, we attempt to find the observer distance, h, from the slab front face; the slab dielectric constant, ε_r; and the slab thickness, d. We consider both the ray-optic and complex-resonance solutions.

Object Features from Ray-optics. The characteristic feature of the ray-optics solution (Eq. 13-17) is that sequential consideration of the terms is equivalent to sequential time consideration. By noting the time of arrival of the first reflection, t_1, we have the location h of the slab front face with respect to the observer. This result is nothing more than that which is normally obtained in radar ranging. We next consider the solution after the arrival of the reflection from the front face but before the arrival of the reflection from the back face. For this time period, Eq. 13-17 gives the following:

$$e_{x1}(z,t) = f(t) - Rf\left(t - \frac{2z}{c}\right) \qquad (13\text{-}39)$$

which is exact until the back-face reflection arrives. Note that this solution would be the total solution if the slab were infinite in thickness. Note also that we have normalized time so that the value, $t = 0$, corresponds to the arrival of the initial pulse at the observer. Since we have taken the point of view of the data collector, we know the field $e_{x1}(z,t)$ at the observation point. In addition, we know the input pulse, $f(t)$, and the distance of the observer from the front face. We may, therefore, solve Eq. 13-39 for the coefficient R. We then substitute Eq. 13-12 and solve for the dielectric constant, with the result,

$$\varepsilon_r = \left[\frac{f\left(t - \dfrac{2h}{c}\right) - f(t) - e_{x1}}{f\left(t - \dfrac{2h}{c}\right) - f(t) + e_{x1}} \right]^2 \tag{13-40}$$

This determination of the dielectric constant, ε_r, not only gives us a parameter we were seeking but also gives us the phase velocity, v, in the slab. We next measure the time elapsed, Δt, between the reflection from the front face and the first reflection from the back face. The final parameter d is then given by the equation, $d = v\Delta t/2$.

We should comment that the appearance of the time variable in Eq. 13-40 is deceptive. Since ε_r is a constant, we would expect to obtain the same value of dielectric constant regardless of the point in time where the measurement is made, provided we operate in the time range of validity in Eq. 13-39. This is indeed the case, provided that the system is noisefree. Certainly the signal-to-noise ratio will affect the accuracy of the result in a real measurement.

Object Features from Complex Resonances. In the above discussion of object features from ray-optics, we commented on the time-sequential nature of the solution. With the arrival of the return from the back face of the slab, this time-sequential characteristic disappears in the complex-resonance solution form. Indeed, the solution for $t > z/c + 2d/v$ contains a series of particular ω_m, as given by Eq. 13-29. Provided we can determine ω_m, we shall now show that we can solve for the object characteristics, ε_r and d. We separate Eq. 13-29 into its real and imaginary parts, viz:

$$Re(\omega_m) = \frac{v}{2d}(2m + 1)\pi \tag{13-41}$$

$$Im(\omega_m) = \frac{v}{2d}\ln R \tag{13-42}$$

We treat Eqs. 13-41 and 13-42 as two equations in the two unknowns, ε_r and d. Dividing Eq. 13-42 by Eq. 13-41 and solving for R, we obtain the following:

$$R = e^{-(2m+1)\pi\tan\Delta_m} \tag{13-43}$$

where

$$\tan\Delta_m = \frac{Im(\omega_m)}{Re(\omega_m)} \tag{13-44}$$

Substituting Eq. 13-12 into Eq. 13-43 and solving for ε_r, we have the following:

$$\varepsilon_r = \operatorname{ctn} h^2[(2m + 1)\pi \tan \Delta_m] \tag{13-45}$$

Once ε_r is known, substitution into Eq. 13-41 yields the slab thickness d. The problem of determination of the parameters that classify the slab has now been solved, provided we can find the complex resonances, ω_m. Indeed, for the slab, a determination of ω_m for a single value of index m is sufficient.

To attempt to determine the complex resonances, we shall examine the solution form in the frequency domain with careful attention to what is measurable and what is assumed known in an experiment. In Eq. 13-24, the term on the left side is the Fourier transform of the data that would be obtained in an observation at a point exterior to the slab. The first term on the right is the incident field and is assumed known. The second term on the right is the reflection from the front face of the slab. The final term is the term containing the complex resonances. In the time domain, this contribution begins at the value, $t = z/c + 2d/v$. We rewrite Eq. 13-24, isolating the complex resonance term as follows:

$$E_{x1}(z,\omega)F(\omega) - F(\omega)e^{ikz} + RF(\omega)e^{-ikz} = \tilde{\rho}_0(\omega)F(\omega)e^{-i(kz+2\beta d)} \tag{13-46}$$

The left side of Eq. 13-46 represents new data obtained by subtracting the incident field and the front face contribution from the original data. We interpret this new data as the output $G(\omega)$ from a single-input, single-output linear system with $F(\omega)$ as the input (see Fig. 13-5). We have

$$G(\omega) = \tilde{\rho}_0(\omega)F(\omega)e^{-i\omega\tau} \tag{13-47}$$

where

$$\tau = \frac{z}{c} + \frac{2d}{v} \tag{13-48}$$

We remark that $G(\omega)$ is not observable in practice for two reasons. First, there is always noise present. Second, the experimental equipment is always bandlimited. We shall consider the noise problem in detail in a subsequent section and shall ignore it for the present. Concerning the bandwidth limitation of the mea-

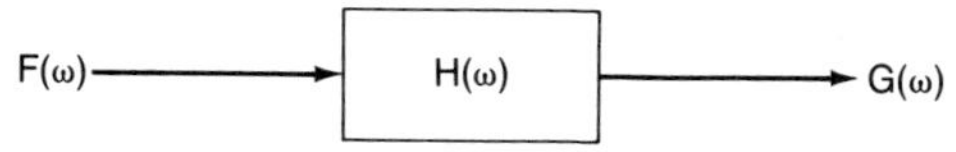

Fig. 13-5. Single-input, single-output linear system.

surement, we model the effect by a filter $T(\omega)$ and obtain a filtered version $G_f(\omega)$ of the output, given by

$$G_f(\omega) = \tilde{\rho}_0(\omega)F(\omega)T(\omega)e^{-i\omega\tau} \tag{13-49}$$

We define the system transfer function, $H(\omega)$, as $G_f(\omega)/F(\omega)$ and obtain

$$H(\omega) = \tilde{\rho}_0(\omega)T(\omega)e^{-i\omega\tau} \tag{13-50}$$

To complete the analysis, we choose for $T(\omega)$ a first-order Butterworth filter[9] given by

$$T(\omega) = \frac{1}{1 + i\dfrac{\omega}{\omega_c}} \tag{13-51}$$

We substitute Eq. 13-51 into Eq. 13-50 and, using the contour integration techniques previously described, take the inverse Fourier transform with the result that

$$h(t) = \sum_{n=-\infty}^{\infty} R_n e^{s_n\tau} + R_f e^{-\omega_c\tau} \tag{13-52}$$

where

$$s_n = i\omega_n \tag{13-53}$$

$$R_n = -\frac{(1 - R^2)v\omega_c}{2d(\omega_c + s_n)} \tag{13-54}$$

$$R_f = -\frac{(1 - R^2)v\omega_c}{Re^{2d\omega_c/v} + 1} \tag{13-55}$$

In Eq. 13-52, we have taken the inverse Fourier transform of the system transfer function, $H(\omega)$. The result, $h(t)$, is called the *impulse*, or *delta function, response* of the system. We note that, since $\tilde{\rho}_0(\omega)$ fails to vanish on C_a as $a \to \infty$, the convergence of the complex exponential series is controlled by the system bandwidth limits, in this case modeled by the $1/|\omega|$ asymptotic characteristic of filter $T(\omega)$. Each term in the series contains a specific complex resonant frequency, ω_n. In addition to the series, Eq. 13-52 contains a term representing the contribution of the filter pole. The identification problem can now be stated as follows: Given data $h(t)$, find the residues R_n and R_f and the resonances s_n and ω_c. If this can be accomplished, the classification of the slab by complex resonance methods is complete.

Relevance to Identification and Classification. In this section on "Object Descriptions," we have used the dielectric slab as an example to illustrate the two solution forms, namely, ray-optic and complex-resonance. In the ray-optic solution, we showed that identification of certain transit times led to the classification of the slab by the parameters, ε_r and d. In the complex-resonance solution, we showed that identification of the complex-resonant frequencies, ω_n, led to the determination of ε_r and d and thereby classified the object.

It should be noted, however, that no method was proposed to identify the ω_n. It has only been suggested that the relevant parameters be somehow identified by "fitting" data $h(t)$ in the complex exponential series in Eq. 13-52. This identification problem will be considered in detail in the next section.

As a postscript of this section, we pointed out several crucial factors inherent in the development. It has been implicitly assumed in the discussion that we know the geometric form-factor as a prelude to classification. Indeed, the parameter-identification procedure can proceed in ignorance but the classification problem cannot. It is important to stress this distinction because it has far-reaching consequences in application to inverse problems such as target determination and geophysical remote sensing. In the ray-optic case, we were able to identify certain transit times from an examination of the data. In order to interpret the times physically, however, it was necessary to assume that the data came from scattering from a slab. Once we made this assumption, we were able to determine the slab parameters. In the complex-resonance case, we could proceed with the identification of the complex-resonant frequencies by fitting the data to an exponential series. This could be done with no knowledge of the form factor. (Indeed this fitting problem will be considered in the next section with no regard to the physical nature of the problem.) Once the ω_n have been determined, however, it becomes necessary to use Eq. 13-29 to proceed with the slab classification.

13.2 EXPONENTIAL SERIES IDENTIFICATION

The result in Eq. 13-52, where the delta-function response is expressed by an exponential series, is not unusual in wave scattering. Lax and Phillips[10] have shown that for the scalar wave equation, classical scattering from objects of finite extent with the Dirichlet boundary condition is, with certain restrictions, a meromorphic function of complex frequency. By the Mittag-Leffler theorem,[11] this result means that the scattered field is given by a sum over poles representing the complex resonances of the scatterer plus an entire function. The inverse transform of the pole series is a complex exponential series. The Lax and Phillips result has been extended by Marin[12] to the vector electromagnetic case for perfectly conducting objects of finite extent. A notable outgrowth of these theoretical results has been the success of the *Singularity Expansion Method* (SEM), proposed by Baum[13] and discussed in depth in a special issue of *Electromagnetics.*[14] Because the exponential series occurs naturally in large classes of problems, it is logical to inquire if the parameters that describe the series can be identified from experimental data. It is also important to determine what role the entire function plays in the description of the scattered field.

Exponential modeling is not a new subject, but it remains an elusive one for the following reason: Mathematically, the series has no "nice" characteristics. The terms are not orthogonal over any interval, nor do they form a basis for any complete linear space. It is therefore difficult to say anything about mathematical convergence and difficult to obtain a "feeling" for numerical convergence. The series, however, is so important physically, certainly even beyond electromagnetics, that we persist in its study.

In this section, we consider the identification of the parameters in the exponential series. We begin with a description of difference-equation models and show their relation to the exponential series. We next briefly display the models in the format of modern system identification. We follow with a description of a nonlinear least-squares algorithm that we have used with some success in exponential parameter identification and give some examples of its use. Finally, we point out some of the difficulties in making the bridge between exponential identification and object classification.

Difference-Equation Models

In Eq. 13-52, we found that the delta-function response was given by a complex exponential series. Since Marin[12] has shown that this result generalizes to a wide class of electromagnetic scattering problems, we now investigate in some depth the determination of the parameters in the exponential series.

Consider the following model of the delta-function response:

$$h(t) = \sum_{n=1}^{N} R_n e^{s_n t} + \tilde{e}(t) \tag{13-56}$$

where $\tilde{e}(t)$ is an error term arising from three sources. First, the scattering problem may not be completely modeled by an exponential series. Such is certainly the case when an entire function is present. Second, in our application here, $h(t)$ is not the output from the model but from the actual electromagnetic process and therefore must contain errors (measurement error, for example).

Third, the order of the exponential series has been truncated to N and is therefore an approximation to the infinite-order complex exponential series.

The identification problem is now stated as follows: Given input data $x(t)$ and output data $y(t)$, identify the residues, R_n, and the poles, s_n, in the delta function response, $h(t)$, such that the error is minimized in some sense. As discussed by Dudley,[15,16] this problem can be cast in a form compatible with modern system identification, where difference-equation modeling of linear systems is a well-understood technique. Consider the difference equation described in the previous chapter (Eq. 12-19) and repeated here for convenience, viz:

$$\sum_{n=0}^{N} a_n y(k-n) = \sum_{n=1}^{M} b_n x(k-n) + e(k) \tag{13-57}$$

where $y(k)$, $x(k)$, and $e(k)$ are sampled versions of $y(t)$, $x(t)$, and $e(t)$, respectively. By convention, we have suppressed the sampling interval T. That is, $y(k)$ is actually $y(kT)$. Taking the z-transform of both sides of Eq. 13-57 gives

$$Y(z) \sum_{n=0}^{N} a_n z^{-n} = X(z) \sum_{n=1}^{M} b_n z^{-n} + E(z) \tag{13-58}$$

where $X(z)$, $Y(z)$, $E(z)$ are the z-transforms of $x(t)$, $y(t)$, and $e(t)$, respectively. We define the system-transfer function, $H(z)$, thus:

$$H(z) = \frac{Y(z)}{X(z)} \tag{13-59}$$

and obtain the following:

$$H(z) = \frac{\sum_{n=1}^{M} b_n z^{-n}}{\sum_{n=0}^{N} a_n z^{-n}} + \tilde{E}(z) \tag{13-60}$$

where the tilde in the error term indicates its alteration during the cross-product operation. Equation (13-60) is a quotient of two polynomials in z^{-1}, with an added error term. Using standard methods,[15] we factor the denominator polynomial into its roots, perform a partial fraction expansion, and take the inverse z-transform to obtain the exponential series,

$$h(k) = \sum_{n=1}^{N} R_n e^{s_n kT} + \tilde{e}(k) \qquad (13\text{-}61)$$

where T is the sample rate. Note that Eq. 13-61 is simply a uniformly sampled version of Eq. 13-56. The identification problem can now be reformulated as follows: Given input data $x(k)$ and output data $y(k)$ from a transient electromagnetic process, identify the residues, R_n, and poles, s_n, in Eq. 13-61 by the following steps:

1. Identify coefficients a_n and b_n in the difference equation, Eq. 13-57.
2. Using the identified coefficients, factor the denominator polynomial in Eq. 13-60 and perform a partial fraction expansion.
3. Take the inverse z-transform to produce R_n and s_n in Eq. 13-61.

Since this procedure contains many possibilities for error,[15] it is crucial that the first step, the identification of the coefficients in the difference equation, be done accurately. This point is discussed at some length in the previous chapter. Its consideration has led to the construction of an algorithm to be described shortly.

Modern System Identification

The ideas inherent to difference equation modeling discussed in the previous section can be cast in the form of modern system identification. Define the time-stepping operator, q^{-1}, by

$$q^{-1}x(k) = x(k - 1) \qquad (13\text{-}62)$$

Define the polynomial stepping operators, A, B, C, D, and F, based on this time-stepping operator. Typically,

$$A(q^{-1}) = \sum_{n=0}^{n_a} a_n q^{-n} \qquad (13\text{-}63)$$

We introduce the generalized difference equation model of Ljung and Söderstrom,[17] given by the following:

$$A(q^{-1})y(k) = \frac{B(q^{-1})}{F(q^{-1})}x(k) + \frac{C(q^{-1})}{D(q^{-1})}e(k) \qquad (13\text{-}64)$$

This model contains most of the well-known single-input, single-output models as special cases. We consider only two, the equation-error model and the output-error model. For more general discussions, the reader is referred to Ljung and Söderstrom.[17]

The equation-error model is obtained from Eq. 13-64 by setting C = D = F = 1 to give the following:

$$Ay(k) = Bx(k) + e(k) \tag{13-65}$$

A special case of Eq. 13-65 occurs when $x(k)$ is the delta function. Dudley[15] has shown that this case is actually Prony's method, as discussed in the previous chapter.

The general output-error model is obtained from Eq. 13-64 by setting $A = 1$ so that

$$y(k) = \frac{B}{F}x(k) + \frac{C}{D}e(k) \tag{13-66}$$

The sequence, $e(k)$, is assumed to be white noise. The first term on the right of the equation models the input-output relationship and the second term models the disturbances. Note that unlike models that use the A polynomial, the output-error model handles these items separately. Consequently, the output-error model is robust in the sense that if B and F are of an order at least that of the actual system, then even if the noise model is incorrect, the technique we will discuss will yield estimates of these polynomials that converge (as the amount of data becomes large) to values that include, respectively, the zeros and poles of the actual system as factors.[17]

There are good reasons for assuming that the noise is white and appears only at the output. The output-noise assumption is reasonable because usually the exciting signal used in transient electromagnetic experiments is usually high-amplitude and completely dominates any measurement noise. The whiteness assumption is reasonable because the data is usually collected by a sampling oscilloscope or similar device so that each sample in a data record is obtained from a different repetition of the same experiment. Under this assumption, the model becomes the following:

$$y(k) = \frac{B}{F}x(k) + e(k) \tag{13-67}$$

The obvious way to fit this model to the data is to substitute the observed data records, $[x(k)]$ and $[y(k)]$, into Eq. 13-67 and use an optimization algorithm to minimize the cost function, V, given by the following:

$$V = \sum_{k=1}^{K} e^2(k) \tag{13-68}$$

over the coefficients of B and F. If the actual disturbance term is Gaussian and white and the degrees of B and F are those of the actual system, then minimizing Eq. 13-68 yields the maximum likelihood estimates of B and F.[18] Maximum-likelihood estimates are attractive because they are consistent (i.e., the estimates converge to the actual values as the amount of data gets large) and asymptotically efficient (i.e., they have the lowest variance among the class of consistent estimates).[19]

If the noise is not white, maximum-likelihood estimates can be acquired by using a noise model as in Eq. 13-66. Another approach, however, seems more suited to transient problems. Records of the noise alone are obtained by collecting data at the output with the excitation system turned off. One of the techniques discussed in the previous chapter is used to obtain an estimate of the noise spectrum, $\phi_e(\omega)$. A "prewhitening" filter with magnitude response, $\phi_e^{-.5}(\omega)$, is then applied to both the input and output records of the transient experiment. Because both records have passed through the same filter, the input-output relationship has not been altered, but the noise has become white. If the polynomial degrees and the estimate of the noise spectrum are correct, then by fitting Eq. 13-67 to the filtered data we obtain maximum-likelihood estimates in the Gaussian noise case. Even in the non-Gaussian case, this prefiltering should reduce the variance of the estimates.

To avoid possible confusion, we wish to point out that maximum-likelihood identification is often associated with the ARMAX model of the last chapter. However, the phrase "maximum likelihood" applies only if the actual model is of this form. We have argued that output error models are more suitable for our purposes, and this has been verified by many simulations. Note that in Eq. 12-25 both noise and input signal pass through the system poles given by the A polynomial. If we fitted this model to a system whose actual form was given by Eq. 13-67, then both the A and the C polynomials would converge to F, and we would have a nonparsimonious representation of the actual system. Consequently, the parameter estimates obtained from the ARMAX model will have a higher variance than if the correct model were used. Unfortunately, although the error sequence in Eq. 13-67 is a linear function of the coefficients of B, it is a highly nonlinear function of the coefficients of F, and minimizing Eq. 13-68 is a difficult nonlinear least-squares problem. The approach we will discuss is the result of a considerable amount of experimental effort by ourselves and several colleagues at the Lawrence Livermore National Laboratory. The conclusion to which our experiments point is that the nonlinear method is superior to any of the several other methods we have tried. In the next section, we will discuss our algorithm, and, in the following section, we will discuss its use. A more thorough discussion is contained by Goodman.[20]

The Nonlinear Least-Squares (NLS) Algorithm

For notational convenience, we define the parameter vector,

$$\boldsymbol{\theta}_m = [f_m(1), \ldots, f_m(I), b_m(0), \ldots, b_m(J)]^T \tag{13-69}$$

where the subscript gives the number of the iteration and integers I and J are the orders of F and B, respectively. Because $f(0)$ is fixed at 1, it is not included in $\boldsymbol{\theta}$. As a function of $\boldsymbol{\theta}$, Eq. 13-68 becomes

$$V(\boldsymbol{\theta}_m) = \sum_{k=1}^{K} e_m^2(k) = \sum_{k=1}^{K} [y(k) - \hat{y}_m(k)]^2 \tag{13-70}$$

where $[y(k)]$ is the output data record and $[\hat{y}_m(k)]$ is the output of the model when it is driven by the input record, i.e.,

$$\hat{y}_m(k) = \frac{B}{F}x(k) \tag{13-71}$$

The gradient of $V(\boldsymbol{\theta})$ is the row vector,

$$\nabla V(\boldsymbol{\theta}) = -2\sum_{k=1}^{K} e(k)[\nabla\hat{y}(k)] \tag{13-72}$$

and its Hessian is the matrix,

$$\nabla^2 V(\boldsymbol{\theta}) = 2\sum_{k=1}^{K} [\nabla\hat{y}(k)]^T[\nabla\hat{y}(k)] - 2\sum_{k=1}^{K} e(k)[\nabla^2\hat{y}(k)] \tag{13-73}$$

The first-order sensitivity derivative, $[\nabla\hat{y}(k)]$, is a sequence of row vectors whose p^{th} components form the sequence $[\partial\hat{y}(k)/\partial\theta_p]$. Similarly, the second-order sensitivity derivative, $[\nabla^2\hat{y}(k)]$, is a sequence of matrices whose $(pq)^{\text{th}}$ entries form the sequence $[\partial^2\hat{y}(k)/\partial\theta_p\partial\theta_q]$. At first glance, the cost of computing $[\nabla^n\hat{y}(k)]$ appears to be of order $K(I + J)^{n+1}$. Fortunately, the cost for both sensitivity derivatives can be made linear in $(I + J)$; our discussion is based on Åström.[21] Let the sequence $[w(k)]$ be the input to the difference equation,

$$u(k) = w(k) - \sum_{i=1}^{I} f(i)u(k - i) \tag{13-74}$$

where the output sequence $[u(k)]$ is computed with zero initial conditions. If $[u_B(k)]$ is the output for input $[x(k)]$; $[u_F(k)]$, the output for input $[-\hat{y}(k)]$; $[u_{FF}(k)]$, the output for input $[-2u_F(k)]$; and $[u_{FB}(k)]$, the output for input $[-u_B(k)]$, then one can show that

$$\begin{aligned}
\frac{\partial \hat{y}(k)}{\partial f(i)} &= u_F(k - i) \\
\frac{\partial \hat{y}(k)}{\partial b(j)} &= u_B(k - j) \\
\frac{\partial^2 \hat{y}(k)}{\partial f(i)\partial f(j)} &= u_{FF}(k - i - j) \\
\frac{\partial^2 \hat{y}(k)}{\partial f(i)\partial b(j)} &= u_{FB}(k - i - j) \\
\frac{\partial^2 \hat{y}(k)}{\partial b(i)\partial (b)(j)} &= 0
\end{aligned} \tag{13-75}$$

If other more obvious simplifications are used as well, then the cost of computing both the gradient and the Hessian will also be linear functions of $(I + J)$.

Most popular optimization algorithms use only the gradient because the cost of computing the Hessian is prohibitive, but the low cost for our Hessian led us to use Newton's method. It is argued in Gill et al[22] that Newton's method is superior if computing the Hessian is cheap; our own experience with minimizing $V(\boldsymbol{\theta})$ agrees with this finding. The classical iteration equation for Newton's method is as follows:

$$\boldsymbol{\theta}_{m+1} = \boldsymbol{\theta}_m + \mathbf{d}_m \tag{13-76}$$

where

$$\mathbf{d}_m = -[\nabla^2 V(\boldsymbol{\theta}_m)]^{-1} [\nabla V(\boldsymbol{\theta}_m)]^T \tag{13-77}$$

Because $V(\boldsymbol{\theta})$ is usually not convex, $\mathbf{d}_m$ is not guaranteed to be a descent direction. In order to assure a descent direction, it is necessary to replace the Hessian with a matrix that approximates it but is positive definite. One method[23] uses the following:

$$\nabla^2 V(\boldsymbol{\theta}) = E \Lambda E^T \tag{13-78}$$

where the columns of E are the orthonormal eigenvectors of the Hessian and Λ is the corresponding diagonal matrix of eigenvalues. A positive definite matrix is obtained by replacing the diagonal elements of Λ with

$$\bar{\lambda}_i = \max[|\lambda_i|,\delta] \tag{13-79}$$

where δ is a small positive number. This method works well on a per-step basis, but computing the decomposition is expensive if there are many parameters.

Another method, based on Moré and Sorensen,[24] first computes the symmetric indefinite factorization,

$$\nabla^2 V(\boldsymbol{\theta}) = U D U^T \tag{13-80}$$

where U is a product of upper-triangular and permutation matrices and D is block-diagonal with blocks of order 1 or 2. Greenstadt's method is then applied to the blocks of D. This method costs much less and is competitive on a per-step basis as well. Furthermore, if LINPACK[25] is available, it can be implemented by modifying only a few lines of the subroutine DSISL. Both methods are available as options in NLS.

Another option is the Gauss-Newton approximation in which only the first term in Eq. 13-73 is used; its advantage is the time savings gained by not calculating the second term. The standard argument for its use is that the second term is small near the minimum because $[e(k)]$ is small. For our problem, one can also show that if the model order is large enough so that $[e(k)]$ is white at the global minimum, then the expected value of the second term will be zero there. The first term is always positive semidefinite but is often ill-conditioned. The usual solution is to add δI to it before it is inverted, where $\delta > 0$,[26] but we have observed that the same methods we used to modify the full Hessian are also good methods for regularizing the first term. Because there are two decisions with two options each, there are four possible search directions. Our experience has been that applying the symmetric-indefinite-factorization method to the full Hessian both close to and far from the minimum is usually the fastest and most reliable. There are plenty of exceptions; however, when the algorithm is converging slowly, changing directions will often speed it up.

In addition to the Newton step, there is an auxiliary step that optimizes the parameters of B for fixed F because, with F fixed, minimizing $V(\boldsymbol{\theta})$ with respect to B is a linear least-squares problem whose global minimum can be determined noniteratively. Using this step every few iterations can greatly speed convergence.

Another necessary modification to Newton's method is the inclusion of a step-length algorithm so that Eq. 13-76 becomes

$$\boldsymbol{\theta}_{m+1} = \boldsymbol{\theta}_m + s_m \mathbf{d}_m \tag{13-81}$$

where $s_m > 0$. The best value of s_m is obviously the one that minimizes $V(\boldsymbol{\theta}_{m+1})$, but determining it could require many evalutions of $V(\boldsymbol{\theta})$. Because of the scaling properties of Newton's method that, near the minimum at least, make the best value of s close to 1, and because computing the derivatives of $V(\boldsymbol{\theta})$ does not cost that much more than computing $V(\boldsymbol{\theta})$ itself, we decided to use Armijo's method,[27] which, rather than seeking the absolute minimum, decreases the size of s until a certain criterion has been satisfied. The initial value of s should be approximately 1.

Another issue that had to be addressed in determining step length was stability because at some point a pole of the model of Eq. 13-67 could cross the unit circle, causing numerical inaccuracies or overflows. Stability is equivalent to the condition that all the reflection coefficients corresponding to F have magnitudes less than 1.[22] Using a constrained nonlinear program[22] to apply magnitude constraints is unsatisfactory, however, because the reflection coefficients of F are highly nonlinear functions of the polynomial coefficients of F. Consequently, such an algorithm is difficult to implement and expensive computationally. The usual, simpler method is merely to reduce the step size whenever the magnitude of a reflection coefficient becomes greater than 1. Unfortunately, this can produce situations in which the algorithm becomes static at a point on the stability boundary that is not optimum. This problem can become severe if, as frequently happens with electromagnetic transient experiments, the response of the device or object being tested includes several lightly damped resonances (corresponding to z-domain poles near the unit circle). The solution we used in an early version of NLS[20] was to relax the stability test by comparing the magnitudes with a number larger than 1. Our reasoning was that $V(\boldsymbol{\theta})$ will tend to become large as the stability boundary is crossed. The algorithm will therefore not allow the model to become too unstable and will eventually drive $\boldsymbol{\theta}$ back across the boundary. This approach worked well in many instances. The reflection coefficients are also highly nonlinear functions of the poles, however, and are therefore not good measures of the degrees of instability. A better measure is the size of $[u_B(k)]$, defined previously. If $[\hat{y}(k)]$ is computed using the following:

$$\hat{y}(k) = \sum_{j=0}^{J} b(j) u_B(k - j) \tag{13-82}$$

then $[u_B(k)]$ is always available. The step size is reduced if its maximum magnitude exceeds a large number; we used 10^{10}.

Running the NLS Algorithm

NLS is an iterative algorithm and requires an initial point the selection of which is important on two accounts. First, the closer the starting point is to a minimum, the fewer will be the number of iterations. Second, $V(\boldsymbol{\theta})$ may have local minima. If the model order is less than that of the actual system, local minima are likely. On the other hand, in the zero-noise case, or, in the limit, as the amount of data becomes large, if the model order is too large, the global minimum will be distributed (any point on some contour in the parameter space will be a global minimum). Under restricted conditions, if the model order is correct, there is a unique global minimum and no local minima. For a discussion of these issues, see Söderström[28] and Stearns.[29]

Our approach to locating the global minimum begins with fitting a high-order model of the form Eq. 13-65. (The closed-form solution to this problem is given in the previous chapter. Ideally, the A and B polynomials will have a common factor that models the noise, and the remaining factors will be the correct model of the actual system. Of course, this usually will not be the case, but often A and B contain factors close enough to those of the actual system for NLS to converge to the global minimum when it uses them as an initial point. As a rule, we have found that the best procedure is to let NLS start with the high-order model (the initial value of F is A) and run until either a minimum is reached or convergence slows. Slow convergence is often a consequence of overmodeling because a distributed global minimum leads to a Hessian that is singular or nearly singular. This situation is usually a good indication that the minimization problem is ill-conditioned.

The obvious way to remove extraneous poles from a model is to throw away those with the smallest residues. This method is incorrect, however, particularly for transient problems in which the exciting pulse may not have energy at all frequencies. When nonparametric models were discussed in the previous chapter, it was shown that such a pulse could lead to high variance estimates. The frequency-domain interpretation for this problem is that if the input energy is almost zero at some frequency, then the estimate of the transfer function at that frequency will have a high variance. (Its value has little effect on the fit to the output record.) Similarly, in the parametric-output-error problem, if the input signal does not excite a particular pole, then the estimate of its residue will have a high variance. (Its value will have little effect on Eq. 13-68.) The correct approach is to discard poles that contribute little to minimizing Eq. 13-68.

The first step in eliminating the extraneous poles is to compute the I roots of $F(z)$, $P_1 \ldots ,P_I$, which are, of course, the poles of the model. If these poles are ordered so that the first $2I_c$ poles are the complex poles grouped in pairs, and the remaining $I - 2I_c$, are the real poles, then in the z-transform domain,

$$\hat{Y}(z) = \sum_{i=1}^{I_c} \frac{\beta_{2i-1} + \beta_{2i}z^{-1}}{1 - 2Re[p_{2i}]z^{-1} + |p_{2i}|^2z^{-2}} X(z) + \sum_{i=2I_c+1}^{I} \frac{\beta_i}{1 - p_iz^{-1}} X(z) \tag{13-83}$$

The coefficients of B are linear functions of the β_i's and can be recovered by recombining the above partial fraction expansion. If in the z-transform domain, we define for $i = 1, \ldots, I_c$ the sequences,

$$H_{2i-1}(z) = \frac{1}{1 - Re[p_{2i}]z^{-1} + |p_i|^2z^{-2}} X(z) \tag{13-84}$$

and

$$H_{2i}(z) = z^{-1}H_{2i-1}(z) \tag{13-85}$$

and for $i = 2I_c + 1, \ldots, I$ the sequence,

$$H_i(z) = \frac{1}{1 - p_iz^{-1}} X(z) \tag{13-86}$$

then, in the time domain,

$$e(k) = y(k) - \sum_{i=1}^{I} \beta_ih_i(k) \tag{13-87}$$

Because $[e(k)]$ is a linear function of the β_i's, minimizing Eq. 13-68 with respect to thee coefficients is a linear regression (linear least-squares) problem. If it is desirable to keep only $L < I$ poles, then the best subset can be obtained by trying all possible combinations of L poles and selecting the one that yields the smallest value of Eq. 13-68. Unfortunately, since this involves solving $I!/(I - L)!L!$ linear least-squares problems, it is usually computationally prohibitive. Selecting a subset of regressors is a standard problem in linear-regression theory, and many suboptimal methods that are not computationally expensive have been developed.[30] We used a backward-elimination method that begins with the solution for all I poles and then removes one pole at a time from the solution until L remain. At each step, the pole that contributes the least to the solution is removed.

If the original model order was too high, the near collinearity of the sequences, $[h_i(k)]$ in Eq. 13-87 can result in very large estimates of the β_i's, which in turn can result in a model whose impulse response exhibits the same kind of bizarre behavior we observed in the nonparametric modeling section of the last chapter. Selecting a subset of regressors can be viewed as a method for regularizing a regression problem (recall that we used "ridge regression" for the nonparametric estimate). Often the reduced-order model will give a much better estimate of the impulse response. In order to improve the estimates of the poles in the reduced-order model, this model should be used as the starting point for another NLS run. Typically, the parameters won't change dramatically, and only a few iterations will be required.

Interpreting the Results

Considerable caution must be exercised in interpreting the results of any method that fits exponentials to data or identifies the response of a system with a linear difference equation model. To begin with, the idea of representing a function with an infinite exponential series is somewhat nonsensical because this series is not unique. In particular, on a finite interval, any reasonable function can be represented by its Fourier series, which, of course, is a sum of undamped exponentials equally spaced in frequency. Usually there is an implicit assumption that the system is approximately finite-order so that only a few of the terms in the expansion are significant. Selecting the correct model order is itself a complex problem, particularly because the finite-order assumption is only an approximation. A good discussion of some of the issues in order selection is contained in Akaike.[31]

Even if the actual system is finite-order, difficulties remain because the estimation problem is often highly ill-conditioned. At a minimum of $V(\boldsymbol{\theta})$, the gradient is zero and the sensitivity of $V(\boldsymbol{\theta})$ to changes in $\boldsymbol{\theta}$ is measured by the Hessian. If the Hessian has small eigenvalues, then $V(\boldsymbol{\theta})$ will be insensitive to changes in $\boldsymbol{\theta}$ along the directions of the corresponding eigenvectors. Consequently, the variances of the estimates will be large along these directions. In almost all of our simulations, the Hessian has been observed to have some small eigenvalues at the minimum; ones in the range 10^{-10} were not at all uncommon. In the Gaussian noise case, we can be more precise because at the minimum the Hessian can be used to obtain an estimate of the Fisher information matrix, J. In particular,

$$\hat{J} = (2\sigma^2)^{-1}\nabla^2 V(\boldsymbol{\theta}) \tag{13-88}$$

where σ^2 is the noise variance. By the Cramer-Rao inequality,[19] the inverse of the Fisher information matrix is a lower bound on the covariance matrix of any unbiased estimate of $\boldsymbol{\theta}$. Therefore, if the Hessian has small eigenvalues, the inverse of the Fisher information matrix will have large ones, and we should expect high-variance estimates.

A general rule of thumb is that it is easier to identify a set of highly underdamped complex poles than to identify a set of real poles. Furthermore, in the case of complex poles, the estimates of the imaginary parts (frequencies are usually much better than the estimates of the real parts (damping factors).[15] Some understanding of why these observations should be so can be obtained by keeping Parseval's theorem in mind and observing frequency-domain behavior. The frequency responses on two real poles will always overlap to a considerable degree. Similarly, so will the responses of two pairs of complex poles having the same frequencies but different damping factors. On the other hand, there will be little overlap between two pairs of complex poles that have different frequencies and small damping factors.

Identification Using NLS: An Example

Consider the following example of the identification of the complex resonances of a lossless dialectric slab. The input signal is the double exponential pulse with $\alpha_1 = 10^7$ and $\alpha_2 = 10^8$, given previously in Fig. 13-3. The signal consists of 1024 equally spaced samples with a sampling interval of 8.494×10^{-10} s. The slab has a dielectric constant of 5.5 and a thickness of 5 m. We process the data using the signal processing algorithm, SIG.[32] We subtract the incident field and the reflection from the front face from the total E-field and shift the time axis so that the value, $t = 0$, corresponds to the initial arrival of the first return from the back face (Fig. 13-6). We examine the input and output spectrum (Figs. 13-7 and 13-8, respectively) and filter both with a tenth-order analog-derived Butterworth low-pass filter at 90 MHz. We find that, after filtering, the data is oversampled with respect to the Nyquist rate. We therefore decimate the data by 4:1, resulting in 256 point input and output records with a sampling interval of 3.3976 ns. Note in Fig. 13-9 that eight resonances are visible in the output spectrum before the combined effects of the input spectrum and the low-pass filter drive the spectrum below the range of the linear plot. We comment that the identification of eight complex resonances (16 poles) requires a model of at least the sixteenth order. The filtering operation is therefore essential in order to keep the order of the system within a reasonable size. In an actual experiment, natural filtering is supplied by the bandwidth limits of the measuring system. Filtering, however, is still required to eliminate high-frequency noise, which can also increase model order unnecessarily.

The filtered and decimated data are next processed by NLS. The algorithm

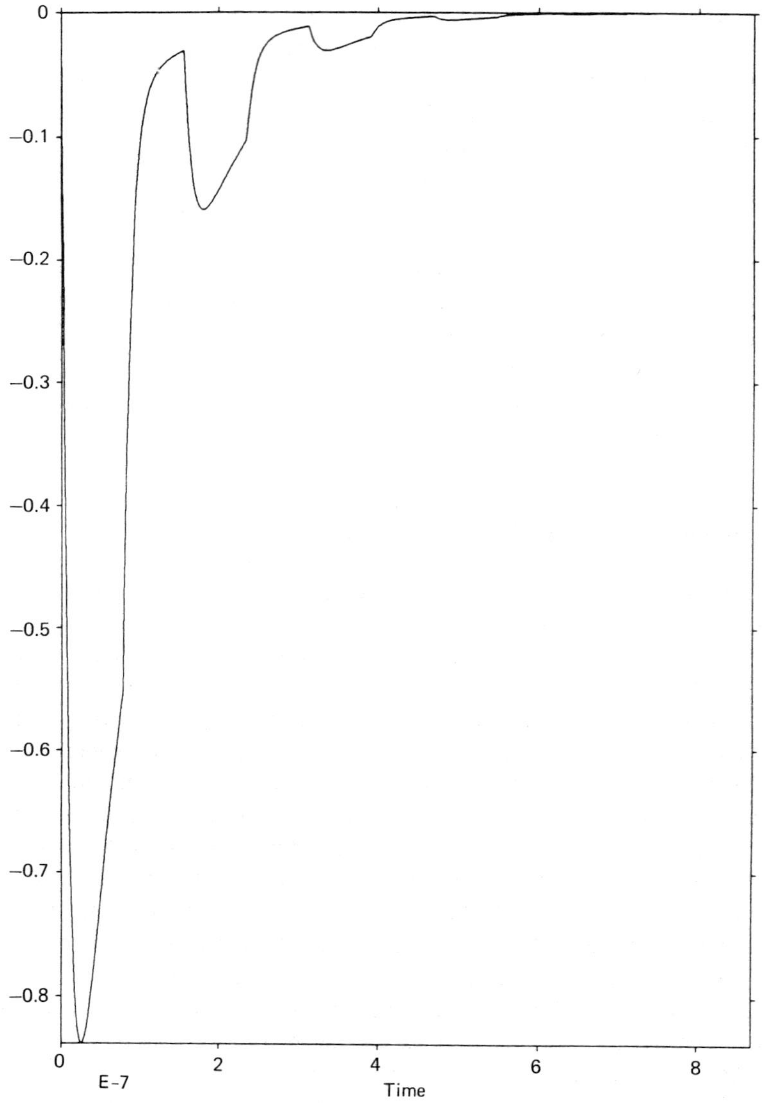

Fig. 13-6. Output E-field, incident field, and front-face reflection subtracted for 1024 samples, where $\Delta t = 8.494 \times 10^{-10}$ s.

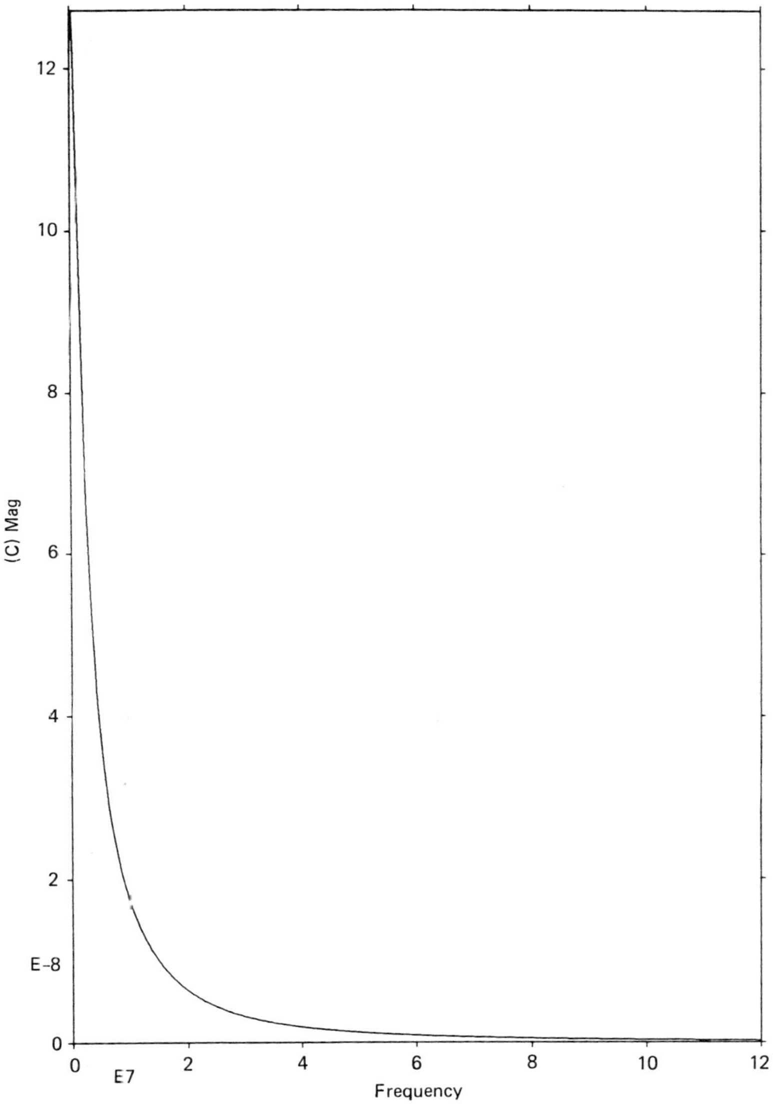

Fig. 13-7. FFT of input.

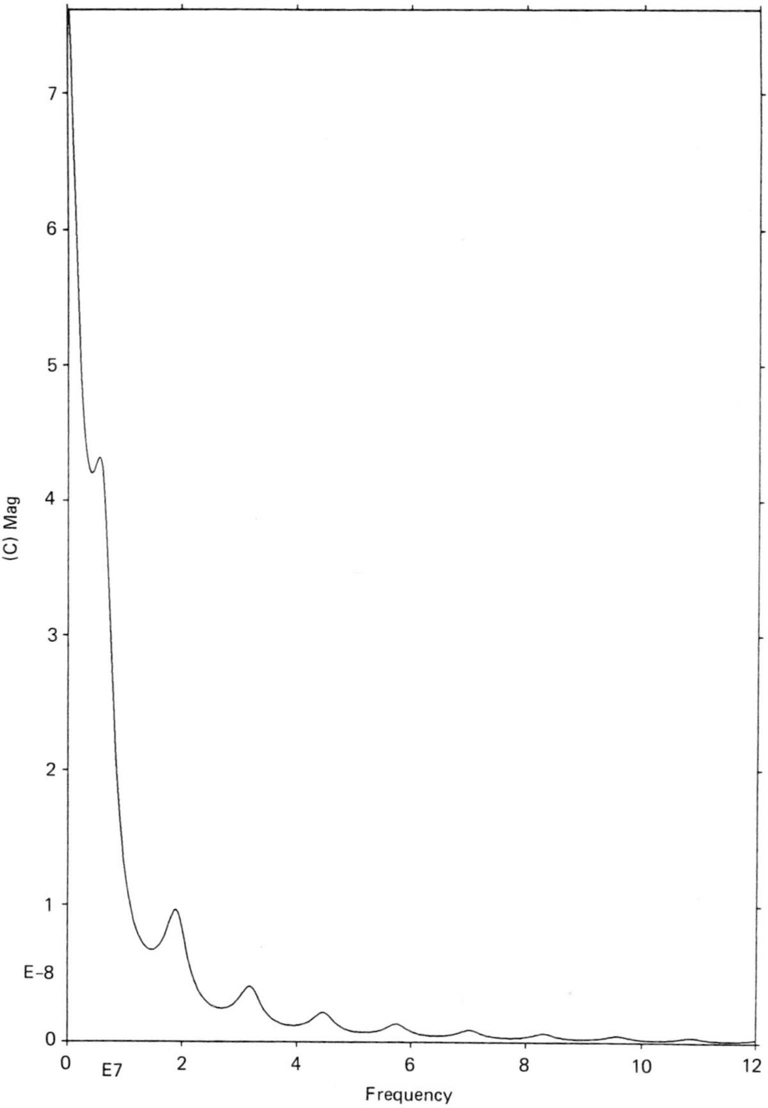

Fig. 13-8. FFT of output.

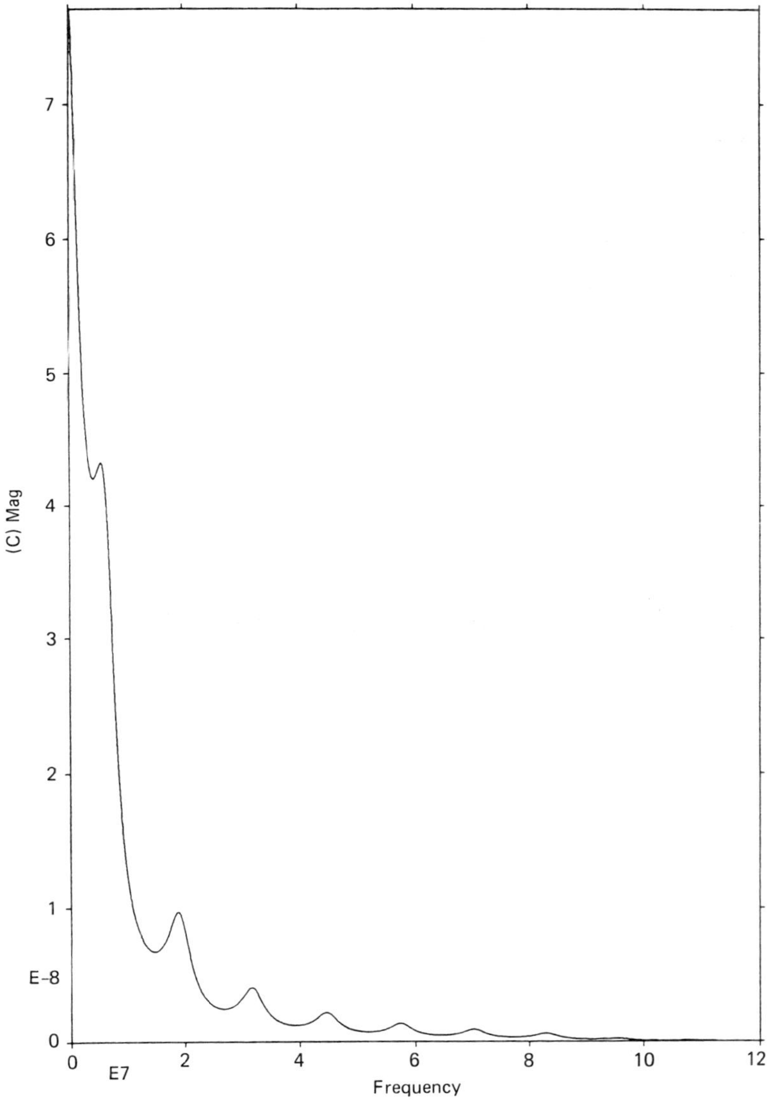

Fig. 13-9. FFT of output after filtering at 90 MHz and decimating 4:1.

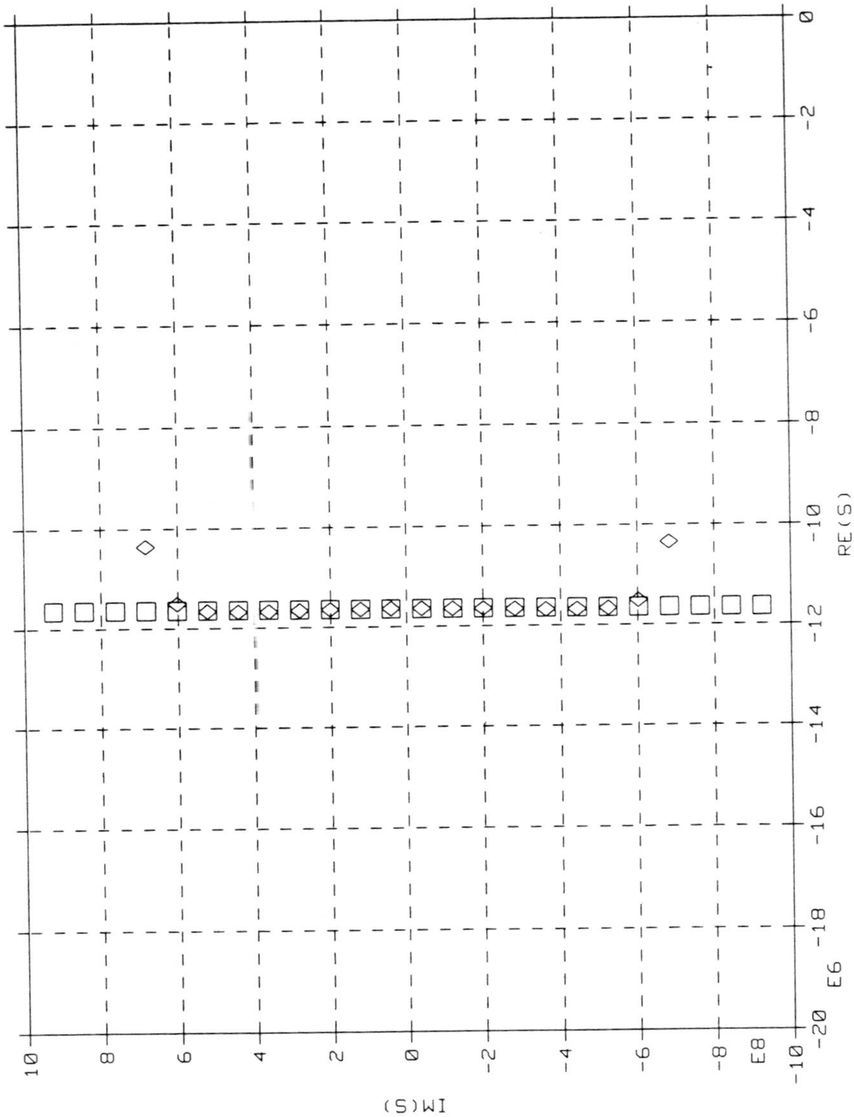

Fig. 13-10. Theoretical (□) and model (△) poles (s-plane)

begins with an ordinary linear, least-squares fit to the data (see Eq. 13-65). We select a model order of 20 with 19 zeros. The poles are presented in the s-plane (Fig. 13-10), with the results of the identification (Δ) displayed along with the theoretical poles from Eq. 13-29 ($s_n = i\omega_n$).

Note that the first seven identified pole pairs match their theoretical counterparts to within the accuracy of the plot. The eighth and ninth pairs begin to deviate, particularly in the real part.[15] These deviations are caused by the deterioration in signal-to-noise ratio at the edge of the low-pass filter ($\omega = 5.655 \times 10^8$).

The identified complex resonances obtained above can be used as discriminators to determine the physical parameters of the slab, as described in the section on Object Features from Complex Resonances. We emphasize that we are assuming that we know the form of the structure but not the specific values of the parameters, ε_r and d. This knowledge is often not available in an actual identification experiment or, at best, is known only in vague, nonscientific terms. For example, we might know only that the output data represents scattering from a "flying object" or a "buried inclusion." In the case of the slab, however, knowledge of the form of the structure leads to the complete classification of the object by the procedure in the section just mentioned.

The preceding identification sequence was carried out on "perfect" data, in the sense that the only noise is the roundoff inherent to a VAX-11/750 computer. Using a Gaussian random number generator in SIG, we may study the effects of noise on the identification process. Concentrating on the system output, we compute the maximum of the output sequence, $\sup[y(k)]$, where $k = 1,2, \ldots, 1024$. If σ is the standard deviation of the Gaussian distribution, we define δ by

$$\sigma = \delta \sup[y(k] \tag{13-89}$$

To the output (Fig. 13-6), we add zero-mean Gaussian noise with $\delta = 0.01$ (see Fig. 13-11). The effect of the noise on the resonances is evident in the frequency domain (see Fig. 13-12) when compared with the noisefree case in Fig. 13-8. One would expect difficulties in the identification procedure, beginning with the second resonance (ordered by increasing frequency). Indeed, if we repeat the identification procedure using a twentieth-order system, the poles in Fig. 13-13 begin to deviate from their theoretical positions after the first pole pair. Again, the initial effect is in the real parts. It is to be noted that a noise level of 1-percent of the peak in the output signal is sufficient to cause serious errors in the pole positions. Even so, in this case the first pole pair gives all the information necessary for complete classification of the slab.

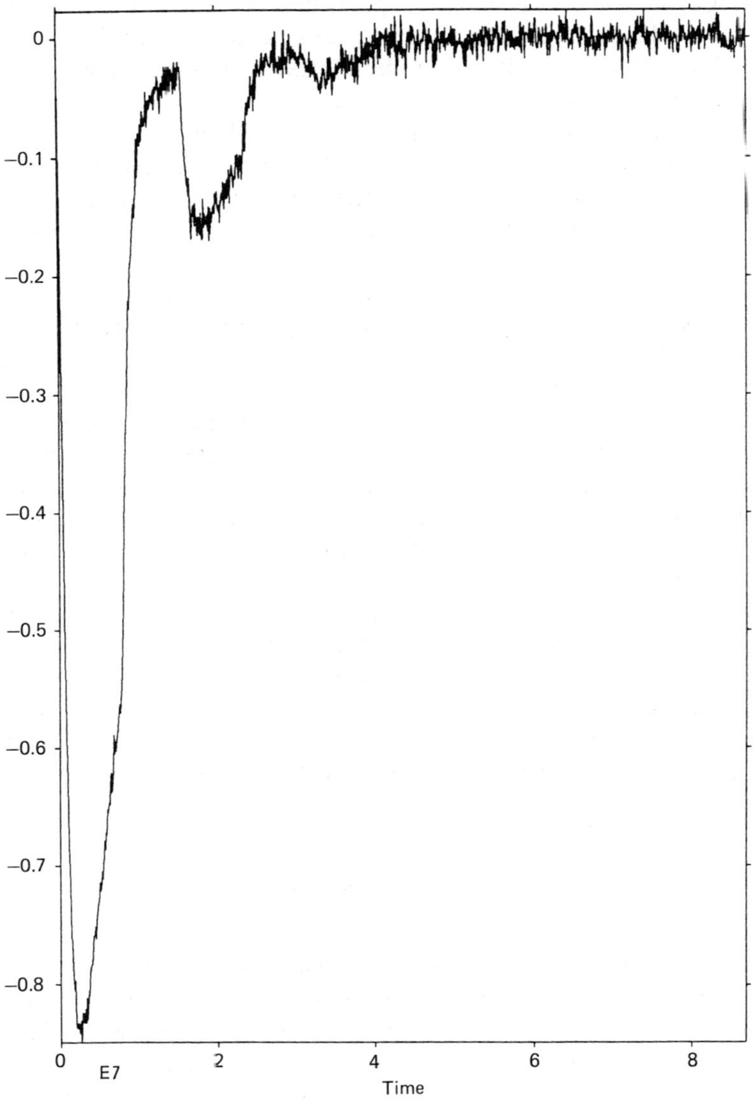

Fig. 13-11. Output with 1-percent Gaussian noise.

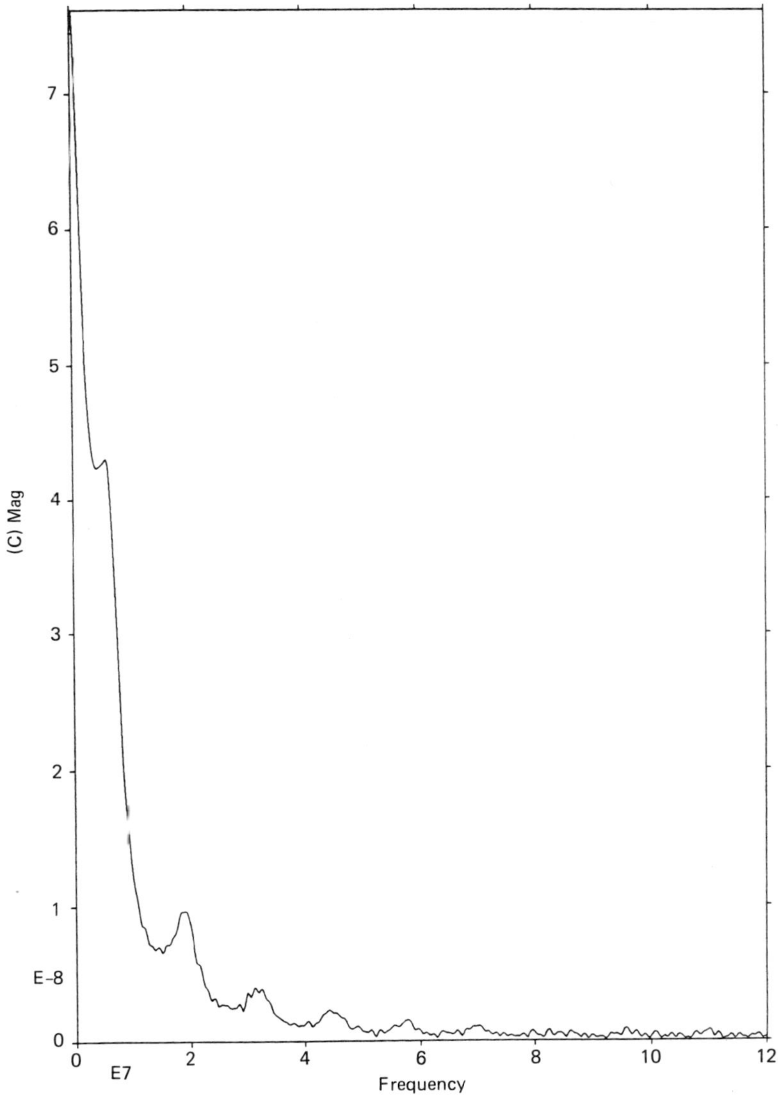

Fig. 13-12. FFT of output with 1-percent Gaussian noise.

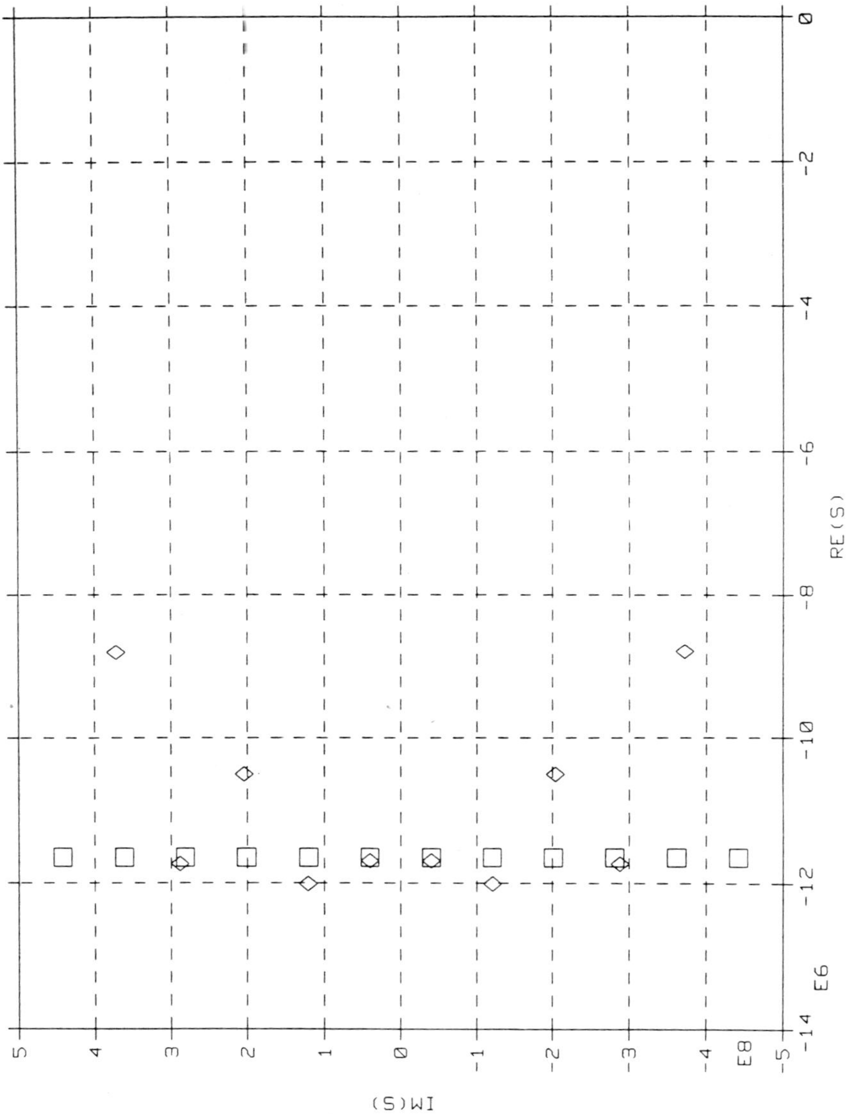

Fig. 13-13. Theoretical (□) and model (△) poles with 1-percent Gaussian noise (s-plane)

13.3 DISCUSSION AND CONCLUSIONS

In this chapter, we have discussed some methods for obtaining characteristics of scattering objects from transient electromagnetic data. We have drawn a distinction between obtaining parameters in an identification procedure (system identification) and relating the parameters to parameters giving physical characteristics of the object (object classification). We have presented an example, the dielectric slab, that has allowed us to describe the object in terms of both a ray-optic and a complex-resonance solution. The ray-optic solution emphasizes local features of the object, whereas the complex-resonance solution emphasizes global features.

Although the concept of "global" versus "local" generalizes to a large class of scattering objects,[1] it has not been possible as yet to produce systematic identification and classification procedures that encompass both descriptions. Indeed, most of the effort to date has been concerned with global identification.

We have discussed in some detail a global identification method based upon the use of a nonlinear least-squares algorithm (NLS) and an interactive signal-processing package (SIG). We have included an example based on simulated data for the dielectric slab. It is now important to undertake a systematic study of scattering from canonical objects on an experimental transient range. Recent improvements on the transient range at Lawrence Livermore National Laboratory make such a program possible.

Transient identification and object classification are still emerging sciences. With the advent of the NLS algorithm and improvements in transient range instrumentation and noise absorption, understanding the capabilities and limitations in complex-resonance identification is at hand. This understanding, however, is only the beginning. As pointed out in the introduction to Sec. 13-2, the scattering from perfectly conducting objects consists of both the complex-resonance series and an entire function. Any attempt at modeling the scattering process by complex resonances alone, therefore, ignores the entire function contribution. Although the role of the entire function in the transient waveform is well-understood,[33,34] its role in identification and classification is controversial.[35,36] Indeed, whereas the complex exponential series is parametrized by the poles and their residues, the entire function contribution is not parametrized at all. This fact has unfortunate consequences in attempts to obtain complex resonances of scattering objects that exhibit high radiation damping. For such structures, the pole positions contain relatively high damping coefficients, and the transient signal is therefore dominated by the entire function. An excellent example of the difficulties is found in the thesis by Simon,[37] which involves identification and classification simulations with a conducting circular cylinder of infinitely axial length.

In conclusion, obtaining scattering-object characteristics by parametric methods involves system identification, transient scattering theory, experimental transient range methods, and signal processing. System identification is a science that enables us to obtain parameters once a model is established. Transient scattering theory provides the model. A good transient range enables us to obtain data in a controlled laboratory environment. Intelligent signal processing allows us to provide good, parsimonious data for the identification algorithms. Progress in parametric methods for object classification depends on all four areas. Particularly important, however, is a better understanding of transient scattering theory, particularly in the parametrization of the entire function contribution.

As a final note, most efforts to date in transient electromagnetic identification have been confined to single-input, single-output cases. It is clear, however, from many studies in inverse scattering that a single aspect angle provides only limited information concerning characteristics of three-dimensional scatterers. It is therefore important to generalize studies to include inputs and outputs at multiple locations around the scattering object. Dudley[38] has proposed a state-space formulation of transient electromagnetic scattering. The model is for the single-input, multiple-output case and includes the invariance of pole position with spatial location. The method, however, remains to be tested with both simulated and range data.

REFERENCES

1. Heyman, E., and Felsen, L. B. 1983. Creeping waves and resonances in transient scattering by smooth convex objects. *IEEE Trans. Antennas Prop.* AP-31:426–437.
2. Tijhuis, A. G., and Blok, H. 1984. SEM approach to the transient scattering by an inhomogeneous, lossy dielectric slab, Part 1: The homogeneous case. *Wave Motion* 6:61–78.
3. Lytle, R. J., and Lager, D. L. 1976. Using the natural frequency concept in remote probing of the earth. *Radio Science* 11:199–209.
4. Miller, E. K., and Lager, D. L. 1982. Inversion of one-dimensional scattering data using Prony's Method. *Radio Science* 17:211–217.
5. Plonus, M. A. 1978. *Applied electromagnetics.* New York: McGraw Hill, pp. 478–481.
6. Pennisi, L. L. 1963. *Elements of complex variables.* New York: Holt, Reinhart, & Winston, pp. 237.
7. Ibid., Theorem 7.1.1.
8. Ibid., Theorem 7.6.3.
9. Oppenheim, A. V., and Willsky, A. S. 1983. *Signals and systems.* Englewood Cliffs, N.J.: Prentice-Hall, pp. 422–427.
10. Lax, P. D., and Phillips, R. S. 1967. *Scattering theory.* New York: Academic Press.
11. Knopp, K. 1947. *Theory of functions,* Part II. New York: Dover.
12. Marin, L. 1973. Natural mode representation of transient scattered fields. *IEEE Trans. Antennas Prop.* AP-21:809–818.
13. Baum, C. E. 1971. On the singularity expansion method for the solution of electromagnetic interaction problems. *AFWL Interaction Notes,* Note 88.
14. Pearson, L. W., and Marin, L., guest eds. 1981. Special issue on the singularity expansion method. *Electromagnetics* 1:349–511.

15. Dudley, D. G. 1979. Parametric modeling of transient electromagnetic systems. *Radio Science* 14:387–396.
16. Dudley, D. G. 1979. Parametric identification of transient electromagnetic systems. *Wave Motion* 5:369–384.
17. Ljung, L., and Söderström, T. 1983. *Theory and practice of recursive identification*. Cambridge, MA: MIT Press.
18. Goodwin, G. C., and Payne, R. L. 1977. *Dynamic System Identification: Experiment Design and Data Analysis*. New York: Academic Press.
19. Kendall, M. G., and Stuart, A. 1973. *The advanced theory of statistics*. vol. 2. London: Griffin.
20. Goodman, D. M. 1983. NLS: A system identification package for transient signals. Lawrence Livermore National Laboratory, Report UCID-19767.
21. Åström, K. J. 1980. Maximum likelihood and prediction error methods. *Automatica* 16, No. 5:551–574.
22. Gill, P. E., Murray, W.; and Wright, M. H. 1981. *Practical optimization*. New York: Academic Press.
23. Greenstadt, J. L. 1967. On the relative efficiencies of gradient methods. *Mathematics of computation* 21:360–367.
24. More, J. J. and Sorenson, D. C. 1979. On the use of directions of negative curvature in a modified Newton method. *Math. Prog.* 15:1–20.
25. Dongarra, J. J.; Moler, C. B.; Bunch, J. R.; and Stewart, G. W. 1979. *LINPACK Users Guide*, SIAM.
26. Marquardt, D. 1963. An algorithm for least-squares estimation of nonlinear parameters. *SIAM J. Appl. Math.* 11:431–441.
27. Polak, E. 1971. *Computational methods in optimization*. New York: Academic Press.
28. Söderström. T. 1975. On the uniqueness of maximum likelihood identification. *Automatica* 11:193–197.
29. Stearns, S. D. 1980. Error surfaces of recursive adaptive filters. Sandia National Laboratory, Report SAND80-1348.
30. Kenney, W. J., and Gentle, J. E. 1980. *Statistical Computing*. New York: Dekker.
31. Akaike, H. 1974. A new look at the statistical model identification. *IEEE Trans. Auto. Control* AC-19, No. 6:716–723.
32. Lager, D. L., and Azevedo, S. G. 1983. User's manual, SIG, A general-purpose signal processing program. Lawrence Livermore National Laboratory, Report UCID-19912.
33. Pearson, L. W. 1984. A note on the representation of scattered fields as a singularity expansion. *IEEE Trans. Antennas Prop.* AP-32:520–524.
34. Morgan, M. A. 1984. Singularity expansion representations of fields and currents in transient scattering. *IEEE Trans. Antennas Prop.* AP-32:466–473.
35. Felsen, L. B. 1985. Comments on early time SEM. *IEEE Trans. Antennas Prop.* AP-33:118–119.
36. Dudley, D. G. 1985. Comments on SEM and the parametric inverse problem. *IEEE Trans. Antennas Prop.* AP-33:119–120.
37. Simon, J. I. 1984. Identification of the complex resonances of a dielectric slab and a circular cylinder. Master's thesis, E. E. Dept., University of Arizona.
38. Dudley, D. G. 1984. A state-space formulation of transient electromagnetic scattering. *IEEE Trans. Antennas Prop.* AP—33:1127–1130.

Abbreviations and Keywords

Abbreviation or Keyword	Definition	Section
A	Area of sensor plate or loop	11.1
ACD	Asymptotic-Conical Dipole	6.1
ADC	Analog-to-Digital Converter	6.4
A_e, $\vec{A}_{eq}$	Equivalent area	6.1, 8.1
A_{eff}	Effective area	8.1
Aperiodic pulses	Pulses that recur in some nonperiodic manner	2.2
ARMAX	Auto Regressive Moving-Average model	12.1, 12.2
$A(\omega)$	Amplitude spectrum	1.0
BAR	BAseband Radar	1.3
BCD	Binary-Coded Decimal	11.5
BLUE	Best Linear Unbiased Estimate	12.1
BW	Passband	11.3
C	Capacitance	8.1, 11.1
C&C	Command and Control	6.7
CCD	Charge-Coupled Devices	6.4
ccw	Counterclockwise	1.2
CFAR	Constant-False-Alarm-Rate	
CML	Cylindrical Moebius Loop	6.1
Coherent pulse	One of a sequence of pulses with the same functional time dependence over the pulse duration	2.2
CPM	Circular Parallel Mutual inductance sensor	6.1
cw	Clockwise	1.2
DFT	Discrete Fourier Transform	12.1
Distal	Farthest from the waveform baseline	2.2
Duration	Finite time interval	2.2
DUT	Device Under Test	10.0
E	Electric field strength	11.1
EBS	Electron Bombarded Semiconductor	6.4
ECL	Emitter-Coupled Logic	4.1

Abbreviations and Keywords (*continued*)

Abbreviation or Keyword	Definition	Section
EM field	ElectroMagnetic field	5.0
EMP	ElectroMagnetic Pulse	1.2, 5.0, 6.0, 6.6, 12.1
FDHM	Full Duration at Half Maximum	2.5
FET	Field-Effect Transistor	4.5, 11.1
FFT	Fast Fourier Transform	10.0, 12.1
FMM	Flush Moebius Mutual inductance sensor	6.1
FO	Fiber-Optic	6.2
$f(t)$	Excitation or signal	1.1, 2.4
FWHM	Full Width at Half Maximum	4.8
$g(\tau)$	Correlation function	2.4
h_e	Effective height of an antenna	11.1
h_{eff}	Effective height of small sensors	8.1
HFW	High Frequency Window	9.1
HSD	Hollow Spherical Dipole	6.1, 8.1
$h(t)$	Impulse response	1.0
IDFT	Inverse Discrete Fourier Transform	12.1
IEC	International Electrotechnical Commission	2.0
IMM	Inside Moebius Mutual inductance sensor	6.1
I_{sc}	Short-circuit current	8.1
I1I	*I*-dot One-turn *I*nsertion unit	6.1
JJ	*J*osephson Junction	2.5
L	Inductance	8.1
LED	Light-Emitting Diode	6.2
LFW	Low Frequency Window	9.1
LINAC	LINear ACcelerator	4.9
ℓ	Line length	10.0
ℓ_e	Effective length	11.1
ℓ_{eq}	Equivalent length	6.1, 8.1
Mesial	Toward the middle of the waveform baseline	2.2
MGL	Multi-Gap Loop	6.1, 8.1
MMT	Matched-Monopole Type	1.2
MTL	Multi-Turn Loop	6.1

Abbreviations and Keywords *(continued)*

Abbreviation or Keyword	Definition	Section
NLS	Nonlinear Least-Squares	12.2
Noncoherent pulse	A pulse that does not provide stationary time dependence	2.2
OMM	Outside Moebius Mutual inductance sensor	6.1
PCS	Plastic-Clad Silica	6.2
Periodic pulses	Pulses that recur with a fixed period of T seconds	2.2
PPD	Parallel-Plate Dipole	6.1
PRF	Pulse Repetition Frequency	1.1
Proximal	Nearest the waveform baseline	2.2
$p(t)$	Pulse-modulated input	1.2
PTD	Programmable Trigger Delay	6.3
Pulse	A physical modification of state	2.2
Pulse train	Sequence of pulses	2.2
RELS	Recursive Extended Least-Squares	12.2
RLS	Recursive Least-Squares	12.2
$r(t)$	Response	1.1
S	Sensor plate area, or speed of rotation in revolutions per second	8.1, 11.1
SCR	Silicon-Controlled Rectifier	4.6
SEM	Singularity Expansion Method	13.2
SFANA	Stepped-Frequency Automatic Network Analyzers	10.2
Single event	Single-shot case of a single pulse	2.2
SNR	Signal-to-Noise Ratio	1.2, 10.0
SRD	Step-Recovery Diode	4.3
$s(t)$	Sampling pulse	2.4
T	Sampling interval	10.0
TD	Tunnel Diode	4.4
TDANA	Time-Domain Automatic-Network Analyzer	10.0, 10.1
TDR	Time-Domain Reflectometry	10.1
TDS	Time-Domain Spectroscopy	10.2
TEM	Transverse Electromagnetic	5.0
TLIR	Time-Limited Impulse Response	1.2

Abbreviations and Keywords (*continued*)

Abbreviation or Keyword	Definition	Section
Transition duration	Difference between the *proximal* and *distal* points	2.2
TTL	Transistor-Transistor Logic	4.1
T_w	Time window interval	10.0
v	Velocity of propagation	10.0
V_L	Output voltage	8.1
V_{oc}	Open-circuit voltage	8.1
Waveform	Observed or measured shape of a pulse	2.2
Waveform duration	Difference between the mesial points (in time) of the first and last transitions	2.2
WTT	Wave-Trapping Type	1.2
Y_T	Transfer admittance	6.8
Z	Characteristic impedance	1.2
Z_c	(1) Characteristic impedance of a cable, or (2) recording system matched impedance	8.1, 11.1
Z_T	Transfer impedance	6.1, 6.8
$Z(\omega)$	Driving point impedance of a dipole	1.1
$\delta(t)$	Dirac delta function	1.1
ε_0	Permittivity of free space	8.1
ε_r	Relative dielectric constant	8.1
μ	Permeability	1.2
$\phi(\omega)$	Phase function	1.0

Appendix

Typical Time Domain Measurement Configurations

Jeremy A. Landt

Los Alamos National Laboratory

This appendix contains examples of typical time-domain measurement configurations that are intended to be illustrative of techniques employed in the 500-MHz to (about) 5-GHz range or for risetimes longer than (about) 50 ps. At lower frequencies and for risetimes longer than several nanoseconds, time-domain techniques are still of interest, and a variety of oscilloscopes and transient recorders can be employed. For risetimes less than 50 ps, optical-electronic techniques can be used to measure transient signals. The specific pieces of equipment used in these examples are presently commercially available but not officially endorsed. In many cases, equivalent equipment or alternative techniques can be used to perform the same measurements. Sampling techniques are treated exclusively.

The ideas and techniques illustrated here have come from many sources. A number of special-purpose systems have been developed in the last several years, for example, on automatic automotive braking system using base-band radar and monopulse subterranean radar (see Chap. 1). The interested reader is referred to the literature for discussion of these and similar systems.

The chosen examples illustrate the utility of time-domain measurements. As has been mentioned in various places throughout this book, their advantages include the following:

1. Clear and natural representation of transient wave phenomena that permits the physics of propagation to be easily grasped and gives a qualitative understanding of transient phenomena.
2. Broadband measurements that allow high-Q characteristics to be measured without limitations imposed on sampling in the frequency domain.
3. Time-range gating can be used to eliminate the need for expensive anechoic chambers.
4. Usable time-domain data can be obtained without precision-calibration standards or techniques. Note that without careful calibration, frequency-domain data has limited utility. One problem encountered when sweeping the frequency is that changes in phase caused by cable lengths give phase errors. Automated techniques

are often required for meaningful swept measurements as well as good calibration procedures.

5. The "big picture" is easily grasped in time-domain data. Resonant frequencies, element Q, and broadband impedance are easily estimated.
6. The time domain is natural for many applications such as EMP and range-delay radar.
7. Scaling laws and scale models can be used to reduce the resources required to test large objects.
8. Equipment faults are located easily in the time domain. These include bad connectors, cable faults, impedance mismatches, and the like.

Time-domain measurements have various limitations, however, including:

1. Specialized equipment is required, and such equipment is often not included in the standard laboratory inventory.
2. Frequency-domain results can be of limited accuracy. The FFT can be used to calculate frequency-domain data from time-domain measurements, but data obtained in this fashion usually do not have the accuracy of carefully made frequency-domain measurements. Signal averaging and careful calibration procedures can improve accuracy.
3. Measurements are usually limited to frequencies less than (about) 10 GHz. Measurements of open systems made below (about) 100 MHz require large facilities or use of absorbing material (an anechoic chamber). The upper frequency limit can be extended with exotic pulsers or by pulsed carrier techniques.
4. Most coaxial and strip lines are lossy and dispersive. Rapid risetimes are lost quickly on excessive cable runs.
5. Trigger jitter often makes it difficult to obtain data. Triggering drift can cause problems, especially when attempts are made to improve signal-to-noise by averaging.
6. External noise can be picked up in open systems.
7. Impedance mismatches produce reflections. Thus, to reduce unwanted reflections, all components should be well matched to the impedance of the measuring system (usually 50 ohms).
8. Accuracy will be impaired by nonlinear sweep, nonlinear deflection (i.e., nonlinear horizontal and vertical deflections), and inaccuracies in A/D converters.

The examples covered here include TDR (time-domain reflectometry) methods and applications, transmission measurements, and base-band radar. Bounded and unbounded applications will be considered. A "bare-bones" approach will be presented as well as the techniques used to automate and collect data in digital format.

TIME-DOMAIN REFLECTOMETRY

Time-domain reflectometry (TDR) is a technique used to measure the reflection of a pulse from an object under test. The finite velocity of light is used to eliminate unwanted reflections occurring during the time that measurements are being made and also to determine the distance to reflecting perturbations. The quantity of interest is the input reflection coefficient. In bounded systems, TDR can be used to measure impedances on transmission lines, to measure load impedances, and to find faults on transmission lines. Coaxial and microstrip components can be measured. Discontinuities caused by faults, mismatches, bad connectors, and changes in impedance can be easily located. In open or radiating systems, antenna input impedance can be measured. In many cases, direct time-domain information is useful, such as finding line faults or determining transmission-line impedance. In some cases—for example, to obtain antenna input impedance—transformation to the frequency domain is desirable. The key to TDR measurements is the use of delay times to separate the reflected signal of interest from unwanted signals in the system. Doing so usually requires well matched components to minimize unwanted reflections.

A basic TDR system consists of several components:

1. A pulser
2. A signal sampler to measure the signal on a transmission line
3. Transmission lines that connect the pulser, sampler, and device to be tested
4. A means of displaying or recording the signal obtained by the sampler.

The equipment chosen depends on the application and the range of frequencies of interest. For example, above 1 GHz, a sampling oscilloscope is the most common form of signal display. For TDR applications, pulsers of modest voltage (<5 V) are adequate. The risetime of the pulser is determined by the spatial resolution required or by the highest frequencies of interest. To be useful, the pulser output should not experience changes in voltage after the primary pulse or step.

A self-contained, portable, but not nonexpandable TDR system is shown in Fig. A-1. The Tektronix 1502 TDR Cable Tester has a risetime less than 140 ps and a jitter less than 40 ps, values that are adequate for many applications. This self contained unit is easy to use, but better used for demonstration of TDR measurements and applications than for illustration of TDR equipment principles. Measurement of rapid transients depends upon jitter-free triggering. Although this problem is minimized in the 1502, it affects sampling systems in general. The pulser in the 1502 generates a step voltage of about 220 mV. The signal sampler is close to the pulser, which is internally matched to

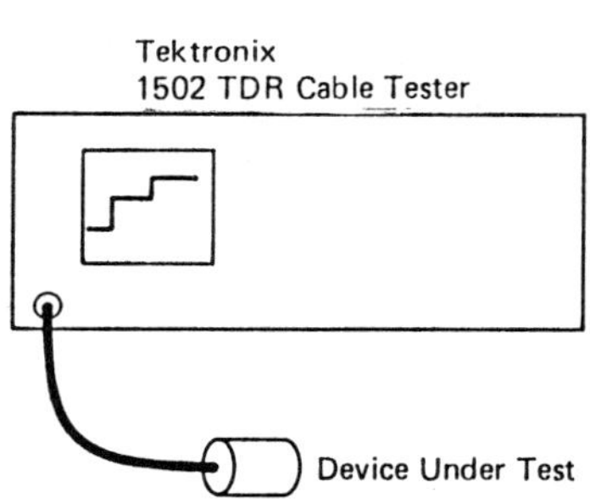

Fig. A-1. Low-cost TDR.

absorb signals reflected from the load. As with any high-frequency system, care must be taken; high-quality connectors should be used and the length of cable connecting the unit to the load minimized.

The 1502 was originally intended for use in finding the location of faults in cables. The distance to a fault is determined by multiplying one-half the product of the signal velocity of the cable by the round-trip time to the fault. The 1502 has a dial that eliminates the need to do the division manually.

The 1502 can be used as a general purpose TDR, however, for example, to determine the qualitative characteristics of an antenna as a function of depth away from a planar interface and as a function of the conductivity of the media. To perform this kind of experiment, connect a 15 cm-long dipole to the 1502 by a length of coaxial cable. This antenna is resonant in air at about 1 GHz. Since the period of 1-GHz signal is 1 ns, the clear time in air is 1 ns per 15 cm, where distance is measured to the closest unwanted scatterer. In water, the clear time is about 1 ns per 2 cm. Thus, a 1-m-square ground plane will give a clear time in air of 3 1/3 ns. For the present example, this is a convenient size that does not require an elaborate antenna range.

The experiment is performed by using an unbalanced dipole, or a monopole driven against a ground plane. In either case, insulated wire should be used; number 12 insulated copper wire used in house wiring works nicely. As the antenna is lowered toward water, changes in the reflection coefficient will be seen. The reflections change dramatically when the antenna is lowered into the water. (The resonant frequency is about 300 MHz at the air-water interface.) When salt is added to the water to increase conductivity, the antenna Q decreases, and the dipole loses much of its resonant character when the salt-water solution becomes very conductive.

A similar experiment in the frequency domain would require an elaborate antenna range, and the raw data would contain considerably less intuitive information about the physics of the interaction of the antenna with the conductive media.

The 1502 is adequate to demonstrate TDR techniques. It can produce output data with a built-in X-Y recorder or record data with an optional strip-chart recorder. Since all of the TDR components are contained within the instrument and not accessible, however, the 1502 cannot be easily expanded, modified, or tailored to specific requirements and is difficult to use for transmission measurements.

These limitations can be overcome by the expandable TDR system manufactured by Tektronix shown in Fig. A-2. Included are the 7S12 TDR/Sampler with an S52 Pulser and an S6 Sampling Head. These accessories are able to turn any 7000-series mainframe into a sampling system with a risetime of less than 40 ps. The system of Fig. A-2 uses a 7603 oscilloscope mainframe since this is the least expensive of the 7000 series, and many laboratories will already have it. Optional transmission equipment includes the 7S11 sampling unit and the S4 sampling head.

All of the TDR measurements previously discussed can be made with this system. To maintain rapid risetimes, the pulse generator and sampling head must be placed as close as possible to the device under test. Risetime is lost rapidly on excessive cable runs. Tektronix accessories include extenders that permit the pulser and sampling heads to be located remotely from the oscilloscope mainframe; these are highly recommended.

The sampling head risetime is 30 ps; the pulser risetime, 25 ps with an 800-ns width;

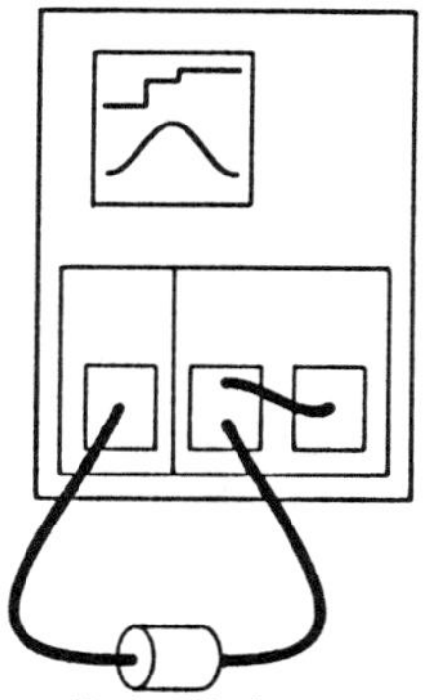

Fig. A-2. Expandable 30-psec TDR and transmission.

and the pulser repetition rate, 40 kHz. One sample of the signal is obtained for each pulse produced by the pulser.

Consequently, a high pulse-repetition rate permits data to be obtained quickly and aids in flicker-free and easily viewable displays.

The system in Fig. A-2 can be expanded and automated, and such features will be discussed later. For this reason, it is preferred over the system in Fig. A-1 if a need for additional capabilities is anticipated.

TRANSMISSION MEASUREMENTS

Transmission measurements are required to determine the radiating properties of antennas and the response of components such as filters, couplers, amplifiers, and the like. For some applications, the low voltage provided by the S-52 pulse generator is adequate. The system in Fig. A-2 can be adapted for transmission measurements by adding a 7S11 sampling unit and an S4 sampling head. Alternatively, if TDR measurements are not required, the S6 sampling head can be used to provide transmission measurements by connecting the pulser directly to the device under test, connecting the output from the device under test to the S6 sampling head, and terminating the other port of the S6 head with a 50-ohm load. Both approaches work well, and obtaining a stable trigger or display encounters minimal trouble.

The voltage provided by the S-52 pulse generator is inadequate for transmission measurements in radiating systems (for example, transmission from one antenna to another) or in devices with very high insertion loss. The transient input impedance of most radiating systems is poorly matched to 50 ohms. The radiating characteristics of most radiating systems is poorly matched to the frequency content of the step voltage produced by the S-52 pulser. Because of these effects, considerable energy is lost in reflection of the pulser output from the radiating antenna. For example, the typical range for the input surge impedance of wire and conical radiators is in the range of several hundred to one thousand ohms. The reflected energy varies as the square of the reflection coefficient. In a 50-ohm system, a 1000-ohm radiating antenna will reflect 82 percent of the incidence energy. Once launched, the radiated field strength will generally fall off inversely as distance from the radiator. The receiving antenna will develop an open circuit voltage that is equal to the product of the incident field strength and the receiving antenna effective height. An insertion loss of 40 dB is typical for a long-wire radiating antenna and a 1-GHz receiving antenna separated by 1 meter. Thus, the 200 mV provided by the S-52 pulser results in received signals of 2 mV, which is an inadequate amount.

The solution is to use a pulser with more output voltage. For most pulser techniques available, this increase is usually accompanied by an increase in risetime (photo conductive switches can produce very high voltages with subnanosecond risetimes, but these pulsers usually are not available commercially).

A typical transmission system for radiating systems is shown in Fig. A-3. This equipment is the same as that of Fig. A-2 with the addition of a Picosecond Pulse Labs Model 1500A pulser and a Tektronix S-53 Trigger Recognizer that minimizes triggering problems. The 1500A pulser output is a Gaussianlike pulse with a peak voltage of 60 V and a half width of 300 ps. A variety of alternative arrangements can be used, but the one in Fig. A-3 provides stable triggering and is easy to set up. In this configuration, the pulse generator must have a stable trigger signal that precedes the output pulse by at least 50 ns.

If more output voltage is required, the IKOR pulser can be used. This pulser has a peak output voltage of about 1 kV but is free running, lacks a trigger output, and has a repetition rate of only 300 Hz. The low repetition rate results in an oscilloscope display that is not easily viewed. To use the IKOR pulser, a Tektronix 7M11 delay line is required to provide a trigger preceding the signal. This delay line increases system risetime to 175 ps. As an alternative time base, the 7T11 can be used. The combination of the 7T11 and the IKOR pulser can give some problems in obtaining reliable triggering, however. Some improvement in triggering performance is possible by using an external trigger not obtained from the signal to be measured but rather from a signal sampler or parasitic antenna near the radiating antenna.

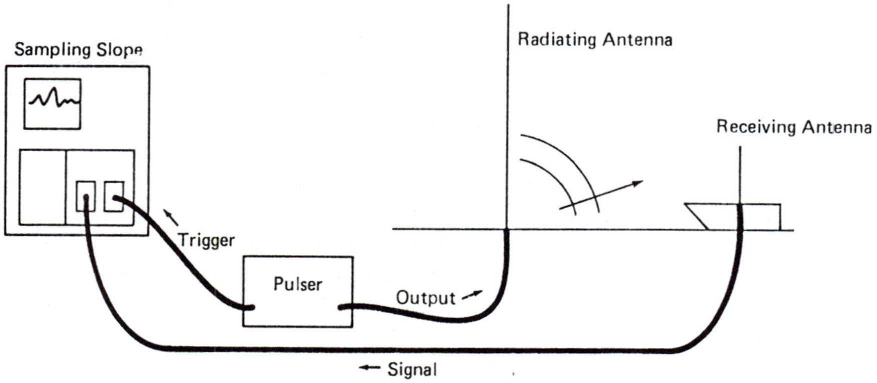

Fig. A-3. High loss transmission.

AUTOMATION AND DATA ACQUISITION

Data quality can be improved by signal averaging. Frequency-domain data can be obtained by transformation of transient data. The output corresponding to a different input signal can be found by convolution with the system impulse response or by multiplication of frequency-domain responses followed by a transform to the time domain (see Chap. 12). These operations require data acquisition. Photography is one form of data recording but is inconvenient for other than simple archival purposes. Horizontal and vertical analog signals are available on the front panel of the 7S12 TDR/Sampler of the systems shown in Figs. A-2 and A-3. These signals are applicable to data-acquisition systems using either analog-to-digital converters and a computer or a digitizing oscilloscope.

A more straightforward approach is that shown in Fig. A-4. The Tektronix 7000 series plug-ins are used with a Tektronix 7854 digitizing-oscilloscope mainframe and a computer with an IEEE-488 interface. The 7854 can perform signal averaging before data is transferred to the computer. The Hewlett-Packard HP-85 computer is a good choice, with well designed and easily used I/O procedures, a built-in tape drive and printer, and adequate memory. In general, the software will have to be written by the user. A simple data-acquisition program is given in Listing 1. Figure A-4 illustrates a TDR application, but the transmission measurements of Figs. A-2 or A-3 could be automated as well.

As an example, the measurement of the insertion loss of a directional coupler will be treated here in detail. Table A-1 lists the required equipment. Equipment setup is straightforward. The mode/address switch on the back of the 7854 GPIB is set to an address of 10, on line, and talk/listen with an LF as end of message (10101010). The 7S12 is installed in the center two compartments of the 7854 (right vertical and A horizontal), the 6-ft extenders are plugged into the 7S12, and the S-6 and S-52 heads are placed at the end of the extenders. The ROMs, ROM drawer, memory card, and HPIB interface are plugged into the back of the HP-85 computer. The 7470 plotter, HP-85, and 7854 are connected by the HPIB cable of the HPIB interface and one HPIB jumper cable. One

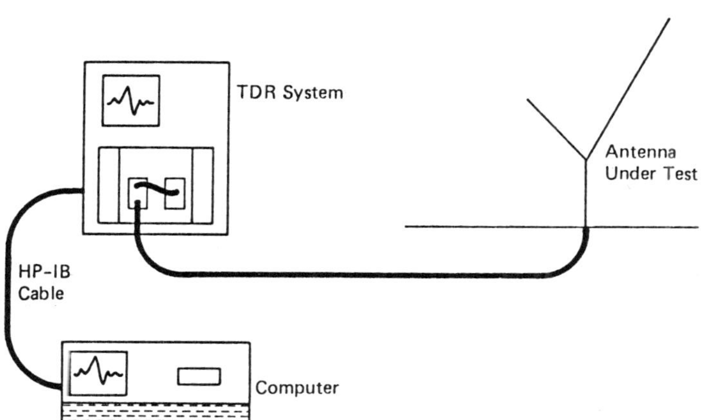

Fig. A-4. Data acquisition.

Listing 1 A short program to transfer data from the Tektronix 7854 digitizing oscilliscope to the HP-85 computer.

```
10 !       PROGRAM SHORT
20 !    2-29-84   J A Landt
30 ! TO TRANSFER DATA FROM THE
40 ! TEKTRONIX 7854 TO THE
50 ! HEWLETT-PACKARD HP-85
60 DIM W$[8200]
70 CLEAR 710
80 OUTPUT 710 ;"O WFM SENDX"
90 ENTER 710 ; N,Z1,X,Z2,Y
100 DISP N,Z1,X,Z2,Y
110 ENTER 710 ; W$
120 PRINT N,Z1,X,Z2,Y
130 PRINT W$
140 END
```

port of the S-6 is terminated with a 50-ohm load. The system is readied for use by energizing all equipment, and, when the 7854 self-test has been completed, by pressing the right-vertical and left-horizontal mode buttons and placing the 7S12 in the repetitive triggering mode. The short program in Listing 1 transfers data from 7854 to the HP-85. The 7854 provides the number of data points, the time increment between data points, the voltage per division, and the data (normally 512 numbers).

Table A-1. Equipment Required for Automated Data Acquisition and Plotting.

		List Price, 1983
A. *Time-domain sampling oscilloscope (Tektronix)*		
1. Tektronix 7854 mainframe		$13750
2. 7S12 TDR/Sampler		3390
3. S52 pulse generator		1655
4. S6 sampling head		2295
5. Two 6-ft extenders		890
6. Assorted SMA fittings (50-Ω loads, coaxial cables, SMA-N and SMA-BNC adapters, etc.)		
B. *Microcomputer and plotter (Hewlett-Packard)*		
1. HP-85A computer		$ 2750
2. 16K memory module	82903A	195
3. ROM drawer	82936A	45
4. I/O ROM	00085-15003	295
5. Printer/plotter ROM	00085-15002	145
6. Matrix ROM	00085-15004	145
7. Advanced progr. ROM	00085-15005	145
8. HP-IB interface	82937A	395
9. Plotter	7470A	1550
10. HP-IB cable	10833B	90
11. Accessories (Data cassettes, printer paper, plotter pens and paper, etc.)		

The wideband-coupling characteristics of an HP 764D directional coupler are to be measured. The calibration of this coupler is -20 ± 2 dB from 225 to 450 MHz. If a scan rate of 2 ns/div is chosen, 10 divisions give a total time span of 20 ns. This will correspond to a frequency step of 50 MHz when the data is processed by a fast Fourier transform. The upper frequency limit is set by the time step, and the Nyquist limit will be 25.6 GHz. The practical upper frequency limit will be about a factor of 10 less, or about 2.56 GHz. Using a faster sweep speed will extend the high-frequency limit of the measurement but will sacrifice the low-frequency resolution. Slowing the sweep speed will do just the opposite. Some trial and error is required to match the sweep speed to the device to be tested. Thus, a 2-ns/div sweep speed is reasonable for the HP 764D coupler.

To perform the measurement, first attach the S-6 directly to the S-52 to measure the applied voltage waveform. The data is acquired with the program given in Listing 2; it is plotted in Fig. A-5 using the program given in Listing 3. It was obtained on the 7854 by pressing the AVE 100 button, followed by its transfer to the HP-85. The fast Fourier transform of the input voltage is shown in Fig. A-6. The FFT was found using the program of Listing 4 and plotted using the program of Listing 5.

The next step is to attach the S-52 to the input of the directional coupler and the S-6 to the coupled port. The remaining ports on the coupler are terminated in 50 ohms. The coupled signal is shown in Fig. A-7 and on an expanded time scale in Fig. A-8. The rapid oscillations in Fig. A-7 are not noise but stable high-frequency oscillations, as shown in Fig. A-8. Note that using the 7854 to average the data has reduced the noise to about 0.05 mV, or about twice the digitization error. The FFT of Fig. A-7 is shown in Fig. A-9. The pulselike waveform of Fig. A-7 transforms into a spectrum similar to $\sin(x)/x$.

The insertion loss is found by dividing the spectrum of Fig. A-9 by the spectrum of Fig. A-6. The program of Listing 6 is used, and the result is shown in Figs. A-10 and A-11. The entire spectrum is given in Fig. A-10. Note that the data is erratic for frequencies with fewer than about 10 samples per cycle. Also plotted in Fig. A-11 are the calibration data given by Hewlett-Packard. The agreement between these two sets of data is ± 0.2 dB, or ± 2 percent. This is very good, considering that no attempt was made to calibrate or fine-tune the time-domain equipment. The 7854, 7S12, S-6, and S-52 were used essentially "out of the box."

These data are very informative. The input pulse can be approximated by a step function, as follows:

$$V_i(t) = A\,U(t) \tag{A-1}$$

The output is the sum of two waves, both replicas of the input but reduced in amplitude, and one delayed and reversed in sign, as follows:

$$V_o(t) = B[U(t) - U(t - T)] \tag{A-2}$$

Thus, a delay line is one of the fundamental building blocks of the coupler. This observation is obvious in the time domain (compare Figs. A-5 and Fig. A-7) but is obscured

Listing 2 HP-85 program to transfer data from Tektronix 7854 digitizing oscilliscope and save on tape.

```
10 !      PROGRAM AQUIRE
20 !    3-18-84  J A Landt
30 ! TO TRANSFER DATA FROM 7854
40 ! SCOPE TO HP-85 AND STORE
50 ! ON TAPE IN FILE 'DATA 1'
60 OPTION BASE 1
70 DIM W(512),W$[8200],N$[10],R$[100]
80 CLEAR
90 DISP "FILE NAME TO STORE DATA ="
100 INPUT D$
110 CLEAR
120 GOSUB 850
130 CLEAR @ DISP "WAITING FOR HP-IB BUS"
140 CLEAR 710
150 OUTPUT 710 ;"O WFM SENDX"
160 ENTER 710 ; N,Z1,X,Z2,Y
170 ENTER 710 ; W$
180 RESET 7
190 PRINT N,Z1,X,Z2,Y
200 CLEAR @ DISP "CONVERTING DATA"
210 I3=0 @ I=1 @ N1=N-1
220 IF W$[I,I]=" " THEN 250
230 I=I+1
240 GOTO 220
250 I1=I+1
260 I=I1
270 IF W$[I,I]="," THEN 300
280 I=I+1
290 GOTO 270
300 I2=I-1
310 I3=I3+1
320 W(I3)=VAL(W$[I1,I2])
330 IF I3<N1 THEN 250
340 I3=N @ I=I2+2 @ W(I3)=VAL(W$[I])
350 ! NOW DATA IN W ARRAY
360 GOSUB 510
370 CLEAR
380 DISP "WRITE DATA TO TAPE NEXT"
390 DISP "" @ DISP "WRITE TO FILE : ";D$
400 DISP ""
410 DISP "HIT 'CONT' TO CONTINUE"
420 PAUSE
430 ASSIGN# 1 TO D$
440 PRINT# 1 ; N,X,Y,W(),R$
450 ASSIGN# 1 TO *
460 CLEAR
470 DISP "DONE"
480 PRINT "DATA WRITTEN TO FILE : ";D$
490 PRINT "" @ PRINT ""
500 STOP
510 GCLEAR @ BEEP @ WAIT 200 @ BEEP
520 CLEAR @ DISP "ENTER COMMENT"
530 INPUT R$
540 DISP ""
550 DISP "PLOT NEXT"
560 DISP "ENTER TIME/DIV ON SCOPE"
570 INPUT T3
```

```
580 DISP ""
590 T1=0 @ T2=10*T3
600 DISP ""
610 DISP "ENTER VOLTS/DIV"
620 INPUT Y3
630 Y2=4*Y3 @ Y1=-4*Y3
640 LOCATE 0,133,0,100
650 SCALE T1,T2,Y1,Y2
660 PEN 1
670 XAXIS Y2,T3,T1,T2
680 XAXIS Y1,T3,T1,T2
690 YAXIS T2,Y3,Y1,Y2
700 YAXIS T1,Y3,Y1,Y2
710 PENUP
720 FOR I=1 TO N
730 X9=(I-1)*X
740 Y9=Y*W(I)
750 PLOT X9,Y9
760 NEXT I
770 PENUP
780 PRINT ""
790 PRINT R$
800 PRINT ""
810 PRINT Y3;" VOLTS/DIV"
820 PRINT T3;" SECONDS/DIV"
830 PRINT ""
840 RETURN
850 DISP "TO TRANSFER DATA FROM"
860 DISP "TEKTRONIX 7854 SCOPE TO"
870 DISP "HP-85 AND STORE ON TAPE"
880 DISP "IN FILE ";D$
890 DISP " "
900 DISP "HIT 'CONT' TO CONTINUE"
910 PAUSE
920 CLEAR
930 DISP "IF FILE ";D$;" DOES NOT"
940 DISP "EXIST, OR OTHER NAME IS"
950 DISP "DESIRED, STOP PROGRAM AND"
960 DISP "CREATE DATA FILE BEFORE"
970 DISP "CONTINUING"
980 DISP " "
990 DISP "TO CREATE A DATA FILE :"
1000 DISP "SIZE IS #*8 IE 512*8=4096+256"
1010 DISP " "
1020 DISP "EXECUTE 'CREATE '";D$;"',1,4352"
1030 DISP "WHERE THE ' IS REALLY A QUOTE"
1040 DISP ""
1050 DISP "HIT 'CONT' TO CONTINUE"
1060 PAUSE
1070 CLEAR
1080 DISP "ADJUST 7854 AND GET DESIRED DATA"
1090 DISP "IN X REGISTER"
1100 DISP ""
1110 DISP "DO THIS USING PANEL BUTTONS"
1120 DISP "AQR, AVERAGE 10, 100, OR 1000"
1130 DISP ""
1140 DISP "THEN HIT 'CONT' TO CONTINUE"
1150 PAUSE
1160 RETURN
1170 END
```

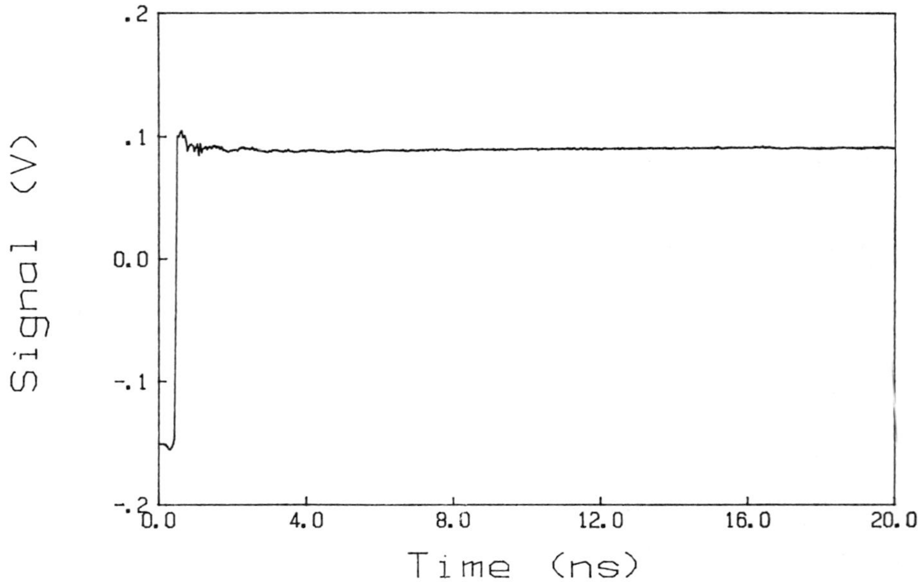

Fig. A-5. Pulse produced by S-52 pulse generator.

in the frequency-domain data (Fig. A-9). Some of the physics of the internal construction of the coupler has been determined by time-domain probing. Note that the flat-topped portion of Fig. A-7 is about 0.0124 V above the base line, a value that corresponds to a reduction of the input voltage by a factor of about 20. An uninformed calculation of insertion loss gives -26 dB, which is in disagreement with Fig. A-11. Thus, while the transient response gives insight concerning the construction of the device, the performance at a particular frequency is not obvious.

Frequency-domain performance should not be estimated by simple division of time-domain voltages. One is tempted to do this, but it is mathematically incorrect. Instead, the transforms should be divided. Doing so is often not difficult. The Laplace transform of Eq. A-1 is expressed as

$$V_i(s) = A/s \tag{A-3}$$

Listing 3 Program to read HP-85 tapes and plot transient data.

```
10 !     PROGRAM TPLOT
20 ! 3-16-84    J A LANDT
30 !    TO READ DATA STORED ON
40 ! TAPE AND PLOT
50 DEG
60 OPTION BASE 1
70 PLOTTER IS 705
```

```
 80 PEN 0
 90 DIM E5(512),R$[100]
100 M=9
110 CLEAR @ DISP "INPUT FILE NAME TO BE READ"
120 INPUT D$
130 PRINT "READING DATA"
140 GOSUB 600
150 D1=X9
160 T6=N*D1
170 C9=T6
180 PENUP
190 GCLEAR
200 D8=-4*Y9 @ D9=4*Y9 @ D7=2*Y9
210 PRINT "INPUT TIME WAVE" @ PRINT " "
220 PRINT "MAG FROM ";D8;" TO ";D9
230 PRINT "TIME FROM 0 TO ";T6;" SEC" @ PRINT " "
240 GOSUB 390
250 M9=1
260 CLEAR @ DISP "RENORMALIZING  --  WAIT"
270 FOR I=1 TO N
280 E5(I)=Y9*E5(I)
290 NEXT I
300 PEN 1
310 FOR I=1 TO N
320 T3=(I-1)*D1
330 T=E5(I)
340 PLOT 1000000000*T3,T
350 NEXT I
360 PENUP @ PEN 0
370 MOVE 0,0
380 STOP
390 LOCATE 25,130,25,95
400 SCALE 0,T6*1000000000,D8,D9
410 GCLEAR @ FXD 1,1 @ PEN 1 @ CSIZE 4
420 LAXES -T6*200000000,(D9-D8)/4,0,D8,1,1,2
430 MOVE T6*1000000000/2,D9-1.1*(D9-D8)
440 LORG 6
450 CSIZE 7,.8
460 LABEL "Time (ns)"
470 MOVE T6*1000000000/2.5,D9-1.18*(D9-D8)
480 LORG 6 @ CSIZE 4,.8
490 LABEL D$;" : ";R$
500 CSIZE 7,.8
510 MOVE -T6*1000000000/6,.5*(D9+D8)
520 LORG 4
530 LDIR 90
540 LABEL "Signal (V)"
550 FRAME @ PEN 0
560 PENUP
570 RETURN
580 STOP
590 ! NUMERICAL INPUT
600 ASSIGN# 1 TO D$
610 READ# 1 ; N,X9,Y9,E5(),R$
620 ASSIGN# 1 TO *
630 PRINT " " @ PRINT " "
640 PRINT "DATA FILE '";D$;"' READ"
650 PRINT " " @ PRINT R$ @ PRINT " "
660 PRINT "N, DT, V/DIV = ";N,X9,Y9
670 E1=E5(1)
680 PRINT "FOR PLOTTING ONLY"
690 RETURN
700 END
```

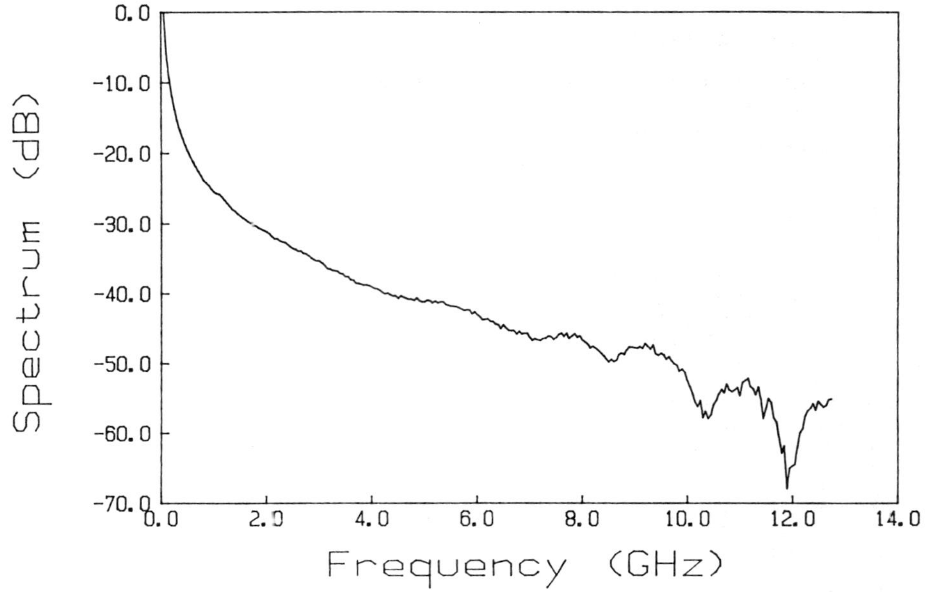

Fig. A-6. Fast Fourier transform of the pulse produced by the S-52 pulse generator (0 dB = 7.56×10^{-10} V/Hz).

and the transform of Eq. A-2 as

$$V_o(s) = \frac{B}{s}[1 - \exp(-sT)] \tag{A-4}$$

The insertion loss is the ratio of Eqs. A-3 and A-4, as follows:

$$I(s) = \frac{B}{A}[1 - \exp(-sT)] \tag{A-5}$$

Let $s = j\omega$. Then Eq. A-5 reduces easily to the following:

$$I(\omega) = j\frac{2B}{A}\sin(\omega T/2) \tag{A-6}$$

Equation A-6 compares well with the measured response shown in Fig. A-11. From Fig. A-5, $A = 0.242$ V, and from Fig. A-7, $B = 0.0124$ V, and T = 1.46 ns. Thus,

$$I(f) = 0.102\sin(f/6.85{*}10^8) \tag{A-7}$$

Listing 4 Program to read HP-85 tapes, calculate the fast Fourier transform, and save on tape.

```
10 !   PROGRAM FORT
20 !  3-16-84   J A LANDT
30 ! FAST FOURIER TRANSFORM
40 ! READS TAPE, TAKES FFT, AND STORES FFT DATA
50 RAD
60 OPTION BASE 1
70 PLOTTER IS 1
80 DIM A(1024),S(256),K(11),E5(512),R$[100]
90 MAT A=ZER @ MAT S=ZER @ MAT E5=ZER
100 M9=9
110 D$="DATA1"
120 F$="FFT1"
130 I8=-1
140 N=2^M
150 PRINT "READING FILE : ";D$
160 GOSUB 2260
170 Y1=-8*Y9 @ Y2=8*Y9 @ Y3=2*Y9
180 D1=X9
190 F9=1/(2*D1)
200 T6=N*D1
210 C9=T6
220 T9=.125*T6 ! TMAX FOR STEP
230 A0=5/T9 ! A FOR STEP
240 PENUP
250 GCLEAR
260 PRINT "INPUT TIME WAVE" @ PRINT " "
270 PRINT "MAG FROM ";Y1;" TO ";Y2
280 PRINT "TIME FROM 0 TO ";T6;" SEC" @ PRINT " "
290 SCALE 0,T6,Y1,Y2
300 YAXIS 0,Y3 @ YAXIS T6,Y3 @ XAXIS Y1,.1*T6 @ XAXIS Y2,.1*T6
310 M9=1
320 CLEAR @ DISP "RENORMALIZING  --  WAIT"
330 FOR I=1 TO N
340 I2=2*I
350 GOSUB 2200
360 A(I2-1)=T @ A(I2)=0
370 NEXT I
380 E9=A(2*N-1)
390 FOR I=1 TO N
400 I2=2*I
410 T3=(I-1)*D1
420 T=A(I2-1)
430 PLOT T3,T
440 ! SUBTRACT STEP FUNCTION
450 T8=-2*A0*(T3-T9)
460 T8=(1-EXP(T8))/(1+EXP(T8))
470 T8=.5*E9*(1+T8)
480 T=T-T8
490 A(I2-1)=T
500 NEXT I
510 PENUP
520 FOR I=1 TO N
530 PLOT (I-1)*D1,A(2*I-1)
540 NEXT I
550 PENUP @ COPY @ PRINT " "
560 PRINT "" @ PRINT "FFT CALC"
570 GOSUB 930
```

```
580 PENUP
590 PRINT "FOURIER COEFFICIENTS"
600 PRINT "  I   FREQ          REAL                    IMAGINARY"
610 CLEAR @ GCLEAR
620 DISP "ADDING BACK STEP"
630 T4=N*D1
640 F4=1/T4
650 FOR I=1 TO N/2
660 F=(I-1)*F4
670 I2=2*I
680 I1=I2-1
690 IF I=1 THEN 710
700 GOSUB 2490
710 S(I)=SQR(A(I1)*A(I1)+A(I2)*A(I2))
720 ! PRINT USING 306 ; I,F,A(I1)
730 ! PRINT USING 307 ; A(I2)
740 IMAGE 19X,MD.DDDDE
750 NEXT I
760 S9=AMAX(S)
770 SCALE 0,F9,0,S9
780 YAXIS F9 @ YAXIS 0
790 XAXIS S9,F9/10 @ XAXIS 0,F9/10
800 PENUP
810 FOR I=1 TO N/2
820 F=(I-1)/(N*D1)
830 PLOT F,S(I)
840 NEXT I
850 PRINT "ZERO STENGTH=";A(1)
860 PRINT "FREQ FROM 0 TO ",F9," HZ" @ PRINT "MAG FROM 0 TO ";S9
870 PRINT USING "4/"
880 GRAPH @ COPY
890 PRINT USING "4/"
900 BEEP @ GOSUB 2360
910 STOP
920 ! START OF FFT SUBROUTINE
930 N0=N @ M0=M @ N1=N/4 @ M1=M-2 @ T0=PI/4 @ K7=N1 @ K8=N1/2
940 S(K8)=SIN(T0)
950 FOR L=2 TO M1
960 T0=T0/2
970 S2=K7
980 K7=K8
990 K8=K8/2
1000 S(K8)=SIN(T0)
1010 C1=N1-K8
1020 S(C1)=COS(T0)
1030 K9=N1-S2
1040 IF K9-K7<0 THEN GOTO 1100
1050 FOR J=K7 TO K9 STEP K7
1060 C=N1-J
1070 D=J+K8
1080 S(D)=S(J)*S(C1)+S(K8)*S(C)
1090 NEXT J
1100 NEXT L
1110 !          ANALYSIS NEXT
1120 K(1)=2*N
1130 FOR L=2 TO M
1140 K(L)=K(L-1)/2
1150 NEXT L
1160 FOR L=M TO 9
1170 K(L+1)=2
```

```
1180 NEXT L
1190 I7=2
1200 FOR J1=2 TO K(10) STEP 2
1210 FOR J2=J1 TO K(9) STEP K(10)
1220 FOR J3=J2 TO K(8) STEP K(9)
1230 FOR J4=J3 TO K(7) STEP K(8)
1240 FOR J5=J4 TO K(6) STEP K(7)
1250 FOR J6=J5 TO K(5) STEP K(6)
1260 FOR J7=J6 TO K(4) STEP K(5)
1270 FOR J8=J7 TO K(3) STEP K(4)
1280 FOR J9=J8 TO K(2) STEP K(3)
1290 FOR J0=J9 TO K(1) STEP K(2)
1300 IF I7-J0>=0 THEN GOTO 1370
1310 T=A(I7-1)
1320 A(I7-1)=A(J0-1)
1330 A(J0-1)=T
1340 T=A(I7)
1350 A(I7)=A(J0)
1360 A(J0)=T
1370 I7=I7+2
1380 NEXT J0
1390 NEXT J9
1400 NEXT J8
1410 NEXT J7
1420 NEXT J6
1430 NEXT J5
1440 NEXT J4
1450 NEXT J3
1460 NEXT J2
1470 NEXT J1
1480 IF I8>0 THEN GOTO 1540
1490 F0=1/D1
1500 FOR I=1 TO N
1510 A(2*I-1)=A(2*I-1)/F0
1520 A(2*I)=-A(2*I)/F0
1530 NEXT I
1540 FOR I=1 TO N STEP 2
1550 T=A(2*I-1)
1560 A(2*I-1)=T+A(2*I+1)
1570 A(2*I+1)=T-A(2*I+1)
1580 T=A(2*I)
1590 A(2*I)=T+A(2*I+2)
1600 A(2*I+2)=T-A(2*I+2)
1610 NEXT I
1620 L1=2
1630 L0=8
1640 N2=2^M1
1650 FOR L=2 TO M
1660 FOR I=2 TO K(1) STEP L0
1670 I1=I+L1
1680 I2=I1+L1
1690 I3=I2+L1
1700 T=A(I-1)
1710 A(I-1)=T+A(I2-1)
1720 A(I2-1)=T-A(I2-1)
1730 T=A(I)
1740 A(I)=T+A(I2)
1750 A(I2)=T-A(I2)
1760 T=-A(I3)
1770 T1=A(I3-1)
```

```
1780 A(I3-1)=A(I1-1)-T
1790 A(I3)=A(I1)-T1
1800 A(I1-1)=A(I1-1)+T
1810 A(I1)=A(I1)+T1
1820 NEXT I
1830 IF L-2<=0 THEN GOTO 2100
1840 K5=K(1)-L0
1850 K6=N2
1860 FOR J=4 TO L1 STEP 2
1870 N3=N1-K6
1880 U1=S(N3)
1890 U2=S(K6)
1900 I6=J+K5
1910 FOR I=J TO I6 STEP L0
1920 I1=I+L1
1930 I2=I1+L1
1940 I3=I2+L1
1950 T=A(I2-1)*U1-A(I2)*U2
1960 T1=A(I2-1)*U2+A(I2)*U1
1970 A(I2-1)=A(I-1)-T
1980 A(I2)=A(I)-T1
1990 A(I-1)=A(I-1)+T
2000 A(I)=A(I)+T1
2010 T=-A(I3-1)*U2-A(I3)*U1
2020 T1=A(I3-1)*U1-A(I3)*U2
2030 A(I3-1)=A(I1-1)-T
2040 A(I3)=A(I1)-T1
2050 A(I1-1)=A(I1-1)+T
2060 A(I1)=A(I1)+T1
2070 NEXT I
2080 K6=K6+N2
2090 NEXT J
2100 L1=2*L1
2110 L0=2*L0
2120 N2=.5*N2
2130 NEXT L
2140 IF I8>0 THEN RETURN
2150 FOR I=1 TO N
2160 A(2*I)=-A(2*I)
2170 NEXT I
2180 RETURN
2190 ! DATA CONVERSION
2200 T=Y9*(E5(I)-E5(1))
2210 RETURN
2220 M9=M9-1
2230 IF M9<1 THEN M9=1
2240 STOP
2250 ! NUMERICAL INPUT
2260 ASSIGN# 1 TO D$
2270 READ# 1 ; N9,X9,Y9,E5(),R$
2280 ASSIGN# 1 TO *
2290 PRINT " " @ PRINT " "
2300 PRINT "DATA FILE '";D$;"' READ"
2310 PRINT " " @ PRINT R$ @ PRINT " "
2320 E9=E5(N9)
2330 PRINT "DC LEVEL REMOVED FOR FFT AND THEN ADDED BACK"
2340 PRINT "STEP=.5*[1+TANH(T-T9)]"
2350 RETURN
2360 CLEAR
2370 DISP "IF FILE ";F$;" DOES NOT EXIST, EXECUTE :"
```

```
2380 DISP "CREATE '";F$;"',1,2304" @ DISP ""
2390 DISP "HIT [CONT] TO STORE MAG OF FFT" @ PAUSE
2400 ASSIGN# 1 TO F$
2410 PRINT# 1,1 ; F4,N,S(),R$
2420 ASSIGN# 1 TO *
2430 CLEAR
2440 DISP "DATA IN FILE ";F$
2450 PRINT "" @ PRINT "FFT DATA WRITTEN TO FILE '";F$;"'"
2460 DISP "FINISHED"
2470 PAUSE
2480 RETURN
2490 W=2*PI*F
2500 T8=PI*W/(2*A0)
2510 T8=EXP(T8)-EXP(-T8)
2520 T8=PI*E9/(A0*T8)
2530 A(I1)=A(I1)-T8*SIN(W*T9)
2540 A(I2)=A(I2)-T8*COS(W*T9)
2550 RETURN
2560 END
```

Listing 5 Program to read FFT file and plot.

```
10 !   PROGRAM FPLOT
20  3-16-84   J A LANDT
30 DEG
40 OPTION BASE 1
50 DIM A(256),R$[100]
60 PLOTTER IS 705
70 PEN 0
80 CLEAR
90 N=256
100 CLEAR @ DISP "MAXIMUM FREQ ON PLOTS=?" @ INPUT F9
110 CLEAR @ DISP "DELTA FREQ FOR TICS=?" @ INPUT F8
120 CLEAR @ DISP "FILE NAME=?" @ INPUT F$
130 ASSIGN# 1 TO F$
140 READ# 1,1 ; F1,N9,A(),R$
150 ASSIGN# 1 TO *
160 A9=AMAX(A)
170 GOSUB 320
180 PRINT "FREQ (Hz)   REL MAG (dB)"
190 PEN 1
200 N9=INT(F9/F1)+1
210 IF N9<N THEN 230
220 N9=N
230 FOR I=2 TO N9
240 F=(I-1)*F1
250 X=20*LGT(A(I)/A9)
260 ! PRINT USING "D.DDDDDE,1X,D.DDDDDE" ; F,X
270 PLOT F*.000000001,X
280 NEXT I
290 PENUP
300 PEN 0 @ MOVE 0,0
```

```
310 STOP
320 LOCATE 25,130,25,95
330 PRINT "MAG REL TO ";A9
340 PRINT "FREQ FROM 0 TO ";F9
350 SCALE 0,F9/1000000000,-70,0
360 GCLEAR @ FXD 1,1 @ PEN 1 @ CSIZE 4
370 LAXES -F8/1000000000,10,0,-70,1,1,2
380 MOVE F9/2000000000,-77
390 LORG 6
400 CSIZE 7,.8
410 LABEL "Frequency (GHz)"
420 MOVE F9/2500000000,-83
430 LORG 6 @ CSIZE 4,.8
440 LABEL F$;" : ";R$
450 CSIZE 7,.8
460 MOVE -F9/6000000000,-35
470 LORG 4
480 LDIR 90
490 LABEL "Spectrum (dB)"
500 FRAME @ PEN 0
510 PENUP
520 RETURN
530 END
```

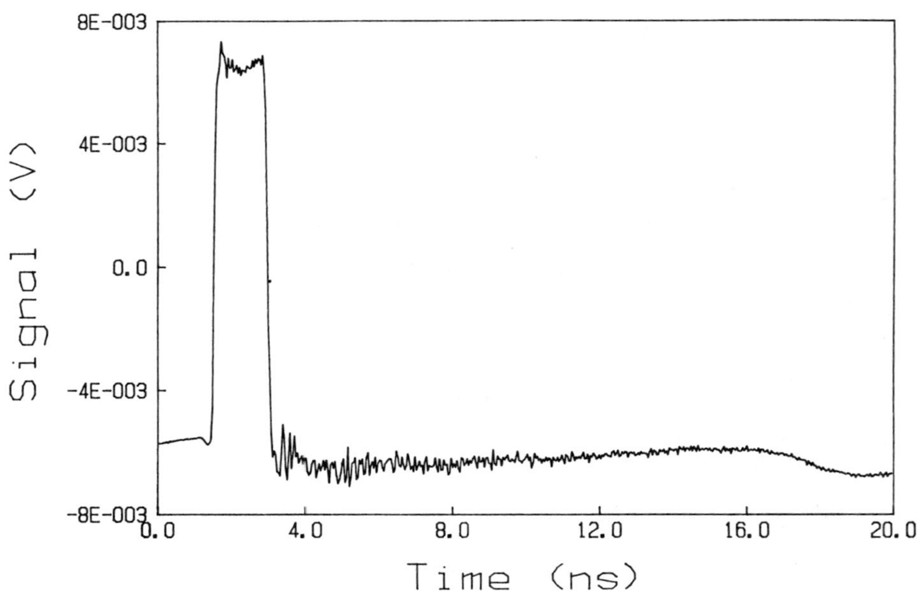

Fig. A-7. Signal measured at the coupled port in the forward direction of a Hewlett-Packard 20-dB directional coupler Model 764D.

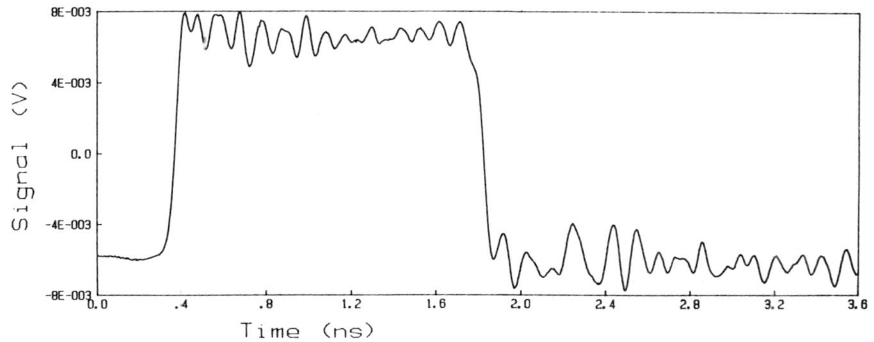

Fig. A-8. Data of Fig. A-7 obtained with a faster sweep rate.

Equation A-7 closely approximates the actual curve in Fig. A-11. From it the center frequency is seen to be 342 MHz with an insertion loss of 19.8 dB, compared to the calibration data of (225 + 450)/2, or 337.5 MHz, and 20 dB. The factor of 2 can be heuristically explained because the output is the sum of two waves. When these waves cancel (at 685 MHz, 1270 MHz, etc.), the output is zero. When they add constructively (at 342 MHz, 1027 MHz, etc.), the output is double that of one of them.

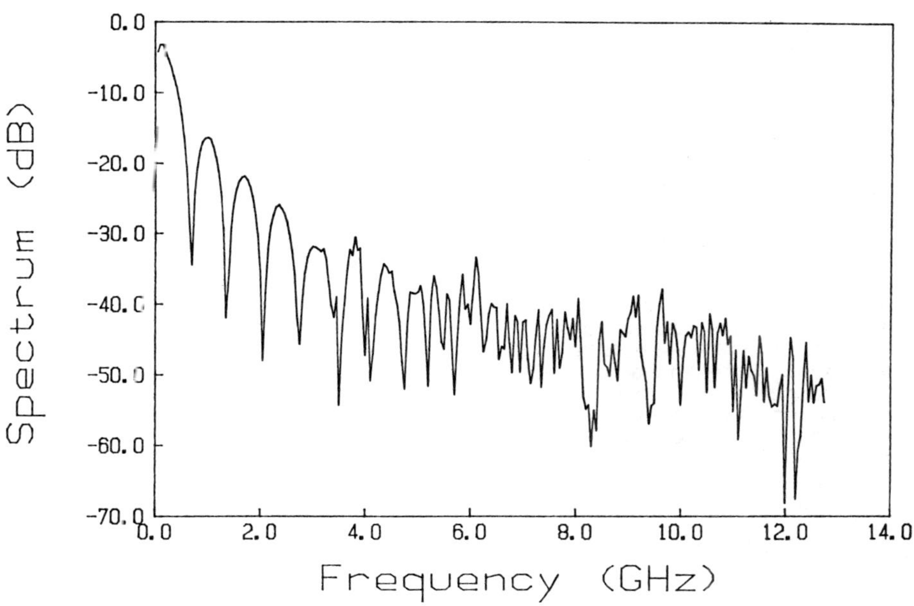

Fig. A-9. Fast Fourier transform of the data of Fig. A-7 (0 dB = 2.60×10^{-11} V/Hz).

Listing 6 Program to find and plot ratio of FFT's (insertion loss).

```
10 !     PROGRAM INSERT
20 !   3-14-84   J A LANDT
30 ! TO READ FILES FFT1 AND FFT2 AND PLOT RATIO
40 DEG
50 OPTION BASE 1
60 DIM A(256),B(256),R$[100],S$[100]
70 PLOTTER IS 1
80 PEN 0
90 CLEAR
100 N=256
110 CLEAR @ DISP "MAXIMUM FREQ ON PLOTS=?" @ INPUT F9
120 CLEAR @ DISP "DELTA FREQ FOR TICS=?" @ INPUT F8
130 CLEAR @ DISP "FOR SIGNAL, FILE NAME=?" @ INPUT F$
140 ASSIGN# 1 TO F$
150 READ# 1,1 ; F4,N9,A(),R$
160 ASSIGN# 1 TO *
170 CLEAR @ DISP "FOR REFERENCE, FILE NAME=?" @ INPUT F$
180 ASSIGN# 1 TO F$
190 READ# 1,1 ; F5,N8,B(),S$
200 ASSIGN# 1 TO *
210 FOR I=1 TO N @ A(I)=A(I)/B(I) @ NEXT I
220 IF F5<>F4 THEN 640
230 IF N9<>N8 THEN 640
240 A9=AMAX(A)
250 A9=1
260 F1=F4
270 GOSUB 430
280 PRINT "FREQ (Hz)   REL MAG (dB)"
290 PEN 1
300 N9=INT(F9/F1)+1
310 IF N9<N THEN 330
320 N9=N
330 FOR I=2 TO N9
340 F=(I-1)*F1
350 X=20*LGT(A(I)/A9)
360 ! PRINT USING "D.DDDDDE,1X,D.DDDDDE" ; F,X
370 PLOT F*.000000001,X
380 NEXT I
390 PENUP
400 PEN 0 @ MOVE 0,0
410 PRINT R$;" / ";S$
420 STOP
430 LOCATE 25,130,25,95
440 PRINT "MAG REL TO ";A9
450 PRINT "FREQ FROM 0 TO ";F9
460 SCALE 0,F9/1000000000,-50,0
470 GCLEAR @ FXD 1,1 @ PEN 1 @ CSIZE 4
480 LAXES -F8/1000000000,10,0,-50,1,1,2
490 MOVE F9/2000000000,-55
500 LORG 6
510 CSIZE 7,.8
520 LABEL "Frequency (GHz)"
530 MOVE F9/2500000000,-59
540 LORG 6 @ CSIZE 4,.8
550 LABEL R$
560 CSIZE 7,.8
570 MOVE -F9/6000000000,-25
580 LORG 4
590 LDIR 90
600 LABEL "Insertion (dB)"
```

```
610 FRAME @ PEN 0
620 PENUP
630 RETURN
640 DISP ""
650 DISP "1ST FILE N, DF =";N9;F4
660 DISP "2ND FILE N, DF =";N8;F5
670 DISP "TROUBLE--STOP"
680 STOP
690 END
```

As a second example, a Mini-Circuits Model ZMDC-20-1 directional coupler was measured. The specifications of this coupler include a 20-dB insertion loss from 25 to 400 MHz. The measured transient response is approximated by a decaying exponential as follows:

$$V(t) = B \exp(-at)\, U(t) \tag{A-8}$$

where $B = 0.025$ V and $1/a = 2.3*10^{-8}s$. In the frequency domain, the insertion loss is determined by the equation,

$$I(\omega) = j\frac{B}{A}(\omega/a)/[1 + j(\omega/a)] \tag{A-9}$$

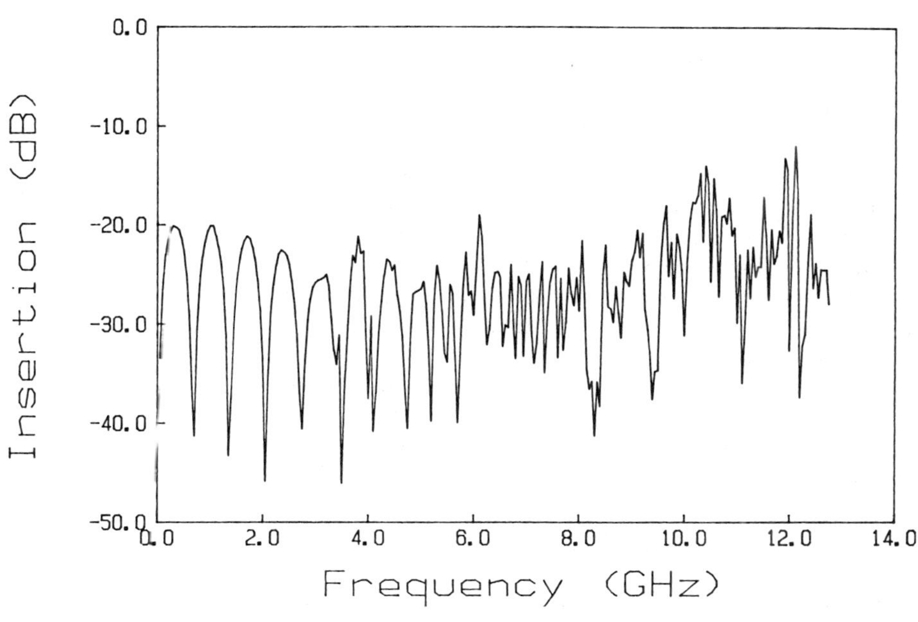

Fig. A-10. Irsertion loss to the coupled port of a Hewlett-Packard directional coupler Model 764D.

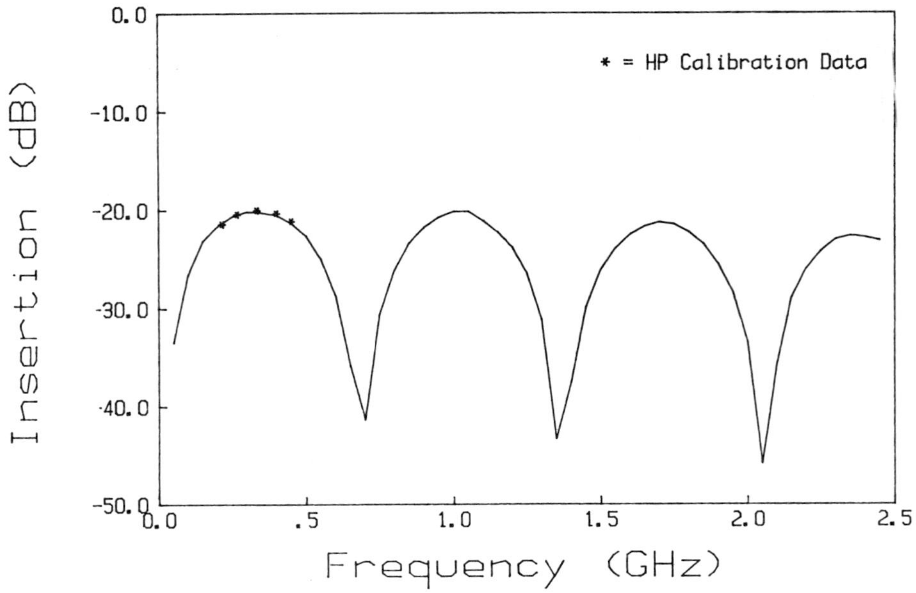

Fig. A-11. Data of Fig. A-10 on an expanded scale, including manufacturer's calibration data.

This coupler is basically a high-pass filter, and the ratio *B/A* accurately represents the insertion loss (19.7 dB). The lower-frequency 3-dB point is about 6.9 MHz. The upper-frequency response is limited by risetime. The construction of this coupler is fundamentally different from that of the HP 764D.

The 7854 has a 10-bit resolution that corresponds to 1024 points, or about 102 points per division, on the face of the CRT. Several alternative methods can be used to acquire the data. A low-end range of equipment includes the 8-bit (256 points) Wavesaver Digital Waveform Recorder produced by EPIC Instruments, Inc., Foster City, California. The Apple and Apple IIe personal computers can be used as digital memory oscilloscopes with the plug-in Model 85 Scope produced by Northwest Instruments Systems, Inc., Beaverton, Oregon. A wide variety of commercial equipment can also be used.

An early Lawrence Livermore National Laboratory approach is shown in Fig. A-12. A 7T11 sampling sweep unit permits external control of sweep position by a *D* to *A* converter driven by a computer. This system works quite well but contains many more subsystems and more elaborate software than the 7854-based system.

SINGLE-EVENT TRANSIENT RECORDING

Commercial high-speed oscilloscopes and transient digitizers can be used for recording single-event transients below (about) 1 GHz. Above 1 GHz, commercially available equipment is usually not available. Special high-speed oscilloscopes have been constructed at Los Alamos National Laboratory and in France to extend the upper frequency limit about 5 GHz.

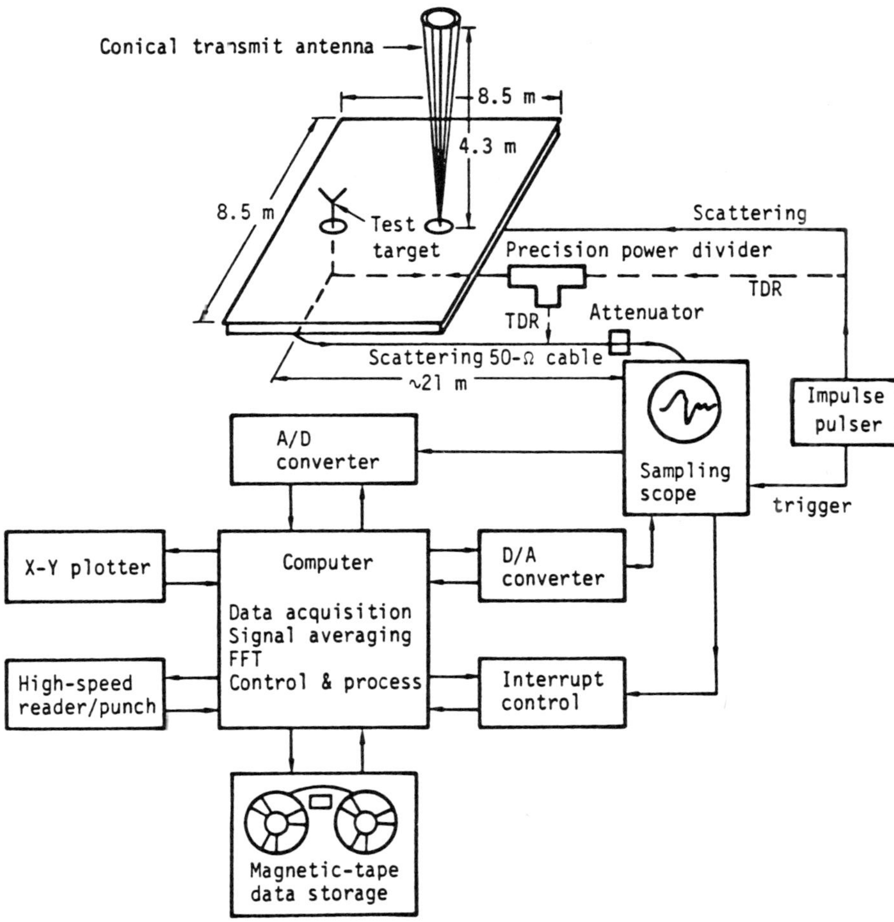

Fig. A-12. Schematic of measurement facility: Impulse pulser sends narrow pulse either through a precision power divider to test target, as in TDR experiment, or directly to conical transmitting antenna, as in scattering experiment. In TDR mode, reflected signal returns to sampling scope via power divider; in scattering mode, signal received by test target returns directly to scope. From these, signal passes to data-processing equipment where it is averaged over repetitive pulses, Fourier-transformed to the frequency-domain, plotted, and stored on magnetic tape.

RADAR

The system shown in Fig. A-3 can be configured to operate as a time-delay radar. In this application, radiating and receiving antennas are located in close proximity. Resistively loaded dipole antennas and crossed bowtie antennas are normally used. Care must be taken so that the 5-V input-voltage limit of the samplers is not exceeded. For this application, it is often useful to slave two pulse generators of opposite polarity, drive a balanced antenna, and provide a balanced receiving antenna. Two samplers are required, and the difference of their signals are displayed by the oscilloscope. EG&G, Santa Barbara, has developed a system based on this approach, and they have also used a seismiclike display that facilitates data interpretation. More complete descriptions of other base-band radar systems can be found elsewhere in this book (Chaps. 1, 2, 4, 5, and 6).

Index